Springer Series in Solid-State Sciences

Volume 205

The Springer Series in Solid-State Sciences features fundamental scientific books prepared by leading and up-and-coming researchers in the field. They strive to communicate, in a systematic and comprehensive way, the basic principles as well as new developments in theoretical and experimental solid-state science.

We welcome submissions for monographs or edited volumes from scholars across this broad domain. Topics of current interest include, but are not limited to:

- Semiconductors and superconductors
- Quantum phenomena
- Spin physics
- Topological insulators
- Multiferroics
- Nano-optics and nanophotonics
- Correlated electron systems and strongly correlated materials
- Vibrational and electronic properties of solids
- Spectroscopy and magnetic resonance

Zhong Lin Wang

Elastic and Inelastic Scattering in Electron Diffraction and Imaging

Second Edition

Springer

Zhong Lin Wang
Beijing Institute of Nanoenergy
and Nanosystems
Chinese Academy of Sciences
Beijing, China

ISSN 0171-1873 ISSN 2197-4179 (electronic)
Springer Series in Solid-State Sciences
ISBN 978-3-031-90818-7 ISBN 978-3-031-90819-4 (eBook)
https://doi.org/10.1007/978-3-031-90819-4

This Springer imprint is published by the registered company Springer Nature Switzerland AG
The registered company address is: Gewerbestrasse 11, 6330 Cham, Switzerland

Preface to the Second Edition

The first edition of this book, *Elastic and Inelastic Scattering in Electron Diffraction and Imaging*, was published in 1995. At the time, it aimed to provide a comprehensive foundation for understanding the fundamental principles and applications of electron scattering in microscopy. Since then, the field of electron microscopy has undergone transformative advancements, driven by breakthroughs in experimental instrumentation, computational methods, and theoretical frameworks. These developments have not only expanded the boundaries of what is possible in electron microscopy but have also opened new avenues for exploring materials at atomic and subatomic scales.

Although my own research interests have shifted away from electron microscopy in the decades following the publication of the first edition, the enduring relevance of this book has been a source of both surprise and inspiration. During a lecture series at Uppsala University in 2019, I was deeply moved when several young faculty members approached me to share that they regarded the book as a "bible" in the field. This sentiment was echoed by many readers and colleagues, including distinguished scientists such as Profs. J. C. H. Spence, C. Colliex, and J. Hren, whose endorsements have been a tremendous honor. The first edition was reviewed by professional journals and their comments are quoted as:

> "a noteworthy achievement and a valuable contribution to the literature", *American Scientist*, 1996
>
> "This is an excellent and comprehensive book ... If you are interested in electron scattering by crystals, in the theory underlying the interpretation of electron micrographs ... you should buy this book. It is comprehensive and right up to date", *J. Microscopy*, 1996

With the encouragement along with the continued demand for the book, motivated me to undertake the task of revising and updating it for a new generation of researchers.

The second edition of this book reflects the remarkable progress made in the field over the past three decades. Chapter 13 has been entirely rewritten to incorporate modern developments, while two new chapters—Chaps. 14 and 15—have been added to address emerging topics and techniques. Chapter 16 has also been significantly revised to ensure the book remains a comprehensive resource, particularly in its treatment of inelastic scattering. These updates aim to bridge the gap between the foundational principles established in the first edition and the cutting-edge advancements that have since reshaped the field.

The impact of electron microscopy on science and technology cannot be overstated. Over the past 30 years, the field has played a pivotal role in advancing our understanding of materials, biological systems, and nanoscale phenomena. Techniques such as aberration-corrected microscopy, ultrafast electron imaging, and advanced spectroscopy have revolutionized the way we observe and interpret the atomic world. This book seeks to contribute to this ongoing progress by providing a updated and thorough exploration of elastic and inelastic scattering, which remain at the heart of electron microscopy.

I would like to express my sincere gratitude to Dr. Weihua Han, Dr. Ning Wu, and Dr. Tao Jiang for their invaluable assistance in preparing the manuscript for this second edition. Their expertise and dedication have been instrumental in ensuring the accuracy and relevance of the content. I am also deeply thankful to the readers of the first edition, whose support and feedback have been a constant source of motivation. It is my hope that this new edition will continue to serve as a valuable resource for students, researchers, and practitioners in the field, inspiring future discoveries and innovations for decades to come.

March 2025

Zhong Lin Wang, Ph.D.
Chief Scientist, Chair Professor
and Director, Beijing Institute
of Nanoenergy and Nanosystems
Chinese Academy of Sciences
Beijing, China

Regents' Professor and Hightower
Chair (Emeritus)
Georgia Institute of Technology
Atlanta, USA

Preface to the First Edition

Elastic and inelastic scattering in transmission electron microscopy TEM are important research subjects. For a long time, I wished to summarize systematically various dynamic theories associated with quantitative electron microscopy and their applications in simulations of electron diffraction patterns and images. This wish now becomes reality. The aim of this book is to explore the physics in electron diffraction and imaging and related applications for materials characterizations. Particular emphasis is placed on diffraction and imaging of inelastically scattered electrons, which, I believe, have not been discussed extensively in any existing books. This book assumes that readers have some preknowledge of electron microscopy, electron diffraction, and quantum mechanics. I anticipate that this book will be a guide to approaching phenomena observed in electron microscopy from the prospects of diffraction physics.

The SI units are employed throughout the book except for Angstrom (Å) which is used occasionally for convenience. To reduce the number of symbols used, the Fourier transform of a real space function $\Psi(\mathbf{r})$, for example, is denoted by the same symbol $\Psi(\mathbf{u})$ in reciprocal space except that $\mathbf{r}$ is replaced by $\mathbf{u}$. Upper and lower limits of an integral in the book are $(-\infty, \infty)$ unless otherwise specified. The $(-\infty, \infty)$ integral limits are usually omitted in a mathematical expression for simplification.

I very much appreciate opportunity of working with Drs. J. M. Cowley and J. C. H. Spence (Arizona State University), J. Bentley (Oak Ridge National Laboratory), R. F. Egerton (University of Alberta), C. Colliex (University of Paris-Sud) and A. Howie (University of Cambridge), who gave me advice and encouragement on many topics. Particular thanks go to Dr. Nea S. Wheeler for her careful and critical review of the manuscript. Thanks to Drs. J. Bentley, S. Dudarev, C. Fanidis, A. V. Latyshev,

J. Liu, R. F. Loane, Ping Lu, L. Reimer, C. J. Rossouw, D. Shechtman, A. E. Smith, D. J. Smith, J. C. H. Spence, B. Viguier, M. H. Yao, and J. M. Zuo, who have kindly allowed me to use their illustrations and results in the text.

March 1995

Zhong Lin Wang
Georgia Institute of Technology
Atlanta, USA

Competing Interests The author has no competing interests to declare that are relevant to the content of this manuscript.

Introduction

Diffraction and imaging of transmitted high-energy electrons are pivotal experimental techniques for determining crystal structures. The continuous advancements in Transmission Electron Microscopy (TEM), Scanning Transmission Electron Microscopy (STEM), and their associated analytical methods have enabled structural determination at atomic resolution, reaching the sub-Angstrom level. The recent commercialization of Cs-corrected TEM/STEM instruments and electron energy filters has ushered in a new era of quantitative electron microscopy. These innovations allow for the separation and digital recording of images and diffraction patterns formed by electrons with distinct energy losses. In practice, quantitative data analysis relies heavily on theoretical simulations. This book delves into the theories and techniques of elastic and inelastic electron diffraction and imaging in quantitative electron microscopy. It aims to summarize and expand upon the various dynamic theories developed over the past 80 years, particularly those arising from studies of elastic and inelastic electron multiple scattering, and to apply these theories to explore new experimental techniques.

Numerous excellent books have been dedicated to the diverse techniques associated with TEM. Hirsch et al. (1977) provide a foundational text on conventional TEM, with a focus on diffraction contrast imaging. Reimer (1984) offers a more physics-oriented perspective on TEM, while Cowley (1981) presents a systematic description of the unified theory of electron, neutron, and X-ray diffraction. Buseck et al. (1988) provide a comprehensive overview of high-resolution TEM and its applications across various fields. Spence and Zuo (1992) and Cowley (1992, 1993) focus on electron diffraction techniques. Books on electron energy-loss spectroscopy (Egerton, 1986) and energy-dispersive X-ray microanalysis delve into the analytical applications of TEM. Given the vast scope of TEM-related subjects, each book typically covers only a fraction of the field. Recent publications, such as those by Hawkes and Spence (2019), Kirkland (1998), Peng et al. (2004), Zuo and Spence (2017), and Williams and Carter (2009), have further enriched the literature on the fundamentals and applications of electron microscopy.

Theories of diffraction and imaging of transmitted electrons are central to TEM. Most existing books focus on elastic scattering theory, often employing the kinematic scattering approach. Humphreys (1979) provides a comprehensive review of Bethe's elastic scattering theory, while Van Dyck (1985) and Amelinckx and Van Dyck (1993) extensively summarize the elastic-scattering multislice theory. Inelastic scattering, however, is typically treated as an absorption effect, incorporated into the theory through an imaginary potential, with thermal diffuse scattering represented by a Debye-Waller factor. Ohtsuki (1983) dedicates more attention to inelastic excitations of charged particles but does not extensively cover diffraction and imaging of inelastically scattered electrons. Consequently, there is a pressing need for a comprehensive summary of the following topics, which are crucial for quantitative data analysis:

1. Dynamic Elastic Scattering Theories: Elastic electron diffraction is often approached using theoretical frameworks tailored to specific subjects, such as Bloch wave theory for convergent beam electron diffraction and multislice theory for high-resolution TEM image simulation. A systematic description of all existing elastic-scattering theories is currently lacking.
2. Reflection High-Energy Electron Diffraction (RHEED): RHEED is a vital technique for observing in-situ surface structure evolution during thin film growth. However, the quantitative analysis of RHEED patterns remains inadequate. It is essential to summarize and compare dynamical approaches for quantitative RHEED data analysis.
3. Absorption Effects: Inelastic scattering is often treated as an absorption effect in the context of elastically scattered electrons, represented phenomenologically by an imaginary potential. However, there is no systematic description of the theoretical basis or calculations of the absorption potential arising from various inelastic-scattering processes. These calculations are critical for quantitative electron diffraction and imaging.
4. Diffraction of Inelastically Scattered Electrons and Kikuchi Patterns: The theory of inelastic electron diffraction is complex due to the incoherent nature of inelastic scattering, making it a rarely discussed topic in existing literature. Despite over half a century of research on inelastic electron scattering, there is a need for a book that consolidates existing theoretical achievements. While energy filters can remove electrons with energy losses greater than sub-eV, phonon-scattered electrons with minimal energy loss (< 0.1 eV) remain in the diffraction pattern.
5. New Imaging Techniques Using Inelastically Scattered Electrons: Recent years have seen growing interest in structural determination using inelastically scattered electrons. For instance, high-angle annular dark-field imaging in STEM relies on signals from high-angle phonon-scattered electrons. This technique offers atomic-number-sensitive structural information with superior resolution compared to conventional bright-field imaging. It enables quantitative chemical mapping at atomic resolution, facilitating column-to-column electronic and chemical analysis.

The ultimate goal of electron microscopy is to quantitatively determine the structure of materials. Achieving this requires the best possible theoretical frameworks and energy-filtered high-resolution diffraction patterns and images. It is anticipated that electron diffraction and imaging will become as indispensable as X-ray and neutron diffraction in materials research, a status they have yet to fully attain. This book focuses on the dynamic diffraction and imaging of elastically and inelastically scattered electrons, emphasizing their applications in structural determination and addressing the aforementioned topics in detail.

The book covers the following principles. First, since elastic-scattering theory is described in various books, diffraction and imaging of inelastically scattered electrons are the main emphasis of this book. Second, all theoretical derivations are initiated from first principles wave mechanics for high-energy electrons ($E_0 > 20$ keV). Elastic scattering is initiated from Schrödinger equation, and inelastic scattering is determined by the solution of Yoshioka's coupled equations. Third, since the interaction between incident electrons and crystals is very strong, all the approaches presented are dynamic theories. Fourth, all theories are linked to one another and presented in the form best suited for numerical calculations. All mathematical operations are given in detail so that the book is self-contained and self-derivable. The entire text is linked together so it forms a unity based on first principles. Finally, all physical quantities are defined in SI units except where stated for example, Angstrom (Å) is used occasionally for convenience.

This book is divided into two main parts: Part I focuses on the diffraction and imaging of elastically scattered electrons, while Part II delves into the theory and applications of inelastically scattered electrons.

Part I: Elastic Scattering of Electrons

Chapter 1 introduces foundational concepts of electron diffraction using the kinematic scattering approach, defining and illustrating key quantities. Chapters 2 and 3 provide a comprehensive quantum mechanical treatment of elastically transmitted electron scattering. Chapter 2 explores the Bloch wave theory (Bethe theory) and its applications in convergent beam electron diffraction, diffraction contrast imaging, and weak-beam imaging. Chapter 3 presents the wave mechanics basis of the Cowley-Moodie multislice theory (1957) and its use in simulating high-resolution electron microscopy images. It also discusses methods for incorporating high-order Laue zone reflections into multislice calculations and introduces real-space multislice theory. A comparison of existing theories highlights their unique strengths and limitations for specific problems. Chapter 5 is dedicated to the dynamic theories of Reflection High-Energy Electron Diffraction (RHEED) from bulk crystal surfaces, offering a detailed derivation and comparison of various theoretical approaches. This chapter stands as the most comprehensive review of RHEED theories to date.

Part II: Inelastic Scattering of Electrons

This part is the first of its kind to provide a thorough discussion of inelastic scattering in electron diffraction. Key topics include:

- Chapter 6: A formal derivation of the imaginary potential used in numerical calculations, based on Yoshioka's coupled equations. It details inelastic scattering processes such as phonon excitation, valence excitation, and atomic inner-shell (single-electron) excitation, calculating their contributions to the absorption potential. The chapter also explains the fundamental features of inelastic scattering and its impact on X-ray emission, making it essential for quantitative data analysis.
- Chapter 7: Presents the classical diffraction theory of thermal diffusely scattered electrons, using the frozen lattice model. It thoroughly discusses multi-phonon and multiple-phonon excitations and introduces diffuse scattering caused by Huang scattering.
- Chapter 8: Extends the Bloch wave theory to inelastically scattered electrons, including cases of imperfect crystals. This approach simplifies calculations for Kikuchi patterns and covers double inelastic scattering.
- Chapter 9: Introduces the reciprocity theorem, which is applied in Chap. 10 to prove the equivalence of Transmission Electron Microscopy (TEM) and Scanning Transmission Electron Microscopy (STEM). It also derives an equivalent theorem for inelastically scattered electrons.
- Chapter 10: Explores Green's function method as a formal theory for solving electron scattering problems. This versatile approach, applicable to crystals of any shape, is particularly effective for treating phonon, atomic inner-shell, and double inelastic scattering. It enables the evaluation of scattering intensities from crystal lattices with varying thermal vibration configurations prior to numerical calculations, making it ideal for thermal diffuse scattering.
- Chapter 11: Applies the multislice theory and real-space multislice theory to inelastically scattered electrons, with detailed examples for simulating atomic-number-sensitive images in STEM and TEM. It extensively discusses Z-contrast imaging techniques and demonstrates multislice calculations for both perfect and imperfect crystals.
- Chapter 12: Introduces a simplified multislice theory for calculating diffraction patterns of phonon-scattered (thermal diffusely scattered) electrons. By simplifying models of elastic rescattering after inelastic excitation, it significantly reduces computational requirements. The theory also explains streak directions in electron diffraction patterns due to phase coupling between atom vibrations.
- Chapter 13: Addresses the contribution of inelastic scattering to high-resolution TEM images, focusing on valence loss and phonon scattering. It illustrates various cases involving surfaces and interfaces, examines the role of thermal diffuse scattering (TDS) in high-angle annular-dark-field STEM imaging, and investigates TDS effects in high-resolution off-axis electron holography.

- Chapter 14: Provides a systematic description of the dynamic scattering of thermal diffusely scattered electrons. It proves the equivalence of the "frozen" lattice model and quantum mechanical phonon scattering theory, enabling its use for high-energy electron diffraction calculations. A comprehensive optical potential method is introduced to describe TDS in electron diffraction and imaging, with rigorous theoretical proofs showing how high-order diffuse scattering contributions are automatically included in dynamical calculations. Practical numerical calculation procedures are also outlined, making this the most complete and accurate treatment of TDS in electron diffraction.
- Chapter 15: Develops a formal statistical dynamical theory to calculate diffuse scattering caused by short-range order (SRO) in distorted crystal structures, accounting for atomic thermal vibrations. This scattering introduces fine details in diffraction patterns and a non-local imaginary potential function that reduces Bragg reflection intensities. The chapter analytically performs statistical structure averaging on imperfections and correlates the results with Cowley's short-range order parameters.
- Chapter 16: Tackles the problem of multiple inelastic scattering in thick specimens, introducing the one-particle density matrix theory as an effective method for handling cases involving multiple elastic and inelastic scattering.
- Chapter 17: Examines the effect of heat exchange between the crystal and its environment on electron scattering behavior, a practical consideration given that crystals are always in contact with the microscope. It highlights potential effects of energy fluctuations on TDS.

  This book synthesizes and advances the theoretical framework for the diffraction and imaging of both elastically and inelastically scattered electrons. It emphasizes advanced topics not systematically covered in existing literature, making it an invaluable resource for scientists in condensed matter physics, materials science, surface science, solid-state chemistry, mineralogy, and biological sciences. It serves as a comprehensive reference for the physics of electron diffraction and imaging.

The author emphasizes the critical importance of quantitative analysis of energy-filtered elastic and inelastic scattering patterns and images, asserting that dynamic calculations are essential. While the literature presents a vast array of theories in diverse forms, unit systems, and approximations, this book aims to unify these approaches in a coherent manner. Although some examples may appear simplified to experts, the goal is to elucidate the underlying principles and provide practical, consistent methodologies. Despite its heavy reliance on mathematics and physics, the book is designed to be self-explanatory, allowing readers to work through the equations step-by-step with patience.

Contents

Part I
Diffraction and Imaging of Elastically Scattered Electrons

Chapter 1
Basic Kinematic Electron Diffraction

In 1986, E. Ruska was awarded the Nobel Physics Prize for his pioneering work of building the world's first transmission electron microscope (TEM) in the late 1920s. The TEM was originally based on the physical principle that charged particles can be focused by magnetic lenses. The strong interaction of incident electrons with electrons and nuclei in crystals makes it feasible to analyze structures of solids based on electron micrographs. The rapid development of TEM techniques in the last 30 years has made it possible to image crystal structures at atomic resolution. Thus TEM has become one of the key research tools in characterizing condensed matter.

Since electrons are charged particles, their interaction with a solid is rather strong in comparison to either X-ray or neutrons, so that multiple scattering effects are always present in electron diffraction. This means that electron diffraction has to be described by dynamic scattering theory especially when quantitative structural analysis is necessary. Dynamical theories, as shown in Chaps. 2–5, are usually based on some fundamental quantities, defined in kinematic scattering theory, for describing the basic characteristics of electron diffraction. Chapter 1 outlines some basic concepts of kinematic electron-scattering theory and introduces a few mathematical operations used in this book. A systematic kinematic treatment of electron diffraction for perfect and imperfect crystal has been given by Cowley (1981).

1.1 Wave Properties of Electrons

Although TEM was based on the optical behavior of charged particles, the electron diffraction is due purely to the wave property of particles. The wavelength λ of an electron is related to its momentum p by the de Broglie relation,

$$\lambda = \frac{h}{p} \tag{1.1}$$

Z. L. Wang, *Elastic and Inelastic Scattering in Electron Diffraction and Imaging*, Springer Series in Solid-State Sciences 205, https://doi.org/10.1007/978-3-031-90819-4_1

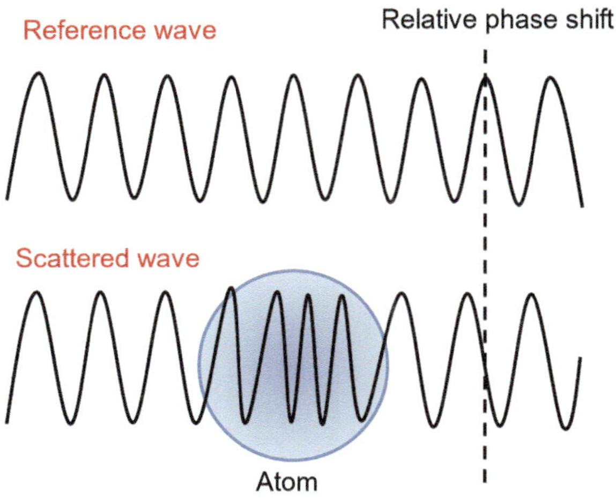

Fig. 1.1 A schematic diagram showing the phase perturbation effect of an incident high-energy electron wave by the scattering of a single atom. The relative phase shift of the scattered wave with respect to the unscattered wave (or reference wave) is indicated by a dashed line

In TEM, if the electron is accelerated to a kinetic energy of eU_0, relativity theory indicates that the momentum p satisfies

$$eU_0 = \left(m_0^2c_0^4 + p^2c_0^2\right)^{1/2} - m_0c_0^2 \tag{1.2}$$

where $m_0c_0^2$ is the rest energy of an electron. Substituting the solution of p from (1.2) into (1.1), the electron wavelength is

$$\lambda = \frac{h}{\left\{2m_0eU_0\left[1 + eU_0/2m_0c_0^2\right]\right\}^{1/2}} \tag{1.3}$$

Substituting the numerical values of the fundamental physical constants, yields

$$\lambda = \frac{1.226}{\left[U_0\left(1 + 0.9788 \times 10^{-6}U_0\right)\right]^{1/2}}\ \text{nm} \tag{1.4}$$

where U_0 is measured in volts. For 100 kV electrons, $\lambda = 0.0037$ nm. For such a small wavelength, a significant phase shift can be created if there is a small change in the scattering potential. A slight modification in crystal potential due to structural evolution can dramatically affect the scattering behavior of the electron. Since the atom size is less than 0.05 nm , a large phase variation can be created after the electron interacts with the atom (Fig. 1.1). This is approximately the origin of atomic resolution phase contrast lattice images of thin crystals using transmitted high-energy electrons (see many articles in Buseck et al. 1988). Wavelengths for electrons accelerated to different energies are listed in Appendix A.

1.2 Plane Wave

In electron imaging and diffraction, an incident electron is expressed as a plane wave. In free space and a nonrelativistic case, the propagation of an electron satisfies the time-dependent Schrödinger equation

$$-\frac{\hbar^2}{2m_0}\nabla^2\Psi = i\hbar\frac{\partial\Psi}{\partial t} \tag{1.5}$$

For an infinity large free space, the plane wave solution of this equation is

$$\Psi = \exp\left(2\pi i \mathbf{K}_0 \cdot \mathbf{r} - i\omega t\right) \tag{1.6}$$

where

$$\hbar\omega = \frac{h^2 K_0^2}{2m_0} = E$$

is the electron kinetic energy and $K_0 = 1/\lambda$ is the electron wave number. The scattering of the electron by a crystal is treated as the scattering of the plane wave by the potential field of the solid. Since electron diffraction is usually a time-independent scattering process, the time-dependent factor $\exp(i\omega t)$ is dropped in the following discussion. In this book, $\exp\left(2\pi i \mathbf{K}_0 \cdot \mathbf{r}\right)$ is taken as a plane wave with wave vector $\mathbf{K}_0$. The sign of the phase factor $(2\pi i)$ is kept consistent in the entire book.

1.3 Single-Atom Scattering

The easiest case in electron diffraction is scattering by a single atom. In this case, the electron is scattered due to its interaction with the electrostatic potential of the atom. In high-energy electrons, if the scattering potential $V(\mathbf{r})$ is weak, the amplitude of an incident plane wave $\exp\left(2\pi i \mathbf{K}_0 \cdot \mathbf{r}\right)$ to be scattered to an exit plane wave of wave vector $\mathbf{K}$ is calculated by the first Born approximation,

$$f(\mathbf{u}) = -\frac{m_0}{2\pi\hbar^2}\int d\mathbf{r}\, \exp(-2\pi i \mathbf{u}\cdot\mathbf{r})V(\mathbf{r}) \tag{1.7}$$

where $\hbar\mathbf{u} = \hbar\left(\mathbf{K} - \mathbf{K}_0\right)$ is the momentum transfer of the incident electron. The Born approximation assumes single scattering; that is, the electron is scattered only once. This assumption is the basis of kinematic scattering theory. If the atomic potential is spherical symmetric, i.e., $V(\mathbf{r}) = V(r)$, (1.7) becomes

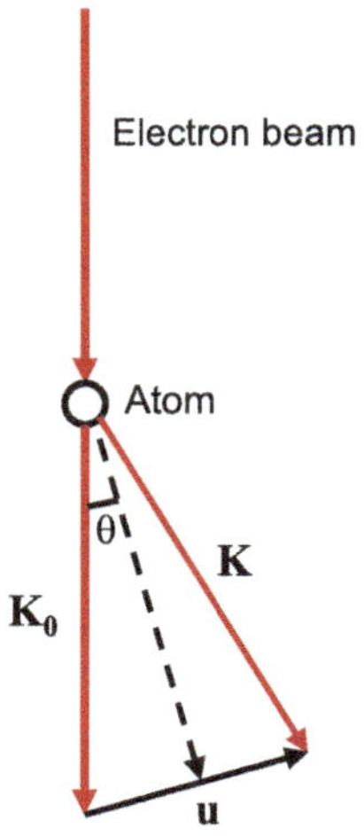

Fig. 1.2 Kinematic scattering of an incident plan wave by a single atom under the Born approximation. The change of electron momentum if $\hbar\mathbf{u}$

$$f(\theta) = -\frac{2m_0}{\hbar^2}\int_0^\infty dr \frac{\sin 2\pi ur}{2\pi ur} V(r) r^2 \tag{1.8}$$

where 2θ is the scattering angle shown in Fig. 1.2 and $\mathbf{u} = \mathbf{K} - \mathbf{K}_0$ with $u = 2K\sin\theta$. The scattering power of the atom is determined by the Fourier transform (**FT**) of its electrostatic potential; thus the electron-scattering factor of the atom is defined as

$$f^e(\mathbf{u}) = \int d\mathbf{r} \exp(-4\pi i\mathbf{s}\cdot\mathbf{r}) V(\mathbf{r}) \tag{1.9}$$

where the scattering vector is defined as $s = \sin\theta/\lambda$, or $\mathbf{s} = \mathbf{u}/2$, with $\mathbf{u}$ a reciprocal space vector (see Sect. 1.6). The electron-scattering factor is usually given as a function of **s**. For the convenience of mathematical expressions in this book, we use **u** as the variable of $f^e(\mathbf{u})$. The electron-scattering factor defined in (1.9) is a quantity that characterizes the scattering power of an atom, and it is independent of the accelerating voltage.

1.4 Mott Formula

The electron-scattering factor is usually difficult to measure experimentally because electron scattering in crystals is a dynamic scattering process. Thus, the scattering of an electron is determined by a solution of the Schrödinger equation. The X-ray diffraction, on the other hand, is usually treated kinematically for thin crystals, and quantitative data analysis is possible using kinematic scattering theory. This is why X-ray diffraction is one of the most accurate methods for structural determination. Since the interaction of X rays (or electromagnetic waves) with a crystal is determined by the charge distribution in the crystal, it is possible to correlate the potential function of an atom with its charge distribution using Maxwell's equations. For an atom of atomic number Z_0, the electrostatic potential function and electron charge density function are related by the Poisson equation

$$\nabla^2 V_\kappa(\mathbf{r}) = -\frac{e}{\varepsilon_0}\left[Z_0\delta(\mathbf{r}) - \rho_\kappa(\mathbf{r})\right] \tag{1.10}$$

where $\delta(\mathbf{r})$ is the Dirac delta function indicating the position of the nucleus. Taking a Fourier transform of (1.10), then using (1.9) and the X-ray scattering factor defined by

$$f_\kappa^x(\mathbf{s}) = \int \mathrm{d}\mathbf{r}\exp(-4\pi i\mathbf{s}\cdot\mathbf{r})\rho_\kappa(\mathbf{r}) \tag{1.11}$$

we have

$$(4\pi i s)^2 f_\kappa^e(\mathbf{u}) = -\frac{e}{\varepsilon_0}\left[Z_0 - f_\kappa^x(\mathbf{s})\right]$$

or

$$f_\kappa^e(\mathbf{u}) = \frac{e}{16\pi^2\varepsilon_0}\frac{\left[Z_0 - f_\kappa^x(\mathbf{s})\right]}{\mathrm{s}^2} \tag{1.12}$$

This is the well-known Mott formula, from which the electron-scattering factor can be directly calculated using the experimentally measured X-ray scattering factor. For general purpose, f^x can be written in analytic form in terms of the Gaussian functions (Doyle and Turner 1968),

$$f^x(s) = \sum_{i=1}^{4} a_i \exp\left(-b_i s^2\right) + c_i \tag{1.13}$$

where the fitting parameters a_i and b_i have been determined for most of the elements in neutral and ionized states. For high-energy electrons, the correction introduced by the exchange effect of valence electrons on the atomic-scattering factor is negligible, so that the scattering factor defined in (1.13) is independent of the incident energy of the electrons. This approximation may not hold if the energy of the incident electron is lower than a few keV (Qian et al. 1993). For atoms with spherical symmetry, the atomic-scattering factor is a real function. The atomic scattering factor could be a complex function if the atomic potential were not absolutely spherically symmetric due to charge redistribution or exchange resulted from solid bonding.

The asymptotic behavior of the scattering factors for large and small values of s must be considered (Peng and Cowley 1988). The electron-scattering factor must converge to the Rutherford-scattering form at high angles. The convergence of f^e at

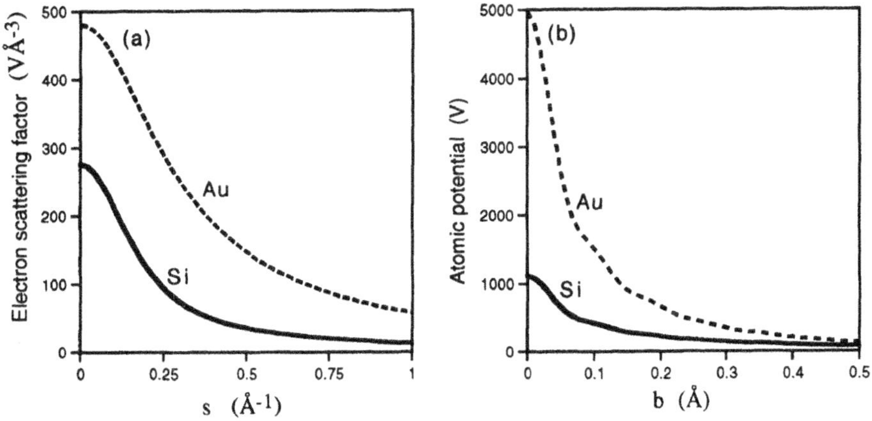

Fig. 1.3 A comparison of **a** electron-scattering factors and **b** atomic potentials of gold and silicon atoms, showing the increase in atom-scattering power with an increased atomic number, where $s = \frac{\sin\theta}{\lambda}$ and $b = \left(x^2 + y^2\right)^{1/2}$

$s = 0$ requires that $f^x(0) = Z_0$ according to the Mott formula or equivalently

$$f^x(0) = Z_0 = \sum_{i=1}^{4} a_i + c_i$$

The empirical analytic expression given by (1.13) tends to deviate from the measured scattering factor when $s > 2\text{Å}^{-1}$. In this case, the expression proposed by Weickenmeier and Kohl (1991) gives better fit especially for $1.5 < s < 3\text{Å}^{-1}$

$$f^e(\mathrm{s}) = s^{-2}\sum_{i=1}^{6} A_i\left[1 - \exp\left(-B_i s^2\right)\right]$$

where A_i and B_i are fitting parameters.

Figure 1.3a compares electron-scattering factors of silicon and gold, and Fig. 1.3b shows the corresponding atomic potentials. It is apparent that the scattering power of Au is much stronger than that of Si. The height of the potential peak is a direct measure of the high-angle scattering power of the atom. The full width of the half-maximum of the potential peak is about 0.2Å.

1.5 Kinematic Electron Diffraction in Thin Crystals

If a crystal is very thin and the crystal potential is weak, then electron scattering can be treated kinematically; i.e., the electron is scattered only once, and there are no multiple-scattering effects. We first consider a three-dimensional periodic crystal

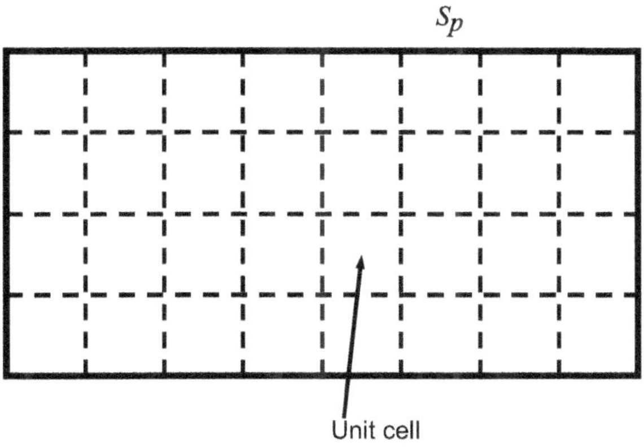

Fig. 1.4 A finite-size periodically structured crystal defined by a shape function S_p. The blocks enclosed by dashed lines represent the distribution of crystal unit cells

of finite size. The shape of this crystal is described by the shape function $S_p(\mathbf{r})$, as illustrated in Fig. 1.4. Under the rigid-body approximation, the potential distribution in real space $\mathbf{r}$ is written as a superposition of the atomic potentials

$$V(\mathbf{r}) = \left\{ \sum_n \sum_\alpha V_\alpha (\mathbf{r} - \mathbf{R}_n - \mathbf{r}_\alpha) \right\} S_p(\mathbf{r}) \tag{1.14}$$

where n is summed over all unit cells filling the entire space (i.e., $-\infty < n < \infty$) and α is summed over all atoms within a unit cell. The kinematic-scattering amplitude of the crystal is a Fourier transform of (1.14) according to the Born approximation,

$$\begin{aligned}
V(\mathbf{u}) = \mathbf{FT}[V(\mathbf{r})] &= \left\{ \sum_n \sum_\alpha \int d\mathbf{r} V_\alpha (\mathbf{r} - \mathbf{R}_n - \mathbf{r}_\alpha) \exp(-2\pi i \mathbf{u} \cdot \mathbf{r}) \right\} \otimes S_p(\mathbf{u}) \\
&= \left\{ \sum_n \exp(-2\pi i \mathbf{u} \cdot \mathbf{R}_n) \sum_\alpha \int d\mathbf{r} V_\alpha(\mathbf{r}) \exp(-2\pi i \mathbf{u} \cdot \mathbf{r}) \exp(-2\pi i \mathbf{u} \cdot \mathbf{r}_\alpha) \right\} \otimes S_p(\mathbf{u}) \\
&= \left\{ \sum_g \delta(\mathbf{u} - \mathbf{g}) \sum_\alpha V_\alpha(\mathbf{u}) \exp(-2\pi i \mathbf{u} \cdot \mathbf{r}_\alpha) \right\} \otimes S_p(\mathbf{u}) \\
&= \left\{ \sum_g \delta(\mathbf{u} - \mathbf{g}) \sum_\alpha V_\alpha(\mathbf{g}) \exp(-2\pi i \mathbf{g} \cdot \mathbf{r}_\alpha) \right\} \otimes S_p(\mathbf{u}) \\
&= \left\{ \sum_g \delta(\mathbf{u} - \mathbf{g}) V_g \right\} \otimes S_p(\mathbf{u})
\end{aligned} \tag{1.15}$$

where $\otimes$ indicates a convolution calculation (see Sect. 1.8) and the Fourier coefficient of the crystal potential is defined as

$$V_g = \frac{1}{\Omega}\sum_{\alpha} f_\alpha(g) < \exp\left[2\pi i\mathbf{g}\cdot(\mathbf{r}_\alpha + \mathbf{u}_\alpha)\right] >= \sum_{\alpha} V_\alpha(\mathbf{g})\exp\left(-2\pi i\mathbf{g}\cdot\mathbf{r}_\alpha\right) \tag{1.16a}$$

the scattering power of the αth atom in the unit cell is

$$V_\alpha(\mathbf{g}) = \frac{1}{\Omega} f_\alpha^e(\mathbf{g})\exp\left(-W_\alpha(g)\right) \tag{1.16b}$$

where the time average $< >$ is introduced to take into account the thermal vibration effect of the atom near its equilibrium position; $\mathbf{u}_\alpha$ is the time-dependent displacement of the atom; and $\exp\left(-W_\alpha(g)\right) = \exp\left(-2\pi^2 < A_\alpha^2 > g^2\right)$ is the Debye-Waller (DW) factor, which is included in all elastic-scattering calculations to characterize weakening of the scattering power of an atom due to its thermal vibration. In deriving (1.15), an identity

$$\sum_n \exp\left(-2\pi i\mathbf{u}\cdot\mathbf{R}_n\right) = \sum_g \delta(\mathbf{u}-\mathbf{g}) \tag{1.17}$$

was used, where $\mathbf{g}$ is defined as a reciprocal lattice vector satisfying

$$\mathbf{g}\cdot\mathbf{R}_n = \mathbf{g}\cdot(n_1\mathbf{a} + n_2\mathbf{b} + n_3\mathbf{c}) = \text{ integer} \tag{1.18}$$

n_1, n_2, and n_3 are integers; **a**, **b**, and **c** are the base vectors of the unit cell. The Dirac delta function $\delta(\mathbf{u}-\mathbf{g})$ allows only reflections that satisfy the Bragg's law in reciprocal space **u**. The intensity of each Bragg reflection is scaled by $\left|V_g\right|^2$, conditions under which $V_g = 0$ for some **g**'s give extinction rules for forbidden reflections under the kinematic-scattering approximation. The forbidden reflections, however, may appear in the diffraction pattern if the multiple-scattering effect is dominant (Gjønnes and Moodie 1965). The sharp Bragg reflections are broadened due to the convolution of the crystal shape factor S_p. Here we have introduced the concept of reciprocal space, Bragg's law, the convolution operation, and the Dirac delta function. Detailed discussions of these are given in the next few sections.

It is very important to point out that (1.17) is a key equation for converting a summation over unit cells in real space into a summation of reciprocal lattice vectors. This equation is frequently used in the following mathematical operations.

To illustrate the nature of the terms appearing in (1.15), Fig. 1.5 shows two electron diffraction patterns of a thin MgO crystal oriented along [100] and [110]. Each Bragg reflection is positioned at a reciprocal lattice point **g**, and the intensity of this reflection is scaled (in kinematic scattering theory) according to $\left|V_g\right|^2$. The intensity at non-Bragg reflection positions is simply zero. Diffraction spots that are absent in diffraction patterns are those where the structure factors vanish. The (001) and (011) reflections of MgO are such examples.

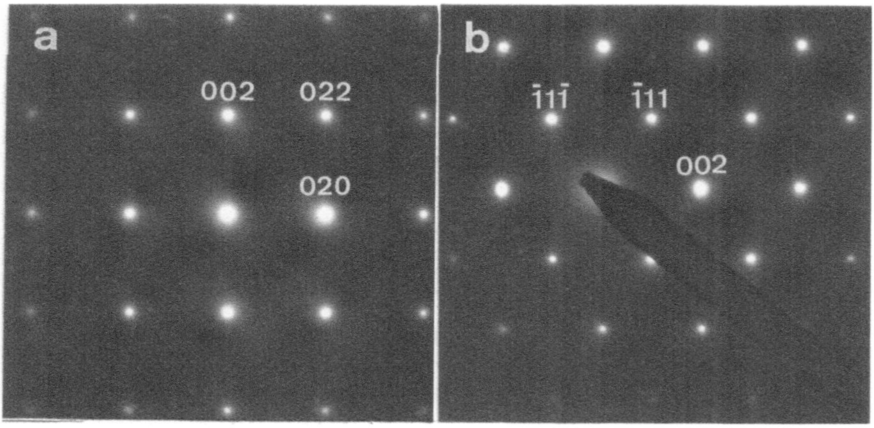

Fig. 1.5 Transmission electron diffraction patterns recorded at 100 keV from an MgO crystal oriented along **a** [100] and **b** [110] directions. The central transmitted (000) beam in (**b**) was blocked by a beam stop

1.6 Reciprocal Space

As shown in (1.18), reciprocal space is defined by a vector $\mathbf{g} = \ell\mathbf{a}^* + m\mathbf{b}^* + n\mathbf{c}^*$, where (ℓ, m, n) are the Miller indices of $\mathbf{g}$ and satisfy

$$\left(\ell\mathbf{a}^* + m\mathbf{b}^* + n\mathbf{c}^*\right) \cdot (n_1\mathbf{a} + n_2\mathbf{b} + n_3\mathbf{c}) = \text{ integer} \tag{1.19}$$

This equation can be satisfied by choosing

$$\mathbf{a}^* = \frac{\mathbf{b} \times \mathbf{c}}{\Omega} \quad \mathbf{b}^* = \frac{\mathbf{c} \times \mathbf{a}}{\Omega} \quad \mathbf{c}^* = \frac{\mathbf{a} \times \mathbf{b}}{\Omega} \tag{1.20}$$

where $\Omega = (\mathbf{a} \times \mathbf{b}) \cdot \mathbf{c}$ is the volume of the unit cell. It can easily be proved

$$\mathbf{a}^* \cdot \mathbf{b} = \mathbf{a}^* \cdot \mathbf{c} = \mathbf{b}^* \cdot \mathbf{a} = \cdots = 0 \tag{1.21a}$$

and

$$\mathbf{a}^* \cdot \mathbf{a} = \mathbf{b}^* \cdot \mathbf{b} = \mathbf{c}^* \cdot \mathbf{c} = 1 \tag{1.21b}$$

The space defined by base vectors ($\mathbf{a}^*$, $\mathbf{b}^*$, $\mathbf{c}^*$) is called reciprocal space (or momentum space). A point in reciprocal space specifies the electrons traveling in the same direction in real space. A high-resolution lattice image in real space is the interference result of electron beams located at different points in reciprocal space. Therefore the angular distribution of scattered electrons is more conveniently described in reciprocal space. Figure 1.6 shows a crystal unit cell and corresponding reciprocal lattice vectors.

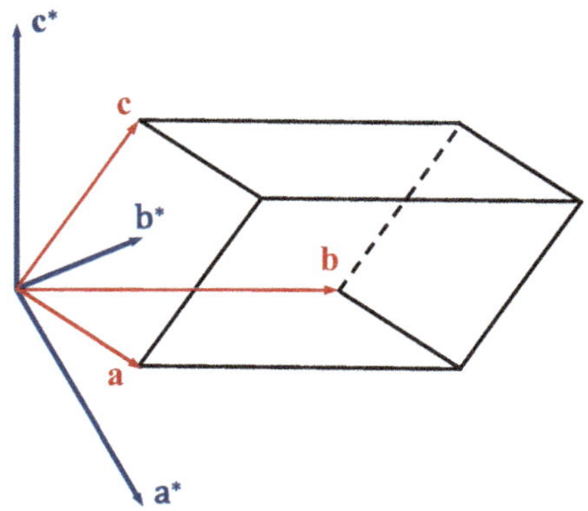

Fig. 1.6 Relationships between the lattice vectors of a crystal unit cell and the corresponding reciprocal lattice vectors

1.7 Bragg's Law

As shown in (1.15), the diffraction of a perfect crystal forms a set of sharp peaks distributed at the reciprocal lattice sites defined by $\mathbf{g} = \ell\mathbf{a}^* + m\mathbf{b}^* + n\mathbf{c}^*$. A strong reflection peak is generated if the change in electron wave vector $\mathbf{u} = \mathbf{K} - \mathbf{K}_0$ equals the reciprocal space vector **g**,

$$\mathbf{u} = \ell\mathbf{a}^* + m\mathbf{b}^* + n\mathbf{c}^* \tag{1.22}$$

This is exactly the result of conservation of momentum in elastic electron scattering. The meaning of (1.22) can be clearly stated using the Ewald sphere construction in reciprocal space, as shown in Fig. 1.7 For elastic scattering, the conservation of energy requires $K = K_0 = 1/\lambda$; thus a sphere of radius K is drawn in reciprocal space with the wave vector of the incident electron beam ending at the origin. The wave vector of the diffracted beam is drawn from the center of the sphere, and the diffracted intensity at **u** is $|V(\mathbf{u})|^2$. The possible beams generated in diffraction pattern are the reciprocal lattice vectors intersecting the Ewald sphere. Circles made by intersections of the Ewald sphere with the planes of reciprocal lattice vectors are called the Laue circles. The reflections are thus classified as the zero-order Laue zone (ZOLZ), the first-order Laue zone (FOLZ), and higher order Laue zones (HOLZs). The ZOLZ is usually defined as the plane of reciprocal lattice points that passes through the origin. An HOLZ is any other reciprocal lattice plane parallel to the plane of the ZOLZ, not passing through the origin, as illustrated in Fig. 1.7. The Ewald sphere construction is a simple and direct method for illustrating the diffracting conditions of electron diffraction. As demonstrated later, classifying Bragg beams as ZOLZ and HOLZ reflections makes theoretical treatment very different.

Figure 1.8 shows an experimental diffraction pattern of [0001] $\alpha - Al_2O_3$. The HOLZs are visible up to the fourth order. If the beam direction is parallel to c^*, the radius of the m th Laue zone ring is $u_m = (2c^* m K_0)^{1/2}$. Thus

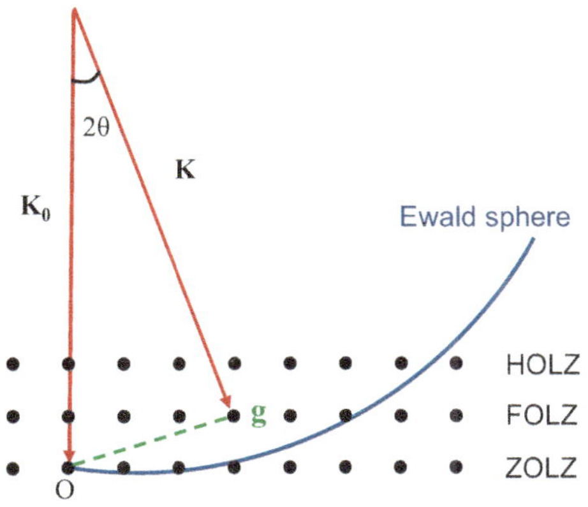

Fig. 1.7 Construction of the Ewald sphere in reciprocal space, illustrating the electron-diffracting condition. Bragg reflections belonging to different Laue zones are indicated by dark spots

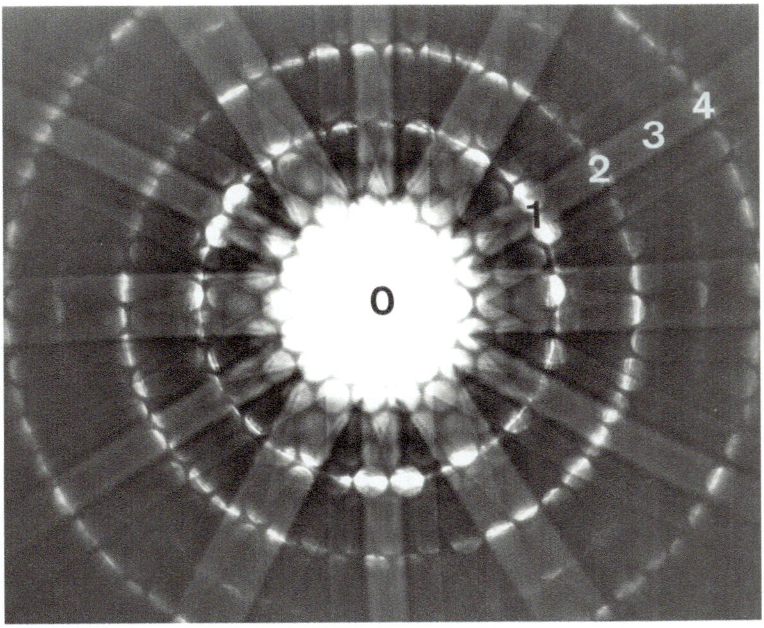

Fig. 1.8 HOLZ reflections of an $\alpha - Al_2O_3$ crystal viewed along [0001]. Numbers indicate corresponding orders of Laue circles. The diffraction pattern was recorded at 100 keV with a converged beam to enhance the intensity of the HOLZ rings

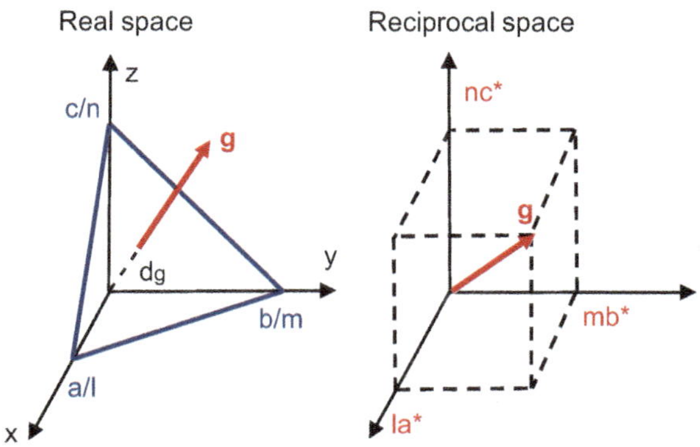

Fig. 1.9 Relationships between **a** real-space atomic planes and the corresponding **b** reciprocalspace lattice vectors for an orthogonal crystal structure

$$u_{m+1}^2 - u_m^2 = 2c^* K_0$$

which can be used to measure the reciprocal lattice vector c^*. Besides HOLZs, Kikuchi bands are also observed. These are the results of electron inelastic scattering, and they are discussed in Part II of this book.

We now return to diffracting conditions. From (1.22), the projections of **u** on real space axes are

$$\mathbf{u} \cdot \mathbf{a} = \ell, \quad \mathbf{u} \cdot \mathbf{b} = m, \quad \mathbf{u} \cdot \mathbf{c} = n \tag{1.23}$$

which represent the well-known Laue conditions for diffraction. In reciprocal space, each reciprocal-space lattice point (ℓ, m, n) represents a set of equally spaced, parallel atomic planes in real space. When one plane is drawn through an atom in the unit-cell origin, the intercepts of the next parallel plane of the set on the axes are a/ℓ, b/m, and c/n. For an orthogonal unit-cell system, as shown in Fig. 1.9, the perpendicular distance between the adjacent planes of the set is

$$\frac{1}{d_g^2} = \frac{\ell^2}{a^2} + \frac{m^2}{b^2} + \frac{n^2}{c^2} \tag{1.24}$$

On the other hand, the distance from the reciprocal lattice origin to the (ℓ, m, n) reciprocal lattice point is the inverse of the distance between the adjacent atomic planes,

$$g = \frac{1}{d_g} \tag{1.25a}$$

and the vector perpendicular to the $(\ell, m, n\ )$ lattice planes is the vector from the origin to the (ℓ, m, n) reciprocal lattice point in reciprocal space. These relationships hold also for nonorthogonal axes. Using (1.25), the modulus of (1.22) is $u = 1/d_g$. Using the scattering angle defined in Fig. 1.2, Bragg's law is explicitly written as

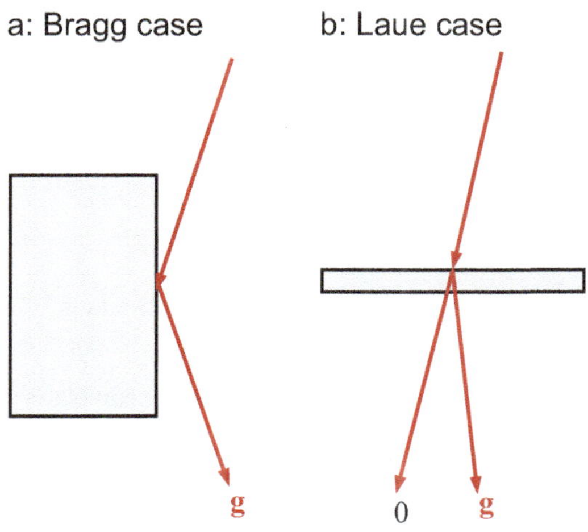

Fig. 1.10 Diagrams showing the diffraction of electrons in the **a** Bragg and **b** Laue cases

$$2d_g \sin\theta_g = \lambda \tag{1.25b}$$

This condition is similar to the $2d_g \sin\theta_g = \lambda$ law for light reflection from a semi-infinity medium. Thus the well-defined diffracted beam given by (1.25b) is conveniently referred to as the Bragg reflection. Equation (1.25b) is usually called the Bragg condition, under which the corresponding reciprocal lattice point intersects the Ewald sphere with zero excitation error.

The diffraction of electrons is classified as a Bragg case or a Laue case. In the Bragg case, an incident beam is diffracted from planes parallel or nearly parallel to the flat surface of a crystal that may be considered semi-infinite. Thus diffracted beams observed are those that reenter the vacuum on the same side of the crystal as the incident beam (Fig. 1.10a). Reflection high-energy electron diffraction (RHEED) from a bulk crystal surface and low-energy electron diffraction (LEED) are typical examples of Bragg cases. Hence the Bragg case is basically the case of backscattering. The dynamical diffraction of reflected electrons in RHEED is described in Chap. 5.

The Laue case involves transmission diffraction (without backscattering) through a parallel-sided crystal plate of infinite extent in two dimensions (Fig. 1.10b). This is the situation in TEM. Most of the following chapters discuss electron diffraction in the Laue case.

1.8 Abbe's Imaging Theory

In TEM micrographs, image contrast is the result of electron scattering by atoms in the specimen and transfer properties of the optic system. The image formation is usually described by Abbe's theory, which is schematically shown in Fig. 1.11. The specimen is illuminated by a plane wave. The transmitted wave function $\Psi(x, y)$ at

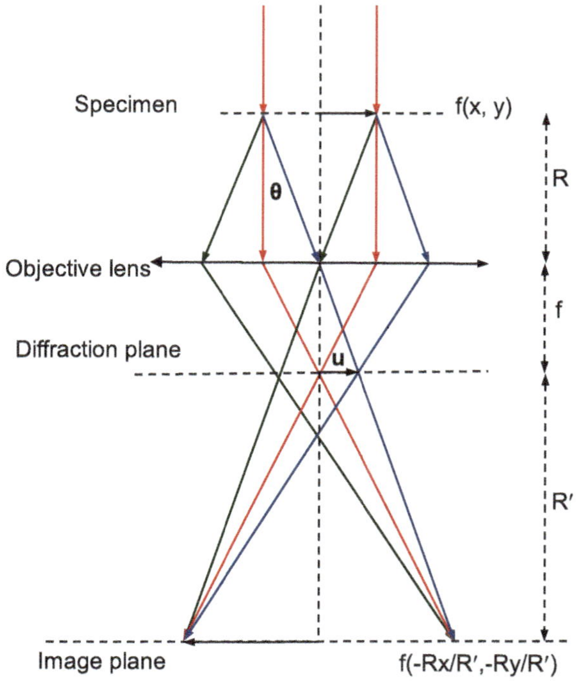

Fig. 1.11 Abbe's theory of image formation in a TEM

the exit face of the specimen can be assumed to be composed of a forward-scattered, axial, transmitted wave and other scattered waves that proceed in directions slightly inclined to it.

An ideal thin lens brings the parallel transmitted waves to a focus on the axis in the back focal plane. Waves leaving the specimen in the same direction (or angle θ with the optic axis) are brought together at a point on the back focal plane (or diffraction plane) at a distance $X = f \tan\theta$ from the axis, where f is the focal length. Thus on the back focal plane, waves from all parts of the illuminated regions propagating in a given direction are added. This is equivalent to the case in which the observation point is at infinity. So in the back focal plane, a Fraunhofer diffraction pattern is formed. The variable used for the distribution of the diffracted amplitude in the diffraction pattern for the one-dimension case is $u = 2K\sin(\theta) \approx 2\theta/\lambda$. Mathematically the formation of the Fraunhofer diffraction pattern is described by a Fourier transform (Cowley 1981),

$$\Psi(\mathbf{u}) = \mathbf{FT}[\Psi(x, y, d)] \tag{1.26}$$

The formation of an image is then described by the inverse Fourier transform of the diffracting amplitude $\Psi(\mathbf{u})$

$$\Psi_i(x, y, d) = \mathbf{FT}^{-1}[\Psi(\mathbf{u})] = \Psi(-x/M, -y/M, d) \tag{1.27}$$

where the minus sign means the inversion of the object by the objective lens and M is the magnification of the lens. Equation (1.27) is for the case of an ideal objective lens. In general, the chromatic and spherical aberrations of the lens, as well as the size of the objective aperture, introduce a transfer function $T_{\text{obj}}(\mathbf{u})$ for the optic system,

$$T_{\text{obj}}(\mathbf{u}) = A_{\text{obj}}(\mathbf{u}) \exp\left(\frac{1}{2}\pi i C_s \lambda^3 u^4 + \pi i \Delta f \lambda u^2\right) \tag{1.28a}$$

where A_{obj} is the shape function of the objective aperture, C_s is the spherical aberration coefficient of the lens, and Δf is the defocus of the lens. Thus (1.27) is modified as

$$\begin{aligned}\Psi_i(x, y, d) &= \mathbf{FT}^{-1}\left[\Psi(\mathbf{u}) T_{\text{obj}}(\mathbf{u})\right] \\ &= \Psi(\mathbf{r}) \otimes \mathbf{FT}^{-1}\left[T_{\text{obj}}(\mathbf{u})\right] = \int d\mathbf{b}' \Psi\left(\mathbf{b}', d\right) T_{\text{obj}}\left(\mathbf{b} - \mathbf{b}'\right)\Big]\end{aligned} \tag{1.28b}$$

The diffraction pattern is given by

$$|\Psi(\mathbf{u})|^2 = |\mathbf{FT}\left[\Psi_i(x, y)\right]|^2 = \left|\int d\mathbf{b} \exp(-2\pi i \mathbf{u} \cdot \mathbf{b}) \Psi_i(\mathbf{b}, d)\right|^2 \tag{1.28c}$$

As a summary, Abbe's imaging theory is a two-step process-taking a Fourier transform of the electron wave function at the exit face of the crystal, then multiplying the lens transfer function and inversely Fourier-transforming the diffraction amplitude back into real space. The electron diffraction pattern is a modulus square of the Fourier transformed wave function at the exit face of the crystal. Applications of Abbe's theory to approaching different imaging techniques are described in Chaps. 11–13.

1.9 Some Mathematical Operations

Several mathematical operations have been introduced, and they are frequently used in diffraction and imaging theories. This section outlines some basic features of these mathematical operations.

1.9.1 Fourier Transformation

The Fourier transform is mostly used to correlate electron waves in reciprocal space (or the diffraction plane) with those in real space (or the image plane), which is defined as

$$\Psi(\mathbf{u}) = \mathbf{FT}[\Psi(\mathbf{b})] = \int d\mathbf{b} \exp(-2\pi i \mathbf{u} \cdot \mathbf{b})\Psi(\mathbf{b}) \quad (1.29a)$$

and the inverse Fourier transform is

$$\Psi(\mathbf{b}) = \mathbf{FT}^{-1}[\Psi(\mathbf{u})] = \int d\mathbf{u} \exp(2\pi i \mathbf{u} \cdot \mathbf{b})\Psi(\mathbf{u}) \quad (1.29b)$$

In this book, integral limits are $(-\infty, \infty)$ unless otherwise specified. The $(-\infty, \infty)$ integral limits are usually omitted in a mathematical expression for simplification.

The Fourier transformation is an important mathematical calculation in describing imaging and diffraction theories (Cowley 1981). A useful relation in imaging theory is the Fourier transform of a plane wave

$$\delta(\mathbf{u} - \mathbf{K}_b) = \int d\mathbf{b} \exp(-2\pi i \mathbf{u} \cdot \mathbf{b}) \exp(2\pi i \mathbf{K}_b \cdot \mathbf{b})$$

It is important to point out that the Fourier transform of a function is denoted by the same symbol in this book, but the variable is replaced by **u**. The Fourier transform of $\Psi(\mathbf{b})$, for example, is denoted by $\Psi(\mathbf{u})$. It is also necessary to indicate that the sign of the phase factor in the Fourier transform is consistent in the entire book. Appendix B gives the general properties of the Fourier transform.

1.9.2 Convolution Calculation

A convolution calculation, denoted by $\otimes$, is usually applied in imaging theory, and it is defined by

$$F(\mathbf{b}) \otimes G(\mathbf{b}) = \int d\mathbf{b}' F(\mathbf{b} - \mathbf{b}') G(\mathbf{b}') = \int d\mathbf{b}' F(\mathbf{b}') G(\mathbf{b} - \mathbf{b}') \quad (1.30)$$

The convolution of two functions commutes,

$$F(\mathbf{b}) \otimes G(\mathbf{b}) = G(\mathbf{b}) \otimes F(\mathbf{b}) \quad (1.31)$$

The effect of convolution is to broaden the distribution profile of one function by that of another. This feature can be illustrated by the convolution of two Gaussian functions

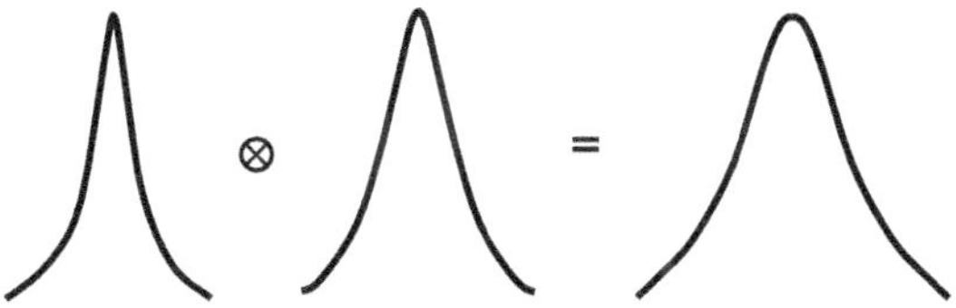

Fig. 1.12 A schematic diagram showing the convolution result of two real functions

$$\exp\left(-\frac{x^2}{a^2}\right) \otimes \exp\left(-\frac{x^2}{b^2}\right) = \frac{\pi^{1/2}ab}{\left(a^2+b^2\right)^{1/2}} \exp\left(-\frac{x^2}{a^2+b^2}\right) \quad a, b > 0$$

A numerical plot of this calculation is illustrated in Fig. 1.12. The convolution calculation is applied in inelastic-scattering theory to characterize the angular redistribution and broadening of electrons due to inelastic excitations.

In electron diffraction calculation, the convolution theorem in (1.32a, b) is often used. The Fourier transform of a product of two functions is the convolution of their Fourier transforms

$$\mathbf{FT}[F(\mathbf{b})G(\mathbf{b})] = \mathbf{FT}[F(\mathbf{b})] \otimes \mathbf{FT}[T(\mathbf{b})] = F(\mathbf{u}) \otimes G(\mathbf{u}) \tag{1.32a}$$

or

$$\mathbf{FT}^{-1}[F(\mathbf{u}) \otimes G(\mathbf{u})] = F(\mathbf{b})G(\mathbf{b}) \tag{1.32b}$$

This relation simply means that the convolution of two functions in reciprocal space is the result of a Fourier transform of the product of the two functions in real space. Equation (1.32b) can easily be proved using the variable substitution method

$$\begin{aligned}
\mathbf{FT}^{-1}[F(\mathbf{u}) \otimes G(\mathbf{u})] &= \int d\mathbf{u} \exp(2\pi i \mathbf{u} \cdot \mathbf{b}) \int d\mathbf{u}' F\left(\mathbf{u} - \mathbf{u}'\right) G\left(\mathbf{u}'\right) \\
&= \int d\mathbf{u} \int d\mathbf{u}' \exp\left[2\pi i \left(\mathbf{u} - \mathbf{u}'\right) \cdot \mathbf{b}\right] F\left(\mathbf{u} - \mathbf{u}'\right) \exp\left(2\pi i \mathbf{u}' \cdot \mathbf{b}\right) G\left(\mathbf{u}'\right) \\
&= \int d\mathbf{u}'' \exp\left[2\pi i \mathbf{u}'' \cdot \mathbf{b}\right] F\left(\mathbf{u}''\right) \int d\mathbf{u}' \exp\left(2'\pi \mathbf{u}' \cdot \mathbf{b}\right) G\left(\mathbf{u}'\right) \\
&= F(\mathbf{b})G(\mathbf{b})
\end{aligned}$$

The convolution theorem is normally used in the multislice theory (Chap. 3) to convert the electron wave from real space to reciprocal space, and vise versa.

1.9.3 Dirac Delta Function

The δ function defined in (1.33a, b) is frequently used in diffraction theory. For the two-dimensional case,

$$\delta(\mathbf{b}) = \begin{cases} 0 & \mathbf{b} \neq 0 \\ \infty & \mathbf{b} = 0 \end{cases} \tag{1.33a}$$

and

$$\int d\mathbf{b}\delta(\mathbf{b}) = 1 \tag{1.33b}$$

The delta function can be considered the limit of a set of real continuous functions, such as Gaussians

$$\delta(x) = \lim_{\chi \to 0} \left\{ \frac{1}{\pi\chi} \exp\left(-\frac{x^2}{\chi^2}\right) \right\} \tag{1.34}$$

An important definition of the delta function, as applied to scattering, is

$$\delta\left(\mathbf{u} - \mathbf{u}'\right) = \int d\mathbf{b} \exp\left[2\pi i \left(\mathbf{u} - \mathbf{u}'\right) \cdot \mathbf{b}\right] \tag{1.35}$$

A very useful relation for converting a summation over unit cells in real space into a summation of reciprocal lattice vectors is

$$\sum_{n=-\infty}^{\infty} \exp\left(2\pi i \mathbf{R}_n \cdot \mathbf{u}\right) = \sum_{g} \delta(\mathbf{u} - \mathbf{g}) \tag{1.36}$$

The basic features of the delta function are given in Appendix C. The delta function has two important properties: The delta function is a symmetric function

$$\delta(\mathbf{b}) = \delta(-\mathbf{b}) \tag{1.37a}$$

The convolution of the delta function with a continuous function is the function itself

$$\int d\mathbf{b}' F\left(\mathbf{b}'\right) \delta\left(\mathbf{b} - \mathbf{b}'\right) = F(\mathbf{b}) \otimes \delta(\mathbf{b}) = F(\mathbf{b}) \tag{1.37b}$$

Applications of these properties are illustrated in subsequent chapters.

Kinematical scattering theory, as the base of analyzing electron diffraction patterns, is important for understanding the dynamical scattering theory. In Chap. 1, based on the first Born approximation, the electron scattering factor, reciprocal space and reciprocal lattice vectors are introduced. The Ewald sphere is a simple, elegant geometrical interpretation of electron diffraction. Many of the imaging and diffraction techniques to be introduced in Chap. 2 can be understood from the Ewald sphere construction. Fourier transform, convolution calculation and delta function are very useful mathematical tools for describing electron diffraction. Chapter 1 is a preliminary chapter that is indispensable for describing the dynamical theories in the remaining chapters of this book.

Chapter 2
Dynamic Elastic Electron Scattering I: Bloch Wave Theory

The kinematic scattering theory illustrated in Chap. 1 holds only for crystals thinner than about 2 nm (Hoerni 1956). The general characteristics, such as the positions of Bragg beams, can be precisely determined by the kinematic theory. The intensity of each reflection, however, is largely affected by multiple scattering effects among existing beams, because the interaction between the electron and the crystal is so strong that multiple-scattering effects are unavoidable. Therefore quantitative data analysis has to be performed based on dynamic scattering calculations. For this reason, the present and following chapters are devoted to the dynamic scattering behavior of electrons.

Diffraction theories of high-energy electrons have been established for many years. One of the most classical approaches is the Bloch wave theory, proposed by Bethe (1928). This theory has taken a dominant role in calculating convergent beam electron diffraction (CBED) and diffraction contrast imaging. In this chapter, the Bloch wave is introduced and its general properties are illustrated. Some simplified cases are discussed to illustrate applications of the theory. Finally the theory is extended to imperfect crystals.

2.1 Relativistic Corrections in Single-Electron Scattering Theory

Based on the first principles approach, we consider the fundamental equation that governs high-energy electron scattering in crystals. Before we show the mathematical description, it is important to consider the nature of the events we are studying. The average distance between successive electrons that strike the crystal in a TEM is about 0.2 mm (for 100 − keV electrons) if the electron flux is on the order of 10^{12} e/s. This distance is much larger than the thickness (typically less than 0.5 μm) of the specimen; thus the interaction between successive incident electrons is extremely

Z. L. Wang, *Elastic and Inelastic Scattering in Electron Diffraction and Imaging*, Springer Series in Solid-State Sciences 205, https://doi.org/10.1007/978-3-031-90819-4_2

weak. Therefore interaction between the incident beam and the crystal can be treated as a case of one electron a time. In other words, electron diffraction theory is basically a single-electron scattering theory. The entire discussion of this book is based on this assumption.

Strictly speaking, high-energy electron scattering obeys the Dirac equation. The Dirac equation contains not only the relativistic effects but also electron spin. It has been shown by Fujiwara (1961) and Howie (1962) that electron spin is negligible in transmission electron diffraction, but relativistic corrections in the electron mass and wavelength have to be considered in both kinematic and dynamic scattering theories (see Gevers and David 1982, for a review). In general, the solution of the Dirac equation for high-energy electron diffraction is quite complex (Fujiwara 1961). We intend to derive a relativistically corrected Schrödinger-like equation. The solution of the equation should be sufficient for accuracy in quantitative electron microscopy. The relativistic energy momentum conservation equation for an electron accelerated to a kinetic energy of $e(U_0 + V)$ is

$$W^2 = p^2c_0^2 + m_0^2c_0^4 = \left[e(U_0 + V) + m_0c_0^2\right]^2 \tag{2.1a}$$

and

$$W = m_e c_0^2 = \frac{m_0 c_0^2}{\left[1 - (v/c_0)^2\right]^{1/2}} = e(U_0 + V) + m_0 c_0^2 \tag{2.1b}$$

where V is the electrostatic potential field of the crystal and c_0 is the speed of light. Neglecting the V^2 term, combining (2.1a, b) yields

$$p^2c_0^2 - 2m_e c_0^2 eV \approx eU_0\left(eU_0 + 2m_0c_0^2\right) \tag{2.2}$$

Replacing p by the operator $-i\hbar\nabla$ and applying (2.2) on Ψ, a Schrödinger-like equation (Humphreys 1979; Spence 1988a) is derived

$$-\frac{\hbar^2}{2m_0}\nabla^2\Psi - \gamma eV\Psi = E\Psi \tag{2.3}$$

where

$$E = eU_0\left[1 + \frac{eU_0}{2m_0c_0^2}\right]$$

and the relativistic factor $\gamma = m_e/m_0$. E is electron energy with relativistic correction. Therefore under the first-order approximation, the Schrödinger equation can be used to describe high-energy electron scattering if relativistic corrections are properly considered. For $V = 0$, the free-space solution of (2.3) gives exactly the same wavelength as that originally based on de Broglie's relation in (1.3). By defining an effective crystal potential

$$U(\mathbf{r}) = \frac{2\gamma m_0 e}{h^2} V(\mathbf{r}) \tag{2.4a}$$

and the electron wave number as

$$K = \frac{(2m_0 E)^{1/2}}{h} \tag{2.4b}$$

Equation (2.3) is rewritten as

$$\left\{\nabla^2 + 4\pi^2 \left[U(\mathbf{r}) + K^2\right]\right\} \Psi(\mathbf{r}) = 0 \tag{2.5}$$

This is the fundamental equation that governs the scattering behavior of electrons in crystals. Discussions in this chapter and Chaps. 3–5 show how to solve this equation using various methods.

2.2 Bethe Theory

2.2.1 Basic Equations

The solution of (2.5) was first given by Bethe (1928). The full solution is written as a linear superposition of Bloch waves

$$\Psi(\mathbf{r}) = \sum_i \alpha_i B_i(\mathbf{r}) \tag{2.6}$$

each Bloch wave $B_i(\mathbf{r})$ is an eigen solution of (2.5), and coefficients α_i are determined by boundary conditions. Equation (2.6) means that each Bloch wave is an eigenstate of the electron crystal system, and the electron wave function is a linear superposition of Bloch waves. The probability that the i th Bloch wave is excited is determined by the superposition coefficient α_i. Although there are many Bloch wave states in the crystal, boundary conditions determine which waves are excited. This is similar to the selection of Bragg beams by the diffracting condition in electron diffraction. Even though there are many possible Bragg reflections, those in the diffraction pattern are selected by the initial diffracting condition, as illustrated by the Ewald sphere in Fig. 1.7.

The Bloch wave theory is usually convenient for examining the diffraction of a periodically structured crystal. In this case, the modified crystal potential U can be expanded as a Fourier series based on reciprocal lattice vectors,

$$U(\mathbf{r}) = \sum_g U_g \exp(2\pi i \mathbf{g} \cdot \mathbf{r}) \tag{2.7a}$$

with

$$U_g = \frac{2\gamma m_0 e}{h^2} \sum_\alpha V_g \exp\left(-2\pi i \mathbf{g} \cdot \mathbf{r}_\alpha\right) \tag{2.7b}$$

Similarly the Bloch wave within the crystal may also be written as a Fourier series

$$B(\mathbf{r}) = \sum_g C_g \exp[2\pi i (\mathbf{k} + \mathbf{g}) \cdot \mathbf{r}] \tag{2.8}$$

where C_g is the Bloch wave coefficient for Bragg reflection **g**. For high-energy electron diffraction, a Bloch wave is neither a spherical wave nor a single plane wave but a linear superposition of plane waves with wave vectors $(\mathbf{k} + \mathbf{g})$. A Bloch wave contains many plane wave components.

Equation (2.8) is introduced to convert a second-order differential equation into a set of linear algebra equations whose solution can easily be obtained from matrix diagonalization. Substituting (2.7) and (2.8) into (2.5) yields

$$\sum_g \left\{ \left[K^2 - (\mathbf{k} + \mathbf{g})^2\right] C_g + \sum_h U_{g-h} C_h \right\} \exp[2\pi i (\mathbf{k} + \mathbf{g}) \cdot \mathbf{r}] = 0 \tag{2.9}$$

where **h** is the reciprocal lattice vector and the interaction between **g** and **h** beams is determined by U_{g-h}. Equation (2.9) holds for all **r** in the crystal. Hence the coefficient of each exponential term must be zero. For the **g** reflection,

$$\left[K^2 - (\mathbf{k} + \mathbf{g})^2\right] C_g + \sum_h U_{g-h} C_h = 0 \tag{2.10a}$$

Equation (2.10a) is actually a set of coupled equations for different **g** reflections. By choosing

$$\mathbf{k} = \mathbf{K} + \nu \mathbf{n} \tag{2.10b}$$

where **n** is a unit vector inward and normal to the crystal slab surface, as shown in Fig. 2.1. We have

$$\left[K^2 - (\mathbf{k} + \mathbf{g})^2\right] = 2\,K S_g - 2(\mathbf{K} + \mathbf{g}) \cdot \mathbf{n}\nu - \nu^2 \tag{2.11a}$$

where the excitation errors S_g are defined as

$$2\,K S_g = K^2 - (\mathbf{K} + \mathbf{g})^2 \tag{2.11b}$$

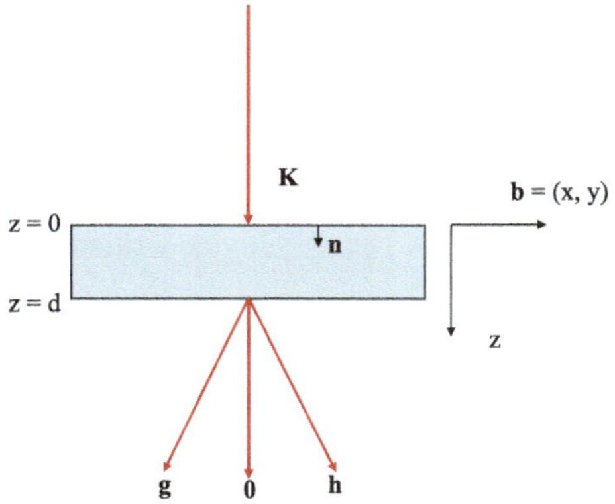

Fig. 2.1 The coordinate system used in describing the transmission electron diffraction

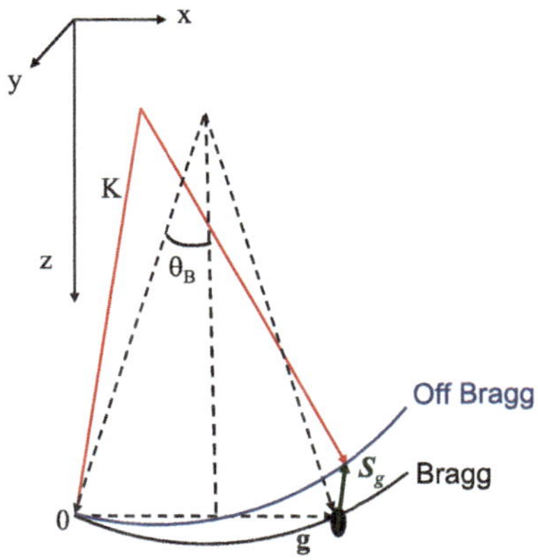

Fig. 2.2 The Ewald sphere construction of electron diffraction at on-Bragg (dashed lines) and off-Bragg (solid lines) conditions. The excitation error S_g is indicated

The S_g is the distance from a reciprocal lattice point **g** to the intersection point of a line, drawn from **g** in parallel to **K**, with the Ewald sphere (see Fig. 2.2). The Bragg condition is defined by $S_g = 0$, which means the Bragg beam is located just on the surface of the Ewald sphere. By neglecting the v^2 term for high-energy transmission electron diffraction (the error introduced by this approximation is very small according to Kim and Sheinin 1982), substituting (2.11a) into (2.10a) yields

$$2\,KS_gC_g + \sum_h U_{g-h}C_h = 2(\mathbf{K}+\mathbf{g})\cdot\mathbf{n}vC_g \tag{2.12}$$

This is the fundamental equation of Bloch wave theory. Characteristics of (2.12) are discussed in the next section.

2.2.2 *Characteristics of Bloch Waves*

We now consider transmission electron diffraction (i.e., the Laue case) by assuming that all the **g**s lie in a plane perpendicular to **n**, i.e., $\mathbf{g} \cdot \mathbf{n} = 0$. This condition restricts the following solution to the ZOLZ reflections if **n** is antiparallel to the z axis. Equation (2.12) can readily be written as

$$2KS_g C_g + \sum_h U_{g-h} C_h = 2\mathbf{K} \cdot \mathbf{n} \nu C_g \tag{2.13}$$

or expressed in a matrix form

$$\mathbf{AC}^{(i)} = 2K_n \nu_i \mathbf{C}^{(i)} \tag{2.14}$$

where elements of the **A** matrix are $A_{gh} = 2KS_g\delta_{gh} + U_{g-h}$ and $K_n = \mathbf{K} \cdot \mathbf{n}$. If m beams are considered in the scattering, **A** is an $(m \times m)$ matrix, and there are m eigenvalues ν_i and m Bloch waves, each Bloch wave having m plane wave components. Equation (2.14) is just an eigen equation of $C_g^{(i)}$ with eigenvalue ν_i. Each eigen solution gives the superposition coefficients $C_g^{(i)}$ of the i th Bloch wave. The number of Bloch waves equals the number of beams or the dimension of the **C** matrix.

The set of equations represented by (2.12) is exact provided an infinite number of **g** values is considered. In practice the number of beams can be reduced depending on the required accuracy of numerical calculations.

Since $U(\mathbf{r})$ is real for elastic scattering without absorption, its Fourier coefficients satisfy $U_g = U_{-g}^*$, and the matrix **A** is thus Hermitian. Hence the eigenvalues ν_i are real (Landau and Lifshitz 1977), although the eigenvectors $\mathbf{C}^{(i)}$ are, in general, complex. The eigenvector **C** matrix is unitary (Rez 1976; Humphreys 1979), i.e., the inverse of **C** is the complex conjugate and the transpose of **C**

$$\mathbf{C}^{-1} = \mathbf{C}^{+} \tag{2.15}$$

Explicitly writing this equation yields

$$\sum_i C_g^{(i)} C_h^{(i)*} = \delta_{g,h} \tag{2.16a}$$

and

$$\sum_g C_g^{(i)} C_g^{(j)*} = \delta_{i,j} \tag{2.16b}$$

where superscriptions i and j are added to represent the i th and the jth Bloch waves. This means that eigenvectors form a complete orthonormal set if the crystal potential is a real function. The following relations also hold among the eigenvectors and eigenvalues (Fukuhara 1966; Spence 1988a):

$$C_g^{(i)}(\mathbf{k}+\mathbf{h}) = C_g^{(i)}(\mathbf{k}) \tag{2.16c}$$

$$v_i(\mathbf{k}+\mathbf{h}) = v_i(\mathbf{k}) + S_h \tag{2.16d}$$

and

$$C_g^{(i)}(\mathbf{k}) = C_{-g}^{(i)*}(\mathbf{k}) \tag{2.16e}$$

A comprehensive description of the symmetry of Bloch waves is given by Gjønnes and Taftø (1993).

We now determine the superpersition coefficients α_i of Bloch waves. The α_i coefficients are easily obtained only if the incident wave is a plane wave. In the Laue case, the boundary condition at $z = 0$ requires $\Psi(\mathbf{K}, \mathbf{b}) = \exp[2\pi i \mathbf{K} \cdot \mathbf{b}]$ or

$$\sum_i \alpha_i \sum_g C_g^{(i)} \exp[2\pi i(\mathbf{K}+\mathbf{g}) \cdot \mathbf{r}] = \exp[2\pi i \mathbf{K} \cdot \mathbf{b}]$$

This equation is satisfied if $\alpha_i = C_0^{(i)*}$, considering the orthonormal relationship in (2.16a). Thus (2.6) becomes

$$\Psi(\mathbf{K}, \mathbf{r}) = \sum_i C_0^{(i)*} B_i(\mathbf{r}) \tag{2.17a}$$

with

$$B_i(\mathbf{r}) = \sum_g C_g^{(i)} \exp\left[2\pi i(\mathbf{K}+\mathbf{g}) \cdot \mathbf{r} + 2\pi i v_i z\right] \tag{2.17b}$$

where the direction of the z axis is parallel to the direction of the incident beam. It is important to point out that the Bloch wave solution introduced in (2.17) can be applied to calculate only the diffraction of a plane wave. Separated calculations are required if the incident electron probe is a superposition of plane waves.

2.2.3 Orthonormal Relationship of Bloch Waves

For Laue reflections with $\mathbf{g}$ lying in the $x - y$ plane, i.e., $g_z = 0$, we now prove the orthonormal relation of the Bloch waves. From (2.17b),

$$\begin{aligned}\int d\mathbf{r} B_i^*(\mathbf{r}) B_j(\mathbf{r}) &= \sum_{gh} C_g^{(i)*} C_h^{(j)} \int d\mathbf{r} \exp\left[2\pi i(\mathbf{h}-\mathbf{g}) \cdot \mathbf{b} + 2\pi i \left(v_j - v_i\right) z\right] \\ &= \sum_g C_g^{(i)*} C_g^{(j)} \int dz \exp\left[2\pi i \left(v_j - v_i\right) z\right] = \delta_{i,j}\end{aligned} \tag{2.18}$$

This means that Bloch waves are an orthonormal set. But this orthogonal relation holds only in the symmetric Laue case (i.e., $g_z = 0$ for the ZOLZ) under the approximation of high energy (Kim and Sheinin 1987). In this case, the orthonormal relation holds regardless of whether the Bloch function is degenerate or nondegenerate. Thus the solution of the Schrödinger equation can usually be expanded as a linear superposition of Bloch waves. For a general case with $g_z \neq 0$, two Bloch functions are orthogonal if $v_j \neq v_i$. If $v_j = v_i$, however, the two Bloch functions may not be orthogonal. The orthonormal relation of the Bloch waves is useful in solving the Green function solution of the Schrödinger equation for a point electron emitter (see Chap. 10).

2.2.4 Bethe Theory and Band Structure Theory

Bethe's theory for electron diffraction is closely related to the energy band concept in the electronic theory of solids, since both represent the electronic state in terms of Bloch functions (Kambe and Moliere 1970). Morse (1930) pointed out the close connection between the two theories and demonstrated that the Bragg peaks observed in low-energy electron diffraction occur in energy ranges that are forbidden by the band structure of the crystal. For high-energy electron diffraction, the two theories may be compared with each other in the following respects:

In band structure theory, Bloch functions are regarded as representing stationary states of valence electrons in a crystal. The propagation vector **k** is given as a parameter, and the eigenvalue of the energy $E(\mathbf{k})$ is calculated as a function of **k**. In Bethe's theory, on the other hand, Bloch functions are used to construct the wave function of the scattered electron inside the crystal, given the energy of the incident electrons. Thus the dispersion surface is an equal energy surface in reciprocal space similar to the Fermi surface in band theory.

In band theory, using the Hartree approximation, the potential is defined as the average Coulomb potential due to the nuclei, the ion core electrons, and all the valence electrons except the one valence electron under consideration. In this case, the exchange and correlation effects between the electron under consideration and all other electrons are not negligible, and thus these affect the Hartree potential. For electron diffraction, however, the Bethe theory states that the potential is due to the nuclei and all electrons belonging to the crystal, and it is not affected by incident electrons.

In band theory, the plane wave expansion of the Bloch function is usually poorly convergent, especially near the nuclei where the wave function varies rapidly. In the Bethe theory for electron diffraction, the convergence of the plane wave expansion is usually sufficiently rapid. In general, the number of plane waves is the number of beams in the diffraction pattern.

2.3 Two-Beam Theory

The most important advantage of the Bloch wave approach is its clarity when describing the excitation of each Bloch wave. We now consider a case in which only two beams are involved. Equation (2.14) becomes

$$\begin{pmatrix} -2K_n v & U_{-g} \\ U_g & 2KS_g - 2K_n v \end{pmatrix} \begin{pmatrix} C_0 \\ C_g \end{pmatrix} = 0 \tag{2.19}$$

The eigenvalues v are obtained by setting the determinant of the matrix equal to zero, thus leading to the following solutions:

$$v_{1,2} = \frac{\left\{ KS_g \pm \left[(KS_g)^2 + |U_g|^2 \right]^{1/2} \right\}}{2K_n} \tag{2.20a}$$

where superscript 1 refers to + and 2 to −. The difference of $v_1 - v_2$ is

$$\Delta v = v_1 - v_2 = \frac{\left[(KS_g)^2 + |U_g|^2 \right]^{1/2}}{K_n} \tag{2.20b}$$

which corresponds to the gap between the two dispersion surfaces; this is discussed later in Chap. 2. If we set

$$w = \frac{KS_g}{|U_g|} = S_g \xi_g$$

where $\xi_g = K/U_g$ is the extinction distance, the ratio of C_0 and C_g can be found from (2.19)

$$\left[\frac{C_g}{C_0} \right]^{(1,2)} = w \pm \left(1 + w^2 \right)^{1/2} \tag{2.20c}$$

By defining $\gamma_0 = \text{arcctg}\, w$ and considering the normalization relation $|C_0|^2 + |C_g|^2 = 1$, the Bloch wave coefficients are

$$C_0^{(1)} = C_g^{(2)} = \cos(\gamma_0/2) \tag{2.21a}$$

and

$$C_0^{(1)} = -C_g^{(2)} = \sin(\gamma_0/2) \tag{2.21b}$$

corresponding to Bloch waves of

$$B_1(\mathbf{r}) = \exp\left(2\pi i \mathbf{K}\cdot\mathbf{r} + 2\pi i v_1 z\right)\left[\cos\left(\gamma_0/2\right) + \sin\left(\gamma_0/2\right)\exp(2\pi i \mathbf{g}\cdot\mathbf{r})\right] \tag{2.21c}$$

and

$$B_2(\mathbf{r}) = \exp\left(2\pi i \mathbf{K}\cdot\mathbf{r} + 2\pi i v_2 z\right)\left[\sin\left(\gamma_0/2\right) - \cos\left(\gamma_0/2\right)\exp(2\pi i \mathbf{g}\cdot\mathbf{r})\right] \tag{2.21d}$$

The two Bloch waves under the two-beam approximation are useful in qualitatively illustrating the physics involved in some imaging and diffraction processes. Substituting (2.21) into (2.17), the intensity of the diffracted beam at the exit face, where $z = d$, is

$$I_g = \left|\sum_i C_0^{(i)*} C_g^{(i)} \exp\left(2\pi i v_i d\right)\right|^2 = \frac{\left|U_g\right|^2 \sin^2(\pi d \Delta v)}{(K_z \Delta v)^2} \tag{2.22a}$$

and

$$I_0 = 1 - I_g \tag{2.22b}$$

The oscillation of I_0 with the variation of specimen thickness d corresponds to thickness fringes observed in bright-field TEM images, as shown in Fig. 2.3a. The dark-field image of the same area (Fig. 2.3b) recorded using the g reflection shows complimentary contrast as expected theoretically. Two adjacent thickness fringes correspond to a thickness change of $\Delta d = 1/\Delta v = \xi_g \,/\left(1 + w^2\right)^{1/2}$, where ξ_g is g dependent. At the Bragg condition $w = 0$, $\Delta d = \xi_g$. This means that the extinction distance is the thickness variation between two adjacent thickness fringes under the Bragg reflection condition.

Using the two-beam theory, we now show the channeling effect in electron diffraction. For simplicity, we consider the expressions of (2.21c) and (2.21d) under the Bragg condition (i.e., $w = 0$)

$$B_1(\mathbf{r}) = \frac{1}{\sqrt{2}} \exp\left(2\pi i \mathbf{K}\cdot\mathbf{r} + 2\pi i v_1 z\right)\left[1 + \exp(2\pi i \mathbf{g}\cdot\mathbf{r})\right]$$
$$= \sqrt{2}\cos(\pi \mathbf{g}\cdot\mathbf{r})\exp\left(2\pi i \mathbf{K}\cdot\mathbf{r} + 2\pi i v_1 z + \pi i \mathbf{g}\cdot\mathbf{r}\right)$$

with

$$\left|B_1(\mathbf{r})\right|^2 = 2\cos^2(\pi \mathbf{g}\cdot\mathbf{r}) \tag{2.22c}$$

and

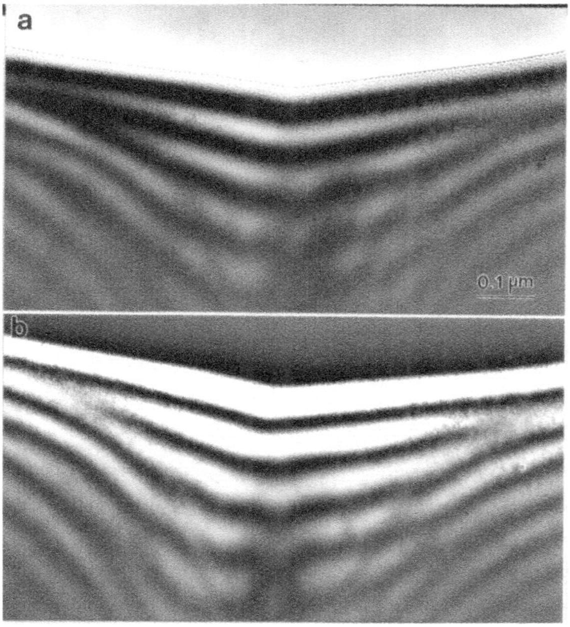

Fig. 2.3 **a** Bright-field image and **b** $g = (004)$ dark-field image of a thin silicon crystal oriented near the [100] zone, showing the thickness fringes. The contrast of the bright-field image is complimentary to that of the dark-field image. Beam energy is 100 keV

$$
\begin{aligned}
B_2(\mathbf{r}) &= \frac{1}{\sqrt{2}} \exp\left(2\pi i \mathbf{K} \cdot \mathbf{r} + 2\pi i v_2 z\right) \left[1 - \exp(2\pi i \mathbf{g} \cdot \mathbf{r})\right] \\
&= i\sqrt{2} \sin(\pi \mathbf{g} \cdot \mathbf{r}) \exp\left(2\pi i \mathbf{K} \cdot \mathbf{r} + 2\pi i v_2 z + \pi i \mathbf{g} \cdot \mathbf{r}\right)
\end{aligned}
$$

with

$$
|B_2(\mathbf{r})|^2 = 2 \sin^2(\pi \mathbf{g} \cdot \mathbf{r}) \tag{2.22d}
$$

The electron current of each Bloch wave flows, on average, parallel to the Bragg planes, but it is modulated across the atom planes, so that for Bloch wave $B_2(\mathbf{r})$ the maximum occurs between the atomic planes (Fig. 2.4). For Bloch wave $B_1(\mathbf{r})$, the situation is reversed, and the current is concentrated on the atom planes, i.e., a stronger channeling effect . The channeling propagation of Bloch wave $B_1(\mathbf{r})$ along atom planes effectively increases the probability of inelastic excitations of the atomic inner shells, resulting in a stronger absorption effect (see Chap. 6).

The channeling effect illustrated here has many important applications in imaging of inelastically scattered electrons (see Chap. 11). Electron channeling along the paths of lowest potential energy (i.e., atomic rows) in crystal enhances the excitations of X-rays and Auger electrons in the atomic rows. The X-ray signals from impurity atoms located within certain crystal planes can be maximized if proper diffracting (or channeling) conditions are set up. This technique is known as atom location by channeling enhanced microanalysis (ALCHEMI) (see Spence 1992, for a review).

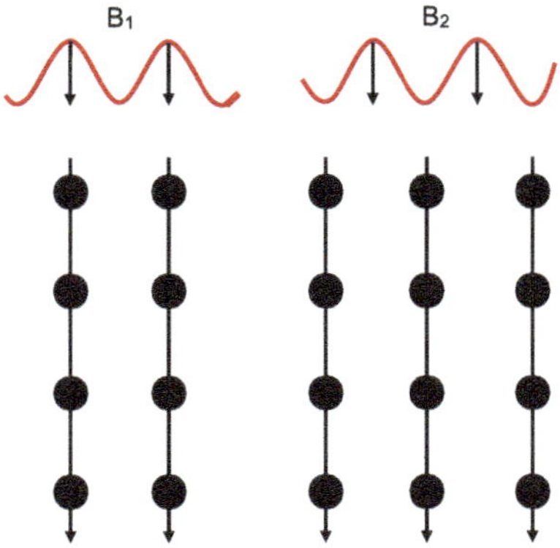

Fig. 2.4 A schematic diagram of two types of Bloch wave fields under two-beam Bragg condition at the reflecting position in a simple cubic lattice. In both cases, the current flow vector (normal to the dispersion surface) is parallel to the reflecting planes. The B_1 wave is more strongly absorbed than B_2 due to a larger probability of inelastic excitation

Equation (2.22a) can be used to estimate the maximum crystal thickness smaller than which the kinematic scattering (or single-scattering) approximation holds (Hoerni 1956). The scattering amplitude of the **g** beam under the kinematic scattering approximation increases linearly with increasing crystal thickness d. In other words, the diffracted intensity increases with d^2. For simplicity, we start with (2.22a), assuming Bragg conditions, so that $I_g \approx \sin^2(\pi d/\xi_g)$. For small thickness satisfying $\pi d/\xi_g \ll 1$

$$I_g = \left(\frac{\pi d}{\xi_g}\right)^2 = \left(\frac{\pi d U_g}{K}\right)^2$$

as expected from kinematic scattering theory. For low-index reflections, the extinction distance ξ_g is a few tens of nanometers. Therefore for kinematic scattering theory to hold, $d \ll \xi_g/\pi$. In this thickness range, electron scattering can be approximately considered the result of single scattering, so that the high-resolution transmission electron microscopy (HRTEM) images of thin crystals are referred to structural images, from which positions of atoms may be directly identified from the image. For crystals with larger thicknesses, multiple scattering effects disturb the localized distribution of the incident electrons, so that the cross scattering of the electrons complicates the image interpretation. In this case, image simulation becomes indispensable.

A more sophisticated two-beam theory has also been derived for wedge-shaped and finite polyhedron crystals (Kato 1952). The major task involved in irregular-shaped crystals is to match the solutions at the boundary. This can be easily done for the two-beam case, but the situation may become very complex if a many-beam theory is involved.

The theory developed here applies to purely elastically scattered electrons without absorption. Inelastic scattering during electron diffraction causes the incident electron to lose not only energy but also become angularly redistributed between Bragg beams. If the objective aperture is used to select one or more Bragg reflected beams to form an image, many of the inelastically scattered electrons are thus excluded from the image. These electrons are effectively absorbed. In general, the absorption effect is characterized by an imaginary potential in elastic-scattering calculations, and this effect decreases the intensity of Braggreflected beams. A systematic introduction of the imaginary potential is illustrated in Chap. 6.

2.4 Dispersion Surfaces

The Bloch wave eigenvalues v_i given by (2.20) depend on both the crystal-scattering potential $\left(U_g\right)$ and the incident electron wave vector. For specific reflections with known U_g s, a relationship between the incident electron wave number K and the Bloch wave eigenvalue v is called the dispersion surface, which is formed by the locus of allowed wave vectors for all beams. All wave vectors inside the crystal are restricted to lie on the dispersion surfaces. Explicitly speaking, the dispersion surface in electron diffraction is a plot of allowed values of the z component of the Bloch wave vector $k_z^{(i)}$ in the crystal. For simplification, we use dispersion relations for the two-beam case to illustrate the construction of dispersion surfaces.

We first consider the zero-order solution of (2.10a), where all Fourier coefficients U_{g-h} (for $\mathbf{g} \neq \mathbf{h}$) are switched off. Then $S_g = 0$, or $\left[\mathbf{k}^{(i)} + \mathbf{g}\right] = K^2$, and this describes a series of free-electron spheres of radius K centered at each of the reciprocal-lattice points $\mathbf{g}$ parallel to the surface (Whelan 1986). One sphere is centered on the origin of reciprocal space and one centered on the reciprocal lattice point

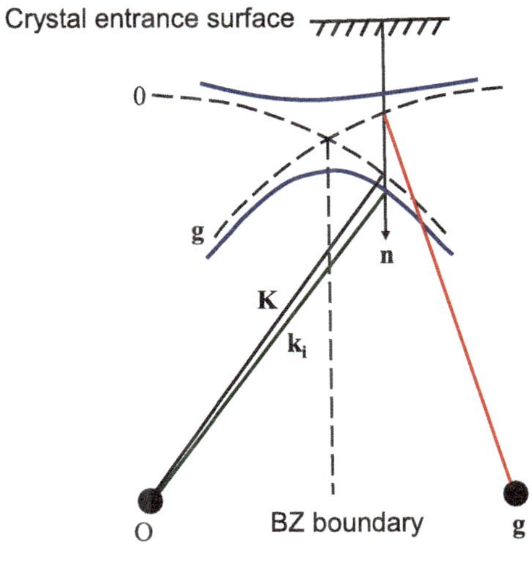

Fig. 2.5 Dispersion surfaces under the two-beam approximation. Here v and S_g are measured in the direction of the surface normal $\mathbf{n}$. The dashed curves 0 and g are parts of the Ewald spheres centered at O and $\mathbf{g}$, respectively, with radius $K = 1/\lambda$

g, as shown in Fig. 2.5. These spheres intersect at the Brillouin zone boundary. A vector is drawn in the direction normal to the entrance surface of the crystal and intersecting the vector **K** at the sphere centered at O. The eigenvalues v_i are calculated from (2.20) using the actual U_g value. The values of v_i are measured along the surface normal direction, starting from a point on that K sphere that is centered at O. Thus the **K** vector must be drawn first. Then points on the dispersion surfaces are drawn at distances v_i measured from the end point of **K** along the surface normal direction **n**. The complete dispersion surfaces are obtained by repeating this procedure for each possible beam direction.

Two-beam diffraction is an ideal case in which the intensities of other beams are zero. In practice, this condition is rarely satisfied. The perturbation effect of weak beams can be reasonably included in the two-beam theory using the Bethe potential, which is discussed in Sect. 2.6. Further studies by Miyake (1959) and Gjønnes (1962a) indicated that the two-beam approximation fails to apply in the range of very short wavelength and small crystal thicknesses, even with the corrected dynamical Fourier potential. This is due to the increased radius of the Ewald sphere, and the shape factor of the crystal for $\lambda \to 0$ (or $K \to \infty$) and $d \to 0$ result in many beam excitations.

2.5 Applications in CBED

The CBED patterns are formed with an electron probe focused on the sample, causing diffraction spots to broaden into disks. The incident probe consists of many plane wave components propagating along different directions, thus forming a converged conical electron probe, as shown in Fig. 2.6. For an incident beam P, the diffraction results in a complete point diffraction pattern consisting of Ps as ruled by the Bragg reflection law. A similar set of point diffraction patterns is formed for another plane wave component Q. Therefore for cases where there are no disk overlaps, a perfect registration is retained between each incident beam direction and the diffracted beams. The intensity profile across the diffracted disk **g** is called a rocking curve, which represents the variation of **g**-reflected intensity with the change of incident beam direction. Thus each point in the central (000) CBED disk corresponds to an incident plane wave component and defines a family of conjugate points differing by reciprocal lattice vectors, one in each CBED disk. The great power of the CBED techniques results from the fact that so much crystal structural information, such as crystal structure factors, charge density distribution, symmetry, strains, unit-cell parameters, and specimen thickness, can be determined quantitatively from analyses of CBED patterns (see Spence and Zuo 1992, and Cowley 1993, for a review).

Figure 2.7 shows a CBED pattern of [110] silicon. Fine details within each diffraction disk are the result of the dynamic diffraction of electrons entering the crystal from different directions. The symmetry of the intensity distribution indicates the projected central-symmetric and the (001) planar mirror-symmetric structure of the

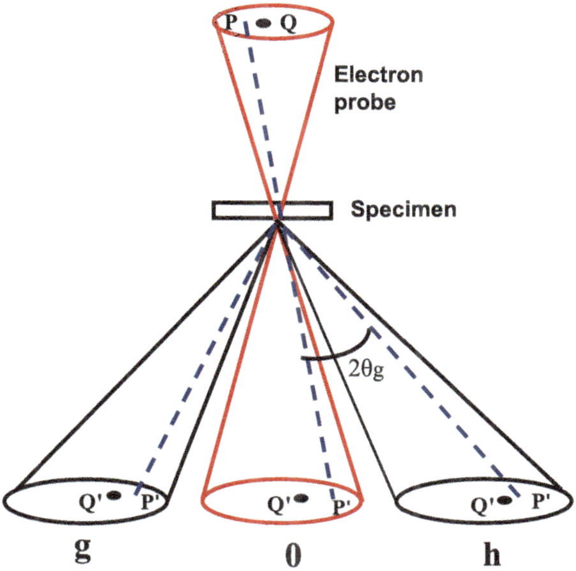

Fig. 2.6 A schematic ray diagram of CBED. If only elastic Bragg scattering is allowed, then source point P gives rise to conjugate points P', one in each disk. Source point Q defines a different incident beam direction and set of diffracted beams Q'

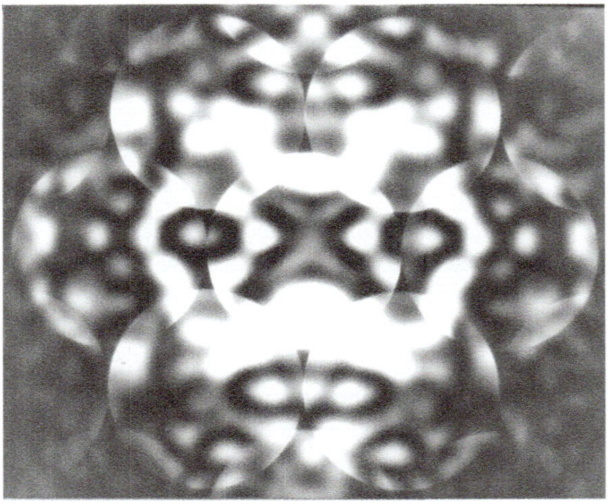

Fig. 2.7 A CBED pattern of silicon viewed along [110] showing structural information contained in the pattern. The pattern was recorded at 100 keV

crystal. Quantification of the intensity distribution in reciprocal space is a powerful method for determining the crystal structure.

Calculating a rocking curve intensity profile in the angular ranges where there is no disk overlap is usually performed by the Bloch wave approach

$$I_g(\mathbf{K}) = \sum_i \sum_j C_0^{(i)*} C_g^{(i)} C_0^{(j)} C_g^{(j)*} \exp\left[2\pi i \left(v_i - v_j\right) d\right] \tag{2.23a}$$

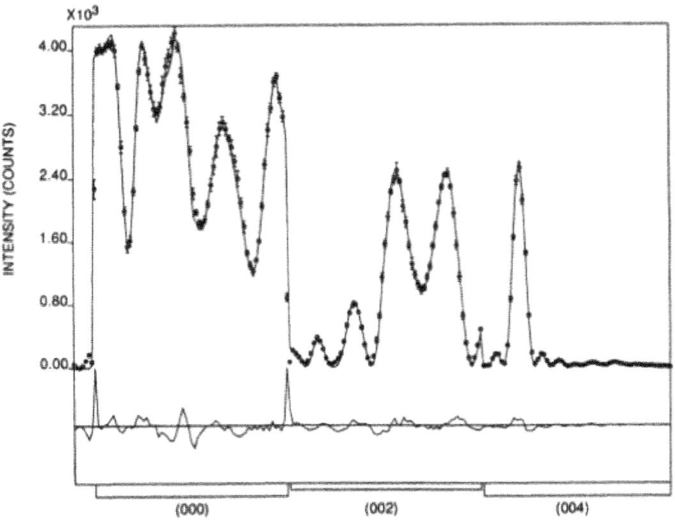

Fig. 2.8 The zero energy loss experimental and computed CBED rocking curve along the (002) row of BeO at 80 kV. The (000), (002), and (004) disks are indicated. The crystal orientation is near the [130] zone axis. Specimen thickness $d = 70.9$ nm. The lower plot shows the difference between calculation and experiment. Courtesy of Spence and Zuo (1992)

In CBED, many parameters are involved in the calculation of $I_g(\mathbf{K})$. Each of these parameters can be determined using different techniques. The specimen thickness is determined from thickness fringes in the diffraction disks under two-beam conditions (Kelly et al. 1975). Unit-cell parameters can be determined by the position of HOLZ line positions in the center disk. Finally, crystal structure factors are refined from the rocking curves of different diffracting disks by comparing the calculated $I_g(\mathbf{K})$ with the experimentally measured I_g. An automated dynamic least-squares refinement program has been developed by Zuo and Spence (1991).

Figure 2.8 shows the experimentally observed zero energy loss CBED pattern (dotted curve) and the many-beam dynamic calculated (solid line) rocking curves for a BeO crystal (Spence and Zuo 1992). The refinement of the calculated curve in comparison with the observed one yields low-index crystal structure factors. For the line scan across the (00n) systematic row, the amplitude and phase of the (002) structure factor at 80 kV are determined as

$$\left|U^{(R)}(002)\right| = 0.039592 \pm 0.00014\ \text{Å}^{-2}, \quad \varphi(002) = -0.88478 \pm 0.017\ \text{rad}$$
$$\left|U^{(I)}(002)\right| = 0.00073 \pm 0.00006\ \text{Å}^{-2}, \quad \varphi'(002) = -1.1 \pm 0.5\ \text{rad}$$

and specimen thickness $d = 71.16$ nm.

If the bonding effect is ignored in a center-symmetric structure, and so that the atom potential is spherical symmetric, the structure factor U_g is a real function. However an imaginary component appears in U_g if the atomic potential is distorted by crystal bonding. Therefore the measurement of $U^{(I)}$ provides important information on charge redistribution in the crystal due to the solid-state effect. Quantitative CBED provides the most accurate method for measuring the charge redistribution of crystals smaller than a few hundred nanometers (Zuo et al. 1988).

If there are disk overlaps in CBED, calculations have to be performed separately for different incident beam directions in the overlapped regions, and the calculated amplitudes are then added coherently, incoherently, or partially coherently depending on the original assumption of the source coherence. For a perfectly coherent beam diffraction, especially if there are substantial disk overlaps, calculations using the multislice theory may be more convenient (see Chap. 3).

2.6 Critical Voltage Effect

The critical voltage effect is one of the typical examples of dynamic scattering in electron diffraction. If a crystal is set at the Bragg condition, normally the diffracted intensity is strong unless the crystal thickness is an integer number of extinction distances. This is due to the constructive interference of waves scattered in the diffracted beam direction. However for a particular incident electron-accelerating voltage, the so-called critical voltage U_{0c}, the diffracted beam intensity of a second-order reflection 2**h** is very small for all thicknesses, due to destructive rather than constructive Bloch wave interference (Nagata and Fukuhara 1967; Uyeda 1968; Watenabe et al. 1968). This voltage depends sensitively on the ratio of first-to-second-order structure factors, and it is used to measure structure factors of low-order reflections that are particularly sensitive to bonding effects in crystals (for reviews, see Cowley 1981; Humphreys 1979; Reimer 1984; Spence and Zuo 1992).

As illustrated in Fig. 2.9, the critical voltage effect occurs at a three-beam diffracting condition, with 2**h** in Bragg condition. This case can be treated either using three-beam theory or two-beam theory with the perturbation of a weak beam (Uyeda 1968; Lally et al. 1972). For simplification, we start from the two-beam theory and consider the perturbation effect of other weak beams.

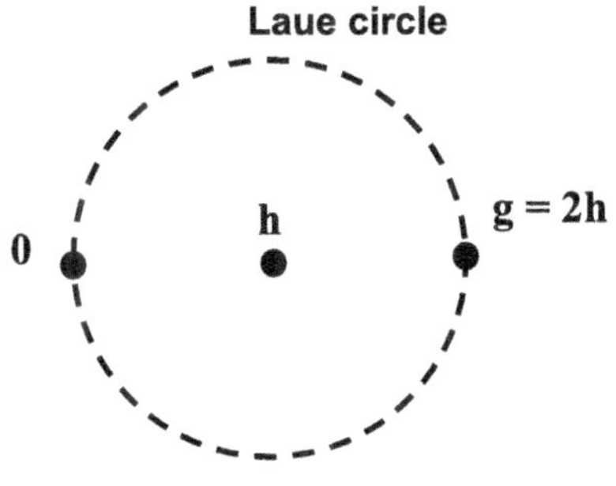

Fig. 2.9 The three-beam diffracting condition for the critical voltage effect

Discussions in Sect. 2.3 were based on an ideal two-beam case in which all other beams were assumed to have zero intensity. One way of including the weak-beam effect is to use the perturbation theory of Bethe and rewrite (2.12) for symmetric Laue cases as

$$2\,K S_h C_h + \sum_{h'} U_{h-h'} C_{h'} = 2 K_n \nu C_h \tag{2.23b}$$

where variable substitutations of **h** by **h**′ and **g** by **h** were introduced. If **h** is a weak beam because $K\,|S_h| \gg |U_g|$, (2.23b) can be solved for the two-beam case as

$$C_h = \frac{\sum_{h'} U_{h-h'} C_{h'}}{2\,K S_h - 2K_n\nu} \approx \frac{U_h C_0 + U_{h-g} C_g}{2\,K S_h} \tag{2.23c}$$

Substituting (2.23c) into (2.12) for the two-beam case, (2.19) is modified as

$$\begin{pmatrix} 2\,K S_0^{eff} - 2K_n\nu & U_{-g}^{eff} \\ U_g^{eff} & 2\,K S_g^{eff} - 2K_n\nu \end{pmatrix} \begin{pmatrix} C_0 \\ C_g \end{pmatrix} = 0 \tag{2.23d}$$

where

$$2\,K S_0^{\text{eff}} = -\sum_h \frac{|U_h|^2}{2\,K S_h}, \quad 2\,K S_g^{\text{eff}} = 2\,K S_g - \sum_h \frac{|U_{g-h}|^2}{2\,K S_h}$$

and

$$U_g^{eff} = U_g - \sum_h \frac{U_h U_{g-h}}{2\,K S_h} \tag{2.24}$$

This effective structure factor is known as the Bethe potential. A more rigorous derivation using the Green's function has been given by Gjønnes (1962b).

We now apply (2.24) to illustrate the critical voltage phenomenon, in which $\mathbf{g} = 2\mathbf{h}$ satisfies Bragg condition, and **h** is weak. For a minimum **g** reflection, we set $U_g^{\text{eff}} = 0$. This condition may be met by varying the accelerating voltage. Using the relation

$$U_{0c} = \frac{m_0 c_0^2 (\gamma - 1)}{e} \tag{2.25}$$

solving γ from $U_g^{\text{eff}} = 0$ (Spence and Zuo 1992) yields

$$U_{0c} = \frac{m_0 c_0^2}{e} \left(\frac{\hbar^2 g^2 V_{2h}}{2 m_0 e V_h} - 1 \right) \tag{2.26}$$

Equation (2.26) gives an approximate value of the critical voltage U_{0c} in terms of the two lowest order structure factors. If the second-order reflection is assumed to depend mainly on the single-atom scattering property, the first-order reflection, which is more sensitive to solid-state bonding effects, may be found with high precision using (2.26) (for a review, see Fox and Fisher 1988). Accuracy of measurements is ultimately limited by the Debye-Waller factor for high-order reflection and the contrast of the intensity minimum at 2**h**.

2.7 Diffraction of Layered Materials

The CBED techniques can be applied to study the strain, composition, and rigid-body displacements across layered materials (or quantum well materials). The Bloch wave theory developed in Sect. 2.2 can be extended to calculate the transmission electron diffraction pattern of layered materials (Rossouw et al. 1991; Peng and Whelan 1990). We now consider a material composed of $m = 1$ to M layers, as shown in Fig. 2.10. The mth layer is located at $z = z_m$, with thickness d_m. For simplicity, we assume there is no rigid shift between the origins of the two adjacent layers. The wave function inside the mth layer is written as

$$\Psi_m(\mathbf{r}) = \sum_i \alpha_m^{(i)} \sum_{g_m} C_{g_m}^{(i)} \exp\left[2\pi i\,(\mathbf{K}+\mathbf{g}_m)\cdot\mathbf{r} + 2\pi i v_{im}\,(z-z_m)\right] \qquad (2.27)$$

where v_{im} and $C_{g_m}^{(i)}$ are the eigenvalues and eigenvectors, respectively, of the Bloch wave solution of the mth crystalline layer and $\mathbf{g}_m$ are the corresponding reciprocal lattice vectors. An analogous relation holds for the ($m+1$)th layer,

$$\Psi_{m+1}(\mathbf{r}) = \sum_i \alpha_{m+1}^{(i)} \sum_{g_{m+1}} C_{g_{m+1}}^{(i)} \exp\left[2\pi i\,(\mathbf{K}+\mathbf{g}_{m+1})\cdot\mathbf{r} + 2\pi i v_{m+1}\,(z-z_{m+1})\right] \qquad (2.28)$$

Matching (2.27) and (2.28) at the interface $z = z_{m+1}$ for the ZOLZ reflections and splitting $\mathbf{r} = (\mathbf{b}, z)$, we obtain

$$\sum_i \alpha_{m+1}^{(i)} \sum_{g_{m+1}} C_{g_{m+1}}^{(i)} \exp\left(2\pi i \mathbf{g}_{m+1}\cdot\mathbf{b}\right) = \sum_i \alpha_m^{(i)} \sum_{g_m} C_{gm}^{(i)} \exp\left(2\pi i \mathbf{g}_m\cdot\mathbf{b} + 2\pi i v_{im} d_m\right)$$

Multiplying this equation by $\exp\left(2\pi i \mathbf{h}_{m+1}\cdot\mathbf{b}\right)$ and integrating over **b**, we obtain

$$\sum_i \alpha_{m+1}^{(i)} C_{h_{m+1}}^{(i)} = \sum_i \alpha_m^{(i)} \sum_{g_m} C_{g_m}^{(i)} \delta\left(\mathbf{h}_{m+1} - \mathbf{g}_m\right) \exp\left(2\pi i v_{im} d_m\right)$$

Multiplying both sides by $C_{h_{m+1}}^{(i)*}$, summing over h_{m+1}, and using the orthonormal relation in (2.16b) yields

Fig. 2.10 A schematic diagram showing layered structures along the beam direction

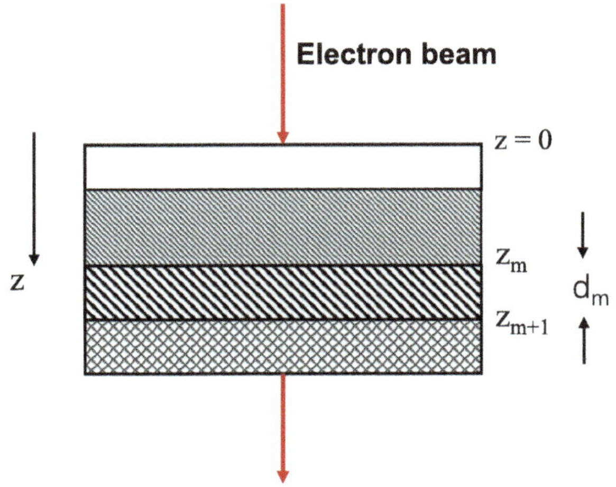

$$\alpha_{m+1}^{(j)} = \sum_i a_m^{(i)} \sum_{g_m} \sum_{h_{m+1}} C_{g_m}^{(i)} C_{h_{m+1}}^{(j)*} \delta\left(\mathbf{h}_{m+1} - \mathbf{g}_m\right) \exp\left(2\pi i v_{im} d_m\right) \tag{2.29}$$

This relation correlates the superposition coefficients of the mth and $(m+1)$th layers. For $m = 1$, the boundary condition is satisfied by choosing $\alpha_1^{(j)} = C_0^{(j)*}$. Using (2.29), it is possible to determine the wave function inside each layer.

Two simple relations can be derived from (2.29). If the two adjacent layers have the same reciprocal lattice base vectors (i.e., same direction and same magnitude), then

$$\alpha_{m+1}^{(j)} = \alpha_m^{(j)} \exp\left(2\pi i v_{jm} d_m\right) \tag{2.30a}$$

If the two layers have different reciprocal lattice base vectors, so that $\mathbf{h}_{m+1} \neq \mathbf{g}_m$ unless $\mathbf{h}_{m+1} = 0$ and $\mathbf{g}_m = 0$, then

$$\alpha_{m+1}^{(j)} = \sum_i \alpha_m^{(i)} C_{0_m}^{(i)} C_{0_{m+1}}^{(j)} \exp\left(2\pi i v_{im} d_m\right) \tag{2.30b}$$

Finally for a case where two adjacent layers have the same set of reciprocal lattice base vectors, a rigid shift $\Delta\mathbf{R}_m$ of the mth layer origin with respect to the origin of the $(m+1)$th layer may be incorporated into the calculation by multiplying the Bloch wave coefficients with a phase factor, i.e.,

$$C_{g_m}^{(i)} \rightarrow C_{g_m}^{(i)} \exp\left(-2\pi i \mathbf{g}_m \cdot \Delta\mathbf{R}_m\right) \tag{2.31a}$$

The preceding discussion makes it possible to apply the Bloch wave theory in calculating CBED patterns of layered materials if the incident beam is assumed to be perpendicular to the surface plane.

2.8 HOLZ Reflections

The theory outlined in the last few sections was based on an approximation of symmetric Laue cases for which $\mathbf{g} \cdot \mathbf{n} = 0$. This condition becomes $g_z = 0$ if the foil normal is antiparallel to the z axis; thus the theory is restricted to ZOLZ reflections. To include HOLZ reflections, we start from (2.12) by defining

$$B_g = \left(1 + \frac{g_n}{K_n}\right)^{1/2} C_g \tag{2.31b}$$

where components normal to the crystal surface are $g_n = \mathbf{g} \cdot \mathbf{n}$, $h_n = \mathbf{h} \cdot \mathbf{n}$, and $K_n = \mathbf{K} \cdot \mathbf{n}$, so that (2.12) is rewritten as

$$\frac{2\,K S_g B_g}{1 + (g_n/K_n)} + \sum_h \frac{U_{g-h} B_h}{\{(1 + g_n/K_n)\,[1 + (h_n/K_n)]\}^{1/2}} = 2K_n v B_g \tag{2.31c}$$

This is a fundamental eigenvalue equation that includes all HOLZ effects and boundary inclination effects. The only approximation made in deriving (2.31c) is to ignore the v^2 term for high-energy electron diffraction in the Laue case. The FORTRAN program provided by Spence and Zuo (1992) calculates (2.31), so that HOLZ reflections are automatically included.

2.9 Real-Space Bloch Wave Theory of ZOLZ Reflections

The Bloch wave theory presented in the last few sections is essentially a reciprocal-space theory in which the differential equation was solved in reciprocal space. The dimension of the matrix is the number of beams involved in the calculation. The numerical calculation of this theory, however, may be quite involved especially for zone axis patterns in which many beams are involved. An alternative Bloch wave theory is thus introduced to solve the Schrödinger equation in real space (Howie 1966; Berry 1971; Buxton et al. 1978; Vincent et al. 1984; for a review, see Bird 1989). This theory is particularly convenient for calculating zone-axis CBED patterns.

2.9.1 Projected Potential Approximation

Before we present the real-space Bloch wave theory, it is important to examine the portion of the crystal potential that is responsible for ZOLZ reflections. In zone-axis patterns, the dominant Bragg reflections are distributed within the ZOLZ. If $\mathbf{g}$ is written as $(\mathbf{g}_b, g_z)$, the crystal potential is written as

$$U(\mathbf{r}) = \sum_g U_g \exp(2\pi i \mathbf{g} \cdot \mathbf{r}) = \sum_g U_g \exp(2\pi i \mathbf{g}_b \cdot \mathbf{b} + 2\pi i g_z z)$$
$$= U^{(0)}(\mathbf{b}) + U^{(H)}(\mathbf{b}, z) \quad (2.32a)$$

where

$$U^{(0)}(\mathbf{b}) = \sum_{g_b} U_{g_b} \exp(2\pi i \mathbf{g}_b \cdot \mathbf{b}) \quad (2.32b)$$

is the projected potential, and

$$U^{(H)}(\mathbf{r}) = \sum_{g_z \neq 0} \exp(2\pi i g_z z) \sum_{g_b} U_g \exp(2\pi i \mathbf{g}_b \cdot \mathbf{b}) \quad (2.32c)$$

which is the potential responsible for HOLZ reflections because $g_z \neq 0$. Therefore the diffraction of ZOLZ reflections is directly determined by the projected potential $U^{(0)}(\mathbf{b})$. This means that the ZOLZ approximation is the projected potential approximation; the projected potential approximation actually restricts the theory to the ZOLZ reflections only.

2.9.2 *ZOLZ Reflections*

Starting from the Schrödinger equation and factoring out the rapid variation in the z direction,

$$\Psi(\mathbf{r}) = \exp(2\pi i \mathbf{K} \cdot \mathbf{r})\Phi(\mathbf{b}, z) \quad (2.33)$$

Substituting (2.33) into (2.5) gives

$$\left[\nabla_b^2 + 4\pi^2 U(\mathbf{r})\right] \Phi(\mathbf{b}, z) \approx -4\pi i K_z \frac{\partial \Phi(\mathbf{b}, z)}{\partial z} \quad (2.34)$$

where the ∇_b^2 operator acts only on the transverse coordinate $\mathbf{b}$; $\partial^2\Phi/\partial z^2$ was dropped under the small angle-scattering approximation; and $\partial\Phi(\mathbf{b}, z)/\partial x$ and $\partial\Phi(\mathbf{b}, z)/\partial y$ are neglected because $K_z \gg K_x$ and $K_z \gg K_y$. Using the potential given in (2.32b) for the ZOLZ reflections, (2.34) becomes

$$\left[\nabla_b^2 + 4\pi^2 U^{(0)}(\mathbf{b})\right]\Phi(\mathbf{b}, z) \approx -4\pi i K_z \frac{\partial \Phi(\mathbf{b}, z)}{\partial z} \quad (2.35)$$

Thus the solution of (2.35) can be directly written (Doyle and Berry 1973) as

$$\Phi = \sum_j \varepsilon^j \exp\left(\zeta s_j z\right) \tau_j(\mathbf{b}) \tag{2.36a}$$

where

$$\zeta = \frac{i}{4\pi K_z} \tag{2.36b}$$

and $\tau_j(\mathbf{b})$ is a two-dimensional Bloch wave solution of

$$\left[\nabla_b^2 + 4\pi^2 U^{(0)}(\mathbf{b})\right] \tau_j(\mathbf{b}) = s_j \tau_j(\mathbf{b}) \tag{2.37}$$

The two-dimensional Bloch wave $\tau_j(\mathbf{b})$ satisfies the following orthonormal relation (Ashcroft and Mermin 1976; also see Sect. 2.2.3):

$$\int d\mathbf{b} \tau_i^*(\mathbf{b}) \tau_j(\mathbf{b}) = \delta_{ij} \tag{2.38}$$

Using the orthonormal relation of the C coefficients, it can be proved directly that

$$\sum_i \tau_i(\mathbf{b}) \tau_i^*\left(\mathbf{b}'\right) = \delta\left(\mathbf{b} - \mathbf{b}'\right) \tag{2.39}$$

This means that $\left\{\tau_j(\mathbf{b})\right\}$ is a complete orthonormal function set. These are important relations that make it possible to expand an arbitrary convergent function in the representation of $\left\{\tau_j(\mathbf{b})\right\}$. This point is demonstrated in Sect. 10.3 for the solution of Green's function. Coefficients ε^j are determined by the boundary condition at $z = 0$. If the incident electron is described by $\Phi(\mathbf{b}, 0)$, the boundary condition requires.

$$\Phi(\mathbf{b}, 0) = \sum_j \varepsilon^j \tau_j(\mathbf{b}) \tag{2.40}$$

Multiplying (2.40) by $\tau_i^*(\mathbf{b})$, integrating over $\mathbf{b}$, and using (2.38) yields

$$\varepsilon^i = \int d\mathbf{b} \Phi(\mathbf{b}, 0) \tau_i^*(\mathbf{b}) \tag{2.41}$$

At the exit face of the crystal slab $z = d$, the amplitude for electrons scattered to $\mathbf{u}$ in the two-dimensional reciprocal space is

$$A(\mathbf{u}) = \sum_j \varepsilon^j \exp\left(\zeta s_j d\right) \int d\mathbf{b} \exp(-2\pi i \mathbf{u} \cdot \mathbf{b}) \tau_j(\mathbf{b}) \tag{2.42}$$

the intensity distribution $\mathbf{I} = |A(\mathbf{u})|^2$.

Compared to the Bloch wave theory in reciprocal space (Sect. 2.2), the most important advantage of the real-space Bloch wave theory is that the diffraction of a finite incident electron probe of $\Phi(\mathbf{b}, 0)$ can be calculated without repeating the calculation for each plane wave component within the incident conical probe. This is convenient for the case of coherent convergent beam electron diffraction, in which disk overlaps may occur. The calculation of Bethe theory, however, has to be performed separately for individual incident plane wave components.

2.9.3 Effects of HOLZ Reflections

The HOLZ reflections can be included in this theory using the perturbation technique (Buxton 1976; Vincent et al. 1984), in which the superposition coefficients of Bloch waves are assumed as z-dependent, i.e.,

$$\Phi = \sum_j \varepsilon^j(z) \exp\left(\zeta s_j z\right) \tau_j(\mathbf{b}) \tag{2.43}$$

and the upper layer potential $U^{(H)}$ is treated as a perturbation. This approach is exactly equivalent to the usual way of treating a time-dependent perturbation, in which z takes the place of time and the upper layer potential $U^{(H)}(\mathbf{b}, z)$ represents the time-dependent perturbation. Substituting (2.43) into (2.5), the first and second-order perturbation terms can be solved by the iteration method (for a review, see Bird 1989).

2.10 Diffraction Contrast Images of Imperfect Crystals

Diffraction contrast imaging, a powerful technique for determining dislocation structures in crystals, is based on the following mechanism (Hirsch et al. 1977). The image is formed by selecting a single Bragg reflected beam using the objective aperture. The intensity of the Bragg beam is perturbed by the variation of local diffracting conditions due to the strain field of the dislocation. Thus this type of imaging is usually referred to diffraction contrast imaging.

This section describes the basic theoretical approach for calculating the diffraction contrast of images. An excellent review has been given by Amelincks and Van Dyck (1993) on the fundamental theories of diffraction contrast imaging and associated applications.

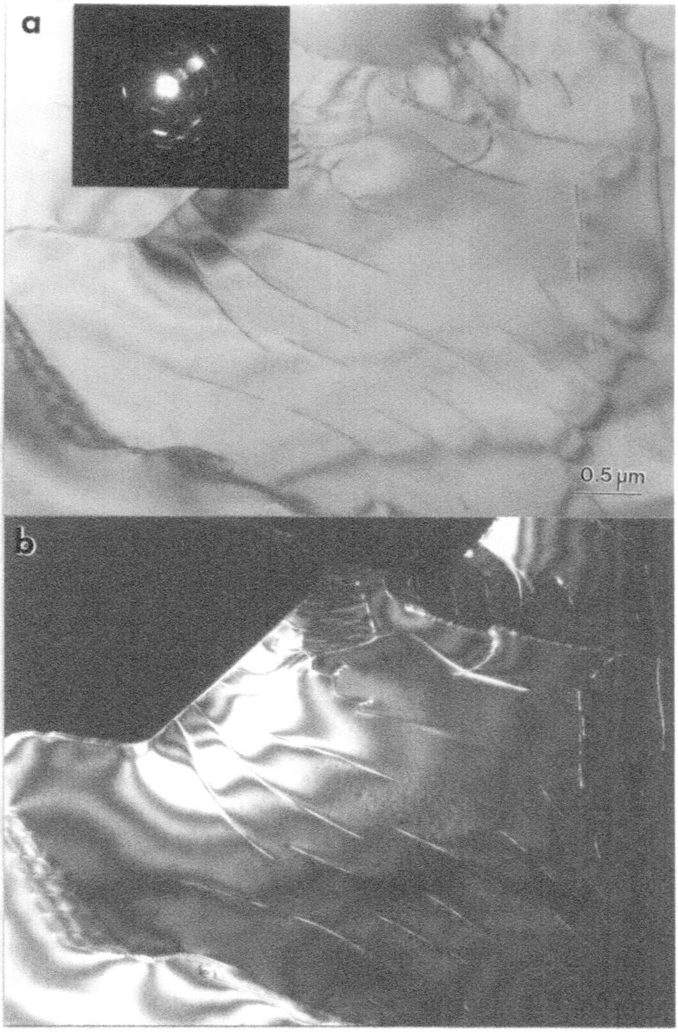

Fig. 2.11 **a** Bright-field and **b** **g** = $(20\bar{2}0)$ dark-field diffraction contrast images of polycrystalline $\alpha - Al_2O_3$ recorded under the two-beam diffracting condition near [0001] zone. The insert is the diffraction pattern

Figure 2.11 shows the bright-field and dark-field diffraction contrast images of dislocations in a polycrystalline $\alpha - Al_2O_3$ specimen. The dislocation lines are clearly resolved in both images. The intensity variation near dislocation lines results from dynamic diffraction from distorted lattices near dislocation cores. The contrast is due to the variation of the local diffracting condition, such as the excitation error, from that of perfectly structured lattices. The contrast is very sensitive to the nature of the dislocation, crystal orientation, incident beam direction, and crystal thickness.

Therefore dynamic simulations are usually necessary if we intend to study detailed lattice distortion near the dislocations. But, as shown at the end of Sect. 2.10.4, no simulation is needed if we are interested only in determining the Burgers vector of dislocations. We now consider the theoretical approach for simulating diffraction contrast images.

2.10.1 Potential of Imperfect Crystals

In practice, atomic displacements can be introduced by defects, dislocations, or strain. For a general case, atom displacement may be described by a displacement vector $\mathbf{R}(\mathbf{r})$, which depends on the position of the atom. Thus the potential distribution in the crystal under the rigid-ion approximation is written as

$$U(\mathbf{r}) = \sum_n \sum_\alpha U_\alpha\left(\mathbf{r} - \mathbf{R}_n - \mathbf{r}_\alpha - \mathbf{R}(\mathbf{r})\right) \tag{2.44}$$

Using (1.17), the reciprocal lattice vector can be introduced as

$$\begin{aligned} U(\mathbf{r}) &= \sum_n \sum_\alpha \int d\mathbf{u} U_\alpha(\mathbf{u}) \exp\left\{2\pi i \mathbf{u} \cdot [\mathbf{r} - \mathbf{R}_n - \mathbf{r}_\alpha - \mathbf{R}(\mathbf{r})]\right\} \\ &= \sum_g \left[\sum_\alpha U_\alpha(\mathbf{g}) \exp\left\{-2\pi i \mathbf{g} \cdot [\mathbf{r}_\alpha + \mathbf{R}(\mathbf{r})]\right\}\right] \exp(2\pi i \mathbf{g} \cdot \mathbf{r}) \end{aligned} \tag{2.45}$$

The Fourier coefficient of the modified potential is

$$U_g(\mathbf{r}) = U_g \exp[-2\pi i \mathbf{g} \cdot \mathbf{R}(\mathbf{r})] \tag{2.46}$$

Therefore the distortion of the crystal lattice by defects introduces a position dependent phase factor in the Fourier coefficient of the crystal potential.

2.10.2 Modified Bloch Wave Theory

For an imperfect crystal, the wave function usually does not obey the Bloch theorem and cannot be expanded into a Fourier series. Alternatively the solution of the Schrödinger equation is written as (Hirsch et al. (1977), Takagi (1962) and Takagi (1969), Howie and Basinski (1968))

$$\Psi(\mathbf{r}) = \sum_g \phi_g(\mathbf{r}) \exp\left[2\pi i \left(\mathbf{K} + \mathbf{g} + \mathbf{S}_g\right) \cdot \mathbf{r}\right] \tag{2.47}$$

where $\left|(\mathbf{K}+\mathbf{g}+\mathbf{S}_g)\right| = K$ and $\mathbf{S}_g = S_g\hat{z}$ is the excitation error. Substituting (2.45)–(2.47) into (2.5), we obtain

$$\sum_g \left\{ \frac{1}{4\pi^2}\nabla^2\phi_g + \frac{i}{\pi}\left(\mathbf{K}+\mathbf{g}+\mathbf{S}_g\right)\cdot\nabla\phi_g + \sum_h \phi_h U_{g-h}\exp[2\pi i(\mathbf{h}-\mathbf{g})\cdot\mathbf{R} + 2\pi i\left(\mathbf{S}_h-\mathbf{S}_g\right)\cdot\mathbf{r}]\right\}\exp\left[2\pi i\left(\mathbf{K}+\mathbf{g}+\mathbf{S}_g\right)\cdot\mathbf{r}\right] = 0 \quad (2.48)$$

Providing that all terms in the curly bracket vary slowly with $\mathbf{r}$, $\mathbf{R}(\mathbf{r})$ does not change significantly in a lattice distance and the $\nabla^2\phi_g$ term is negligible in comparison with the term $\left(\mathbf{K}+\mathbf{g}+\mathbf{S}_g\right)\cdot\nabla\phi_g$, then we have

$$\left(\mathbf{K}+\mathbf{g}+\mathbf{S}_g\right)\cdot\nabla\phi_g \approx \pi i\sum_h \phi_h U_{g-h}\exp\left[2\pi i(\mathbf{h}-\mathbf{g})\cdot\mathbf{R} + 2\pi i\left(\mathbf{S}_h-\mathbf{S}_g\right)\cdot\mathbf{r}\right] \quad (2.49)$$

This is the basic equation of the modified Bloch wave theory. The left-hand side of (2.49) indicates that it is necessary to differentiate ϕ_g in the direction of ($\mathbf{K}+\mathbf{g}$ $+\mathbf{S}_g$). Equation (2.49) was first proposed by Takagi (1962), Takagi (1969) and Taupin (1964) for approaching crystals with defects, dislocation, or strain, and it is thus called the *T* equation. To simplify this equation and make it useful in image calculation, we now discuss the column approximation.

2.10.3 Column Approximation

The column approximation is usually assumed in image simulations of defects and dislocations. Figure 2.12 illustrates the column approximation (Hirsch et al. 1977; Howie and Whelan 1961). A dislocation is situated at *D* inside the foil. An electron wave is incident on the top surface. The dislocation line causes a displacement of an atom in the column along *AB* from its true position by an amount $\mathbf{R}$ that depends on its distance *z* from the upper surface. The column approximation amounts to assuming that the electron wave function at *B* is the same as that at the lower surface of a crystal of infinite lateral extension, with the same displacement $\mathbf{R}(z)$, depending only on *z*, and not on the position of the column. Such a crystal can be considered an assembly of thin crystal rods, each perfect but displaced relative to each other. The wave function at the exit face of each column (or rod) depends only on the scattering of that column, so there is no interaction between columns. The width of the column is about 2 nm. This approximation may not be valid at dislocation cores where a rapid variation of $\mathbf{R}$ is possible. A rigorous discussion of the column approximation was given by Howie and Basinski (1968).

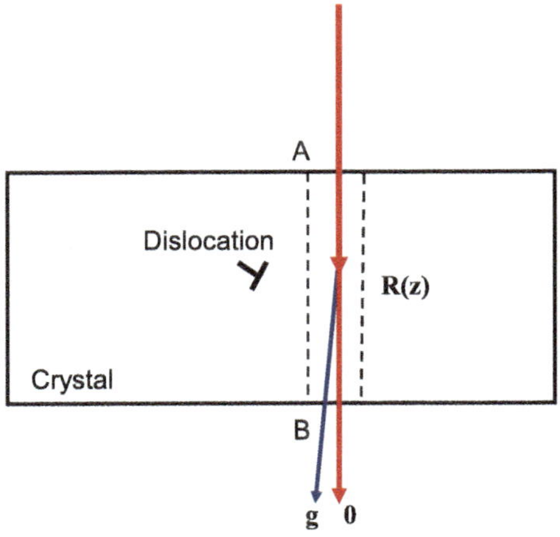

Fig. 2.12 A column approximation used to calculate the diffraction contrast images of dislocations. The crystal is assumed to be composed of parallel columns. Within each column, the local lattice displacement **R** is assumed to depend only on z. The (x, y) dependence of **R** is introduced in calculations for different columns

2.10.4 *Howie-Whelan Equation*

Contrast in dislocation images is produced by the variation of local crystal orientation due to the atom displacement **R**. This results in the intensity variation of local reflected Bragg beams. This type of image can be simulated from column to column using full dynamic diffraction theory. The scattering of each column is equivalent to the scattering of a perfect crystal without considering the effects of its neighbor columns. In this case, ϕ_g is approximated to depend only on z. If the incident beam is along the z axis, so that

$$\left(\mathbf{K} + \mathbf{g} + \mathbf{S}_g\right) \cdot \nabla\phi_g \approx K\frac{d\phi_g}{dz},$$

Equation (2.49) becomes

$$\frac{d\phi_g}{dz} = \sum_h \frac{\pi i}{\xi_{g-h}}\phi_h \exp\left[2\pi i\,(\mathbf{h} - \mathbf{g}) \cdot \mathbf{R} + 2\pi i\left(\mathbf{S}_h - \mathbf{S}_g\right) \cdot \mathbf{r}\right] \tag{2.50}$$

This is a many-beam equation for calculating the contrast of dislocations. Equation (2.50) can be solved exactly if there is no lattice distortion (i.e., $\mathbf{R} = 0$) (Van Dyck 1976).

Under two-beam approximation, (2.50) reduces to the Howie-Whelan equation (Howie and Whelan 1961),

$$\frac{d\phi_0}{dz} = \frac{\pi i}{\xi_0}\phi_0 + \frac{\pi i}{\xi_g}\phi_g \exp\left(2\pi i\mathbf{g} \cdot \mathbf{R} + 2\pi i\mathbf{S}_g \cdot \mathbf{r}\right) \tag{2.51a}$$

$$\frac{d\phi_g}{dz} = \frac{\pi i}{\xi_0}\phi_g + \frac{\pi i}{\xi_g}\phi_0 \exp\left(-2\pi i\mathbf{g} \cdot \mathbf{R} - 2\pi i\mathbf{S}_g \cdot \mathbf{r}\right) \tag{2.51b}$$

Equations (2.51a) and (2.51b) are the basic equation applied to simulate the diffraction contrast images. We now discuss the diffracting condition under which the contrast of dislocations disappears.

From (2.51), it is clear that the dislocation contrast disappears if $\mathbf{g} \cdot \mathbf{R} = 0$. For a general dislocation, the displacement vector $\mathbf{R}$ is directly related to the Burgers vector $\mathbf{b}_B$ and $\mathbf{b}_B \times \mathbf{u}_D$, where $\mathbf{u}_D$ is the direction of the dislocation line. In general, the reflected intensity of $\mathbf{g}$ is not affected if atom displacements are restricted to the reflecting plane of $\mathbf{g}$. Therefore diffracting conditions under which the contrast of dislocations disappears are

$$\mathbf{g} \cdot \mathbf{b}_B = 0 \quad \mathbf{g} \cdot (\mathbf{b}_B \times \mathbf{u}_D) = 0$$

These are the general rules for determining the nature of dislocations in TEM. The condition $\mathbf{g} \cdot \mathbf{b}_B = 0$ means that atom displacements are restricted in the plane perpendicular to $\mathbf{g}$; thus no contrast is produced if the image is recorded using this Bragg reflection.

The Howie-Whelan equation (2.51) is the fundamental equation for simulating diffraction contrast images recorded under two-beam conditions. Figure 2.13 shows a comparison of the experimentally observed stacking fault images in TiAl for two different zone axes (Fig. 2.13a) with simulated images for stacking faults bounded by partial dislocations (Viguier et al. 1994). For TiAl, the majority of stacking faults and their bounding dislocations were observed to be completely out of contrast for $\mathbf{g} = \left[2\ 0\ \bar{2}\right]$ because $\mathbf{g} \cdot \mathbf{R} = 2n\pi$. For $\mathbf{g} = \left[2\ \bar{2}\ 0\right]$, however, the bounding partial dislocations were visible, but faults were invisible. These observations indicate that faults are lying in the (111) plane (i.e., $\mathbf{R} = \pm 1/3\,[111]$) and partial dislocations have a Burgers vector parallel to $\mathbf{g} = \left[\ 1\bar{2}1\ \right]$. Figure 2.13b is simulated for an extrinsic stacking fault with a double Schckley partial dislocation of $\mathbf{b}_B = 1/3\left[1\ \bar{2}\ 1\right]$. Simulated fault images corresponding to intrinsic stacking faults and extrinsic stacking faults (without partial dislocations) are shown in Fig. 2.13c–e. The best fit is found in Fig. 2.13b. Thus the stacking fault is extrinsic and bounded by partial dislocations with Burgers vector $1/3\,[1\bar{2}1]$. For simulations of electron diffraction contrast images of arbitrary displacement fields, the program package SIMCON provided by Janssens et al. (1992) could be very useful.

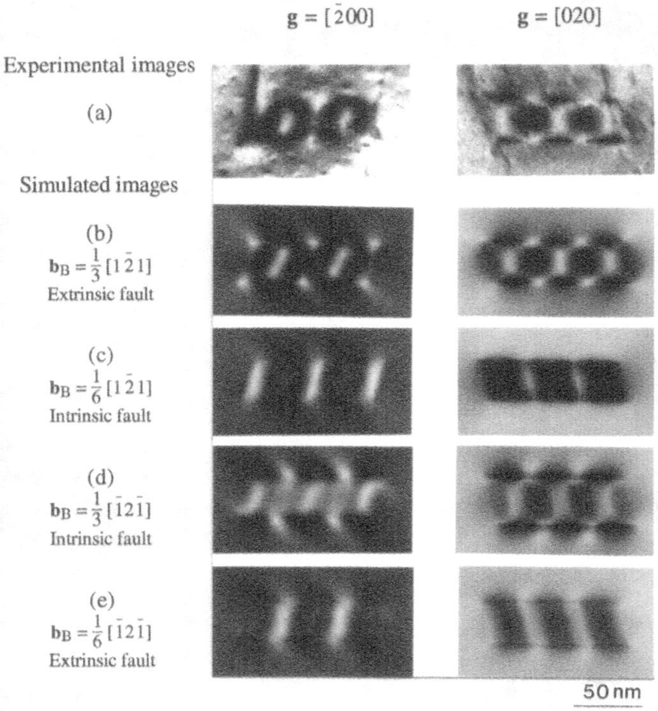

Fig. 2.13 **a** Experimental bright-field images of a stacking fault dipole in TiAl recorded under two-beam diffracting conditions of $\mathbf{g} = [\bar{2}00]$ (left-hand column) and $\mathbf{g} = [020]$ (right-hand column); **b** simulated for an extrinsic stacking fault with a partial dislocation of $b_B = 1/3[1\bar{2}1]$. **c**–**e** Simulated images for extrinsic (without bounding dislocations) and intrinsic stacking faults under identical diffracting conditions: foil normal $\mathbf{n} = [2\bar{1}4]$, $d = 115$ nm, beam direction $\mathbf{B} = [019]$ and $S_g = 0.006$ nm^{-1} for $\mathbf{g} = [\bar{2}00]$; $\mathbf{B} = [506]$ and $S_g = 0.019$ nm^{-1} for $\mathbf{g} = [020]$. The best fit is found in (**b**). Courtesy of Viguier et al. (1994)

2.10.5 α Coefficient Method

We now introduce an alternative treatment of electron diffraction in the distorted crystals under the column approximation. This approach is applied in Sect. 2.11 to calculate the contrast of weak-beam images. We first write the solution of the Schrödinger equation as

$$\Psi(\mathbf{r}) = \sum_i \alpha^{(i)}(z) \sum_g C_g^{(i)} \exp\left[2\pi i(\mathbf{K}+\mathbf{g})\cdot\mathbf{r} + 2\pi i v_i z\right] \exp(-2\pi i \mathbf{g}\cdot\mathbf{R}) \tag{2.52}$$

This theory is expressed in the form of Bloch wave coefficients, which are assumed to depend only on z, based on the column approximation. The comparison of (2.52) with (2.47) yields

$$\phi_g(\mathbf{r}) = \sum_i \alpha^{(i)}(z) C_g^{(i)} \exp\left[-2\pi i \mathbf{S}_g \cdot \mathbf{r} + 2\pi i v_i z\right] \exp(-2\pi i \mathbf{g} \cdot \mathbf{R}) \qquad (2.53)$$

Substituting this equation into (2.50) and using (2.13) and (2.16), we obtain

$$\frac{d\alpha^{(j)}(z)}{dz} = 2\pi i \sum_i \alpha^{(i)}(z) \exp\left[2\pi i \left(v_i - v_j\right) z\right] \sum_g C_g^{(i)} C_g^{(j)*} \frac{d(\mathbf{g} \cdot \mathbf{R})}{dz} \qquad (2.54)$$

Equation (2.54) represents the change in amplitude of the jth Bloch wave in a crystal slice of thickness dz at a depth z, due to the change of local orientation. This theory is more convenient for calculating the weak-beam image.

2.11 Weak-Beam Imaging

As illustrated in the last section, the contrast of dislocation lines is determined by the magnitude of the local lattice distortion, and the image resolution is no better than 2 nm . The resolution of diffraction contrast images may be improved in the dark-field images recorded by using weak diffracted beams (Cockayne et al. 1969). A simple qualitative description of the weak-beam imaging technique follows. If a perfect crystal is oriented at the exact Bragg position, the corresponding diffracted beam is strong (unless the crystal thickness happens to be an integer multiple of the extinction distance). As the crystal is tilted away from the Bragg position, the intensity in the diffracted beam decreases, so that the beam becomes a weak beam. If the crystal contains a dislocation, lattice planes around the dislocation are locally tilted, and hence some of these planes may locally be tilted back into the Bragg position, as shown in Fig. 2.14. If this is the case, locally scattered intensity from perfect crystal regions is weak, but intensity scattered locally from the region satisfying the Bragg condition is strong (Hirsch et al. 1977).

Figure 2.15 b shows a weak-beam image of the same specimen area in Fig. 2.11. It is apparent that the dislocation lines are much thinner than those appearing in either the bright-field or the dark-field images in Fig. 2.11. The grain boundaries, which are hardly seen in the bright-field image, are clearly resolved in the weak-beam image. There appears to be much less dynamic diffraction effect in the weak-beam image.

The weak-beam image is usually taken using a weak $-\mathbf{h}$ beam when a $+\mathbf{g}$ beam is set at Bragg condition (Fig. 2.14). We now use (2.54) to illustrate the dynamic description of weak-beam imaging of imperfect crystals,

$$\frac{d\alpha^{(j)}(z)}{dz} = \sum_i \alpha^{(i)}(z) \exp\left(2\pi i \Delta v_{ij} z\right) \frac{dR^{ij}(z)}{dz} \tag{2.55}$$

where

$$R^{ij}(z) = 2\pi i \sum_g C_g^{(i)} C_g^{(j)*} \mathbf{g} \cdot \mathbf{R}, \quad \Delta v_{ij} = v_i - v_j \tag{2.56}$$

Using (2.55), we perform the following mathematical operation (Cockayne 1972):

$$\int_0^d dz \frac{d}{dz} \alpha^{(j)}(z) \exp\left[R^{jj}(z)\right]$$

$$= \int_0^d dz \sum_{i \neq j} \alpha^{(i)}(z) \frac{dR^{ij}(z)}{dz} \exp\left[2\pi i \Delta v_{ij} z - R^{jj}(z)\right] \tag{2.57}$$

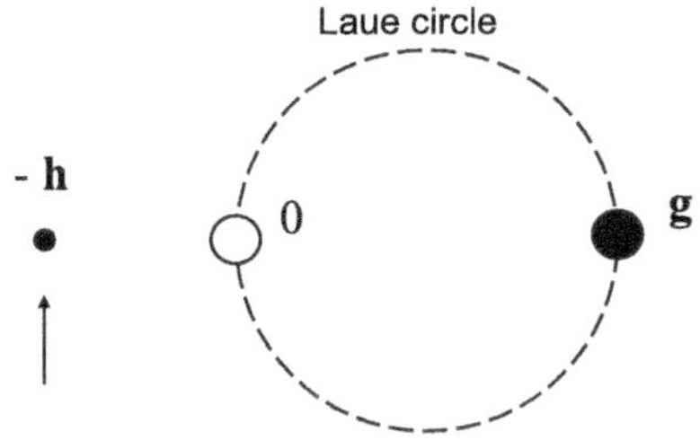

Fig. 2.14 The three-beam diffracting condition for weak-beam imaging

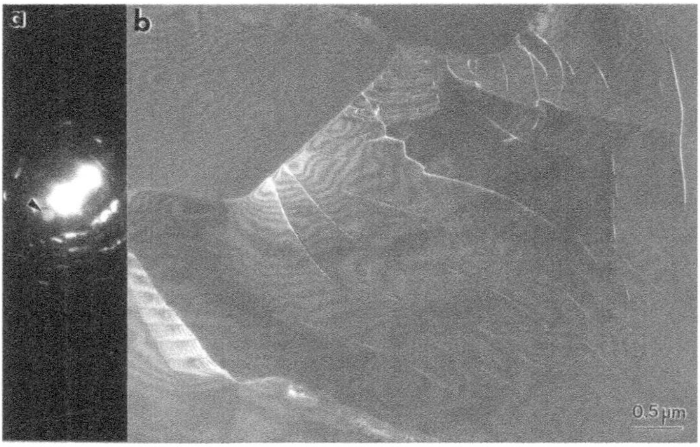

Fig. 2.15 **a** Three-beam electron diffracting condition for recording the dark-field weak-beam image, shown in (**b**), of the polycrystalline $\alpha - Al_2O_3$ specimen used for Fig. 2.11. The beak used for forming the weak-beam image is indicated by an arrowhead

Integrating the left-hand side of this equation, we obtain

$$\alpha^{(j)}(d) = \exp\left[R^{jj}(d)\right] \times \left\{ \int_0^d dz \sum_{i \neq j} \alpha^{(i)}(z) \frac{dR^{ij}(z)}{dz} \exp\left[2\pi i \Delta v_{ij} z - R^{jj}(z)\right] + \alpha^{(j)}(0) \right\} \tag{2.58}$$

The purpose of the preceding mathematical operation is to exclude the contribution of the terms with $i = j$. The first-order solution of (2.58) is obtained by taking $\alpha^{(i)} \approx C_0^{(i)^*}$ on the right-hand side; thus

$$\alpha^{(j)}(d) \approx \exp\left[R^{jj}(d)\right] \times \left\{ \sum_{i \neq j} C_0^{(i)*} \int_0^d dz \frac{dR^{ij}(z)}{dz} \exp\left[2i\pi i \Delta v_{ij} z - R^{jj}(z)\right] + C_0^{(j)^*} \right\} \tag{2.59}$$

If $dR^{ij}(z)/dz$ is a slow-variation function of z, the magnitude of $\alpha^{(j)}(d)$ depends approximately on

$$\int_0^d dz \exp\left[2\pi i \Delta v_{ij} z - R^{jj}(z)\right]$$

The integral is maximum when $[\ 2\pi i \Delta v_{ij} z - R^{jj}(z)]$ is constant (Cockayne 1972) or in an equivalent form

$$2\pi i \Delta v_{ij} - \frac{dR^{jj}(z)}{dz} = 0$$

or

$$v_j - v_i + \sum_g C_g^{(i)} C_g^{(j)*} \frac{d(\mathbf{g} \cdot \mathbf{R})}{dz} = 0 \tag{2.60}$$

Under the two-beam approximation, this condition becomes (Cockayne 1972)

$$S_g + \frac{d(\mathbf{g} \cdot \mathbf{R})}{dz} = 0 \tag{2.61a}$$

This is the condition for dislocations to show bright contrast in a weak-beam image. This relation simply means that the local orientation variation has to reduce the excitation error of the $-\mathbf{g}$ beam to satisfy the Bragg condition. For imperfect crystals and under the column approximation, the effective local excitation error is given by

$$S_g^{(eff)} = S_g + \frac{d(\mathbf{g} \cdot \mathbf{R})}{dz}$$

The condition $S_g^{(eff)} = 0$ means that the local region satisfies the Bragg condition, thus giving strong reflection intensity.

The column approximation was assumed in weak-beam imaging theory. However the large scattering angle between $-\mathbf{h}$ and $\mathbf{g}$ reflections shown in Fig. 2.14 could significantly increase the width of the column in some cases (Howie and Sworn 1970; Lewis and Villagrana 1979), resulting in incorrect positions of the dislocation contrast. It is thus necessary to examine each individual case carefully.

As just pointed out, the weak-beam image is formed due to the variation of the excitation error w with the local orientation. The dynamic diffraction effect in the $-\mathbf{h}$ beam is greatly reduced due to strong excitation of the $\mathbf{g}$ beam. For crystals containing grain boundaries, dislocations at interfacial regions are best seen in the weak-beam image; Fig. 2.16 shows such an example. The grain boundary is hardly seen in the bright-field image (Fig. 2.16a); it becomes relatively easy to see the interfacial dislocations in the dark-field diffraction contrast image (Fig. 2.16b). The fine details of dislocation are clearly seen in the weak-beam image (Fig. 2.16c), but the quantitative interpretation of the contrast is rather difficult.

Weak-beam imaging is best suited for imaging dislocations, but it is almost impossible to apply it to determine the nature of dislocations. The two-beam diffraction contrast images are very useful for determining the Burgers vector of dislocations.

2.12 Absorption Effect in Dynamical Calculations

Interaction between the incident electron and the crystal is quite complex. Besides elastic scattering, inelastic excitations, such as X-ray emission, photon emission, and atom ionization, are usually involved. For thin crystals, the probability of inelastic scattering is usually small. As shown in Chap. 6, inelastic scattering of electrons is equivalent to introducing an imaginary potential into the elastic-scattering equation. This is usually known as the absorption effect. In the Bloch wave picture, taking the two-beam case as an example, one Bloch wave is a maximum between the columns of the atomic row, and the other Bloch wave is a maximum on the atomic rows (see Fig. 2.4). It is thus expected that the former is weakly scattered or absorbed and the latter will be strongly scattered or absorbed due to different inelastic-scattering processes. This is the so-called anomalous absorption effect. We can include absorption by replacing $V(\mathbf{r})$ with $V(\mathbf{r}) + iV'(\mathbf{r})$. In general, $|V(\mathbf{r})| \gg |V'(\mathbf{r})|$, so we can use the first-order perturbation method to estimate the effect of absorption. The perturbation iV' results in a change ΔE in the energy of the Bloch wave B_i given by

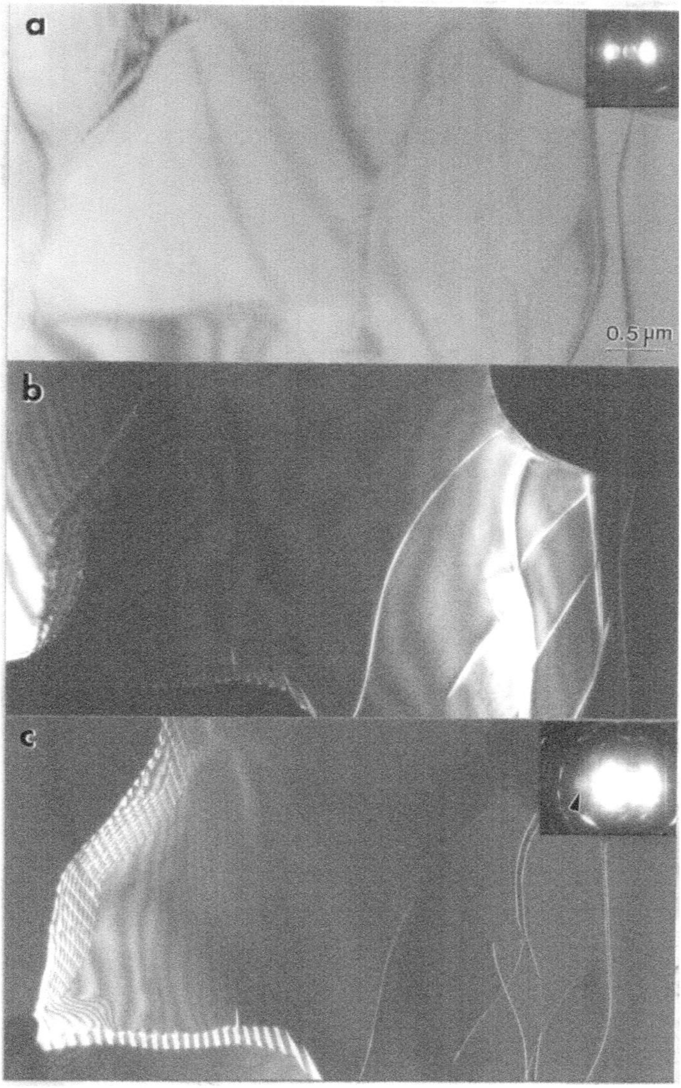

Fig. 2.16 **a** Bright-field, **b** $\mathbf{g} = (20\bar{2}0)$ dark-field, and **c** weak-beam images of a polycrystalline $\alpha - Al_2O_3$ specimen showing the advantages of the weak-beam technique for imaging interface dislocations. The beam used to form the weak-beam image is indicated by an arrowhead

$$\Delta E = \int d\mathbf{r} B_i^*(\mathbf{r}) \left[-ieV'(\mathbf{r})\right] B_i(\mathbf{r}) \tag{2.61b}$$

The change in energy is equivalent to a change in K. Adding an $iU'_{g'h}$ term in (2.13), we have

$$2KS_g C_g + \sum_{g \neq h} U_{g-h} C_h = 2K_z v C_g - i \sum_{g \neq h} U'_{g-h} C_h$$

This is equivalent to adding an imaginary component in the wave vector along the z direction. Thus the perturbation effect to the electron wave vector is

$$\Delta K_z = iv'_i = -\frac{m_e \Delta E}{h^2 K_z} = \frac{m_e i}{h^2 K_z} \int d\mathbf{r} B_i^*(\mathbf{r}) \left[eV'(\mathbf{r})\right] B_i(\mathbf{r})$$

Using the Bloch wave expression in (2.17), the final calculation gives (Humphreys 1979)

$$v'_i = \frac{1}{2K_z} \sum_g \sum_h C_g^{(i)*} U'_{g-h} C_h^{(i)} \tag{2.62}$$

Hence v'_i is the imaginary part of the wave vector for the i th Bloch wave, and it can easily be calculated from the eigenvectors and the U' matrix. The following substitutions are thus made:

$$v_i \to v_i + iv'_i \tag{2.63a}$$

$$U_g \to U_g + iU'_g \tag{2.63b}$$

$$\frac{1}{\xi_g} \to \frac{1}{\xi_g} + \frac{i}{\xi'_g} \tag{2.63c}$$

with $\xi'_g = K/U'_g$. This is the phenomenological treatment of the absorption effect in electron diffraction. The calculation of the imaginary potential is illustrated in Chap. 6.

2.13 Summary

As a summary of this chapter, we have introduced the Bloch wave theories and their applications in electron diffraction and imaging. The Bethe theory is powerful particularly because of its physical insight for evaluation of the contribution of different Bloch waves. This theory is much more convenient for sophisticated analytical expressions when there are only a few beams involved. The Bethe theory is best suited for CBED calculations in reciprocal space if there are no disk overlaps, but the calculations can be easily performed only for three-dimensional perfect crystals. Real-space Bloch wave theory is convenient for simulations of diffraction contrast images and CBED if there are disk overlaps. A detailed comparison of the Bethe theory with other theories, presented in the next few chapters, is given in Sect. 4.5.

Chapter 3
Dynamic Elastic Electron Scattering II: Multislice Theory

For simulations of electron images, the Bloch wave approach is more convenient for dealing with these cases in which either the crystal contains no defects or the image resolution is so low that the column approximation holds. When the image resolution approaches the atomic level, the dominant contrast mechanism is phase contrast, and the effect of diffraction contrast is minimized. In this case, it is more convenient to apply the multislice theory, particularly when defects are present in the crystal. The most important advantage of the multislice theory is that it does not require three-dimensional periodicity of the crystal; thus images of interfaces, surfaces, and dislocations can be simulated. In this chapter outlines the theoretical scheme of the multislice theory and its applications. Methods to improve the theory are illustrated to include HOLZ reflections accurately.

3.1 Physical Optics Approach

The multislice many-beam dynamic electron diffraction theory of Cowley and Moodie (1957) was originally derived from the physical optics approach, in which electron transmission through a crystal is represented by transmission through a set of two-dimensional phase objects, as shown in Fig. 3.1. The crystal is cut into many slices of equal thickness Δz in the direction perpendicular or nearly perpendicular to the incident beam. This treatment is usually convenient if incident electrons are along one of the low-index zone axes. When slice thickness tends to be very small, the scattering of each slice can be based on the projected potential approximation. If the backscattering effect is negligible, the transmission of the electron wave through each slice can be considered a two-step process-the phase modulation of the wave by the projected atomic potential within the slice and the propagation of the modulated wave in a vacuum for distance Δz along the beam direction before striking the next crystal slice (Fig. 3.1). The projected potential approximation was made for each

Z. L. Wang, *Elastic and Inelastic Scattering in Electron Diffraction and Imaging*, Springer Series in Solid-State Sciences 205, https://doi.org/10.1007/978-3-031-90819-4_3

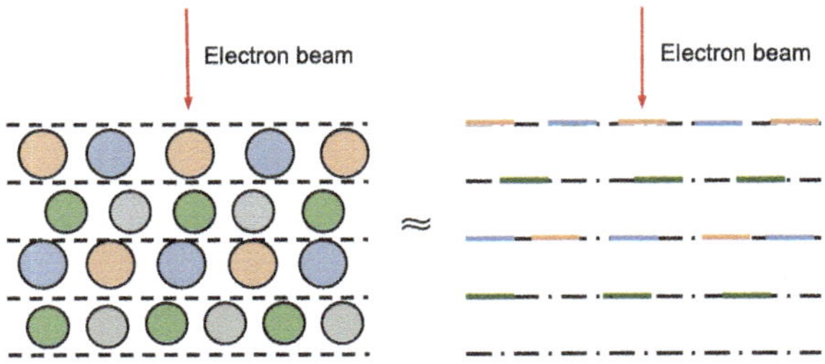

Fig. 3.1 The multislice method. The crystal is cut into many parallel slices; electron beam scattering by the crystal is calculated slice-by-slice without considering backscattering. If the thickness of the crystal slice is sufficiently small, the potential field in each slice can be considered projected on the entrance face of each slice. This is an excellent approximation especially for transmission high-energy electron diffraction

slice. For simplification, we assume that planes of crystal slices are perpendicular to the incident beam direction. A more general case in which the incident beam enters the crystal with a small inclined angle is given in Sect. 3.2.

3.1.1 *Phase Object Approximation*

If an electron of incident energy E is traveling in a potential field $V(\mathbf{r})$, the effective wave vector is related to the effective wavelength of the electron by

$$K_{eff} = \frac{1}{\lambda_{eff}} = \frac{(2m_0)^{1/2}}{h}[E + eV(\mathbf{r})]^{1/2} = \frac{1}{\lambda}\left[1 + \frac{V(\mathbf{r})}{U_0}\right]^{1/2} \tag{3.1}$$

Thus the relative phase shift of the wave traveling in the field relative to the wave traveling in the absence of a field is

$$\int_{z_n}^{z_n+\Delta z} dz 2\pi\left(\frac{1}{\lambda_{eff}} - \frac{1}{\lambda}\right) \approx i\sigma V(\mathbf{b}) \tag{3.2}$$

where $\sigma = \pi/\lambda U_0$, which is derived by ignoring the relativistic effect, and the projected potential of the n_c th slice is

$$V(\mathbf{b}) = \int_{z}^{z+\Delta z} dz V(\mathbf{r})$$

Therefore the effect of the potential field is represented by multiplying the wave function with a phase-grating function

$$Q_{n_c}(\mathbf{b}, \Delta z) = \exp[i\sigma V(\mathbf{b})] \tag{3.3}$$

This is known as the phase object approximation.

3.1.2 *Huygens' Principle*

In the second step, we use Huygens' principle to derive the relationship between the wave $\Phi\left(\mathbf{b}', z_n\right)$ at z_{n_c} with that at $z_{n_{c+1}}$, as shown in Fig. 3.2. According to Huygens' principle, each point on a wave front gives rise to a spherical secondary wave having an initial amplitude proportional to the amplitude of the incident wave. The amplitudes of the secondary wave are added to give the amplitude on a plane of observation. The mathematical expression of the principle (Cowley, 1981) is

$$\Phi\left(\mathbf{b}, z_{n_{c+1}}\right) = -\frac{i}{\lambda}\int d\mathbf{b}'\Phi\left(\mathbf{b}', z_{n_c}\right)\frac{\exp\left(2\pi i K\rho'\right)}{\rho'}\frac{1}{2}\left(1+\frac{\rho'\cdot\mathbf{z}}{\rho' z}\right) \tag{3.4}$$

where $\boldsymbol{\rho}' = \mathbf{b}' - \mathbf{b} + \Delta\mathbf{z}$ is the relative position of the observation point with respect to the source point, as shown in Fig. 3.3. There is a change of phase $\pi/2$, represented by $-i$; a scale factor of λ^{-1}; and an obliquity factor,

$$\frac{1}{2}\left(1+\frac{\rho'\cdot\mathbf{z}}{\rho' z}\right)$$

which ensures that the waves add up with a maximum amplitude in the forward direction and do not propagate backward. The integration of $\mathbf{b}'$ sums over all elements on the wave front that contribute to the wave at $\mathbf{b}$. Under the small-angle scattering approximation, substituting

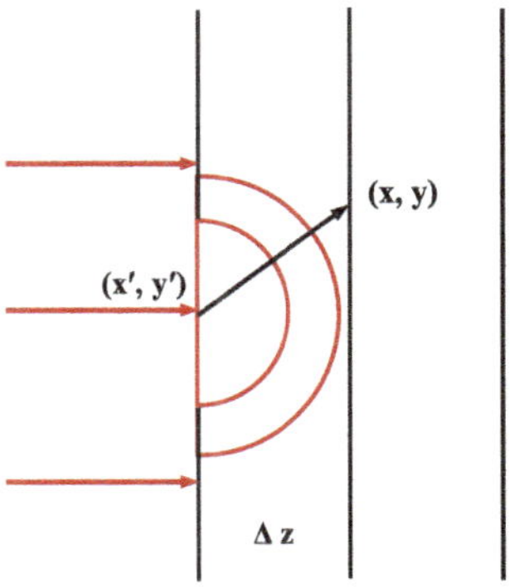

Fig. 3.2 The application of Huygens' principle in electron wave propagation in real space. This step is used to introduce the propagation function in multislice theory

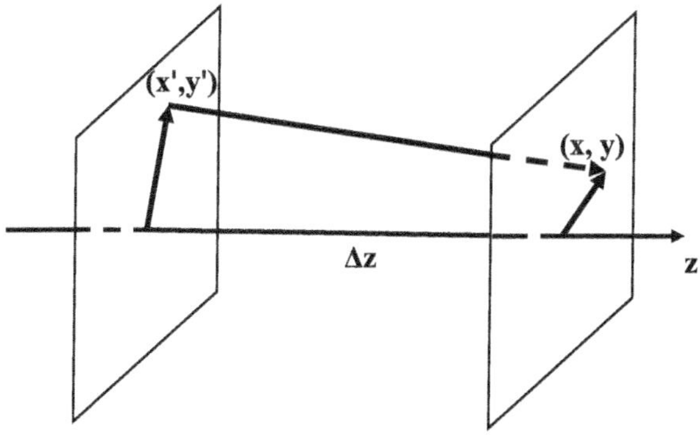

Fig. 3.3 The real-space coordinate system for describing electron propagation in free space

$$\frac{\rho' \cdot z}{\rho' z} \approx 1 \quad \rho' \approx \Delta z$$

in the denominator and in the exponents,

$$\rho' = \left(\left|\mathbf{b}' - \mathbf{b}\right|^2 + \Delta z^2\right)^{1/2} \approx \Delta z + \frac{\left|\mathbf{b}' - \mathbf{b}\right|^2}{2\Delta z} \tag{3.5}$$

Equation (3.4) becomes

$$\begin{aligned}\Phi\left(\mathbf{b}, z_{n_c+1}\right) &= \exp(2\pi i K \Delta z) \int d\mathbf{b}' \Phi\left(\mathbf{b}', z_{n_c}\right) \frac{\exp\left(\pi i K \mid \mathbf{b}' - \mathbf{b} \mid^2 / \Delta z\right)}{i\lambda \Delta z} \\ &= \exp(2\pi i K \Delta z) \Phi\left(\mathbf{b}, z_{n_c}\right) \otimes P(\mathbf{b}, \Delta z)\end{aligned} \tag{3.6}$$

where the propagation function is defined as

$$P(\mathbf{b}, \Delta z) = \frac{\exp\left(\pi i K |\mathbf{b}|^2 / \Delta z\right)}{i\lambda \Delta z} \tag{3.7a}$$

Therefore the propagation of an incident wave for a distance Δz in vacuum is characterized by a propagation function. Since the $\exp(2\pi i K \Delta z)$ term is constant if Δz remains the same for each slice, the phase factor can thus be dropped in the following discussion. It can be proved that the Fourier transform of P is still a Gaussian-like function

$$\int d\mathbf{b} P(\mathbf{b}, \Delta z) \exp(-2\pi i \mathbf{u} \cdot \mathbf{b}) = \exp\left(-\pi i \mathbf{u}^2 \Delta z \lambda\right) \tag{3.7b}$$

This characteristic is used in the following sections.

3.1.3 Multislice Theory

From Sects. 3.1.1 and 3.1.2, the wave function before and after transmitting the n_c th crystal slice is correlated by

$$\Phi\left(\mathbf{b}, z_{n_c+1}\right) = \left[\Phi\left(\mathbf{b}, z_{n_c}\right) Q_{n_c}(\mathbf{b}, \Delta z)\right] \otimes P(\mathbf{b}, \Delta z) \tag{3.8}$$

This is the famous multislice equation of Cowley and Moodie (1957). This equation has three important features. First no assumption was made regarding the arrangement of atoms in the slices, so that the theory can be applied to calculate electron scattering in crystals containing defects and dislocations. Second since there is no restriction on the phase-grating function $Q_{n_c}(\mathbf{b})$, it can be modified slice-to-slice according to the desired structure. The theory can thus be applied to calculate images of complex structures, such as interfaces and multilayers. Finally the equation can be applied to calculate wave propagation through crystals slice-by-slice, so that a real-space picture is given of how the electron wave is built up and the channeling current is formed. The multislice calculation includes contributions from all possible Bloch waves, and boundary conditions are automatically matched.

3.2 Quantum Mechanical Basis of Multislice Theory

As shown in the last section, the multislice equation was first derived from the physical optics approach under the small-angle-scattering approximation without backscattering. The accuracy of the theory depends on the thickness of the crystal slice. It has been shown (Goodman and Moodie 1974) that in the limiting case when the thickness of the slice Δz goes to zero and the number of slices N goes to infinity in such a way that $N\Delta z = d$, the multislice description becomes a rigorous representation of the scattering process, completely consistent with the more conventional quantum mechanical descriptions. Equation (3.8) has been derived directly by using Feynman (1948) path-integral method for high-energy electron diffraction (Van Dyck 1975; Jap and Glaeser 1978). In this section, we derive the multislice equation directly from the Schrödinger equation.

3.2.1 Inclined Incident Beam Case

To illustrate the quantum mechanical basis of the multislice theory, we consider a general case in which the electron strikes the crystal entrance surface at a small angle ϑ, as shown in Fig. 3.4. The crystal slice is usually cut parallel to a crystallographic plane. This case occurs in either dark-field or bright-field HRTEM of small-angle

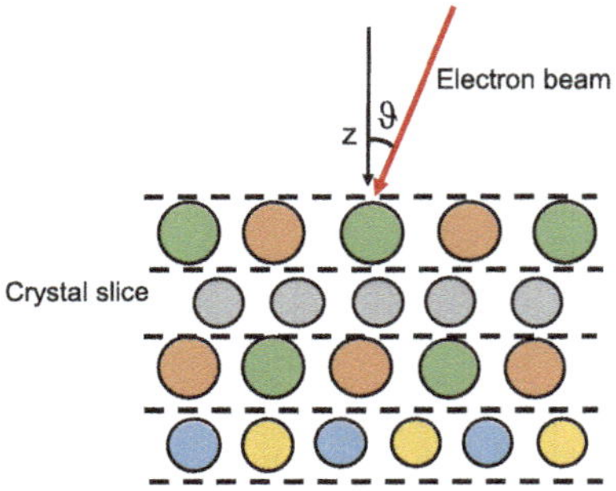

Fig. 3.4 The multislice method for the inclined incident beam case. Slices are cut perpendicular to the z axis, then the crystal potential in each slice is projected onto the $x-y$ plane, but the incident beam enters the surface at an angle ϑ. This approach provides some convenience for dealing with those cases in which the incident beam is slightly off the zone axis

wedge-shaped crystals. We now follow Ishizuka's method to derive the multislice equation based on the first-principles approach (Ishizuka and Uyeda 1977; Ishizuka 1982).

The first approximation that we make is high-energy electron forward scattering. It is always assumed that there is no backscattering in electron diffraction. The validity of this approximation has been examined by Van Dyck (1985), who found that the amplitude ratio of the backscattered wave to the transmitted wave is on the order of $(V/U_0)(1-V/U_0)^{-1}$, which tends to be very small for high-energy electrons. Based on this approximation, (2.5) is first written as an integral equation (see Appendix D)

$$\Psi(\mathbf{r}) = \exp(2\pi i\mathbf{K}\cdot\mathbf{r}) + \pi\int d\mathbf{r}'\frac{\exp\left(2\pi iK\left|\mathbf{r}-\mathbf{r}'\right|\right)}{|\mathbf{r}-\mathbf{r}'|}U\left(\mathbf{r}'\right)\Psi\left(\mathbf{r}'\right) \tag{3.9}$$

where the incident wave is assumed to be a plane wave, although this assumption does not affect the final result even if the incident wave is a converged wave packet. The plane wave component can be removed in the following discussion by defining $\Psi(\mathbf{r}) = \exp(2\pi i\mathbf{K}\cdot\mathbf{r})\Phi(\mathbf{r})$, so that

$$\begin{aligned}\Phi(\mathbf{r}) &= 1 + i\pi\lambda\int d\mathbf{r}'P'\left(\mathbf{r}-\mathbf{r}'\right)U\left(\mathbf{r}'\right)\Phi\left(\mathbf{r}'\right)\\ &= 1 + i\sigma\int d\mathbf{r}'P'\left(\mathbf{r}-\mathbf{r}'\right)V\left(\mathbf{r}'\right)\Phi\left(\mathbf{r}'\right)\end{aligned} \tag{3.10}$$

where P' is a propagation function for a general incident beam case,

$$P'(\mathbf{r}) = \frac{1}{i\lambda}\frac{\exp(2\pi iKr - 2\pi i\mathbf{K}\cdot\mathbf{r})}{r} \tag{3.11}$$

and

$$\sigma = \frac{\pi e \gamma}{\lambda E} = \frac{1}{\hbar v} \quad E = eU_0 \left(1 + \frac{eU_0}{2m_0 c_0^2}\right)$$

The constant σ given here includes the relativistic correction. Before finding the multislice solution of (3.10), we consider two properties of the P' function.

Property 1

$$\int_{\Sigma} d\mathbf{b} P'(\mathbf{b} - \mathbf{b}_1, z - z_1) \, P'(\mathbf{b}_2 - \mathbf{b}, z_2 - z) \approx \frac{K}{K_z} P'(\mathbf{b}_2 - \mathbf{b}_1, z_2 - z_1) \quad (3.12a)$$

or in reciprocal space

$$P'(\mathbf{u}, z - z_1) \, P'(\mathbf{u}, z_2 - z) \approx \frac{K}{K_z} P'(\mathbf{u}, z_2 - z_1) \quad (3.12b)$$

where the plane of integration Σ is arbitrarily located between z_1 and z_2. To prove this relation, we take a two-dimensional Fourier transform of P',

$$P'(\mathbf{u}, z) = \int d\mathbf{b} \frac{1}{i\lambda} \frac{\exp(2\pi i K r - 2\pi i \mathbf{K} \cdot \mathbf{r})}{r} \exp(-2\pi i \mathbf{u} \cdot \mathbf{b})$$

Integrating successively over x and y, the result is Ishizuka (1982)

$$P'(\mathbf{u}, z) = \frac{K}{\left[K^2 - (\mathbf{K}_b + \mathbf{u})^2\right]^{1/2}} \exp\left\{-2\pi i z \left[K_z - \left(K^2 - |\mathbf{K}_b + \mathbf{u}|^2\right)^{1/2}\right]\right\} \quad (3.13)$$

Using (3.13), it can be readily shown

$$P'(\mathbf{u}, z - z_1) \, P'(\mathbf{u}, z_2 - z) = \frac{K}{\left(K^2 - |\mathbf{K}_b + \mathbf{u}|^2\right)^{1/2}} P'(\mathbf{u}, z_2 - z_1) \approx \frac{K}{K_z} P'(\mathbf{u}, z_2 - z_1) \quad (3.14)$$

where an approximation of $K^2 \gg |\mathbf{K}_b + \mathbf{u}|^2$ was made. This requires the angle ϑ between wave vector $\mathbf{K}$ and the normal direction z of the slices to be small. This condition is usually satisfied in dark-field HRTEM imaging. If we inverse Fourier transform (3.14) into real-space, (3.12a) is obtained.

Property 2

$$\int d\mathbf{b} P'(\mathbf{b}, \Delta z) = \frac{K}{K_z} \quad (3.15)$$

For small-angle scattering so that $\left|\mathbf{r} - \mathbf{r}'\right| \approx z - z'$, P' becomes

$$P'(\mathbf{b}, \Delta z) = \frac{1}{i\lambda\Delta z}\exp\left[2\pi i\,(K - K_z)\,\Delta z\right]\exp\left(i\frac{\pi b^2}{\lambda\Delta z} - 2\pi i\mathbf{K}_b \cdot \mathbf{b}\right)$$
$$= \exp\left[2\pi i\,(K - K_z)\,\Delta z\right]\exp\left(-2\pi i\mathbf{K}_b \cdot \mathbf{b}\right)P(\mathbf{b}, \Delta z) \tag{3.16}$$

In general, the constant factor $\exp\left[2\pi i\,(K - K_z)\,\Delta z\right]$ can be omitted for slices of equal thickness.The $\exp\left(-2\pi i\mathbf{K}_b \cdot \mathbf{b}\right)$ factor corresponds to the shift of the electron wave function in the **b** plane due to the inclined propagation of electrons.

3.2.2 Multislice Solution of the Schrödinger Equation

We now return to (3.10). To determine its multislice solution, we need to derive a relationship between the wave $\Phi\,(\mathbf{b}, z_0)$ going into a crystal slice at $z = z_0$, and $\Phi(\mathbf{b}, z)$ after penetrating through the slice at $z = z$. The first step is to separate the entrance wave and the newly scattered wave of the slice.

$$\Phi(\mathbf{b}, z) = 1 + i\sigma\int d\mathbf{b}'\int_{z'=-\infty}^{z'=z} dz'P'\left(\mathbf{b} - \mathbf{b}', z - z'\right)V\left(\mathbf{b}, z'\right)\Phi\left(\mathbf{b}', z'\right)$$
$$= 1 + i\sigma\int d\mathbf{b}'\int_{z'=-\infty}^{z'=z_0} dz'P'\left(\mathbf{b} - \mathbf{b}', z - z'\right)V\left(\mathbf{b}, z'\right)\Phi\left(\mathbf{b}', z'\right)$$
$$+ i\sigma\int d\mathbf{b}'\int_{z'=z_0}^{z'=z} dz'P'\left(\mathbf{b} - \mathbf{b}', z - z'\right)V\left(\mathbf{b}, z'\right)\Phi\left(\mathbf{b}', z'\right) \tag{3.17}$$

Using the following properties of function P' introduced in (3.12a) and (3.14):

$$1 = \frac{K_z}{K}\int d\mathbf{b}_0 P'\left(\mathbf{b} - \mathbf{b}_0, z - z_0\right) \tag{3.18a}$$

$$P'\left(\mathbf{b} - \mathbf{b}', z - z'\right) = \frac{K_z}{K}\int d\mathbf{b}_0 P'\left(\mathbf{b} - \mathbf{b}_0, z - z_0\right)P'\left(\mathbf{b}_0 - \mathbf{b}', z_0 - z'\right) \tag{3.18b}$$

replacing the first and the second terms on the right-hand side of (3.17) by (3.18a) and (3.18b), respectively, yields

$$\Phi(\mathbf{b}, z) = \frac{K_z}{K}\int d\mathbf{b}_0\left[1 + i\sigma\int d\mathbf{b}'\int_{z'=-\infty}^{z'=z_0} dz'P'\left(\mathbf{b}_0 - \mathbf{b}', z_0 - z'\right)V\left(\mathbf{b}, z'\right)\Phi\left(\mathbf{b}', z'\right)\right]$$
$$P'\left(\mathbf{b} - \mathbf{b}_0, z - z_0\right) + i\sigma\int d\mathbf{b}'\int_{z'=z_0}^{z'=z} dz'P'\left(\mathbf{b} - \mathbf{b}', z - z'\right)V\left(\mathbf{b}, z'\right)\Phi\left(\mathbf{b}', z'\right) \tag{3.19}$$

Comparing the terms inside [ ] in (3.19) with (3.17) for $z = z_0$, then (3.19) can be written as

$$\Phi(\mathbf{b}, z) = \frac{K_z}{K}\int d\mathbf{b}_0 \Phi(\mathbf{b}_0, z_0)\, P'(\mathbf{b} - \mathbf{b}_0, z - z_0) + i\sigma \int d\mathbf{b}' \int_{z'=z_0}^{z'=z} dz' P'\left(\mathbf{b} - \mathbf{b}', z - z'\right) V\left(\mathbf{b}', z'\right) \Phi\left(\mathbf{b}', z'\right) \tag{3.20}$$

To solve (3.20), we expand Φ for (i σ),

$$\Phi(\mathbf{b}, z) = \sum_{L=0}^{\infty} (i\sigma)^L f_L(\mathbf{b}, z) \tag{3.21}$$

Substituting (3.21) into (3.20) and collecting terms with the same orders of $(i\sigma)^L$,

$$f_0(\mathbf{b}, z) = \frac{K_z}{K}\int d\mathbf{b}_0 P'(\mathbf{b} - \mathbf{b}_0, z - z_0)\, \Phi(\mathbf{b}_0, z_0) \tag{3.22a}$$

$$f_L(\mathbf{b}, z) = \int d\mathbf{b}' \int_{z'=z_0}^{z'=z} dz' P'\left(\mathbf{b} - \mathbf{b}', z - z'\right) V\left(\mathbf{b}', z'\right) f_{L-1}\left(\mathbf{b}', z'\right) \tag{3.22b}$$

If V varies slowly in the region of $\Delta z = (z - z_0)$, using (3.12a)

$$\begin{aligned} f_1(\mathbf{b}, z) &= \frac{K_z}{K}\int d\mathbf{b}_0 \Phi(\mathbf{b}_0, z_0) \int_{z'=z_0}^{z'=z} dz' \int d\mathbf{b}' P'\left(\mathbf{b} - \mathbf{b}', z - z'\right) V\left(\mathbf{b}', z'\right) \\ &\quad P'\left(\mathbf{b}' - \mathbf{b}_0, z' - z_0\right) \\ &\approx \frac{K_z}{K}\int d\mathbf{b}_0 \Phi(\mathbf{b}_0, z_0) \left[\frac{K}{K_z}\int_{z'=z_0}^{z'=z} dz' V\left(\mathbf{b}_0, z'\right)\right] P'(\mathbf{b} - \mathbf{b}_0, z - z_0) \end{aligned} \tag{3.22c}$$

Applying the same method, it can easily be proved

$$f_L(\mathbf{b}, z) \approx \frac{K_z}{K}\int d\mathbf{b}_0 \Phi(\mathbf{b}_0, z_0) \frac{1}{L!}\left[\frac{K}{K_z}\int_{z'=z_0}^{z'=z} dz' V\left(\mathbf{b}_0, z'\right)\right]^L P'(\mathbf{b} - \mathbf{b}_0, z - z_0) \tag{3.22d}$$

Defining the projected potential as

$$V(\mathbf{b}) = \int_{z'=z_0}^{z'=z} dz' V\left(\mathbf{b}, z'\right) \tag{3.23a}$$

the relation that governs the wave before and after being scattered by a crystal slice is

$$\begin{aligned}\Phi(\mathbf{b}, z) &= \int d\mathbf{b}_0 \Phi\left(\mathbf{b}_0, z_0\right) \exp\left[i\sigma \frac{K}{K_z} V\left(\mathbf{b}_0\right)\right] \frac{K_z}{K} P'\left(\mathbf{b}-\mathbf{b}_0, z-z_0\right) \\ &= \left\{\Phi\left(\mathbf{b}, z_0\right) \exp\left[i\sigma \frac{K}{K_z} V(\mathbf{b})\right]\right\} \otimes \left[\frac{K_z}{K} P'\left(\mathbf{b}, z-z_0\right)\right]\end{aligned} \tag{3.23b}$$

Equation (3.23a) is the multislice solution of the Schrödinger equation for an inclined incident electron beam. This equation is identical to (3.8) if we set $K_z = K$ and $K_b = 0$. The following two substitutions have been made for the inclined incident beam

$$\sigma \to \sigma \frac{K}{K_z} \quad P \to \frac{K_z}{K} P'$$

It is important to point out that the projected potential used in (3.23b) for the inclined incident beam case is projected along the z axis direction (i.e., the normal direction of the slices) instead of the beam direction. The major modifications are contained in the propagation function P'. The extra factor K_z/K is introduced for normalizing P'.

In the multislice calculation, the major error source is the thickness of the slice. The upper limit of the slice thickness can be estimated from the preceding derivation of the multislice formula as follows. The potential distribution should change slowly over a range $\Delta z\theta$, where θ is the scattering semiangle, and a factor of 2 is introduced to consider half-slice thickness. Using the Bragg reflection condition $2\theta g \approx \lambda/\Delta r$, where Δr defines the distance over which the potential does not change appreciably, so that $2\Delta z\theta g < \Delta r$, yielding $\Delta z < 2\Delta r^2/\lambda$. For 300 keV electrons, $\lambda = 0.0019$ nm, if $\Delta r = 0.01$ nm, so that $\Delta z < 0.1$ nm. Therefore good accuracy can be achieved if the slice thickness is chosen as the diameter of an atom.

The upper limit of the inclined angle of the beam is determined by the condition under which the approximated relation $\left(K^2 - |\mathbf{K}_b + \mathbf{u}|^2\right)^{1/2} \approx K_z$ holds. This can be written equivalently as $\sin\vartheta \approx g/2K$, where ϑ is the angle between the wave vector $\mathbf{K}$ and the normal direction of the slices, and g can be considered as the maximum reciprocal lattice vector to be enclosed by the objective aperture.

3.3 Simulations of HRTEM Images and Electron Microdiffraction Patterns

The Cowley-Moodie multislice theory has been widely applied in simulating electron diffraction patterns and images. The book edited by Krakow and O'Keefe (1989) covers various applications of the theory. In this section, a few examples are illustrated to demonstrate applications of the theory.

Observing the defect structure in crystalline materials is an important application of HRTEM. The multislice method is the only theory that can be applied to simulate images recorded for any desired structures. We now use the surface profile image to illustrate its application. We know that all elemental and compound semiconductors exhibit surface atomic reconstruction due to broken bonds. The reconstructed surface can be imaged if it is placed parallel to the incident beam direction in TEM. Figure 3.5a shows an HRTEM image of a clean CdTe (001) surface (Lu and Smith 1991). The beam direction is [110] and $E_0 = 300$keV. The image resolution of the microscope was about 0.2 nm. It is apparent that the surface atoms moved to new positions due to the formation of the surface. From experimental micrographs and comparison with image simulations, a structure model (shown in Fig. 3.5c) for the (3×1) reconstructed (001) surface was developed (Lu and Smith 1991). The (3×1) reconstruction involves both the formation of surface dimers and the presence of vacancies at the surface. Every third atomic pair is missing along the [110] direction, and the remaining two atom pairs at the surface rotate slightly toward each other to form the surface dimer. The multislice theory is then applied to simulate the image based on the structure model given in Fig. 3.5c, and the simulated result is shown in Fig. 3.5b. The simulation was performed for a crystal of thickness 2.8 nm at a defocus value of -35 nm . It is very clear that the simulated image is in close agreement with the experimental observed image.

Studies of nanostructured and nanocomposite materials are important applications of HRTEM, which has the unique capability of determining nanophases and nanocrystallography. The chemisorption ability and catalytic properties of metal particles supported on reducible oxides, for example, are often altered by high temperature reduction in a process known as strong metal support interaction, which is caused by the encapsulation of metal particles by the oxide overlayer species that have migrated from the oxide support. This result has been observed in the Pt/TiO_2 catalyst (Yao and Smith 1994). Figure 3.6a, b show two experimental HRTEM images of small Pt particles (about 3–4 nm in size) supported by TiO_2. On well-faceted edge-on {111} surfaces, crystalline monolayers (indicated by arrowheads) are seen. Most dark dots correspond to atomic columns in the monolayer, but the contrast is elongated parallel to the (100) plane. Image simulations have been made for clean Pt{111} surfaces (Fig. 3.6c), Pt{111} surfaces with monolayer Ti coverage (Fig. 3.6d), and Pt{111} surfaces with monolayer TiO_2 coverage (Fig. 3.6e). The simulation was performed by constructing a supercell that contains the entire faceted particle. In comparison to experimental images shown in Fig. 3.6a, b, e shows the best fit. There-

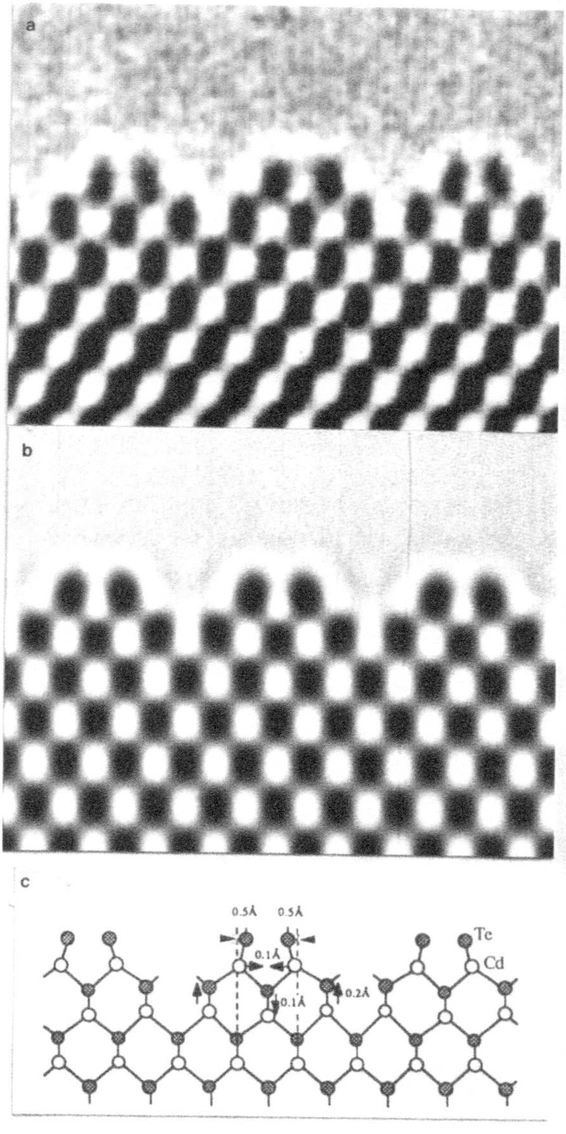

Fig. 3.5 **a** An HRTEM profile image of the CdTe(001) surface in a [110] projection showing the (3 × 1) reconstruction. The image was recorded in situ at a temperature about 240°C. **c** Schematic model for the (3 × 1) reconstructed surface as viewed along the [110] direction. Approximate atom displacements deduced from experimental images are shown. **b** A multislice simulated image for the (3 × 1) reconstructed surface. Courtesy of Drs. Lu and Smith (1991)

fore the decoration layer on the {111} surfaces of Pt particles is TiO_2, and the layer is terminated with oxygen.

The multislice theory can be applied to calculate not only HRTEM images but also coherent nanoprobe electron microdiffraction patterns, recorded in scanning transmsission electron microscopy (STEM) using a probe of diameter smaller than 2 nm (see Cowley 1992, for a review). Calculations of the multislice theory are

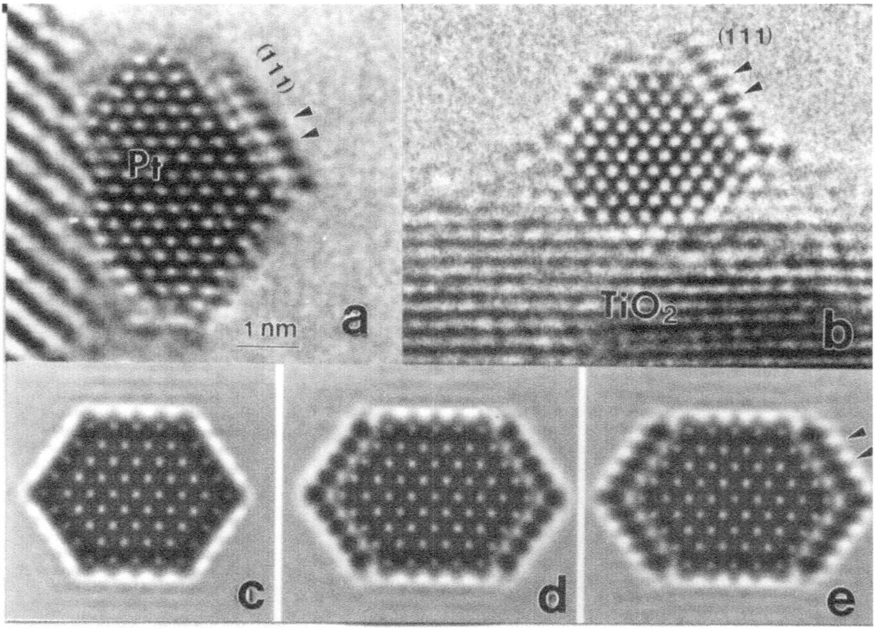

Fig. 3.6 **a**, **b** The HRTEM images of Pt metal particles supported on TiO_2 substrate. The dark contrast at the {111} surface corresponds to the adsorbed TiO_2 layer. **c**–**e** are simulated images of a Pt particle with clean {111} surfaces, {111} surfaces with monolayer Ti coverage, and {111} surfaces with monolayer TiO coverage, respectively; electron beam energy 400 keV, objective lens defocus –51 nm. Courtesy of Dr. M. H. Yao

particularly powerful for crystals containing defects or irregularly shaped crystals, such as small faceted particles (Liu et al. 1993).

3.4 Calculations of HOLZ Reflections

In the multislice theory, the potential distribution in each slice is projected onto a plane, so that the potential distribution within each slice is independent of z. This affects calculations of the HOLZ reflections, since they are the result of periodic crystal structure along the z axis. We now examine the effects of slice thickness on HOLZ calculations. As shown in (2.32a) and (2.32b), the crystal potential can be written as a superposition of potentials responsible for the kinematically reflected intensities of different orders of Laue zones (Self et al. 1983; Kilaas et al. 1987),

$$V(\mathbf{r}) = \sum_{g_z} \left\{ \sum_{g_b} V_g \exp\left(2\pi i \mathbf{g}_b \cdot \mathbf{b}\right) \right\} \exp\left(2\pi i g_z z\right) \tag{3.24}$$

Then the projected potential for a crystal slice is

$$V(\mathbf{b}) = \sum_{g_z} \left[\sum_{g_b} V_g \exp(2\pi i \mathbf{g}_b \cdot \mathbf{b}) \right] \int_{z}^{z+\Delta z} dz \exp(2\pi i g_z z)$$
$$= \sum_{g_z} \left[\sum_{g_b} V_g \exp(2\pi i \mathbf{g}_b \cdot \mathbf{b}) \right] \frac{\sin \pi g_z \Delta z}{\pi g_z} \exp\{2\pi i g_z [z + \Delta z/2]\} \quad (3.25)$$

For orthonormal crystal structures and selecting the slice thickness as the size of the c lattice vector $\Delta z = c$,

$$V(\mathbf{b}) = \Delta z \sum_{g_b} V_{g_b} \exp(2\pi i \mathbf{g}_b \cdot \mathbf{b}) \quad (3.26)$$

This expression contains only contributions from reciprocal lattice vectors belonging to the ZOLZ, and V is independent of z. Hence there is only one phase-grating function for the entire calculation. To include the effect of HOLZ reflections, slice thickness has to be chosen smaller than c, or $\Delta z = c/n_1$, where n_1 is the number of slices per unit cell; thus

$$V(\mathbf{b}) = \sum_{g_z} \left[\sum_{g_b} V_g \exp(2\pi i \mathbf{g}_b \cdot \mathbf{b}) \right] \frac{\sin \pi g_z c/n_1}{\pi g_z} \exp\{2\pi i g_z [z + c/2n_1]\} \quad (3.27)$$

The effects of HOLZ reflections are considered to some extent, but the final calculated results depend on the number of slices per unit cell. In general, one atomic layer per slice is a reasonable choice (Kilaas et al. 1987). More importantly, V depends on z (i.e., the depth of the slice in the crystal), making it impossible to use the precalculated phase-grating function for all slices if we are interested in HOLZ reflections. In this case, a phase-grating function must be calculated for each slice and then stored in a memory unit. This process not only takes large amounts of disk space, but it is also very time consuming. It is thus desirable to explore some new methods that may improve the accuracy of multislice calculations of HOLZ reflections.

3.5 Improved Multislice Approaches

3.5.1 Modified Multislice Theory for ZOLZ Reflections

The conventional multislice equation was derived by assuming that the wave is first modified by a phase-grating function under the projected potential approximation. The wave then propagates for a distance Δz in a vacuum, resulting in broadening of the wave distribution as characterized by the propagation function. To examine errors introduced by these approximations, we start from the exact solution for ZOLZ reflections. This exercise is necessary to propose some new methods that may improve the accuracy of the multislice calculations. We now try to find the solution of (2.35) for a thin crystal slice located at $z = z0$,

$$\frac{\partial \Phi(\mathbf{b}, z)}{\partial z} = \zeta \left[\nabla_b^2 + 4\pi^2 U^{(0)}(\mathbf{b})\right] \Phi(\mathbf{b}, z) \tag{3.28}$$

Since $U^{(0)}$ is independent of z, the solution of (3.28) can be written (Van Dyck and Coene 1984) directly as

$$\Phi(\mathbf{b}, z + \Delta z) = \exp\left\{\zeta \Delta z \left[\nabla_b^2 + 4\pi^2 U^{(0)}(\mathbf{b})\right]\right\} \Phi(\mathbf{b}, z) \tag{3.29}$$

where the exponential of the operator is defined in a power series of

$$\exp(A) = \sum_{n=0}^{\infty} \frac{A^n}{n!}$$

Equation (3.29) is the basic starting point of the real-space multislice theory (see Van Dyck 1985, for a review). The solution given by (3.29) is usually called the real-space multislice theory. In general, for two operators

$$\exp(A + B) = \exp(A) \exp(B) \exp[-(A, B)/2] \tag{3.30}$$

This relation is known as Weyl's formula with $(A, B) = AB - BA$. In general, if two operators A and B are not commulative, then $\exp(A + B) \neq \exp(A) \exp(B)$. The two operators

$$A = \frac{i\Delta z}{4\pi K} \nabla_b^2 \quad B = \frac{\pi i \Delta z}{K} U^{(0)}(\mathbf{b})$$

in (3.29) are such a case. However for $\Delta z \to 0$, the second-order correction $(BA - AB)$ is proportional to Δz^2, and it can be neglected. Thus for a very thin slice, (3.29) becomes

$$\begin{aligned} \Phi(\mathbf{b}, z + \Delta z) &\approx \exp\left(\zeta \Delta z \nabla_b^2\right) \exp\left[\frac{\pi i \Delta z}{K} U^{(0)}(\mathbf{b})\right] \Phi(\mathbf{b}, z) \\ &= \exp\left(\zeta \Delta z \nabla_b^2\right) \exp[i\sigma V(\mathbf{b})] \Phi(\mathbf{b}, z) \\ &= \exp\left(\zeta \Delta z \nabla_b^2\right) \left[Q_{n_c}(\mathbf{b}, \Delta z) \Phi(\mathbf{b}, z)\right] \end{aligned} \tag{3.31}$$

To see the effect of the operator ∇_b^2, we consider the following operation on an arbitrary function $f(\mathbf{b})$, whose Fourier transform is $f(\mathbf{u})$.

$$\begin{aligned}
\exp\left(\zeta\,\Delta z\nabla_b^2\right) f(\mathbf{b}) &= \exp\left(\zeta\,\Delta z\nabla_b^2 \int d\mathbf{u}\, f(\mathbf{u}) \exp(2\pi i \mathbf{u}\cdot\mathbf{b})\right.\\
&= \int d\mathbf{u}\, f(\mathbf{u}) \left[\exp\left(\zeta\,\Delta z\nabla_b^2\right) \exp(2\pi i \mathbf{u}\cdot\mathbf{b})\right]\\
&= \int d\mathbf{u}\, f(\mathbf{u}) \left[\left[\sum_n \frac{\left(\zeta\,\Delta z\nabla_b^2\right)^n}{n!}\right] \exp(2\pi i \mathbf{u}\cdot\mathbf{b})\right]\\
&= \int d\mathbf{u}\, f(\mathbf{u}) \left[\sum_n \frac{\left(-4\pi^2\zeta\,\Delta z u^2\right)^n}{n!}\right] \exp(2\pi i \mathbf{u}\cdot\mathbf{b})\\
&= \int d\mathbf{u}\, f(\mathbf{u}) \exp\left(-4\pi^2\zeta\,\Delta z u^2\right) \exp(2\pi i \mathbf{u}\cdot\mathbf{b})\\
&= \int d\mathbf{u}\, f(\mathbf{u}) \exp\left(-\pi i u^2 \Delta z \lambda\right) \exp(2\pi i \mathbf{u}\cdot\mathbf{b})\\
&= f(\mathbf{b}) \otimes P(\mathbf{b}, \Delta z)
\end{aligned} \tag{3.32}$$

where the Fourier transform of the propagation function (3.7b) and the convolution theorem were used in the last step. Substituting (3.32) into (3.31), the multislice formula is obtained

$$\Phi(\mathbf{b}, z + \Delta z) = \left[Q_{n_c}(\mathbf{b}, \Delta z)\Phi(\mathbf{b}, z)\right] \otimes P(\mathbf{b}, \Delta z) \tag{3.33}$$

The preceding exercise shows that the operation of $\exp\left(\zeta\,\nabla_b^2\right)$ introduces the effect of the propagation function. Therefore $\exp(A)$ is the propagator operator, and $\exp(B)$ is the phase-grating operator. Based on this result, calculation accuracy can be improved if the projected plane of the potential is the geometrical center of the slice (Van Dyck 1983), i.e.,

$$\exp[A + B] = \exp[A/2 + B + A/2] \approx \exp[A/2]\exp[B]\exp[A/2] \tag{3.34a}$$

the multislice formula (3.33) is thus modified as

$$\Phi(\mathbf{b}, z + \Delta z) = \left\{[\Phi(\mathbf{b}, z) \otimes P(\mathbf{b}, \Delta z/2)]Q_{n_c}(\mathbf{b}, \Delta z)\right\} \otimes P(\mathbf{b}, \Delta z/2) \tag{3.34b}$$

or if the potential is projected by half onto each face of the slice so that

$$\exp[A + B] = \exp[B/2 + A + B/2] \approx \exp[B/2]\exp[A]\exp[B/2] \tag{3.35a}$$

The multislice formula (3.33) is thus modified as

$$\Phi(\mathbf{b}, z + \Delta z) = \left\{\left[\Phi(\mathbf{b}, z)Q_{n_c}(\mathbf{b}, \Delta z/2)\right] \otimes P(\mathbf{b}, \Delta z)\right\} Q_{n_c}(\mathbf{b}, \Delta z/2) \tag{3.35b}$$

These methods are expected to improve the accuracy of multislice calculations for the ZOLZ reflections.

3.5.2 *Modified Multislice Theory for HOLZ Reflections*

In conventional multislice approach, the inclusion of HOLZ reflections greatly increases the amount of numerical computation. In this section, we introduce a second-order multislice theory that not only improves the accuracy of multislice calculations for ZOLZ reflections but also takes into account HOLZ reflections with a reasonable efficiency (Van Dyck 1980). We now start with (2.35) without making the projected potential approximation,

$$\frac{\partial \Phi(\mathbf{b}, z)}{\partial z} = \zeta \left[\nabla_b^2 + 4\pi^2 U(\mathbf{b}, z)\right] \Phi(\mathbf{b}, z) \tag{3.36}$$

For mathematical convenience, we define an operator $\Delta_b = \zeta \nabla_b^2$ and consider for a thin slice of thickness $\Delta z = \eta$. Equation (2.35) is converted into an integral equation

$$\Phi(\mathbf{b}, \eta) = \Phi(\mathbf{b}, 0) + \int_0^\eta dz \left[\Delta_b + i\sigma V(\mathbf{b}, z)\right] \Phi(\mathbf{b}, z) \tag{3.37}$$

This equation can be expanded iteratively into powers series of Δ_b and $i\sigma V$

$$\Phi(\mathbf{b}, \eta) = \left\{1 + \int_0^\eta dz \left[\Delta_b + i\sigma V(\mathbf{b}, z)\right] + \int_0^\eta dz \left[\Delta_b + i\sigma V(\mathbf{b}, z)\right] \int_0^z dz' \left[\Delta_b + i\sigma V\left(\mathbf{b}, z'\right)\right] + \cdots \right\} \Phi(\mathbf{b}, 0) \tag{3.38}$$

By defining the projected potential

$$V(\mathbf{b}) = \int_0^\eta dz V(\mathbf{b}, z) \tag{3.39a}$$

and a function

$$Z(\mathbf{b}) = \frac{1}{V(\mathbf{b})} \int_0^\eta dz V(\mathbf{b}, z) z \tag{3.39b}$$

and by using the following two identities:

$$\int_0^\eta dz V(\mathbf{b}, z) \int_0^z dz' V\left(\mathbf{b}, z'\right) = \frac{1}{2} V^2(\mathbf{b}) \tag{3.40a}$$

and

$$\int_0^\eta dz \int_0^z dz' V\left(\mathbf{b}, z'\right) = [\eta - Z(\mathbf{b})]V(\mathbf{b}) \tag{3.40b}$$

Equation (3.38) is rewritten as

$$\begin{aligned}\Phi(\mathbf{b}, \varepsilon) &= \left\{1 + \eta\Delta_b + \frac{(\eta\Delta_b)^2}{2} + i\sigma V(\mathbf{b}) + \frac{[i\sigma V(\mathbf{b})]^2}{2} + \Delta_b[\eta - Z(\mathbf{b})]i\sigma V(\mathbf{b})\right.\\ &\quad \left. + Z(\mathbf{b})i\sigma V(\mathbf{b})\Delta_b + \cdots\right\}\Phi(\mathbf{b}, 0)\\ &= \left\{1 + \eta\Delta_b + \frac{(\eta\Delta_b)^2}{2} + i\sigma V(\mathbf{b}) + \frac{[i\sigma V(\mathbf{b})]^2}{2} + \eta\Delta_b i\sigma V(\mathbf{b})\right.\\ &\quad \left. + Z(\mathbf{b})i\sigma\left[V(\mathbf{b})\Delta_b - \Delta_b V(\mathbf{b})\right] + \cdots\right\}\Phi(\mathbf{b}, 0)\end{aligned} \tag{3.41}$$

Equation (3.41) is the expansion of the exact solution of (3.37) for small-slice thickness. We now examine the expansion of the conventional multislice solution from (3.31)

$$\begin{aligned}\Phi(\mathbf{b}, \eta) &= \exp\left[\eta\Delta_b\right]\exp[i\sigma V(\mathbf{b})]\Phi(\mathbf{b}, 0)\\ &= \left[1 + \eta\Delta_b + \frac{(\eta\Delta_b)^2}{2} + \cdots\right]\left\{1 + i\sigma V(\mathbf{b}) + \frac{[i\sigma V(\mathbf{b})]^2}{2} + \cdots\right\}\Phi(\mathbf{b}, 0)\\ &= \left[1 + \eta\Delta_b + \frac{(\eta\Delta_b)^2}{2} + i\sigma V(\mathbf{b}) + \frac{[i\sigma V(\mathbf{b})]^2}{2} + \eta\Delta_b i\sigma V(\mathbf{b}) + \cdots\right]\Phi(\mathbf{b}, 0)\end{aligned} \tag{3.42}$$

Comparing (3.42) to (3.41), even the first-order term $Z(\mathbf{b})i\sigma\left[V(\mathbf{b})\Delta_b - \Delta_b V(\mathbf{b})\right]$ does not show up completely in (3.42); thus (3.31) does not even cover all the terms from the first-order scattering. This is probably why very small-slice thickness is required to ensure the accuracy of the multislice calculation, i.e., $\left[V(\mathbf{b})\Delta_b - \Delta_b V(\mathbf{b})\right] \approx 0$ if Δz is small. However the small-slice thickness makes numerical calculation rather inefficient. We now introduce a modified multislice equation, which is expected to give the expansion terms shown in (3.41) (Van Dyck 1980),

$$\begin{aligned}\Phi(\mathbf{b}, \eta) = &\exp\left\{i\sigma V(\mathbf{b})\left[1 + \eta'(\mathbf{b})\right]/2\right\}\\ &\times \exp\left[\eta\Delta_b\right]\exp\left\{i\sigma V(\mathbf{b})\left[1 - \eta'(\mathbf{b})\right]/2\right\}\Phi(\mathbf{b}, 0)\end{aligned} \tag{3.43a}$$

or

$$\Phi(\mathbf{b}, z + \Delta z) = \left[\Phi(\mathbf{b}, z)Q^{(1)}(\mathbf{b})\right] \otimes P(\mathbf{b}, \Delta z)Q^{(2)}(\mathbf{b}) \tag{3.43b}$$

with

$$Q^{(1)}(\mathbf{b}) = \exp\left\{i\sigma V(\mathbf{b})\left[1 - \eta'(\mathbf{b})\right]/2\right\} \tag{3.43c}$$

$$Q^{(2)}(\mathbf{b}) = \exp\left\{i\sigma V(\mathbf{b})\left[1 + \eta'(\mathbf{b})\right]/2\right\} \tag{3.43d}$$

and

$$\eta'(\mathbf{b}) = \frac{Z(\mathbf{b}) - \eta/2}{\eta/2}$$

is the relative deviation in (x, y) plane between the center of potential $Z(b)$ and the geometrical center $\eta/2$ of the slice; it is called the potential eccentricity of (x, y). It can be shown that (3.43a) is the only expression which matches (3.41) up to the second order by using only three exponential functions (Van Dyck 1980). Successive application of (3.43b) on each slice gives the wave function at the exit surface of the crystal. The most important advantage of (3.43c) is the inclusion of HOLZ reflections. The numerical calculation efficiency is greatly improved because the phase-grating function does not depend on the depth z of the slice, as discussed in Sect. 3.3 for conventional multislice method.

Another modified multislice theory for the inclusion of HOLZ reflections is called the improved phase-grating method (Van Dyck 1983). The wave function is written as a product of the phase-grating function of the crystal potential with a function $\Theta(\mathbf{r})$

$$\Phi(\mathbf{r}) = \exp\left[i\sigma \int_0^z dz V(\mathbf{b}, z)\right]\Theta(\mathbf{r}) \tag{3.44}$$

where $\Theta(\mathbf{r})$ is a function to be determined. Substituting (3.44) into (3.36), the first-order solution of $\Theta(\mathbf{r})$ can be obtained. The final expression of $\Theta(\mathbf{r})$ is very lengthy and thus is not listed here. Numerical calculation shows that the improved multislice equation (3.43d) gives the most accurate result (Kilaas et al. 1987).

3.6 Effects of a Magnetic Field

In electron microscopy, the magnetic field near the specimen may affect the scattering behavior of the transmitted electrons. This effect can be included in multislice calculations using a modified phase-grating function.

If $\mathbf{A}^{(0)}(\mathbf{r})$ denotes the vector potential of the magnetic field $\mathbf{B}^{(0)}(\mathbf{r})$, with $\mathbf{B}^{(0)}(\mathbf{r}) = \nabla \times \mathbf{A}^{(0)}(\mathbf{r})$ and $\nabla \cdot \mathbf{A}^{(0)}\mathbf{r} = 0$, the Hamiltoniam describing the interaction of an electron with the crystal is

$$H = \frac{\left[p + e\mathbf{A}^{(0)}\right]^2}{2m_0} - e\gamma V \tag{3.45}$$

Then the Schrödinger equation becomes

$$\nabla^2\Psi + 4\pi^2\left(U + K^2\right)\Psi + \frac{4\pi ie}{h}\cdot\mathbf{A}^{(0)}\cdot\nabla\Psi - \frac{4\pi^2e^2}{h^2}\cdot\left[A^{(0)}\right]^2\Psi = 0 \quad (3.46)$$

By substituting $\Psi(\mathbf{r}) = \exp(2\pi iKz)\Phi(\mathbf{r})$, and neglecting the $\partial^2\Phi/\partial z^2$ term for small-angle scattering, we obtain

$$\begin{aligned} -4\pi iK\frac{\partial\Phi}{\partial z} =& \left[\nabla_b^2 + 4\pi^2U\right]\Phi - \frac{8\pi^2eK}{h}A_z^{(0)}\Phi \\ &+ \frac{4\pi ie}{h}\cdot\mathbf{A}^{(0)}\cdot\nabla\Phi - \frac{4\pi^2e^2}{h^2}\cdot\left[A^{(0)}\right]^2\Phi \end{aligned} \quad (3.47)$$

As we showed in (3.32), the operator ∇_b^2 is responsible for propagating electrons in space; thus under the first-order approximation, we can entirely describe the effects of the magnetic field by using the phase-grating function, i.e.,

$$\frac{\partial\Phi}{\partial z} \approx i\sigma V\Phi - \frac{2\pi ie}{h}A_z^{(0)}\Phi - \frac{e\lambda}{h}\cdot A^{(0)}\cdot\nabla\Phi - \frac{\pi ie^2\lambda}{h^2}\cdot\left[A^{(0)}\right]^2\Phi \quad (3.48)$$

There are two cases in which effects of the magnetic field may need to be considered; one is the magnetic lens and the other, the magnetic nature of the sample itself. For magnetic lenses, the term containing $\left[A^{(0)}\right]^2$ is not small. In TEM, components of the magnetic vector potential in the paraxial region are written as

$$A_x^{(0)} = -\frac{1}{2}B_z^{(0)}(z)y, \quad A_y^{(0)} = \frac{1}{2}B_z^{(0)}(z)x, \quad A_z^{(0)} = 0 \quad (3.49)$$

In this case, the phase-grating function may be found after introducing a rotational coordinate system (see Pozzi 1989, for details).

For magnetic domains of $\mathbf{B}^{(0)} = (0, B_0, 0)$, for example, where B_0 is constant, the vector potential can be chosen as $\mathbf{A}^{(0)} = -1/2\left(\mathbf{r}\times\mathbf{B}^{(0)}\right) = (-B_0z/2, 0, B_0x/2)$. If the term containing $\left[A^{(0)}\right]^2$ is small, (3.48) is rewritten as Wang and Spence (1990)

$$\frac{\partial\Phi/\partial z}{\Phi} \approx \frac{i\sigma V - (\pi \mathrm{ie}/h)B_0x}{1 + (e\lambda/2\,h)B_0x} \quad (3.50)$$

Solving this equation for a thin crystal slice yields

$$\Phi(\mathbf{b}, z + \Delta z) = [\Phi(\mathbf{b}, z)Q_B(\mathbf{b})] \quad (3.51a)$$

where the new phase-grating function is

$$Q_B(\mathbf{b}) = \exp\left[\Delta z \frac{i\sigma V(\mathbf{b}, z) - (\pi \mathrm{ie}/h) B_0 x}{1 + (e\lambda/2h) B_0 x}\right] \approx \exp\left[i\sigma V(\mathbf{b}) - \frac{\pi \mathrm{ie} \Delta z}{h} B_0 x\right] \tag{3.51b}$$

Convoluting the propagation function to (3.51a), the multislice formula becomes

$$\Phi(\mathbf{b}, z + \Delta z) = [\Phi(\mathbf{b}, z) Q_B(\mathbf{b})] \otimes P(\mathbf{b}, \Delta z) \tag{3.52}$$

which is the same as the conventional multislice formula except the phase-grating function is replaced by Q_B. Equation (3.52) was applied to calculate the magnetic contrast of a monolayer of Fe on a (110) GaAs surface in reflection electron microscopy (Wang and Spence 1990).

3.7 Summary

To summarize, the multislice theory introduced in Sects. 3.1 and 3.2 is often used in HRTEM image simulations. It is possible to include any type of defects in the calculation, but that requires artificially truncating a supercell large enough to enclose the defect. The entire space is assumed to be filled by supercells, and scattering within each supercell should be approximately independent of its neighbor cells. This is the so-called periodic continuation technique. The real-space multislice theory introduced in Sect. 3.5.1 can be applied to calculate images of defects using the so-called real-space patching method (Coene, et al. 1985), which assumes only one defect in the entire space and is thus believed to be more realistic than the periodic continuation technique. Modified multislice methods for including HOLZ reflections in the dynamic calculations have been introduced.

Chapter 4
Dynamic Elastic Electron Scattering III: Other Approaches

Besides the Bloch wave and multislice theories introduced in Chaps. 2 and 3, there are several other theories on dynamic electron scattering. Each of these methods is convenient for some particular purpose. Chapter 4 introduces these theories, then establishes the equivalence among them.

4.1 Scattering Matrix Theory

The scattering matrix theory was developed from the eigenvalue (2.14), and the result is expressed in a rather simple matrix form (Fujimoto, 1959). We first consider some identities from (2.14)

$$\mathbf{A}\mathbf{C}^{(i)} = 2\,K\,v_i\,\mathbf{C}^{(i)} \tag{4.1}$$

Multiplying matrix **A** to both sides of (4.1) and reusing (4.1),

$$\mathbf{A}^2\mathbf{C}^{(i)} = 2K_n v_i \mathbf{A}\mathbf{C}^{(i)} = [2\,K\,v_i]^2\,\mathbf{C}^{(i)} \tag{4.2}$$

and in general for an integer $n > 1$,

$$\mathbf{A}^n\mathbf{C}^{(i)} = [2\,K\,v_i]^n\,\mathbf{C}^{(i)} \tag{4.3}$$

The **g** component of the ith Bloch wave is

$$[2\,K\,v_i]^n\,C_g^{(i)} = \sum_h \left[\mathbf{A}^n\right]_{gh} C_h^{(i)} \tag{4.4}$$

Z. L. Wang, *Elastic and Inelastic Scattering in Electron Diffraction and Imaging*, Springer Series in Solid-State Sciences 205, https://doi.org/10.1007/978-3-031-90819-4_4

where $\mathbf{A}^n$ is a $m \times m$ matrix and $[\mathbf{A}^n]_{gh}$ are elements in the g column and h row of the $\mathbf{A}^n$ matrix. We now use (4.4) to simplify the expression for the amplitude of the g beam; from (2.17),

$$\begin{aligned}
\Psi_g &= \sum_i C_0^{(i)*} C_g^{(i)} \exp\left[2\pi i v_i z\right] \\
&= \sum_i C_0^{(i)*} \sum_n \frac{1}{n!} \left[2\pi i v_i z\right]^n C_g^{(i)} \\
&= \sum_i C_0^{(i)*} \sum_n \frac{(\pi i z/K)^n}{n!} \left[2\, K v_i\right]^n C_g^{(i)} \\
&= \sum_i C_0^{(i)*} \sum_n \sum_h \frac{(\pi i z/K)^n}{n!} \left[\mathbf{A}^n\right]_{gh} C_h^{(i)} \\
&= \sum_n \frac{(\pi i z/K)^n}{n!} \left[\mathbf{A}^n\right]_{g0} = \left\{\exp\left[\frac{\pi i z}{K}\mathbf{A}\right]\right\}_{g0}
\end{aligned} \tag{4.5}$$

where the matrix

$$\sum_n \frac{(\pi i z/K)^n}{n!} \left[\mathbf{A}^n\right] = \left\{\exp\left[\frac{\pi i z}{K}\mathbf{A}\right]\right\}$$

The amplitude of the **g** beam at the exit face of the crystal is

$$\left\{\exp\left[\frac{\pi i z}{K}\mathbf{A}\right]\right\}_{g0}$$

or in general

$$\begin{pmatrix} \Psi_0 \\ \Psi_1 \\ \vdots \\ \Psi_g \end{pmatrix} = \exp\left[\frac{\pi i z}{K}\mathbf{A}\right] \begin{pmatrix} 1 \\ 0 \\ \vdots \\ 0 \end{pmatrix} \tag{4.6}$$

Thus the propagation of electrons through a crystal of thickness z is represented by the operation of the scattering matrix

$$\mathbf{S} = \exp\left[\frac{\pi i z}{K}\mathbf{A}\right] \tag{4.7}$$

In numerical calculations, the slow convergence of the **S** matrix limits direct calculation of the S matrix for the entire crystal thickness. Alternatively the crystal can be cut into slices of equal thickness ($z = \Delta z\mathbf{N}$), and the **S** matrix for each thin slice can be calculated to a high degree of accuracy. In general, if

$$\exp\left[\frac{\pi i \Delta z}{K}\mathbf{A}\right]$$

requires M' power, then

$$\exp\left[\frac{\pi i \Delta z\,\mathrm{N}}{K}\,\mathrm{A}\right]$$

requires $N \times M'$ powers (Sturkey 1962). The scattering matrix for the entire thickness is a product of the matrixes for each successive slice,

$$\mathbf{S} = \mathbf{S}_N \mathbf{S}_{N-1} \cdots \mathbf{S}_1 \tag{4.8}$$

It is possible to include the absorption effect in the scattering matrix calculation. The theory does not need to solve the dispersion eigenvalue equations, and it is feasible to apply it to many-beam cases (Fisher 1968). More sophisticated matrix theory has been developed by Kogiso and Kainuma (1968) to increase the accuracy of the matrix calculation. Matrix theory can also be applied to calculate electron diffraction from faulted crystals (Sturkey 1962).

For particular cases involving only two or three beams, the scattering matrix may be calculated analytically. Sturkey (1962) has outlined the analytical expressions of the scattering matrixes for some simplified cases. Here we list a few examples

$$\exp\begin{pmatrix} 0 & ire^{id} \\ ire^{-id} & ik \end{pmatrix} = \exp(ik/2)\begin{pmatrix} \cos\xi - ik(\sin\xi)/(2\xi) & ie^{id}(\sin\xi)/\xi \\ ie^{-id}(\sin\xi)/\xi & \cos\xi + ik(\sin\xi)/(2\xi) \end{pmatrix} \tag{4.9}$$

where $\xi = \left[r^2 + (k/2)^2\right]^{1/2}$.

$$\exp\begin{pmatrix} 0 & ia & ib \\ ia & 0 & 0 \\ ib & 0 & 0 \end{pmatrix} = \begin{pmatrix} \cos p & ia(\sin p)/p & ib(\sin p)/p \\ ia(\sin p)/p & 1\text{–}2a^2\left(\sin^2 p/2\right)/p^2 & -2ab\left(\sin^2 p/2\right)/p^2 \\ ib(\sin p)/p & -2ab\left(\sin^2 p/2\right)/p^2 & 1\text{–}2b^2\left(\sin^2 p/2\right)/p^2 \end{pmatrix} \tag{4.10}$$

where $p = \left(a^2 + b^2\right)^{1/2}$.

$$\exp\begin{pmatrix} 0 & iA_1 e^{i\alpha} \\ iA_2 e^{i\beta} & 0 \end{pmatrix} = \begin{pmatrix} \cos\zeta & iA_1 e^{i\alpha}(\sin\zeta)/\zeta \\ iA_2 e^{i\beta}(\sin\zeta)/\zeta & \cos\zeta \end{pmatrix} \tag{4.11}$$

where $\zeta = (A_1 A_2)^{1/2}\exp[i(\alpha + \beta)]$.

4.2 Green's Function Theory

Green's function theory is a standard method for converting a differential equation into an integral equation. This theory can be conveniently applied to prove some of the scattering characteristics of high-energy electrons (see Chap. 9). For convenience, (2.5) is rewritten as

$$\left[\nabla^2 + 4\pi^2 K^2\right] \Psi(\mathbf{r}) = -4\pi^2 U(\mathbf{r})\Psi(\mathbf{r}) \tag{4.12}$$

The solution to this equation can be written in integral form (see Appendix D)

$$\Psi(\mathbf{r}) = \Psi_0(\mathbf{r}) - 4\pi^2 \int d\mathbf{r}'\, G_0\left(\mathbf{r}, \mathbf{r}'\right) U\left(\mathbf{r}'\right) \Psi\left(\mathbf{r}'\right) \tag{4.13}$$

where $\Psi_0(\mathbf{r})$ is a solution of the homogeneous equation $\left[\nabla^2 + 4\pi^2 K^2\right] \Psi_0(\mathbf{r}) = 0$ and G_0 is Green's function determined by

$$\left[\nabla^2 + 4\pi^2 K^2\right] G_0\left(\mathbf{r}, \mathbf{r}'\right) = \delta\left(\mathbf{r} - \mathbf{r}'\right) \tag{4.14}$$

The solution of (4.14) can be found by (1) taking the Fourier transform of the equation,

$$4\pi^2\left(K^2 - u^2\right) G_0\left(\mathbf{u}, \mathbf{r}'\right) = \exp\left(-2\pi i \mathbf{u} \cdot \mathbf{r}'\right)$$

(2) solving for G_0, and (3) taking the inverse Fourier transform

$$G_0\left(\mathbf{r}, \mathbf{r}'\right) = \int d\mathbf{u} \frac{\exp\left[2\pi i \mathbf{u} \cdot \left(\mathbf{r} - \mathbf{r}'\right)\right]}{4\pi^2\left(K^2 - u^2\right)} \tag{4.15}$$

In a spherical coordinate system where $\left(\mathbf{r} - \mathbf{r}'\right)$ is the polar axis,

$$G_0\left(\mathbf{r}, \mathbf{r}'\right) = \int_0^\infty du\, u^2 \int_0^\pi d\theta \sin\theta \int_0^{2\pi} d\varphi \frac{\exp\left[2\pi i u \left|\mathbf{r} - \mathbf{r}'\right| \cos\theta\right]}{4\pi^2\left(K^2 - u^2\right)} \tag{4.16}$$

Performing the integration calculation, we obtain

$$G_0\left(\mathbf{r}, \mathbf{r}'\right) = \frac{1}{4\pi^2 i \mid \mathbf{r} - \mathbf{r}' \mid} \int_{-\infty}^{\infty} du \frac{u \exp\left[2\pi i u \left|\mathbf{r} - \mathbf{r}'\right|\right]}{\left(K^2 - u^2\right)} \tag{4.17}$$

Integrating u can be truncated in a complex plane for an enclosed integral path (see Appendix D); using the residual theorem, Green's function for forward scattering is

$$G_f\left(\mathbf{r},\mathbf{r}'\right) = -\frac{\exp\left[2\pi i K\left|\mathbf{r}-\mathbf{r}'\right|\right]}{4\pi\left|\mathbf{r}-\mathbf{r}'\right|} \tag{4.18a}$$

For backscattering, it is

$$G_b\left(\mathbf{r},\mathbf{r}'\right) = -\frac{\exp\left[-2\pi i K\left|\mathbf{r}-\mathbf{r}'\right|\right]}{4\pi\left|\mathbf{r}-\mathbf{r}'\right|} \tag{4.18b}$$

Therefore the solution of (4.14) can be obtained iteratively in a power series (Fujiwara 1959)

$$\begin{aligned}\Psi(\mathbf{r}) &= \Psi_0(\mathbf{r}) - 4\pi^2\int d\mathbf{r}_1 G_0\left(\mathbf{r},\mathbf{r}_1\right)U\left(\mathbf{r}_1\right)\Psi_0\left(\mathbf{r}_1\right)\\ &+\left[-4\pi^2\right]^2\int d\mathbf{r}_1 G_0\left(\mathbf{r},\mathbf{r}_1\right)U\left(\mathbf{r}_1\right)\int d\mathbf{r}_2 G_0\left(\mathbf{r}_1,\mathbf{r}_2\right)U\left(\mathbf{r}_2\right)\Psi_0\left(\mathbf{r}_2\right)+\cdots,\end{aligned} \tag{4.19}$$

or

$$\Psi(\mathbf{r}) = \Psi_0(\mathbf{r}) + \Psi_1(\mathbf{r}) + \cdots + \Psi_m(\mathbf{r}) + \cdots, \tag{4.20a}$$

where

$$\Psi_m(\mathbf{r}) = \left[-4\pi^2\right]\int d\mathbf{r}' G_0\left(\mathbf{r},\mathbf{r}'\right)U\left(\mathbf{r}'\right)\Psi_{m-1}\left(\mathbf{r}'\right) \tag{4.20b}$$

Equations (4.20a) and (4.20b) are the so-called Born series, which were derived for a general scattering process without making the small-angle scattering approximation. Thus the theory is more convenient than the conventional multislice theory for treating the following problems: (1) low-energy electron scattering in the spherical wave representation; (2) electron back scattering; (3) Fresnel diffraction in which the source is located near the specimen, such as in low-energy point source TEM (Kreuzer et al. 1992); and (4) diffraction patterns of small crystals. For high-energy transmission electron diffraction, Green's function theory is less useful due to the slow convergence of the Born series.

4.3 Semireciprocal Space Method

The semireciprocal space theory is based on the Darwin-type approach initiated for X-ray diffraction by reflection from the face of a large perfect crystal (Darwin 1914), in which transmission and reflection coefficients are calculated from summations of transmitted and diffracted amplitudes at each plane. The application of the Darwin-type approach in electron diffraction was made by Howie and Whelan (1961) and Tournarie (1962). We now derive the formal theory following (2.35) for high-energy transmission electron diffraction,

$$\left[\nabla_b^2 + 4\pi^2 U(\mathbf{b}, z)\right] \Phi(\mathbf{b}, z) \approx -4\pi i K \frac{\partial \Phi(\mathbf{b}, z)}{\partial z} \tag{4.21}$$

The solution of this equation can be written as

$$\Phi(\mathbf{b}, z) = \sum_g \Phi_g(z) \exp[2\pi i \mathbf{g} \cdot \mathbf{r}] \tag{4.22}$$

where the two-dimensional Bloch wave coefficients are assumed to depend only on z. If the crystal potential can be equivalently written as

$$U(\mathbf{b}, z) = \sum_g U_g(z) \exp[2\pi i \mathbf{g} \cdot \mathbf{r}] \tag{4.23}$$

Substituting (4.22) and (4.23) into (4.21), we obtain

$$\frac{d\Phi_g(z)}{dz} = \frac{i\pi}{K}\left[2\,K S_g \Phi_g(z) + \sum_h U_{g-h}(z)\Phi_h(z)\right] \tag{4.24}$$

Equation (4.24) can be written in a matrix form

$$\frac{d\mathbf{\Phi}(z)}{dz} = \frac{i\pi}{K}\mathbf{A}(z)\mathbf{\Phi}(z) \tag{4.25}$$

where $\Phi(z)$ is a single-column matrix with elements $\Phi_g(z)$.

For ZOLZ reflections where $\mathbf{A}$ is independent of z, (4.4) can be solved directly

$$\mathbf{\Phi}(z) = \exp\left(\frac{i\pi}{K}\mathbf{A}z\right)\mathbf{\Phi}(0) \tag{4.26}$$

Equation (4.26) is exactly the same as the solution for the scattering matrix theory (4.6).

The semireciprocal space theory expresses dynamic scattering in a simple equation. But the theory is not convenient in practical calculation due to the slow convergence of the exponential matrix calculation unless the number of beams are less than three.

4.4 Scattering Operator in Electron Diffraction

Based on (4.26), if the crystal thickness is artificially separated, as from 0 to d_1, and from d_1 to d_2, the equation that correlates scattering amplitudes at $z = d_1$ and $z = d_2$ is

$$\mathbf{\Phi}\,(d_2) = \mathbf{T}_c\,(d_2 - d_1)\,\mathbf{\Phi}\,(d_1) \tag{4.27a}$$

where the correlation matrix is

$$\mathbf{T}_c\,(d_2 - d_1) = \exp\left[\frac{i\pi}{K}\mathbf{A}\,(d_2 - d_1)\right] \tag{4.27b}$$

Therefore the scattering amplitudes at d_2 are related to amplitudes at d_1 by a scattering matrix. Equations (4.27a) and (4.27b) is convenient for describing electron diffraction in faulted crystals. We now derive the scattering operator in the Bloch wave representation. For convenience, we start from a general relation of wave amplitudes at z

$$\Psi_g(z) = \sum_i \alpha_i C_g^{(i)} \exp\left[2\pi i\, v_i z\right] \tag{4.28a}$$

or in a matrix form

$$\boldsymbol{\Psi}(z) = \mathbf{C}\left\{\exp\left[2\pi i\, v_i z\right]\right\}_{\mathrm{D}} \boldsymbol{\alpha} \tag{4.28b}$$

where $\Psi(z)$ and α are single-column matrixes with elements $\Psi_g(z)$ and α_i, respectively, and $\{\exp\left[2\pi i\, v_i z\right]\}_{\mathrm{D}}$ is a diagonal matrix. We now consider the amplitudes at $z = z_1$ and $z = z_2$,

$$\boldsymbol{\Psi}\,(z_1) = \mathbf{C}\left\{\exp\left[2\pi i\, v_i z_1\right]\right\}_{\mathrm{D}} \boldsymbol{\alpha} \tag{4.29a}$$

$$\boldsymbol{\Psi}\,(z_2) = \mathbf{C}\left\{\exp\left[2\pi i\, v_i z_2\right]\right\}_{\mathrm{D}} \boldsymbol{\alpha} \tag{4.29b}$$

Solving $\boldsymbol{\alpha}$ from (4.29a) and substituting the solution into (4.29b), we have

$$\boldsymbol{\Psi}\,(z_2) = \mathbf{C}\left\{\exp\left[2\pi i\, v_i\,(z_2 - z_1)\right]\right\}_{\mathrm{D}} \mathbf{C}^{-1}\boldsymbol{\Psi}\,(z_1) = \mathbf{C}\left\{\exp\left[2\pi i\, v_i\,(z_2 - z_1)\right]\right\}_{\mathrm{D}} \mathbf{C}^{\dagger}\boldsymbol{\Psi}\,(z_1) \tag{4.30}$$

where a relation of $\mathbf{C}^{-1} = \mathbf{C}^{\dagger}$ was used. For the gth column of (4.30),

$$\Psi_g\,(z_2) = \sum_h \left[\mathbf{P}_b\,(z_2 - z_1)\right]_{gh} \Psi_h\,(z_1) \tag{4.31a}$$

where the scattering operator

$$\left[\mathbf{P}_b\,(z_2 - z_1)\right]_{gh} = \sum_i C_g^{(i)} \exp\left[2\pi i\, v_i\,(z_2 - z_1)\right] C_h^{(i)*} \tag{4.31b}$$

which correlates the scattering amplitudes at $z = z_1$ with those at $z = z_2$, as shown in Fig. 4.1. This relationship is applied in Sect. 8.7.1 to calculate the images of faulted crystals.

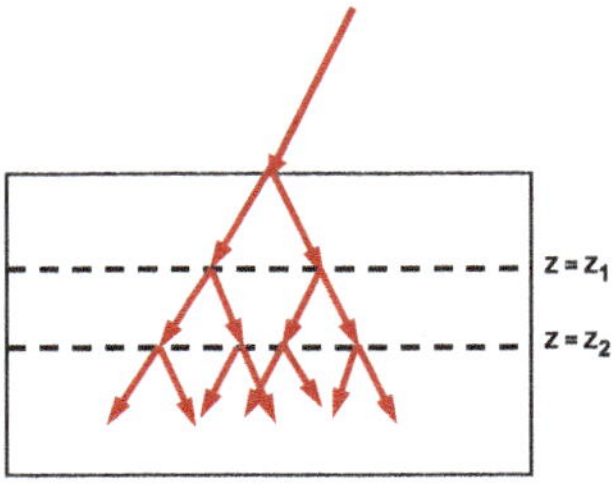

Fig. 4.1 The matrix operator that correlates the scattering amplitudes at $z = z_1$ with those of $z = z_2$. This operator simplifies the theoretical description of the evolution of Bragg beam amplitudes at different depths of the crystal (see text)

4.5 Diffraction in Imperfect Crystals

Electron diffraction in imperfect crystals was briefly described in Sect. 2.10. In this section we outline the original theory of Takagi (1969) for dealing with crystals containing defects. Instead of using (2.47), the electron wave function is written as

$$\Psi(\mathbf{r}) = \sum_g \phi'_g(\mathbf{r}) \exp\left[2\pi i(\mathbf{K}+\mathbf{g})\cdot\mathbf{r} - 2\pi i\mathbf{g}\cdot\mathbf{R}(\mathbf{r}_0)\right] \quad (4.32)$$

where $\mathbf{R}(\mathbf{r}_0)$ is a continuous function representing the displacement of the atom located at $\mathbf{r}_0$, with $\mathbf{r} = \mathbf{r}_0 + \mathbf{R}$, and $\phi'_g(\mathbf{r})$ is the amplitude of the $\mathbf{g}$ reflection. Using the potential expression given by

$$U(\mathbf{r}) = \sum_g U_g \exp\left[2\pi i\mathbf{g}\cdot\mathbf{r} - 2\pi i\mathbf{g}\cdot\mathbf{R}(\mathbf{r}_0)\right] \quad (4.33)$$

substituting (4.32) and (4.33) into (2.5), we obtain

$$\sum_g \Big\{\nabla^2\phi'_g + 4\pi i[\mathbf{K}+\mathbf{g}-\nabla(\mathbf{g}\cdot\mathbf{R})]\cdot\nabla\phi'_g + \Big[4\pi^2K^2 - 2\pi i\nabla^2(\mathbf{g}\cdot\mathbf{R}) - 4\pi^2 \mid \mathbf{K}+\mathbf{g}$$
$$- \nabla(\mathbf{g}\cdot\mathbf{R})|^2\Big]\phi'_g + 4\pi^2\sum_h \phi'_h U_{g-h}\Big\} \exp[2\pi i(\mathbf{K}+\mathbf{g})\cdot\mathbf{r} - 2\pi i\mathbf{g}\cdot\mathbf{R}] = 0 \quad (4.34)$$

The following two assumptions were made to simplify this equation: (1) $\nabla^2\phi'_g$ and $\nabla^2(\mathbf{g}\cdot\mathbf{R})$ terms are neglected on account of the macroscopic character of variations in ϕ' and $\mathbf{g}\cdot\mathbf{R}$; and (2) the remaining terms in the curly bracket also demonstrate macroscopic variation and can thus be omitted from the integration. Multiplying (4.34) by $\exp[-2\pi i(\mathbf{K}+\mathbf{h})\cdot\mathbf{r} + 2\pi i\mathbf{h}\cdot\mathbf{R}]$ and integrating over $\mathbf{r}$, (4.34) becomes

$$i[\mathbf{K}+\mathbf{g}-\nabla(\mathbf{g}\cdot\mathbf{R})]\cdot\nabla\phi'_g + 2\pi KS'_g\phi'_g + \pi\sum_h U_{g-h}\phi'_h \approx 0 \quad (4.35)$$

where the modified excitation error is defined as

$$2KS'_g = K^2 - |\mathbf{K} + \mathbf{g} - \nabla(\mathbf{g} \cdot \mathbf{R})|^2 \tag{4.36}$$

To write (4.35) into a familiar form, an oblique coordinate parallel to $(\mathbf{K} + \mathbf{g})$ is defined as ρ_g; thus

$$\frac{\partial \phi'_g}{\partial \rho_g} \approx 2\pi i S'_g \phi'_g + \frac{\pi i}{K} \sum_h U_{g-h} \phi'_h \tag{4.37}$$

Equation (4.37) was first derived by Takagi (1962, 1969) and Taupin (1964) to approach crystals with defects, dislocations, and strain. This theory has been extensively developed and applied by Kato (1973, 1976a, b, 1979) to approach the diffraction of crystals containing defects, thereby leading to a new field called statistical diffraction theory, which describes the diffraction of a crystal containing a statistical distribution of imperfections (Kato 1980a, b, 1991).

An alternative approach can be proposed based on (4.32). If the electron wave function is expressed as a superposition of Bloch waves

$$\Psi(\mathbf{r}) = \sum_i \alpha_i(z) \sum_g C_g^{(i)} \exp\left[2\pi i\,(\mathbf{k_i} + \mathbf{g}) \cdot \mathbf{r} - 2\pi i \mathbf{g} \cdot \mathbf{R}(\mathbf{r})\right] \tag{4.38}$$

Comparing this equation with (4.27) yields

$$\phi'_g(\mathbf{r}) = \sum_i \alpha_i(z) C_g^{(i)} \exp\left[2\pi i\, v_i z\right] \tag{4.39}$$

Substituting (4.39) into (4.37), multiplying both sides by $C_g^{(j)*}$ and summing over g, we have

$$\frac{d\alpha_j}{dz} \approx 2\pi i \sum_g \sum_i \left[S'_g C_g^{(j)*} C_g^{(i)} \alpha_i - 2\pi i\, v_i \alpha_i \right.$$
$$\left. + \frac{\pi i}{K} \sum_g \sum_h U_{g-h} \sum_i C_g^{(j)*} C_h^{(i)} \alpha_i \right] \exp\left[2\pi i\left(v_i - v_j\right) z\right] \tag{4.40}$$

Since the Bloch wave solution for a perfect crystal satisfies (2.10a),

$$\sum_h U_{g-h} C_h^{(i)} = \left[K^2 - (\mathbf{k}_i + \mathbf{g})^2\right] C_g^{(i)} \tag{4.41}$$

Using (4.36), and under the column approximation, (4.40) becomes

$$\frac{d\alpha_j}{dz} \approx 2\pi i \sum_g \sum_i C_g^{(j)*} C_g^{(i)} \frac{(k_{iz} + g_z)}{k_{jz}} \mathbf{g} \cdot \frac{d\mathbf{R}(z)}{dz} \exp\left[2\pi i \left(v_i - v_j\right) z\right] \alpha_i \quad (4.42)$$

This equation is in an identical form as (2.54), which correlates amplitudes of Bloch waves with the spatial variation of atomic static displacements. Gjønnes (1972) has also proposed a theory for dealing with imperfect crystals using Green's function solutions and the perturbation theory.

4.6 Equivalence Among Various Theories

Theoretical approaches described in Chaps. 2, 3 and this chapter represent a few main schemes for describing dynamic electron diffraction. These theories were developed for various purposes; the Bloch wave theory (or eigenvalue method), for example, is more convenient for calculating rocking curves in electron diffraction patterns, while the multislice theory is very powerful for calculating HRTEM images. These theories appear to be different, but they are equivalent to each other. Figure 4.2 shows how all theories can be derived: Starting with the Schrödinger equation, Green's function solution was derived in the form of the Born series (Sect. 4.2). Application of Green's function solution to a thin crystal slice gave the multislice theory (Sect. 3.2). The semireciprocal space theory is derived from the multislice equation when slice thickness $\Delta z \to 0$ (see below). After applying variable transformations, the Bloch wave (or eigenvalue) solution is readily obtained from the semireciprocal space approach. Finally the scattering matrix theory was derived from the eigenvalue equation (Sect. 4.1).

To derive the semireciprocal space theory from the multislice theory, we first consider the limiting case of the multislice equation for $\Delta z \to 0$ (Goodman and Moodie 1974). When $\Delta z \to 0$, there is no difference if the propagation and phase-grating operators are exchanged. Thus for convenience, (3.8) is written as

$$\Phi(\mathbf{b}, z + \Delta z) = Q(\mathbf{b})[\Phi(\mathbf{b}, z) \otimes P(\mathbf{b}, \Delta z)] \quad (4.43)$$

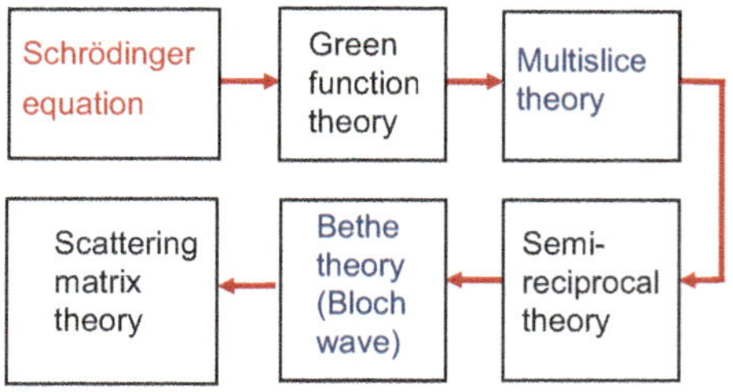

Fig. 4.2 The relationship among different dynamic scattering theories for elastic electron scattering. Each theory can be derived from the previous one. Approximations from which each theory is derived are given in the text.

which means that propagation occurs before phase grating. The convolution operation in this equation can be explicitly written as

$$\Phi(\mathbf{b}, z + \Delta z) = \exp[i\sigma V(\mathbf{b})] \int d\mathbf{b}' \frac{\exp\left(\pi i K b'^2/\Delta z\right)}{i\lambda\Delta z} \Phi\left(\mathbf{b} - \mathbf{b}', z\right)$$
$$= \exp[i\sigma V(\mathbf{b})] \int dx' \int dy' \frac{\exp\left[\pi i K \left(x'^2 + y'^2\right)/\Delta z\right]}{i\lambda\Delta z} \sum_{m=0}^{\infty} \frac{1}{m!}\left[-x'\frac{\partial}{\partial x} - y'\frac{\partial}{\partial y}\right]^m \Phi(\mathbf{b}, z) \tag{4.44}$$

where the Taylor expansion of $\Phi\left(\mathbf{b} - \mathbf{b}', z\right)$ was used. The expansion of $[-x'(\partial/\partial x) - y'(\partial/\partial y)]^m$ leads to integrals of the form

$$\frac{1}{(p+q)!}\frac{\partial^{p+q}}{\partial x^p \partial y^q}\Phi\left(\mathbf{b}, z_n\right) \int dx' \int dy' x'^p y'^q \frac{\exp\left[\pi i K\left(x'^2 + y'^2\right)/\Delta z\right]}{i\lambda\Delta z} \tag{4.45}$$

The only nonzero integrals for the low orders in Δz are

$$\Phi(\mathbf{b}, z) \int dx' \int dy' \frac{\exp\left[\pi i K\left(x^2 + y'^2\right)/\Delta z\right]}{i\lambda\Delta z} = \Phi(\mathbf{b}, z) \tag{4.46a}$$

and

$$\frac{1}{2!}\left[\frac{\partial^2}{\partial x^2} + \frac{\partial^2}{\partial y^2}\right]\Phi(\mathbf{b}, z) \int dx' \int dy' x'^2 \frac{\exp\left[\pi i K\left(x'^2 + y'^2\right)/\Delta z\right]}{i\lambda\Delta z}$$
$$= \frac{i\Delta z}{4\pi K}\left[\frac{\partial^2}{\partial x^2} + \frac{\partial^2}{\partial y^2}\right]\Phi(\mathbf{b}, z) \tag{4.46b}$$

Substituting these values into (4.44) and taking $\exp[i\sigma V(\mathbf{b})] \approx 1 + i\sigma V(\mathbf{b}, z)\Delta z$, the following result is obtained:

$$\Phi(\mathbf{b}, z + \Delta z) \approx \Phi(\mathbf{b}, z) + \frac{\partial}{\partial z}\Phi(\mathbf{b}, z)\Delta z$$
$$\approx [1 + i\sigma V(\mathbf{b}, z)\Delta z]\left[\Phi(\mathbf{b}, z) + \frac{i\Delta z}{4\pi K}\left(\frac{\partial^2}{\partial x^2} + \frac{\partial^2}{\partial y^2}\right)\Phi(\mathbf{b}, z)\right] \tag{4.47}$$

If only the first-order terms of Δz are considered, we have

$$\frac{\partial\Phi}{\partial z} \approx i\sigma V\Phi + \frac{i}{4\pi K}\left(\frac{\partial^2}{\partial x^2} + \frac{\partial^2}{\partial y^2}\right)\Phi \tag{4.48a}$$

$$-4\pi i K\frac{\partial\Phi}{\partial z} \approx 4\pi^2 U\Phi + \left(\frac{\partial^2}{\partial x^2} + \frac{\partial^2}{\partial y^2}\right)\Phi \tag{4.48b}$$

Equation (4.86b) is exactly the same as (4.21). Thus the equivalence of multislice theory with the semireciprocal space theory is established.

To prove the equivalence of the Bethe theory and semireciprocal space theory, we start with the expansion forms of the wave function in the two representations, in the Bloch wave representation

$$\Psi(\mathbf{r}) = \sum_i C_0^{(i)*} \sum_g C_g^{(i)} \exp\left[2\pi i(\mathbf{K}+\mathbf{g})\cdot\mathbf{r} + 2\pi i v_i z\right] \tag{4.49}$$

in the semireciprocal space theory

$$\Psi(\mathbf{r}) = \exp(2\pi i\mathbf{K}\cdot\mathbf{r}) \sum_g \Phi_g(z) \exp[2\pi i\mathbf{g}\cdot\mathbf{r}] \tag{4.50}$$

Comparing (4.49) and (4.50) directly gives

$$\Phi_g(z) = \sum_i C_0^{(i)*} C_g^{(i)} \exp\left[2\pi i v_i z\right] \tag{4.51}$$

Substituting (4.51) into (4.24), assembling terms with the same $\exp\left[2\pi i v_i z\right]$ factor, and assuming A_g is independent of z, yields

$$\mathbf{AC} = 2\,K v_i \mathbf{C} \tag{4.52}$$

This is the eigenvalue equation of the Bethe theory for reflections with $g_z = 0$ (see (2.14)). Thus the equivalence of the semireciprocal theory and the Bethe theory is proved.

4.7 Comparing of Bloch Wave and Multislice Theories

The Bloch wave theory and the multislice theory are equivalent. Numerical calculations made by Ma and Marks (1990) show the complete consistency of the two theories in transmission electron diffraction. The Bloch wave and the multislice theories can be compared in following respects.

For HRTEM image simulations, the multislice theory is much more convenient than the Bloch wave theory, specially when there are imperfections in the crystal, such as point defects and dislocations. The multislice theory is best suited for simulations of images of crystal surfaces and interfaces, but the amount of computation increases linearly with increased crystal thickness. For a perfect crystal containing no defects, on the other hand, the HRTEM phase contrast image calculations using the Bloch wave theory can give a clear picture of how each Bloch wave contributes to the image (Kambe 1982), and the amount of calculation is independent of crystal thickness.

For diffraction contrast image calculations, the Bloch wave theory is much more convenient and efficient if the column approximation is made.

For CBED, the Bloch wave calculation is much easier if there are no disk overlaps. If the beam convergence is so large that there are disk overlaps between different Bragg reflections, diffraction amplitudes of all the **g**s have to be calculated separately for different incident beam directions, then they are added coherently, partially coherently, or incoherently, depending on the assumption of source coherence. This calculation process may be very inefficient for crystals of large unit cells.

For coherent-beam CBED, the calculation can easily be performed using the multislice theory if there are disk overlaps. The two-dimensional rocking surface in the diffraction pattern is obtained by a single multislice calculation. In this case, the multislice calculation may be much faster than the Bloch wave calculation.

The most important advantage of the Bloch wave theory is its explicity for exploring the physics involved in electron diffraction, particularly if there are only a few beams. The contribution of each Bloch wave in the diffraction pattern or image can easily be isolated, leading to a clear interpretation to calculated results (Kambe 1982). In the multislice calculation, contributions of all Bloch waves are combined, but the multislice calculation becomes more effective if there are many beams involved.

The major source of error in the Bloch wave calculation is the number of beams to be included. The error source in the multislice calculation is the thickness of the slice.

For HOLZ reflections, the Bloch wave calculation is more convenient and accurate (Qin and Urban 1990).

Since the multislice theory was developed by ignoring electron backscattering, it cannot be applied to simulate backscattered electron diffraction, but no such restriction prohibits application of the Bloch wave theory.

Table 4.1 gives the theories best suited for certain calculations.

Table 4.1 Recommended dynamic elastic-scattering theories for numerical simulations

Applications	Multislice	Block wave	Green's function	Semireciprocal
High-resolution phase contrast TEM images	✔			
CBED of perfect crystals without disk overlap		✔		
CBED of imperfect crystals	✔			
Coherent CBED with disk overlap	✔	✔		
Microdiffraction of small crystals	✔			
Diffraction of back scattered electrons		✔	✔	
Low-voltage point source TEM ($E_0 < 500$ V)		✔	✔	
Diffraction contrast image		✔	✔	
Weak-beam image		✔		✔
RHEED	✔			
REM image	✔			

Chapter 5
Diffraction and Imaging of Reflected High-Energy Electrons from Bulk Crystal Surfaces

Transmission electron diffraction is a Laue case in which the crystal is so thin that backscattering is ignored. This is the Laue case illustrated in Fig. 1.10a. An alternative application of TEM is to study surfaces of bulk crystals in the Bragg case. In this case, a crystal surface is oriented parallel or nearly parallel to the optic axis of the microscope, and the incident electron beam strikes the surface at a grazing angle. Diffraction patterns recorded in this case are RHEED patterns, which are particularly sensitive to the structure of the surface layer. The dark-field surface image can be formed if one or more Bragg-reflected beams is selected by the objective aperture. This is the reflection electron microscopy (REM) technique, which has been developed for studying surface structures (for a review, see Wang 1993a; Yagi 1993). The RHEED is also a powerful technique for examining the in situ growth of thin films. Applications of RHEED for surface structure determinations are covered in Larsen and Dobson (1988). Techniques, theories, and applications of REM and RHEED received comprehensive coverage in Wang (1996).

In this chapter describes the dynamic scattering theories for RHEED. The main theoretical schemes are outlined and comparisons illustrate characteristics of each approach.

5.1 Geometry of Rheed

Figure 5.1 shows the scattering geometry of RHEED. An incident electron beam strikes the surface at an angle of about $1 - 3°$ with the surface plane. Since the crystal is effectively semi-infinite, there are no transmitted electrons. Part of the electrons are Bragg reflected into the vacuum. The RHEED was first used for oxide film studies by Miyake (1959), and it has been widely applied to monitor thin-film growth during molecular beam epitaxy (MBE) deposition (see for example, Ino, 1987). The small incidence angle makes RHEED a surface-sensitive technique.

Z. L. Wang, *Elastic and Inelastic Scattering in Electron Diffraction and Imaging*, Springer Series in Solid-State Sciences 205, https://doi.org/10.1007/978-3-031-90819-4_5

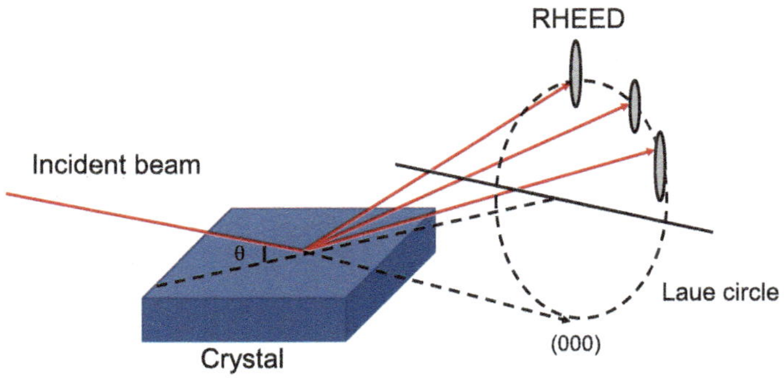

Fig. 5.1 The scattering geometry of RHEED for surface studies

Diffraction of incident electrons by the surface lattice has been used to define surface crystallographic structures. The technique reveals almost instantaneous changes in either the coverage of the specimen surface by adsorbates or the surface structure of the film itself. A review of the RHEED geometry and associated reciprocal space analyses are given by Mahan et al. (1990). Reflected beams may exhibit streaks parallel to the surface normal. This phenomenon results from limited penetration of the electron beam into the crystal surface, and it is equivalent to applying a shape function, as introduced in Chap. 1, perpendicular to the surface.

An example of RHEED patterns and REM images of a reconstructed Si(111) surface during $1 \times 1 \rightarrow 7 \times 7$ transition with a change of specimen temperature is shown in Fig. 5.2. In the inserted RHEED patterns, reflected beams produced by the surface monolayer reconstruction are clearly seen. The dark-field image of the surface formed using the specularly reflected beam(s) shows the distribution of 1 -atom-high surface steps. The surface is clearly covered by the 1×1 structure at temperatures above $T_c = 1083$ K. Below this temperature, the surface starts to show the 7×7 structure. More importantly, the formation of 7×7 structure is clearly associated with 1 -atom-high surface steps. This information cannot easily be provided by other surface-sensitive techniques, such as low-energy electron diffraction. In general, RHEED patterns provide crystallographic structures of the top few surface layers, and REM gives the images of surface domain structures directly.

Figure 5.3 shows a REM image of the Pt(111) surface. The surface exhibits many 1 -atom-high surface steps. The vertical scale is much larger than the horizontal scale due to the foreshortening effect. Due to limited depth of focus, only part of the image is in focus. Thus a single REM image displays a continuous focus series of surface steps. A dark object on the surface shows symmetric mirror image, which resulted from the erected object blocking reflected electrons (*upper half*) and the incident electrons (*lower half*) .

Positions of diffracted beams can be determined from the kinematic diffraction theory if the refraction effect of the crystal surface is considered. The refraction effect results from perturbation of electron the wavelength by the crystal inner potential. The dynamic scattering effect is quite strong in RHEED, although the electron pen-

Fig. 5.2 The RHEED patterns and reflected electron images of an Si(111) surface during a 1 × 1 to 7 × 7 transition. **a** A 1 × 1 surface reconstruction above the critical temperature T_c(= 1083 K). **b** The coexistence of 1 × 1 (*bright areas*) and 7 × 7 (*dark contrast*) surface structures. The transition to a 7 × 7 reconstruction is clearly associated with 1-atom-high surface steps. Courtesy of Dr. Latyshev et al. (1991)

etration depth is limited to a few atom layers because of the long-path interaction of the electron beam with the crystal along the beam direction. Therefore, analysis of RHEED patterns has to be performed by dynamic scattering. The following sections describe the basic methods and associated approximations of several theories that proposed for RHEED calculations. A comprehensive description of RHEED and REM is given by Wang (to be published).

5.2 Bloch Wave Theory

The Bloch wave theory was first applied to RHEED by Colella (1972) and Moon (1972). Figure 5.4 shows the coordinate system and surface model constructed in the Bloch wave calculation. The left-hand half-space is filled with atoms of a periodic

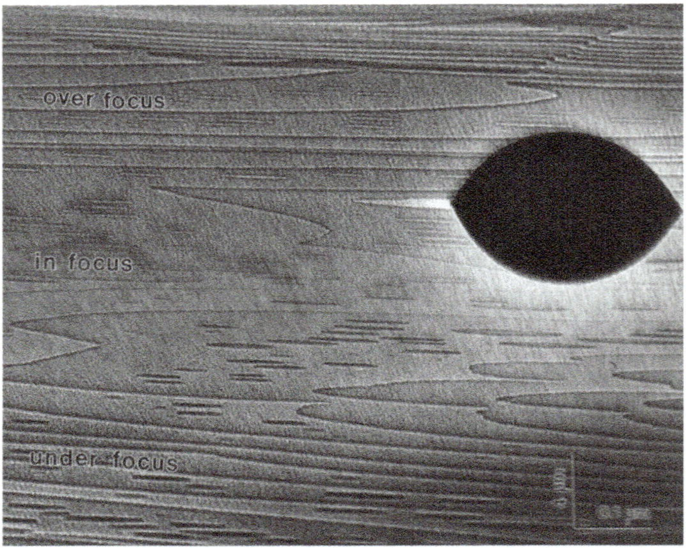

Fig. 5.3 A REM image of a Pt(111) surface recorded at 100 kV using a (666) specularly reflected beam

Fig. 5.4 The construction of a crystal surface for Bloch wave calculations in RHEED. The coordinate system is redefined with the *z* axis perpendicular to the surface and the *x* axis parallel to the surface. The incident beam strikes the surface in a direction nearly parallel to the *x* axis

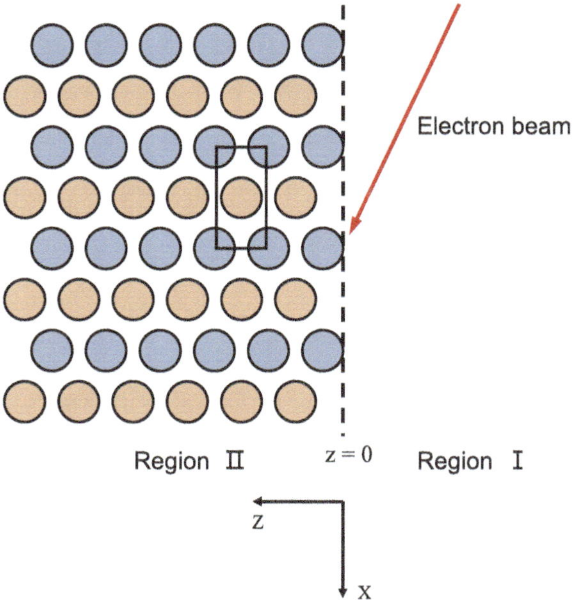

structure (Region II) and the right-hand half is a vacuum (Region I), so that the surface is a sharp termination of the bulk crystal at $z = 0$, and no surface reconstruction is allowed. The z axis direction is defined to be antiparallel to the normal direction of the surface. Thus the potential distribution is $V = V(\mathbf{r})$ for $z > 0$ and $V = 0$ for $z < 0$. The electron wave function inside the bulk crystal can be written as a superposition of Bloch waves

$$\Psi_{\text{II}}(\mathbf{r}) = \sum_i \alpha_i B_i(\mathbf{r}) = \sum_i \sum_g \alpha_i C_g^{(i)} \exp\left[2\pi i(\mathbf{K} + \mathbf{g}) \cdot \mathbf{r} + 2\pi i v_i z\right] \tag{5.1}$$

where α_i are summation coefficients of different Bloch waves determined by boundary conditions. The C_g coefficients are determined by (2.9), which is rewritten as

$$\left[2\,K S_g - 2\,(K_z + g_z)\,v - v^2\right] C_g + \sum_h U_{g-h} C_h = 0 \tag{5.2}$$

where K_z and g_z are components of the electron wave vector $\mathbf{K}$ and reciprocal lattice vector $\mathbf{g}$ in the direction parallel to the surface normal, respectively. In RHEED, since $K_z \ll K$, v may not be small in comparison with K_z, so the v^2 term cannot be neglected (this is a major departure from the Laue case). Thus for the N beam case, (5.2) has $2N$ Bloch wave solutions, which is twice the number for transmission electron diffraction. Equation (5.2) may be written in matrix form as

$$\mathbf{AC} = v[\mathbf{ZC} + v\mathbf{IC}] \tag{5.3}$$

where $[\mathbf{A}]_{gh} = 2\mathbf{KS}_g\delta_{gh} + U_{g-h}$, $[\mathbf{Z}]_{gh} = 2\,(K_z + g_z)\,\delta_{gh}$ and $\mathbf{I}$ is a unit matrix.

Equation (5.3) can be solved exactly using the following method (Kim and Sheinin 1982). If we define a matrix $\mathbf{X}$ to be

$$\mathbf{X} = \mathbf{ZC} + v\,\mathbf{IC} \tag{5.4a}$$

or rewrite it as

$$-\,\mathbf{ZC} + \mathbf{IX} = v\,\mathbf{C} \tag{5.4b}$$

then (5.3) becomes

$$\mathbf{AC} = v\mathbf{X} \tag{5.5}$$

Equations (5.4b) and (5.5) are combined into one matrix form

$$\mathbf{A}_x\mathbf{C}_x = v\mathbf{C}_x \tag{5.6}$$

where

$$\mathbf{A}_x = \begin{pmatrix} -\mathbf{Z} & \mathbf{I} \\ \mathbf{A} & \mathbf{0} \end{pmatrix} \quad \mathbf{C}_x = \begin{pmatrix} \mathbf{C} \\ \mathbf{X} \end{pmatrix}$$

Equation (5.6) can be solved using the conventional method of matrix diagonalization.

For the convenience of the following discussion, we define the tangential components of $\mathbf{K}$ and $\mathbf{g}$ parallel to the surface to be $\mathbf{K}_t$ and $\mathbf{g}_t$, respectively. In a vacuum, the electron wave is a summation of the incident wave and the reflected wave

$$\Phi_{\mathrm{I}}(\mathbf{r}) = \exp[2\pi i\mathbf{K}\cdot\mathbf{r}] + \sum_g R_g \exp\left[2\pi i\left(\boldsymbol{\chi}_{gt} - \boldsymbol{\chi}_{gz}\right)\cdot\mathbf{r}\right] \tag{5.7}$$

where $\boldsymbol{\chi}_{gt}$ and $\boldsymbol{\chi}_{gz}$ are wave vector components of reflected beams in directions parallel and perpendicular to the surface plane, respectively, and they are determined by boundary conditions; the R_g are the amplitudes of the reflected beams. Boundary conditions require

$$\Psi_{\mathrm{I}}(\mathbf{b}, 0) = \Psi_{\mathrm{II}}(\mathbf{b}, 0) \tag{5.8a}$$

and

$$\left.\frac{\partial\Psi_{\mathrm{I}}(\mathbf{r})}{\partial z}\right|_{z=0} = \left.\frac{\partial\Psi_{\mathrm{II}}(\mathbf{r})}{\partial z}\right|_{z=0} \tag{5.8b}$$

Equation (5.8a) and (5.8b) provide a total of $2N$ equations, but the unknown variables contained in (5.1) and (5.7) are R_g, χ_{gz}, and α_i, for a total of $4N(g = 1, 2, \ldots, N$ and $i = 1, 2, \ldots, 2N$). Thus equations we obtained from (5.8a) and (5.8b) are not sufficient to determine required physical quantities uniquely.

To solve this problem, a virtual surface, called the back surface, which is parallel to the front surface, is created at a distance D from the front surface (Colella 1972; Moon 1972). Thus the crystal is an erective slab defined by two parallel faces at $z = 0$ and $z = D$. This arrangement is shown in Fig. 5.5. The artificially created vacuum part is called Region III. Beams propagating in Region III can be transmitted waves only if the crystal slab is sufficiently thick so there are no reflected wave from the front surface; thus

$$\Psi_{\mathrm{III}}(\mathbf{r}) = \sum_g T_g \exp\left[2\pi i\left(\boldsymbol{\kappa}_{gt} + \boldsymbol{\kappa}_{gz}\right)\cdot\mathbf{r}\right] \tag{5.9}$$

where $\boldsymbol{\kappa}_{gt}$ and $\boldsymbol{\kappa}_{gz}$ are components of the reflected beam wave vector in directions parallel and perpendicular to the surface plane, respectively; the T_g are amplitudes of the transmitted beams. Determining various quantities can be divided into the following three steps:

1. To match waves at the boundary, tangential components of wave vectors satisfy

$$\boldsymbol{\chi}_{gt} = \mathbf{K}_t + \mathbf{g}_t = \boldsymbol{\kappa}_{gt} \tag{5.10a}$$

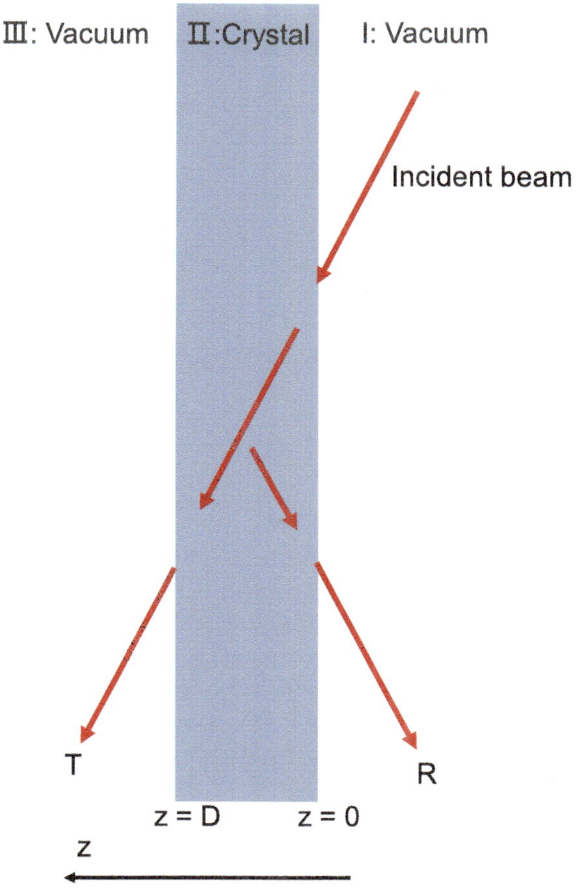

Fig. 5.5 The truncation of a virtual surface located at $z = D$ for uniquely determining the Bloch wave solution. Regions I and III are vacuums, and Region II is a bulk crystal of thickness D

Conservation of energy for elastic scattering also requires

$$\mathbf{K} = \chi_g = \kappa_g \tag{5.10b}$$

Using (5.10a), normal components of wave vectors are

$$\chi_{gz} = \kappa_{gz} = \pm\left[K^2 - (\mathbf{K}_t + \mathbf{g}_t)^2\right]^{1/2} \tag{5.10c}$$

2. Using (5.10a)–(5.10c), wave functions in Region III are rewritten as

$$\Psi_{\text{III}}(\mathbf{r}) = \sum_g T_g \exp\left[2\pi i\left(\boldsymbol{\chi}_{gt} + \boldsymbol{\chi}_{gz}\right)\cdot\mathbf{r}\right] \tag{5.11}$$

where the sign reversal of χ_{gz} with respect to (5.7) indicates the opposite propagation of waves in Region III (along $+z$) as compared to that in Region I (along $-z$).

3. Matching (5.1), (5.7), and (5.11) at the boundaries according to (5.8a) and (5.8b), we obtain

$$\delta_{g0} + R_g = \sum_i \alpha_i C_g^{(i)} \tag{5.12a}$$

$$K_z \delta_{g0} - \chi_{gz} R_g = \sum_i \alpha_i C_g^{(i)} \left(K_z + g_z + v_i\right) \tag{5.12b}$$

$$T_g \exp\left[2\pi i \chi_{gz} D\right] = \sum_i \alpha_i C_g^{(i)} \exp\left[2\pi i \left(K_z + g_z + v_i\right) D\right] \tag{5.12c}$$

$$\chi_{gz} T_g \exp\left[2\pi i \chi_{gz} D\right] = \sum_i \alpha_i C_g^{(i)} \left(K_z + g_z + v_i\right) \exp\left[2\pi i \left(K_z + g_z + v_i\right) D\right] \tag{5.12d}$$

These $4N$ equations are sufficient to give a set of unique solution of all variables.

However it has been pointed out by Marks and Ma (1988) that problems are encountered when the absorption effect is considered. In this case, the electron wave vector has an imaginary component, so that the $\exp\left[2\pi i \left(K_z + g_z + v_i\right) D\right]$ factor damps very quickly, especially for a large slab thickness D. Thus (5.12c) and (5.12d) are not useful for determining the required parameters. The so-called evanescent wave is thus generated, and it propagates in a direction parallel or nearly parallel to the surface. An alternative method for finding a unique solution is to use the energy flow concept (Ma and Marks 1989; Marks and Ma 1988), which follows.

In wave mechanics, the direction of energy flow is characterized by a vector $\mathbf{j}$, which is proportional to the expectation value of the Bloch wave momentum and the group velocity of the Bloch wave. For a single Bloch wave, the energy flow vector can be directly derived by substituting the Bloch wave solution (2.8) into the equation for current flow density $\mathbf{j}$ in quantum mechanics (see Sect. 6.2.3). For the ith Bloch wave,

$$\mathbf{j}_0^{(i)} = \frac{\hbar}{2im_e}\left[\Psi^*\nabla\Psi - \Psi\nabla\Psi^{*}\right] = \frac{h}{m_e}\exp\left[-4\pi i \mathbf{k}_i^I \cdot \mathbf{r}\right] \sum_g \left|C_g^{(i)}\right|^2 \left(\mathbf{k}_i^R + \mathbf{g}\right) \tag{5.13}$$

where $\mathbf{k}_i^R$ and $\mathbf{k}_i^I$ are the real and imaginary components, respectively, of wave vector $\mathbf{k}_i$. Physically equation (5.13) means that the average momentum of a Bloch wave is the sum of the momenta of its plane wave components, weighted by a probability function $\left|C_g\right|^2$. The vector $\mathbf{j}_0^{(i)}$ defines the true path of the electron wave. The z component of the energy flow is

$$j_{0z}^{(i)} = \frac{h}{m_e} \exp\left[-4\pi i \mathbf{k}_i^I \cdot \mathbf{r}\right] \sum_g \left|C_g^{(i)}\right|^2 \left(k_{iz}^R + g_z\right) \tag{5.14}$$

We can now consider two cases. First if the energy flow of a Bloch wave is not toward the crystal, it is nonphysical to consider this Bloch wave in the solution. Second waves whose momenta increase at increased crystal thickness cannot be excited because they vanish with deep penetration due to the absorption effect. Combining the two preceding points, Bloch waves that can be excited satisfy $j_{0z}^{(i)} \geq 0$ and $k_{iz}^I \geq 0$ (note the z axis is antiparallel to the surface normal). Thus we have only N possible Bloch waves that can be excited inside the crystal after applying these selection rules, making it possible to produce a set of unique solutions using only the boundary conditions for the surface located at $z = 0$. In this case, the virtual surface truncated at $z = D$ is unnecessary.

However there are two problems with the Bloch wave approach for RHEED. First the crystal surface is assumed to be a sharp cutoff of the perfect lattice, so that scattering of the extended surface potential in the region $z < 0$ is ignored. The surface potential is particularly important in RHEED because of its high sensitivity to the structure of the topmost surface layer. For densely packed surface structures, such as Au(110), the surface formed by sharply cutting the crystal has unavoidably split the atoms at or very near the cutting plane into two parts, resulting in a nonphysical model of broken atoms. In this case, the Bloch wave method fails (Ma and Marks 1992). Examining the applicability of the Bloch wave theory in RHEED case-by-case is therefore recommended. Second, the Bloch wave method in principle is suitable only for a perfect and periodic crystal. Thus it is difficult to incorporate surface imperfections, such as surface steps, dislocations, reconstruction, or relaxation, into calculations. In practice, the contrast with imperfect structures is the most interesting feature. Therefore, the application of Bloch wave approach in RHEED calculations may be limited. An alternative method is to combine the Bloch wave theory with the parallel-to-surface multislice (PaTSM) method, introduced in Sect. 5.3 (Peng 1994).

5.3 Parallel-to-Surface Multislice Theories

To incorporate surface potential effects and surface relaxation into RHEED calculations, other approaches have been proposed in which a crystal is considered periodic in the plane parallel to the surface, and nonperiodic modulations of the potential occur only in the direction normal to the surface (Maksym and Beeby 1981; Ichimiya 1983; Lynch and Moodie 1972; Zhao and Tong 1988). These are PaTSM theories. In this model, the crystal potential is written as

$$U(\mathbf{b}, z) = \sum_g U_g(z) \exp[2\pi i \mathbf{g} \cdot \mathbf{b}] \tag{5.15}$$

The solution of the Schrödinger equation can be written as

$$\Psi(\mathbf{b}, z) = \sum_g \Psi_g(z) \exp\left[2\pi i \left(\mathbf{K}_t + \mathbf{g}\right) \cdot \mathbf{b}\right] \tag{5.16}$$

Substituting (5.15) and (5.16) into (2.5), we obtain

$$\frac{d^2\Psi_g(z)}{dz^2} + \Gamma_g^2\Psi_g(z) + 4\pi^2 \sum_h U_{g-h}(z)\Psi_h(z) = 0 \tag{5.17}$$

where

$$\Gamma_g^2 = 4\pi^2\left[K^2 - (\mathbf{K}_t + \mathbf{g}_t)^2\right] \tag{5.18}$$

Equation (5.17) is a coupled second-order differential equation whose solutions are the amplitudes of reflected beams. Two methods have been proposed to solve (5.17) for RHEED, and they are based on the same physical approach of cutting the crystal into slices parallel to the surface, but they are slightly different in mathematical treatments. Figure 5.6 shows the slice cutting for RHEED calculations.

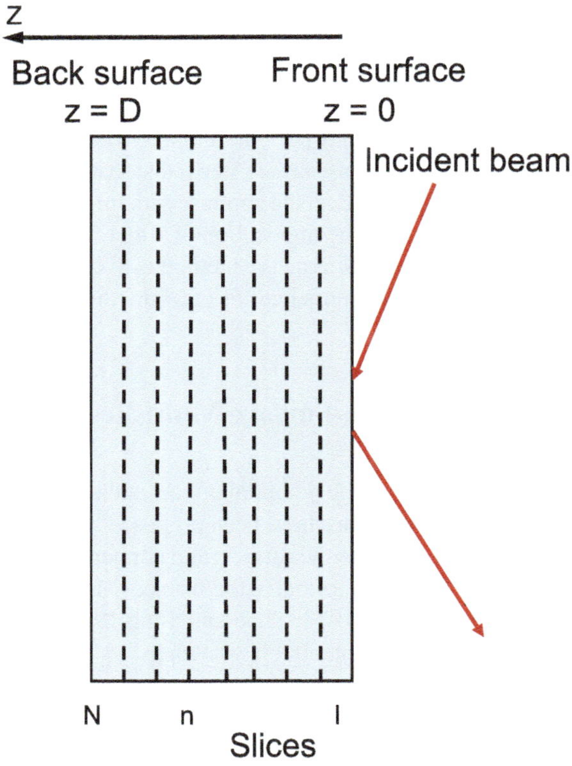

Fig. 5.6 The multislice method with the slice cut parallel to the surface. Again a back surface is truncated to determine uniquely a two-dimensional Bloch wave solution

5.3.1 Method I

The first method was developed independently by Ichimiya (1983, 1990), Maksym and Beeby (1981), and it has generally been applied to the calculation of rocking curves, the relations between the Bragg reflected intensity, and rocking the incident beam direction. In solving (5.17), the first step is to transform the second-order differential equation into a first-order differential equation. For this purpose, the following transformations using G_g and H_g are introduced

$$\left(\frac{d}{dz} + i\Gamma_g\right)\Psi_g(z) = iG_g(z) \tag{5.19a}$$

and

$$\left(\frac{d}{dz} - i\Gamma_g\right)\Psi_g(z) = -iH_g(z) \tag{5.19b}$$

Using (5.19) and (5.17) becomes

$$\frac{dG_g(z)}{dz} - i\Gamma_g G_g(z) - 4\pi^2 i\sum_h \frac{U_{g-h}(z)}{2\Gamma_h}\left[G_h(z) + H_h(z)\right] = 0 \tag{5.20a}$$

and

$$\frac{dH_g(z)}{dz} + i\Gamma_g H_g(z) + 4\pi^2 i\sum_h \frac{U_{g-h}(z)}{2\Gamma_h}\left[G_h(z) + H_h(z)\right] = 0 \tag{5.20b}$$

Equations (5.20a) and (5.20b) are expressed in matrix form as

$$\frac{d\mathbf{G}(z)}{dz} - \mathbf{B}[\mathbf{G}(z) + \mathbf{H}(z)] = 0 \tag{5.21a}$$

and

$$\frac{d\mathbf{H}(z)}{dz} + \mathbf{B}[\mathbf{G}(z) + \mathbf{H}(z)] = 0 \tag{5.21a}$$

where **B** is a $N \times N$ matrix with elements

$$[\mathbf{B}]_{gh} = i\Gamma_g\delta_{gh} + 4\pi^2 i\frac{U_{g-h}(z)}{2\Gamma_h} \tag{5.22}$$

Equation (5.21) can be written in the more convenient form

$$\frac{d\mathbf{W}(z)}{dz} = \mathbf{M}(z)\mathbf{W}(z) \tag{5.23}$$

where

$$\mathbf{W} = \begin{pmatrix} \mathbf{G} \\ \mathbf{H} \end{pmatrix} \tag{5.24}$$

is a supercolumn vector, and $\mathbf{M}$ is a 2×2 supermatrix

$$\mathbf{M} = \begin{pmatrix} \mathbf{B} & \mathbf{B} \\ -\mathbf{B} & -\mathbf{B} \end{pmatrix} \tag{5.25}$$

If the crystal is cut into many thin slices parallel to the surface, then the $\mathbf{M}$ matrix is approximately independent of z within each slice. This is the model to use for solving (5.23), and the accuracy of this method depends on the thickness of the slice. For a thin slice, the solution to (5.23) is

$$\mathbf{W}(z + \Delta z) = \exp\left[\int_{z}^{z+\Delta z} dz'\mathbf{M}\left(z'\right)\right]\mathbf{W}(z) \tag{5.26}$$

$$z$$

The successive application of (5.26) to each slice leads to the solution

$$\mathbf{W}(z) = \mathbf{P}\mathbf{W}(0) \tag{5.27a}$$

where

$$\mathbf{P} = \mathbf{P}_N \cdots \mathbf{P}_n \cdots \mathbf{P}_1 \tag{5.27b}$$

and

$$\mathbf{P}_i = \exp\left[\int_{z_i}^{z_i+\Delta z} dz'\mathbf{M}\left(z'\right)\right] \tag{5.27c}$$

The $\mathbf{P}$ is called the transfer matrix that correlates the scattering amplitude at the surface with that in the bulk.

The next major task is to match the solution (5.27a) to the boundary condition. For simplicity, we assume that the bulk crystal has a thickness D, so that boundary conditions must be matched at $z = 0$ and $z = D$ (Fig. 5.6). If D is sufficiently large, backscattering from the surface located at $z = D$ does not affect the wave distribution at the front surface. From (5.17), vacuum Region I contains only the incident wave and the reflected wave, so that

$$\left[\Psi_g(z)\right]_{z<0} = \exp\left[i\Gamma_0 z\right]\delta_{g0} + R_g \exp\left[-i\Gamma_g z\right] \tag{5.28}$$

where the amplitude of the incident beam is assumed to be unity. In Region III outside the crystal, there are only transmitted waves, so that

$$\left[\Psi_g(z)\right]_{z>D} = T_g \exp\left[i\Gamma_g z\right] \tag{5.29}$$

Substituting (5.28) and (5.29) into (5.19a) and (5.19b) leads to the following relations:

$$G_g(0) = 2\Gamma_0\delta_{g0} \tag{5.30a}$$

and

$$H_g(0) = 2R_g\Gamma_g \tag{5.30b}$$

for the front surface at $z = 0$; and

$$G_g(D) = 2T_g\Gamma_g \exp\left[i\Gamma_g D\right] \tag{5.31a}$$

and

$$H_g(D) = 0 \tag{5.31b}$$

for the back surface located at $z = D$. Therefore solutions of (5.24) and (5.27) must satisfy boundary conditions in (5.30) and (5.31). We now apply the condition in (5.31) first at the last slice N at $z = D$. From (5.27a), the vector matrix is related by

$$\mathbf{W}\left(z_{N-1}\right) = \left(\mathbf{P}_N\right)^{-1}\mathbf{W}(D) \tag{5.32}$$

where $(\mathbf{P}_N)^{-1}$ is the inverse of $\mathbf{P}_N$, written in a 2×2 super matrix form as

$$\left(\mathbf{P}_N\right)^{-1} = \begin{pmatrix} \boldsymbol{\beta}_{11}\left(z_N\right) & \boldsymbol{\beta}_{12}\left(z_N\right) \\ \boldsymbol{\beta}_{21}\left(z_N\right) & \boldsymbol{\beta}_{22}\left(z_N\right) \end{pmatrix} \tag{5.33}$$

If we rewrite (5.31) in a matrix form

$$\mathbf{W}(D) = \begin{pmatrix} \mathbf{0} \\ \mathbf{T} \end{pmatrix} \tag{5.34}$$

where 0 is a zero vector, then substituting (5.33) and (5.34) into (5.32) gives (5.35) after eliminating the **T** matrix

$$\mathbf{W}\left(z_{N-1}\right) = \begin{pmatrix} \mathbf{G}\left(z_{N-1}\right) \\ \mathbf{H}\left(z_{N-1}\right) \end{pmatrix} = \begin{pmatrix} \mathbf{G}\left(z_{N-1}\right) \\ \boldsymbol{\beta}_{21}\left(z_N\right)\left[\boldsymbol{\beta}_{12}\left(z_N\right)\right]^{-1}\mathbf{G}\left(z_{N-1}\right) \end{pmatrix} \tag{5.35}$$

For the $(N-1)$th slice, from (5.27a), we have

$$\mathbf{W}\left(z_{N-2}\right)=\left(\mathbf{P}_{N-1}\right)^{-1}\mathbf{W}\left(z_{N-1}\right) \tag{5.36}$$

where the $\left(\mathbf{P}_{N-1}\right)^{-1}$ is the inverse of $\mathbf{P}_{N-1}$, and it is written in a general form

$$\begin{pmatrix}\mathbf{G}\left(z_{N-2}\right)\\ \mathbf{H}\left(z_{N-2}\right)\end{pmatrix}=\begin{pmatrix}\boldsymbol{\beta}_{11}\left(z_{N-1}\right) & \boldsymbol{\beta}_{12}\left(z_{N-1}\right)\\ \boldsymbol{\beta}_{21}\left(z_{N-1}\right) & \boldsymbol{\beta}_{22}\left(z_{N-1}\right)\end{pmatrix}\begin{pmatrix}\mathbf{G}\left(z_{N-1}\right)\\ \mathbf{H}\left(z_{N-1}\right)\end{pmatrix} \tag{5.37}$$

Solving (5.37) for $\mathbf{G}$ and $\mathbf{H}$ and using (5.35), we have

$$\begin{aligned}\mathbf{G}\left(z_{N-2}\right)&=\left\{\boldsymbol{\beta}_{11}\left(z_{N-1}\right)+\boldsymbol{\beta}_{12}\left(z_{N-1}\right)\boldsymbol{\beta}_{21}\left(z_{N}\right)\left[\boldsymbol{\beta}_{12}\left(z_{N}\right)\right]^{-1}\right\}\mathbf{G}\left(z_{N-1}\right)\\ &\equiv\mathbf{B}_1(N-2)\mathbf{G}\left(z_{N-1}\right)\end{aligned} \tag{5.38a}$$

$$\begin{aligned}\mathbf{H}\left(z_{N-2}\right)&=\left\{\boldsymbol{\beta}_{21}\left(z_{N-1}\right)+\boldsymbol{\beta}_{22}\left(z_{N-1}\right)\boldsymbol{\beta}_{21}\left(z_{N}\right)\left[\boldsymbol{\beta}_{12}\left(z_{N}\right)\right]^{-1}\right\}\mathbf{G}\left(z_{N-1}\right)\\ &\equiv\mathbf{B}_2(N-2)\mathbf{G}\left(z_{N-1}\right)\end{aligned} \tag{5.38b}$$

Solving for $\mathbf{G}\left(z_{N-1}\right)$ from (5.38a), then substituting into (5.38b) and (5.38b) becomes

$$\mathbf{H}\left(z_{N-2}\right)=\mathbf{B}_2(N-2)\left[\mathbf{B}_1(N-2)\right]^{-1}\mathbf{G}\left(z_{N-2}\right) \tag{5.39}$$

For the nth slice, we obtain successively

$$\mathbf{G}\left(z_{n-1}\right)=\left\{\boldsymbol{\beta}_{11}\left(z_n\right)+\boldsymbol{\beta}_{12}\left(z_n\right)\mathbf{B}_2(n)\left[\mathbf{B}_1(n)\right]^{-1}\right\}\mathbf{G}\left(z_n\right)\equiv\mathbf{B}_1(n-1)\mathbf{G}\left(z_n\right) \tag{5.40a}$$

and

$$\mathbf{H}\left(z_{n-1}\right)=\left\{\boldsymbol{\beta}_{21}\left(z_n\right)+\boldsymbol{\beta}_{22}\left(z_n\right)\mathbf{B}_2(n)\left[\mathbf{B}_1(n)\right]^{-1}\right\}\mathbf{G}\left(z_n\right)\equiv\mathbf{B}_2(n-1)\mathbf{G}\left(z_n\right) \tag{5.40a}$$

When we reach the first slice (i.e., the front surface at $z=0$), the boundary condition in (5.30) requires

$$\mathbf{G}(0)=\left\{\boldsymbol{\beta}_{11}\left(z_1\right)+\boldsymbol{\beta}_{12}\left(z_1\right)\mathbf{B}_2(1)\left[\mathbf{B}_1(1)\right]^{-1}\right\}\mathbf{G}\left(z_1\right)\equiv\mathbf{B}_1(0)\mathbf{G}\left(z_n\right) \tag{5.41a}$$

and

$$\mathbf{H}(0)=\left\{\boldsymbol{\beta}_{21}\left(z_1\right)+\boldsymbol{\beta}_{22}\left(z_1\right)\mathbf{B}_2(1)\left[\mathbf{B}_1(1)\right]^{-1}\right\}\mathbf{G}\left(z_1\right)\equiv\mathbf{B}_2(0)\mathbf{G}\left(z_n\right) \tag{5.41a}$$

where

$$[\mathbf{G}(0)]_g=2\Gamma_0\delta_{g0}\text{ and }[\mathbf{H}(0)]_g=2R_g\Gamma_g \tag{5.42}$$

Solving (5.41) yields

$$\mathbf{G}(0)=\mathbf{B}_1(0)\left[\mathbf{B}_2(0)\right]^{-1}\mathbf{H}(0) \tag{5.43}$$

Using (5.42), the gth column of (5.43) is

$$R_g = \frac{\Gamma_0}{\Gamma_g} \sum_h [\mathbf{B}_1(0)]_{gh} \left\{ [\mathbf{B}_2(0)]^{-1} \right\}_{h0} \tag{5.44}$$

Therefore the reflected intensity of **g** is $|R_g|^2$. The numerical calculation starts at the Nth slice to eliminate the transmission matrix **T**. For sufficiently large D, the calculated result of (5.44) is independent of the crystal thickness.

In the preceding discussion, the transfer matrix of each slice can be different. It is possible to modify the theory for calculating the reflected intensity of a layered material, but caution must be exercised to satisfy the expansion form of a crystal potential (5.15) and relevant boundary conditions at the interfaces. In general, for an incident angle of about 3°, the thickness D of the crystal is no more than about 2 nm.

Figure 5.7 shows a comparison of the calculated RHEED rocking curve and the observed curves of an Si(111)1 × 1 surface. This surface is considered to include such disordered structures as a randomly relaxed bulklike structure (bulklike model), a relaxed bulk-like structure with random vacancies (vacancy model), a relaxed bulk-like structure with random adatoms (adatom model), or a random dimer-adatom-stacking-fault structure (random DAS model). In Fig. 5.6, the curve for the adatom model and the random DAS model are in good agreement with the experimental curve. For the vacancy model, the peak position at the (222) reflection of the curve shifts about 0.2° to a lower angle compared with the peak of the experimental one indicated by the arrow in the figure. To finalize the structure model, the crystal was rotated to a different zone axis direction, since we see only the projected structure along the beam direction for the ZOLZ reflections. The result showed that the adatom model fits the experimental observation best (Ichimiya 1993). This example demonstrates clearly the applications of dynamic simulations for determining surface structures.

5.3.2 Method II

The second method was originally proposed for calculating the diffracted intensity of low-energy electrons (Moon 1972; Lynch and Smith 1983), and it has been successfully applied to RHEED. The theory was derived from the same scheme as Method I except for some slightly different mathematical treatments (Zhao and Tong 1988). For convenience, (5.17) is written in matrix form as

$$\frac{d^2 \mathbf{\Psi}(z)}{dz^2} = -\mathbf{L}(z)\mathbf{\Psi}(z) \tag{5.45a}$$

with

$$[\mathbf{L}(z)]_{gh} = \Gamma_g^2 \delta_{gh} + 4\pi^2 U_{g-h}(z) \tag{5.45b}$$

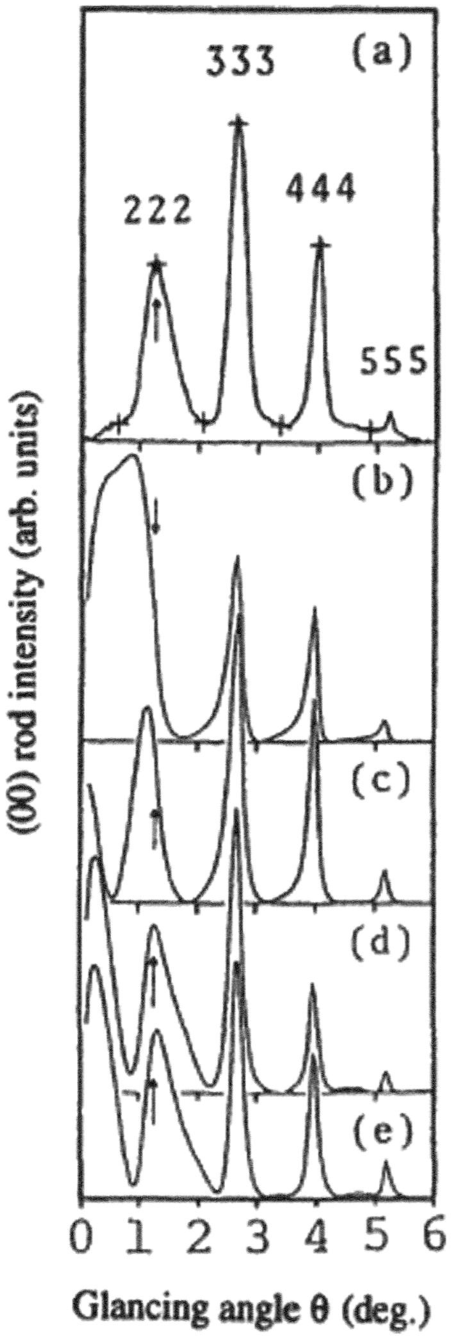

Fig. 5.7 Experimental and dynamic parallel-to-surface multislice calculated rocking curves of the Si(111) surface for beam energy 10 keV. **a** Experimental curve; **b–e** calculated curves based on the bulk-like mode, vacancy model, adatom model, and random DAS model, respectively. Arrows indicate the peak position of the (222) reflection in experimental curve (**a**) (Ichimiya 1983)

Mathematically (5.45a) can be written as two equations

$$\frac{d\boldsymbol{\Psi}(z)}{dz} = \boldsymbol{\Psi}'(z) \tag{5.46a}$$

and

$$\frac{d\boldsymbol{\Psi}'(z)}{dz} = -\mathbf{L}(z)\boldsymbol{\Psi}(z) \tag{5.46b}$$

Equations (5.46a, 5.46b) can be written as a supermatrix

$$\frac{d\boldsymbol{\Psi}''(z)}{dz} = \mathbf{N}(z)\boldsymbol{\Psi}_{(z)} \tag{5.47}$$

where $\boldsymbol{\Psi}''(z)$ is a column vector of order two, whose elements are two infinite column vectors $\boldsymbol{\Psi}(z)$ and $\boldsymbol{\Psi}_{(z)}$ describe the scattering amplitudes and their z direction slopes; $\mathbf{N}$ is a 2×2 super matrix of the form

$$\mathbf{N}(z) = \begin{pmatrix} \mathbf{0} & \mathbf{I} \\ -\mathbf{L}(z) & \mathbf{0} \end{pmatrix} \tag{5.48}$$

and $\mathbf{I}$ is a unit matrix. If the $\mathbf{N}(z)$ matrix is approximated to be independent of z within each slice, (5.47) can be solved in the multislice method as

$$\begin{pmatrix} \boldsymbol{\Psi}(z) \\ \boldsymbol{\Psi}'(z) \end{pmatrix} \approx \boldsymbol{\Omega}(z) \begin{pmatrix} \boldsymbol{\Psi}(0) \\ \boldsymbol{\Psi}'(0) \end{pmatrix} \tag{5.49}$$

where the propagation operator $\boldsymbol{\Omega}(z)$ is a product series

$$\boldsymbol{\Omega}(z) = \boldsymbol{\Omega}_N(\Delta z_N) \cdots \boldsymbol{\Omega}_{n_c}\left(\Delta z_{n_c}\right) \cdots \boldsymbol{\Omega}_1(\Delta z_1) \tag{5.50a}$$

and

$$\boldsymbol{\Omega} n_c\left(\Delta z_{n_c}\right) = \exp\left[\Delta z_{n_c} \begin{pmatrix} \mathbf{0} & \mathbf{I} \\ -\mathbf{L}\left(z_{n_c}\right) & \mathbf{0} \end{pmatrix}\right] \tag{5.50b}$$

is the matrix operator for a thin slice. Calculations for (5.50) can be simplified using the diagonalization transform (Lynch and Moodie 1972; Lynch and Smith 1983). We now match (5.49) to the boundary conditions. For convenience, the general form of the $\boldsymbol{\Omega}$ matrix can be written as

$$\boldsymbol{\Omega} = \begin{pmatrix} \boldsymbol{\Omega}_{11} & \boldsymbol{\Omega}_{12} \\ \boldsymbol{\Omega}_{21} & \boldsymbol{\Omega}_{22} \end{pmatrix} \tag{5.51}$$

Amplitudes of the reflected waves are determined by the boundary conditions. In Region I,

$$\left[\Psi_g(z)\right]_{z<0} = \exp\left[i\Gamma_0 z\right]\delta_{g0} + R_g \exp\left[-i\Gamma_g z\right] \tag{5.52}$$

so that boundary conditions are

$$\boldsymbol{\Psi}(0) = \mathbf{a}_i + \mathbf{R} \tag{5.53a}$$

and

$$\Psi'(0) = i\boldsymbol{\Gamma}\mathbf{a}_i - i\boldsymbol{\Gamma}\mathbf{R} \tag{5.53b}$$

where $[\mathbf{a}_\mathrm{i}]_g = \delta_{g0}$ is the incident wave amplitude vector, $\mathbf{R}$ is the vector of the reflected wave, and $\boldsymbol{\Gamma}$ is the diagonal matrix representation of the z components of the momenta of various waves. In Region III, there are only transmitted waves

$$\left[\boldsymbol{\Psi}_g(z)\right]_{z>D} = T_g \exp\left[i\Gamma_g z\right] \tag{5.54}$$

Boundary conditions are

$$\boldsymbol{\Psi}(D) = \exp[i\boldsymbol{\Gamma}D]\mathbf{T} \tag{5.55a}$$

and

$$\boldsymbol{\Psi}'(D) = i\boldsymbol{\Gamma}\exp[i\boldsymbol{\Gamma}D]\mathbf{T} \tag{5.55b}$$

We now use (5.51), matching (5.49) to (5.52) and (5.54) at $z = 0$ and $z = D$, respectively; the following equations are obtained

$$\exp[i\boldsymbol{\Gamma}D]\mathbf{T} = \boldsymbol{\Omega}_{11}\left(\mathbf{a}_i + \mathbf{R}\right) + \boldsymbol{\Omega}_{12}\left(i\boldsymbol{\Gamma}\mathbf{a}_i - i\boldsymbol{\Gamma}\mathbf{R}\right) \tag{5.56a}$$

and

$$i\boldsymbol{\Gamma}\exp[i\boldsymbol{\Gamma}D]\mathbf{T} = \boldsymbol{\Omega}_{21}\left(\mathbf{a}_i + \mathbf{R}\right) + \boldsymbol{\Omega}_{22}\left(i\boldsymbol{\Gamma}\mathbf{a}_i - i\boldsymbol{\Gamma}\mathbf{R}\right) \tag{5.56b}$$

Finally the reflection vector is

$$\mathbf{R} = (\chi + \nu)^{-1}(\chi - \nu)\mathbf{a}_i \tag{5.57a}$$

with

$$\chi = i\boldsymbol{\Omega}_{22}\boldsymbol{\Gamma} + \boldsymbol{\Omega}_{12}\boldsymbol{\Gamma} \tag{5.57b}$$

and

$$\nu = i\boldsymbol{\Gamma}\boldsymbol{\Omega}_{11} - \boldsymbol{\Omega}_{21} \tag{5.57c}$$

Results have been obtained through various matrix multiplication and inversions. Calculating of (5.50a) is perhaps the most difficult because of the requirement of high-order matrix multiplication. The following identity can be useful in reducing calculations (Lynch and Smith 1983):

$$\exp\left[\Delta z\begin{pmatrix} 0 & \mathbf{I} \\ -\mathbf{L}(z) & \mathbf{0} \end{pmatrix}\right] = \frac{1}{2}\begin{bmatrix} (\mathbf{E}_L + \mathbf{E}_L^{-1}) & -i\mathbf{M}_L^{-1}(\mathbf{E}_L - \mathbf{E}_L^{-1}) \\ i\mathbf{M}_L^{-1}(\mathbf{E}_L - \mathbf{E}_L^{-1}) & (\mathbf{E}_L + \mathbf{E}_L^{-1}) \end{bmatrix} \tag{5.58}$$

where $\mathbf{M}_L$ denotes a matrix that satisfies

$$\mathbf{M}_L\mathbf{M}_L = -\mathbf{L} \tag{5.59a}$$

and

$$\mathbf{E}_L = \exp(i\mathbf{M}_L\Delta z) \tag{5.59b}$$

The theory presented here has been applied to simulating RHEED patterns (Lynch and Smith 1983; Smith et al. 1992). Figure 5.8a is an experimentally observed convergent beam RHEED pattern of Pt(111) near the $[11\bar{2}]$ zone axis, which is called a Kossel pattern (similar to a Kikuchi pattern). This pattern shows the reflected intensity distribution for a large range of beam incidence, and it also exhibits double-resonance parabolas. The arrowhead indicates the (333) reflection beam. Intensity enhancement can be seen where parabolas intersect Bragg beams. Figure 5.8b is a simulated RHEED pattern for the incident beam geometry in Fig. 5.8a; a good agreement has been achieved. Figure 5.8c shows simulations for larger angles of incidence starting with (333) up to approximately (666).

As pointed out at the beginning of Sect. 5.3, the PaTSM theories were formulated on the assumption that the atomic potential can be nonperiodic in the direction perpendicular to the surface, but not within the surface plane, making it possible

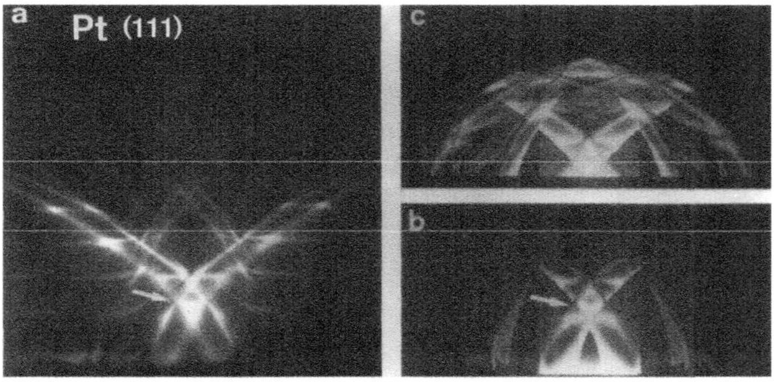

Fig. 5.8 **a** The experimental large-angle convergent beam RHEED pattern from Pt(111) near $[11\bar{2}]$. **b** A simulated RHEED pattern using the PaTSM theory (Method II). **c** A simulation for larger tilt angle; beam energy 100 keV. Courtesy of Dr. Smith et al. (1992)

to include the surface potential effect and atomic relaxation in the surface normal direction. Surface steps, dislocations, and reconstruction within the plane parallel to the surface cannot easily be included in this approach. An alternative method is to introduce an artificial supercell that is so large that surface structures of interests can be incorporated. This has been shown to be possible for calculations of surface step images in REM (McCoy and Maksym 1993).

The Bloch wave and the PaTSM calculations are usually performed in reciprocal space, so that they are convenient for calculating RHEED rocking curves via variation of the beam incident angle. However the major disadvantage of these methods is the difficulty of including surface steps and defects, so they are rarely applied to calculate REM images.

The combined PaTSM method and the Bloch wave theory have been applied to treat cases involving the surface potential effect or surface reconstruction in RHEED (see Peng 1994, for a review). The main scheme is to cut the crystal into thin slices parallel to the surface, so that each slice can be treated as a two-dimensional periodic structure based on the Bloch wave theory. Waves in adjacent slices are uniquely determined by boundary conditions.

5.4 Perpendicular-to-Surface Multislice Theory

The perpendicular-to-surface multislice (PeTSM) theory is based on the multislice formulation of dynamic electron diffraction in TEM, introduced in Chap. 3. The slice is cut normal to the surface and the electron beam azimuth (Peng and Cowley 1986), as shown in Fig. 5.9. The crystal and the vacuum part, in which the incident beam approaches the surface, are defined as a supercell. The entire space is filled with the repetition of the supercell. In RHEED geometry, if we look along the incident beam direction, the scattering angle of the electrons is no more than 5°; thus the assumption that we can ignore backscattering still holds. If crystal slices are cut in the plane perpendicular or nearly perpendicular to the crystal surface plane and the direction of the incident beam, then the transmission of electrons through each slice is analogous to that in TEM. The relationship that correlates the wave before and after penetrating a thin crystal slice of thickness Δx (see Sect. 3.2) is

$$\Phi(x, y, z) = \left\{ \Phi(x_0, y, z) \exp\left[i\sigma \frac{K}{K_x} V(y, z) \right] \right\} \otimes \left[\frac{K_x}{K} P'(y, z, x - x_0) \right] \tag{5.60}$$

where the coordinate system defined for RHEED is shown in Fig. 5.4. The $V(y, z)$ is the projected crystal potential for a slice of thickness Δx. The slice is half filled with atoms, and the other half is a vacuum. Transmitted waves inside the crystal and reflected waves in the vacuum are calculated simultaneously slice-by-slice. In the vacuum, output waves are incident waves and reflected waves. The former can be filtered out by using the Fourier transform technique because they are propagating in different directions. The propagation function is

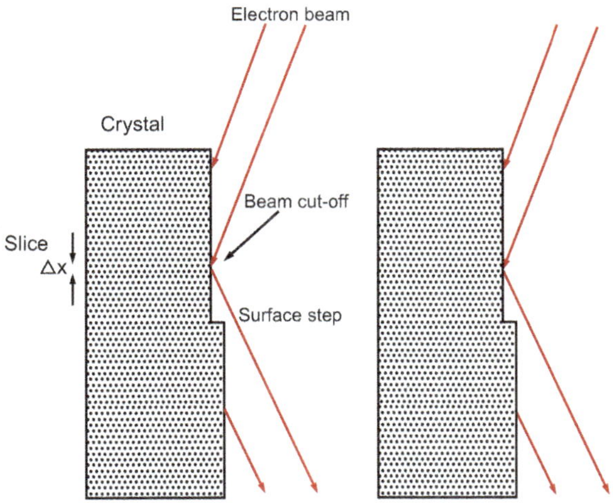

Fig. 5.9 The construction of a supercell for the PeTSM calculation. The slice thickness Δx is usually chosen to be no more than 3Å. A surface step can easily be introduced in the calculation by varying the phase-grating function of the slices

$$P'(y, z, \Delta x) = \frac{1}{\lambda i \Delta x} \exp\left[\frac{\pi i \left(y^2 + z^2\right)}{\lambda \Delta x} - 2\pi i \left(K_y y + K_z z\right)\right] \tag{5.61}$$

This theory has three important advantages. First there is no restriction on the method for constructing the phase-grating function of each slice. It is thus possible, in principle, to include any desired modification of the surface structure, such as steps, defects, and surface potential, in the calculation. Second this real-space method simply provides an easy way of tracking the propagation process of electrons slice-by-slice, given the possibility of viewing the build-up process of electron waves near the surface; this feature is unique to this method. Finally there is no restriction on the form of the incident electron, so that it is feasible to calculate the scattering of a nanoprobe electron beam at the surface. This type of calculation however may not be so easily performed by the theories outlined in Sects. 5.2 and 5.3, in which a plane wave incident beam is assumed in the mathematical derivation.

On the other hand, there are three difficulties with this theory. First a large supercell needs to be truncated to isolate all the scattering occurring in one cell from that in neighboring cells to avoid interference between cells. The steady-state solution of the wave function starts to emerge after sampling thicknesses larger than 70 nm (for GaAs at 100 kV) (Wang 1989a). A surface step contrast can be calculated only after the reflected wave arrives at a steady state, i.e., an intensity distribution that is no longer affected by the edge effect of the incident beam window but purely determined by the properties of the crystal surface. This is a rather time-consuming process. Second the incident electron wave is usually assumed to be a window function, so that the sharp edges of the effective incident wave introduce severe edge effects due to Fresnel diffraction as the wave propagates along the surface (Peng and Cowley 1988). Finally this theory is not effective for calculating RHEED rocking curves because the entire calculation has to be repeated for each incident beam direction.

To overcome the first difficulty, Ma and Marks (1990) examined the numerical consistency between Bloch wave calculations and PeTSM calculations under identical scattering potential and combined the two approaches in the following way. The steady-state wave is calculated first by the Bloch wave method for a perfect periodic surface, then the interaction of the steady-state wave with surface steps is calculated according to PeTSM theory. This combined method attempts to eliminate disadvantages of the two individual methods, but caution has to be exercised in Bloch wave calculations for close-packed surfaces. To overcome the second difficulty, an edge-patching method has been introduced (Ma 1991) to modify the incident wave (in the vacuum part) slice-by-slice by considering a constant phase shift. This method makes it possible to calculate the reflection of a plane wave from the surface, so that calculated results can be directly compared with REM observations. Recently this method has been applied to quantifying RHEED patterns recorded from GaAs(001) $- 2 \times 4$ surfaces grown by molecular beam epitaxy (Ma et al. 1993).

The PeTSM theory is most useful for REM image and fixed-beam RHEED pattern calculations. This real-space approach provides a real-space picture of how the resonance wave is built up along the surface, how much electrons penetrate the surface, and how far they travel along the surface before being rediffracted into a vacuum. In the next section, this theory is applied to the investigation of scattering processes of electrons in RHEED geometry.

The calculation of REM images is usually difficult especially if there are steps on the surface. Calculations by Anstis (1989) using the Bloch wave and PaTSM theory have given some interesting results. This is an open area, and there are many advances are being made.

5.5 Electron Reflection Process in Rheed

In the RHEED geometry, full dynamic calculations are necessary to describe reflections of high-energy incident electrons from surfaces. In particular, it is interesting to investigate the influence of surface defects, such as steps, on the propagation of electrons at the surface. In this section, we use the PeTSM theory to calculate electron scattering from a GaAs (110) surface in RHEED geometry. To track down the scattering process of electrons, a small beam of width $d = 1$ nm is chosen, which strikes on the GaAs (110) surface under the (660) specularly reflected Bragg conditions. The purpose of choosing a small narrow beam is to investigate the propagation of the electrons along the surface after beam cutoff. The incident beam covers the surface area up to the one hundred and fiftieth slice (slice thickness $\Delta x = 0.2827$ nm) in the direction of the beam azimuth. The intensity distribution of electrons at the surface is output in the plane perpendicular to the beam azimuth, so that the buildup of waves at a crystal surface can be seen by comparing intensity distributions at different slices.

Figure 5.10 shows the calculated electron intensity distributions at a GaAs(110) surface under the specular reflection of (660) with an azimuth deviation angle of 8.88 mrad parallel to the surface. After cutting off the incident wave, generated waves are divided into three possible beams. One beam is the transmission wave (T), which propagates in the crystal and is eventually absorbed. The second beam is the surface resonance wave (S), which propagates along the surface. The last beam is the reflected wave (R), which is in the vacuum and propagates away from the surface (this wave does not appear in Fig. 5.10 because of its relatively weak intensity). The surface resonance wave is concentrated in the top two to three atomic layers.

Resonance propagation of electrons at a crystal surface can enhance reflected intensities. Figure 5.11 shows the calculated total reflectance R (i.e., the ratio of the total reflected intensity to the intensity of the entire incident electrons) of electrons from a GaAs(110) surface versus the propagation distance in the beam direction. The incident beam is cut off at the one hundred and fiftieth slice. An important feature in Fig. 5.11 is that the reflected intensity initially increases almost linearly even after the beam is cut off. This indicates that the resonance propagation of the electrons along the surface can generate strong reflection waves. There seems to exist a turning point at which the reflection rate dR/dz decreases significantly. This point occurs at about the three hundredth slice. The mean traveling distance of electrons along the surface is estimated at 50–70 nm. The contribution of direct reflections, i.e., the immediate reflection of electrons from the surface, is small, on the order of one-third total reflected intensity. Most incident electrons travel along the surface for some distance before being scattered back into vacuum.

Differences in reflected intensities from a perfect crystal surface and a surface with a down-step (with respect to the beam) can be seen by comparing calculated RHEED patterns, as shown in Fig. 5.12. In the case of a perfect flat surface, reflected intensities are mainly concentrated at the (660) and the (480) reflections (Fig. 5.12a). When a 1-atom-high down-step is introduced at the two hundredth slice (56.54 nm) away from the beam cutoff position, besides the regular (480) and (660) reflection beams, an extra beam appears at the arrowhead in Fig. 5.12b. If the down-step is set at the eightieth slice (22.62 nm) away from the beam cutoff, this extra beam becomes comparatively strong and splits into two. Splitting this beam in the direction perpendicular to the surface shows that this wave comes from a small region-the top few surface layers (Fig. 5.12c). When the down-step is set at the fiftieth slice (14.14 nm) away from the beam cutoff, the extra beam becomes a single spot, and it is still strong in intensity (Fig. 5.12d). The formation of this extra beam can be considered as the propagation of the resonance waves in vacuum after being interrupted by the step. Line-scan-reflected-intensity distribution across the calculated (660) beam for the several cases addressed in Fig. 5.12 shows that the position of the down-step can also significantly affect the intensity distribution in the (660) disk due to dynamic scattering effects (Wang 1989b).

Theoretically predicted results have been observed in RHEED experiments using a VG HB5^{TM} STEM. The incident beam was chosen as 0.5 nm in diameter to study the small-beam reflection effect from a surface. Figure 5.13 shows a series of RHEED patterns made when the beam scans across a surface down-step. When the beam is

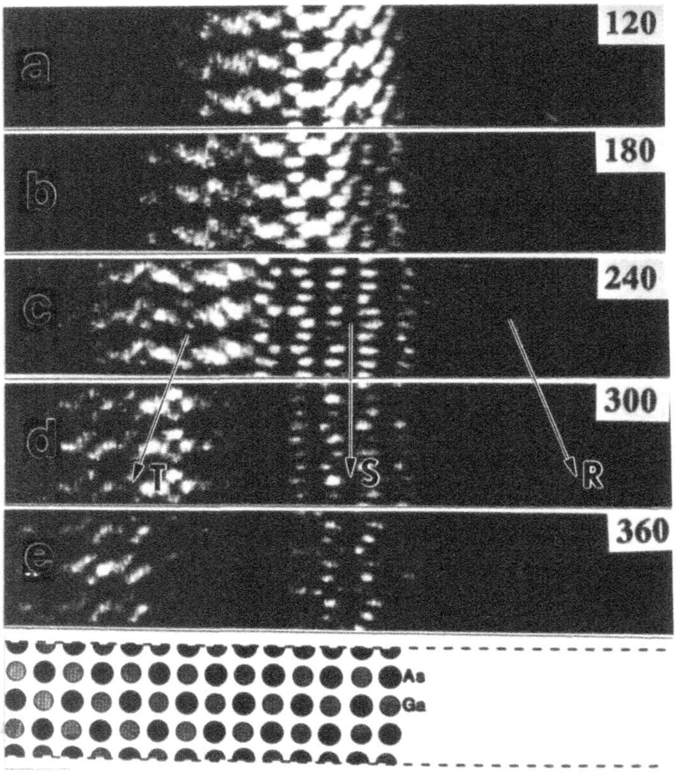

Fig. 5.10 The scattering of high-energy electrons from the surface in RHEED calculated using the PeTSM theory. This figure displays the calculated electron intensity distributions as viewed in the beam direction near a perfect GaAs(110) surface at different slice positions (or depth along the beam direction), as indicated at the corners of a–e. The projected atomic model is shown at the bottom of the output. The left-hand side is the crystal; the right-hand side is a vacuum; and the interface of the two is the (110) surface. The *T*, *S*, and *R* indicate the transmitted wave into the bulk, the resonance wave near the surface, and the reflected wave in a vacuum, respectively. The beam azimuth is close to [001]; the incidence angle is 23.5 mrad; and the azimuth deviation angle in the direction parallel to the surface is 8.88 mrad. The incident beam is cut off at the one hundred and fiftieth slice. The slice thickness is 2.827 Å. The projected atomic model is shown at the bottom of the output

far away from a down-step (Fig. 5.13a), there is almost no indication of the existence of the extra beam, in agreement with the prediction in Fig. 5.12a. Only the (660) and (480) reflections are observed. When the beam approaches the step, an extra beam appears near the edge of the surface, as indicated by an arrowhead in Fig. 5.13b. When the beam moves very close to the step, this beam becomes strong (Fig. 5.13c). This beam disappears suddenly just after the beam crosses the step (Fig. 5.13d). These observations agree with theoretical calculations shown in Fig. 5.12.

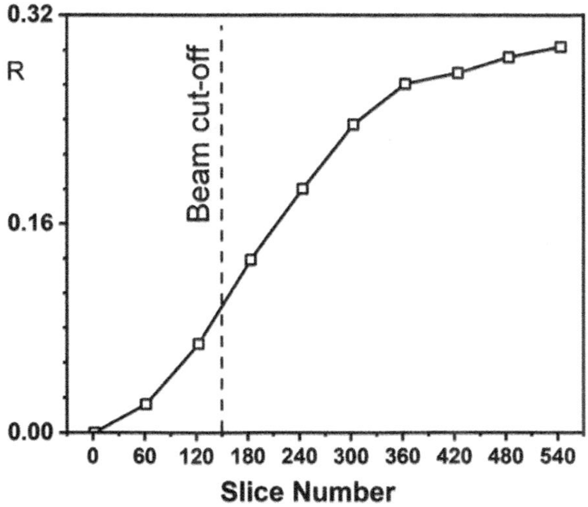

Fig. 5.11 Resonance reflection in RHEED; the calculated total reflectance (R) versus the slice number from the results in Fig. 5.10. The incident beam illuminates only the surface from the first to the one hundred and fiftieth slice, but the total reflected intensity still increases even after the incident beam has been cut off

5.6 Thermal Diffuse Scattering in Rheed

Figure 5.14 shows a RHEED pattern recorded from an annealed $LaAlO_3(100)$ surface. The beam azimuth is [001]. Superlattice reflections observed between bulk reflections are produced by a 5×5 surface reconstruction. The diffuse background observed in the pattern may come from three sources, thermal diffuse scattering (TDS), multiple valence loss scattering, or surface disorder. The contribution made by valence loss electrons can be removed by an energy filter (see Chap. 6), but the TDS remains. It is thus necessary to illustrate some appropriate dynamic theories that can be applied to calculate TDS RHEED patterns.

The theories presented in last few sections deal mainly with elastic scattering from crystal surfaces. Background and Kikuchi patterns observed in RHEED are partially produced by TDS. Green's function theory, introduced in Chap. 10, can conveniently be applied to calculate RHEED of TDS electrons (Dudarev 1991). Here we introduce a modified theory following the PaTSM theory in Sect. 5.3.1 for calculating the TDS pattern under the Einstein model (Korte and Meyer-Ehmsen 1993).

The theory is based on the first-order perturbation method. The crystal potential is written as the sum of a periodic function (V_0) and nonperiodic function (ΔV). Bragg reflections are considered to come only from the periodic potential V_0. Dynamic scattering is assumed to arise from V_0, and scattering from ΔV is treated kinematically. We now consider TDS intensity at a non-Bragg position $\mathbf{u} = \mathbf{g} + \mathbf{q}$. For each $\mathbf{q}$, reflection beams to be included in the calculations are (1) a periodic crystal reciprocal lattice array $\{\mathbf{g}\}$ and (2) the diffuse set $\{\mathbf{u} = \mathbf{g} + \mathbf{q}\}$. For the $N\mathbf{g}$ beam case,

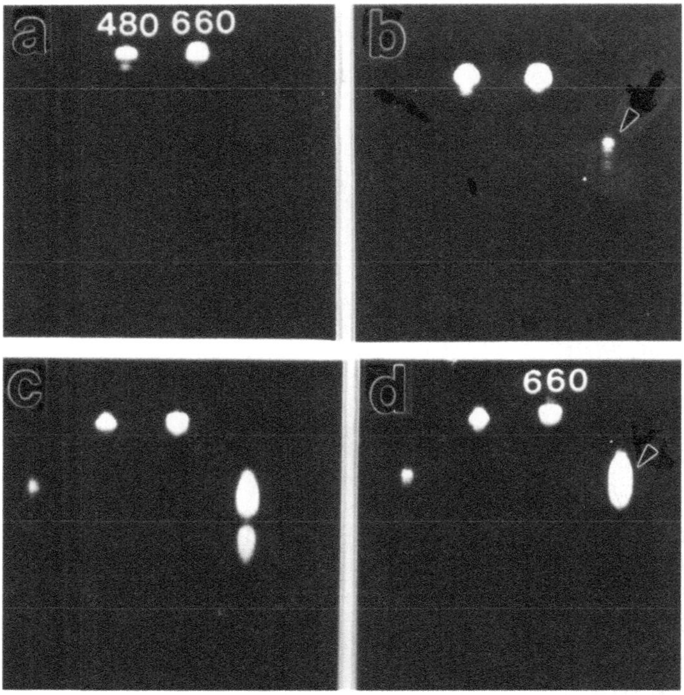

Fig. 5.12 Calculated RHEED patterns of GaAs(110) surfaces for different down-step positions located at **a** ∞, **b** the two hundredth slice (565Å), **c** the eightieth slice (226Å), and **d** the fiftieth slice (141Å) from the beam cut off position. The step height is chosen as a monolayer high. Simulation conditions are the same as those in Fig. 5.10

total beams to be included in the calculations are $2N$, as shown in Fig. 5.15. Waves corresponding to sharp reflections (**g** and **g**′) are coupled by $U_{g-g'}$. Diffuse beams (**u** and **u**′) interact via $U_{u-u'}$. Interaction between **g** beam with the diffuse beam **u** is ΔU_{g-u}, where $\Delta U_u = \mathrm{FT}[\Delta V(\mathbf{r})]$. Therefore the **B** matrix defined in (5.22) is redefined as

$$[\mathbf{B}]_{gh} = i\Gamma_g\delta_{gh} + 4\pi^2 i\frac{U_{g-h}(z)}{2\Gamma_h} \tag{5.62a}$$

$$[\mathbf{B}]_{uu'} = i\Gamma_u\delta_{uu'} + 4\pi^2 i\frac{U_{u-u'}(z)}{2\Gamma_{u'}} \tag{5.62b}$$

and

$$[\mathbf{B}]_{gu} = 4\pi^2 i\frac{\Delta U_{g-u}(z)}{2\Gamma_u} \tag{5.62c}$$

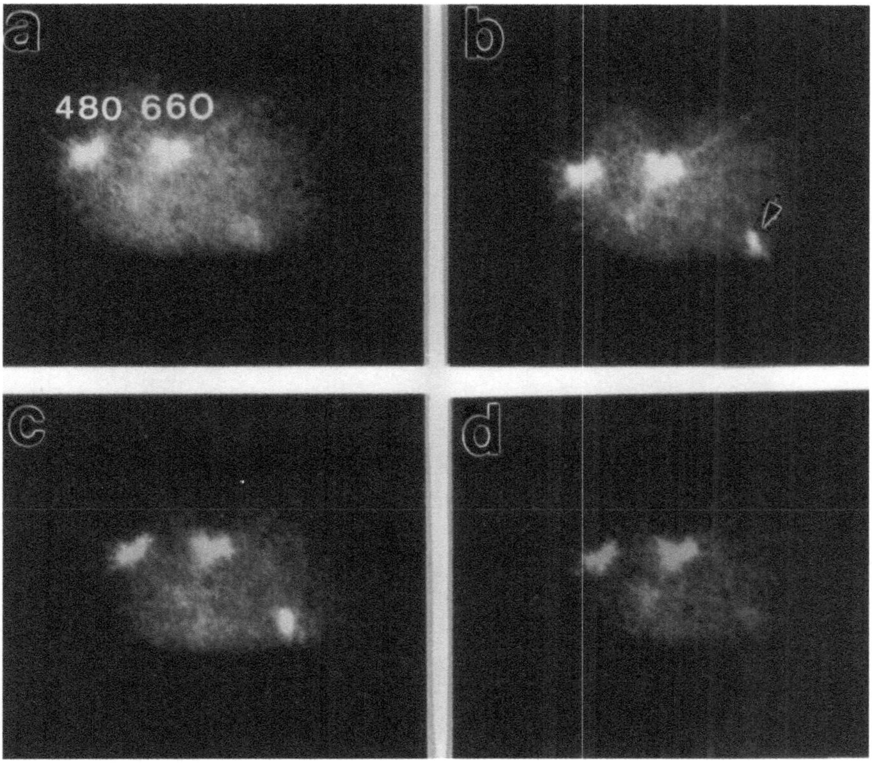

Fig. 5.13 Microdiffraction RHEED patterns with a 5 – Å beam scanning across a down-step on a GaAs(110) surface. These patterns were obtained when the beam was **a** far away from a step, **b** approaching the step, **c** close to the step, and **d** after passing the step. The reflection located near the surface shadow appears only when the incident beam is near the surface down-step (Wang 1989a)

The numerical calculation follows the same equation given in (5.21) or (5.23). The entire calculation has to be repeated for each **q**, analogous to the calculation of a rocking surface in RHEED.

For TDS, the perturbation potential ΔV is time-dependent,

$$\Delta V(t) \approx \sum_{\kappa} \mathbf{u}_{\kappa}(t) \cdot \nabla V_{\kappa}(\mathbf{r} - \mathbf{r}_{\kappa}) \tag{5.63}$$

where $\mathbf{u}_{\kappa}(t)$ is the time-dependent displacement of the κ th atom. Thus the calculated intensity has to be averaged over time. This process normally requires a huge number of calculations. Under the Einstein model, TDS intensity is the incoherent superposition of the scattered intensity of all atoms. Therefore ΔV is approximated as a time-independent function

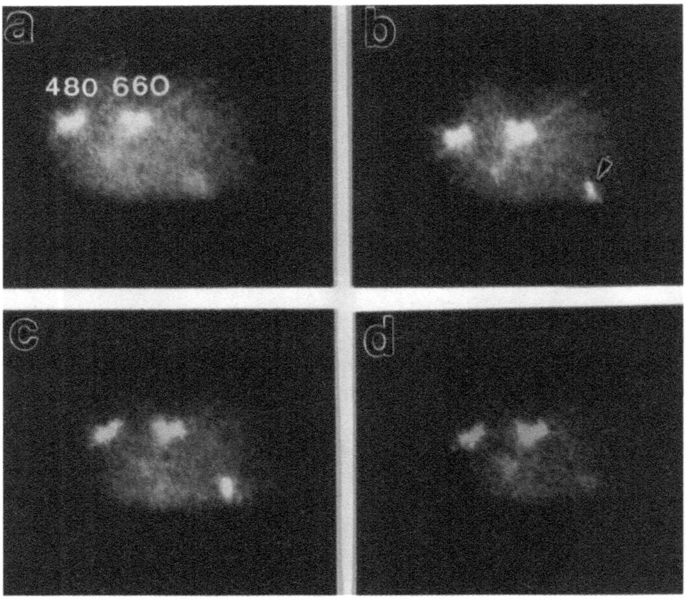

Fig. 5.14 A RHEED pattern recorded at 300 kV from a thermal annealed $LaAlO_3$ (100) surface, exhibiting {100} twin and 5 × 5 surface reconstruction. Besides the Bragg reflections from periodic structures of the bulk and surface, a diffuse background decorated with Kikuchi lines and continuous resonance parabolas is seen

$$\Delta V_u \approx \sum_{\kappa} a_{\kappa} \mathbf{FT} \left[\frac{\partial V_{\kappa} (\mathbf{r} - \mathbf{r}_{\kappa})}{\partial r} \right] \quad (5.64)$$

The TDS intensity at $\mathbf{u} = \mathbf{g} + \mathbf{q}$ can be obtained from a single calculation. This has been shown by Korte and Meyer-Ehmsen (1993), and results show good agreement with experimental observations. In RHEED, the Einstein model is usually an excellent approximation. The perturbation theory introduced here can also be applied to calculate the diffraction of disordered surfaces.

It must be pointed out that the calculation based on this scheme may give the correct distribution of TDS in the diffraction pattern, but the absolute TDS intensity may need be examined in detail because the ΔV_u given in (6.64a) and (6.64b) artificially treats the TDS from all atom sites as coherent. This may greatly limit the application of this modified theory. More rigorous theories are presented in Part II of this book.

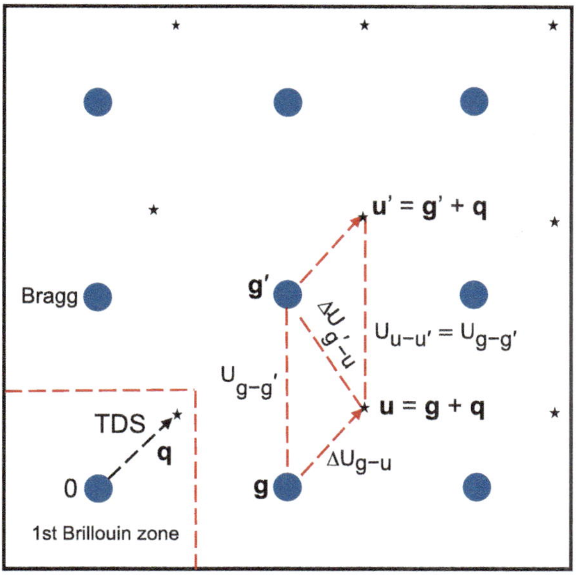

Fig. 5.15 Beam interactions in dynamic calculations of thermal diffusely scattered electrons using the PaTSM theory. $U_g = (2\gamma m_0 e/h^2)\, V_g$ and $\Delta V_u = \mathbf{FT}(\Delta V)$. Nine Bragg beams corresponding to reflections from a periodic structure are accompanied by nine TDS beams. The calculation is performed for a total of $9 + 9 = 18$ beams for each $\mathbf{q}$ point, where $\mathbf{q}$ is restricted to the first Brillioun zone

5.7 Summary

The RHEED is an important technique for examining in situ surface structure evolution in the epitaxy growth of thin films. We introduced several dynamic theories that have been applied to calculate diffraction and image intensities of reflected high-energy electrons. The PaTSM theories are best suited for calculating RHEED rocking curves. The PeTSM theory is powerful for simulating images formed by reflected electrons, particularly where there are steps and defects. Merits and disadvantages of each theory are outlined. Finally we introduced a dynamic theory for calculating the RHEED pattern of thermal diffusely scattered electrons.

Part II
Diffraction and Imaging of Inelastically Scattered Electrons

Chapter 6
Inelastic Excitations and Absorption Effect in Electron Diffraction

6.1 Kikuchi Patterns

As illustrated in Part I, Bragg reflections are the only beams expected to be observed in electron diffraction patterns of crystals. In practice, diffuse scattering, distributed at non-Bragg reflection angles, is seen, and some dark and bright line pairs and bands, so-called Kikuchi lines and bands, also appear in the pattern. These features are observed experimentally, as shown by a diffraction pattern of $\alpha - Al_2O_3$ $[11\bar{2}0]$ in Fig. 6.1. The formation of a Kikuchi pattern is attributed to the inelastic-scattering effect in electron diffraction.

In high-energy electron scattering, various inelastic-scattering events, such as phonons, plasmons (or valence electrons), and atomic inner shell electrons, can be excited because of the interaction of the incident electron with electrons and nuclei in the crystal. The result of each inelastic-scattering process is a transition of the crystal state due to energy and momentum transfers from the incident electron. Figure 6.2 shows a series of energy-filtered electron diffraction patterns of Si [111] of about 50 nm thickness, which exhibit the angular distribution of electrons with different energy losses (Reimer et al. 1990). The zero-loss filtered diffraction pattern (Fig. 6.2b) is much sharper than the unfiltered diffraction pattern (Fig. 6.2a). Bragg beams are apparent, especially in the zero-loss and plasmon filtered diffraction patterns. Increasing energy loss (Fig. 6.2b–d) results in an increased blurring of Bragg spots with the angular distribution of inelastically scattered electrons. Intensities in the system of excess and deficit Kikuchi lines also vary due to the increasing angular width of inelastically scattered electrons with increasing ΔE. When energy loss is increased, the energy-filtered diffraction pattern shows only excess Kikuchi bands. From Fig. 6.2e with $\Delta E = 1800$ eV to Fig. 6.2 f with $\Delta E = 2000$ eV, the intensity and the contrast of excess bands and the deficit high-order Laue zone Kikuchi lines near the center of the [111] pole increase when passing the K ionization edge of Si located at $\Delta E = 1839$ eV. Patterns formed by the atomic inner-shell-scattered electrons contain additional information that may indicate the process of electron inelas-

Z. L. Wang, *Elastic and Inelastic Scattering in Electron Diffraction and Imaging*, Springer Series in Solid-State Sciences 205, https://doi.org/10.1007/978-3-031-90819-4_6

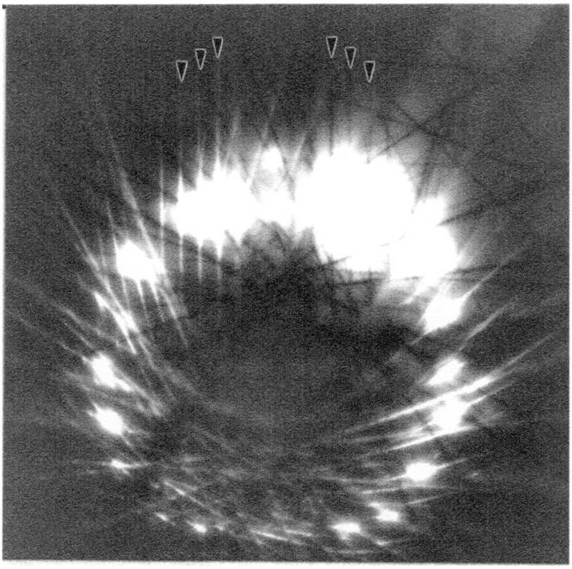

Fig. 6.1 An electron diffraction pattern recorded at 100 kV from an $\alpha - Al_2O_3$ crystal showing the formation of Kikuchi lines and bands. The crystal is oriented near the $[11\bar{2}0]$ zone axis. The paired Kikuchi lines, with excess and deficit intensities, are indicated by arrowheads

tic excitation. Diffraction patterns formed by electrons of different energy losses are angular distribution maps of these electrons in the Kikuchi pattern. Quantitative interpretation of these experimental data depends solely on theories of inelastic electron diffraction. But these theories are mathematically very complex because of convoluted multiple elastic and multiple inelastic scattering. This point can be illustrated by Fig. 6.3 regarding the formation of Kikuchi patterns (Reimer 1984).

6.1.1 Formation of Kikuchi Lines

When an electron beam strikes a crystal foil, various diffracted beams are generated due to elastic scattering. Each of the beams has an equal probability of being inelastically scattered afterward. Inelastic scattering is responsible for the formation of the Kikuchi patterns. The center (000) beam, for example, can be inelastically scattered at some depth within the crystal, so that the electron angular distribution can be approximately described by a cone-shaped function (Fig. 6.3). Part of the inelastically scattered electrons that strike crystal planes at a Bragg angle θ_g are elastically diffracted (i.e., elastic rescattering), resulting in a bright excess line in the observation plane. This line is approximately straight owing to the low value of θ_g. The Bragg reflection decreases the intensity of the beam propagating along $\mathbf{K}_0$, and the incident core intersects with the observation plane, resulting in a dark Kikuchi line. A set of bright-dark Kikuchi line pairs is generated if the mechanism is applied to every set of crystal planes. The cone of a semiconical angle $\pi/2 - \theta_g$, called the Kossel cone, is fixed to the crystal, which means that the Kikuchi lines

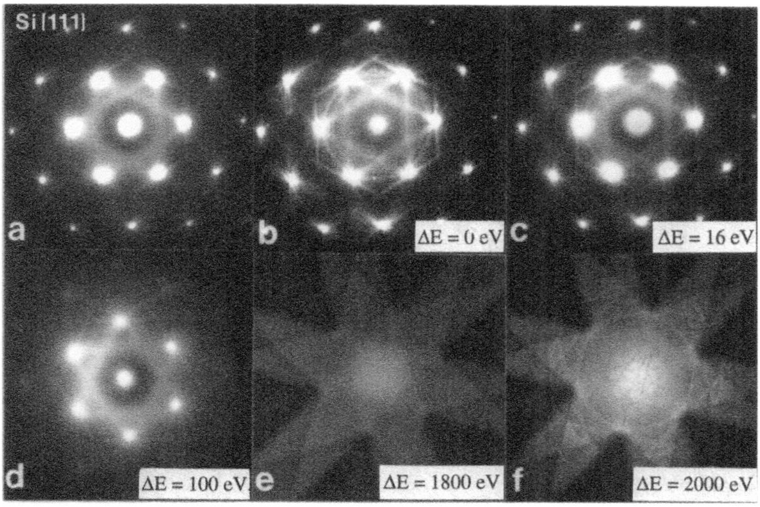

Fig. 6.2 Energy-filtered electron diffraction patterns, recorded at 80 keV, of Si oriented along [111] with thickness $d = 50$ nm. **a** Unfiltered; **b** $\Delta E = 0$ eV; **c** $\Delta E = 16$ eV; **d** $\Delta E = 100$ eV; **e** $\Delta E = 1800$ eV, pre-Si K edge filtered; and **f** $\Delta E = 2000$ eV, after Si K edge filtered. (Courtesy of Dr. Reimer et al. (1990)

move only if the crystal is tilted. This characteristic is employed to orient crystals in TEM. Positions of Bragg-reflected beams, however, move with the variation of the incident beam direction.

For thick crystals, the contrast of bright-dark Kikuchi line pairs can be reversed to dark-bright for the following reasons. A set of bright-dark Kikuchi lines can be observed so long as more electrons hit the lattice planes from one side than from the other. With increasing foil thickness, this is the case for large scattering angles, for which the scattered intensity distribution decreases with increasing scattering angle. For small scattering angles, however, the angular distribution of scattered electrons is so diffuse and uniform that an equal number of electrons hit the lattice planes from both Kossel cones (Fig. 6.3), and a cancellation of the pattern of bright-dark lines is expected.

6.1.2 Inelastic Excitations in Crystals

Kikuchi patterns are formed by the processes of elastic, inelastic and reelastic scattering. Figure 6.4 indicates the main inelastic processes that may be excited in high-energy electron scattering. First plasmon (or valence) excitation, which characterizes transitions of electrons from the valence band to the conduction band, involves an energy loss in the range of 1–50 eV and an angular spreading of less than 0.2 mrad for high-energy electrons. The decay of plasmons results in the emission of ultra-

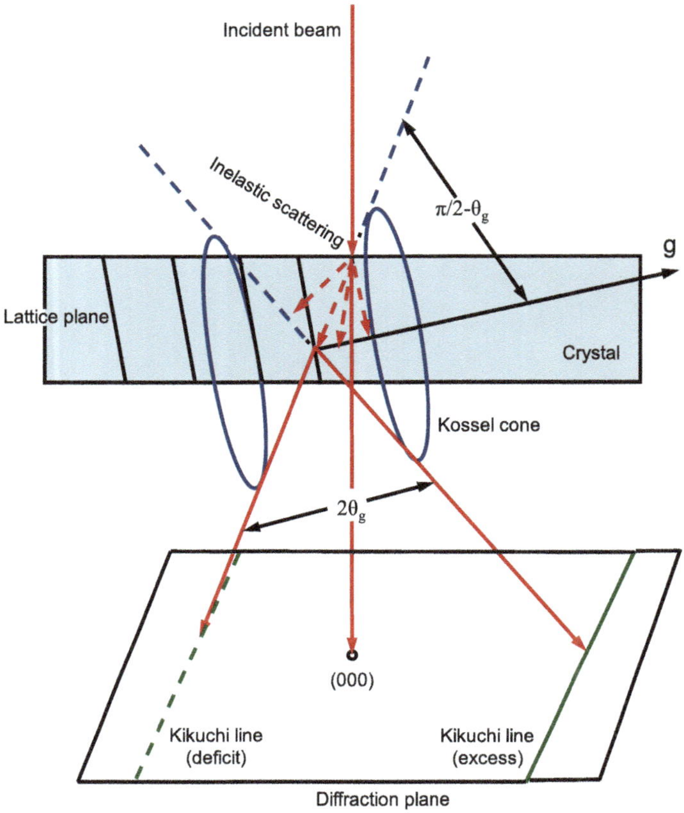

Fig. 6.3 Formation of the Kikuchi pattern in electron diffraction due to inelastic scattering (see text)

violet light. The cathodoluminescence (C.L.) technique is based on the detection of visible light, which is emitted when an electron in a higher energy state (usually at an impurity) fills a hole in a lower state that was created by the fast electron. Second thermal diffuse scattering or phonon scattering is the result of atomic vibrations in crystals. This process does not introduce significant energy loss, but it produces large momentum transfer, which can scatter the incident electron out of the selecting angular range of the objective aperture. This is a localized inelastic-scattering process. The electrons scattered to high angles are primarily due to TDS. Finally atomic inner shell ionization is excited by the energy transfer of incident electrons, resulting in an ejected electron from the deep-core states (Fig. 6.5a). Since only one inner shell electron is involved in the excitation, this process is also called single-electron excitation. Excitation introduces an energy loss in the range of a few tens to thousands eV with an angular spreading of the order $\vartheta_E = \Delta E/2E_0$, where ΔE is the electron energy loss and E_0 is the incident electron energy. Analogous to C.L., holes created at deep-core states tend to be filled by core-shell electrons from higher energy levels;

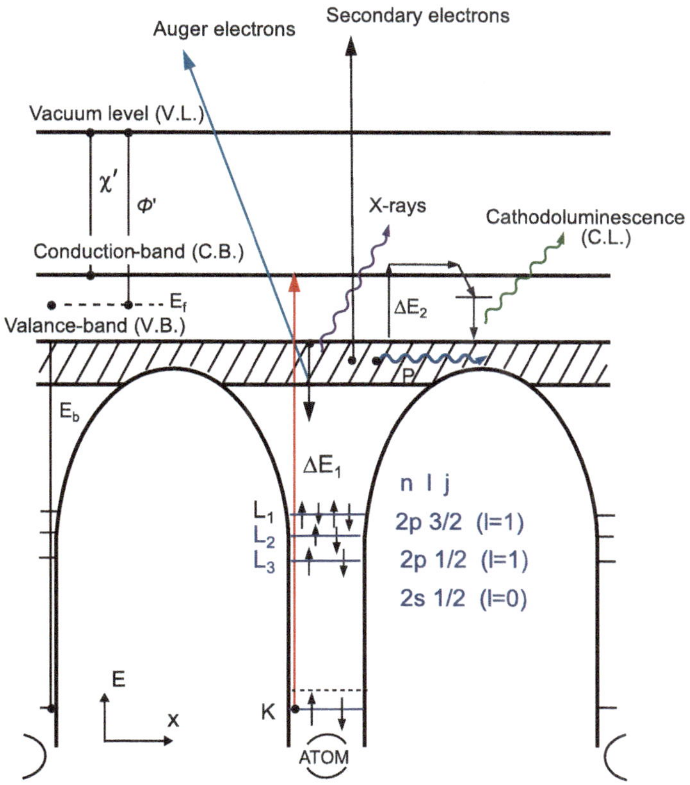

Fig. 6.4 One-electron energy level diagram plotted against positions of atoms showing important excitations by an incident electron in a semiconductor material. Here ε_f is the Fermi level, E_b the binding energy. The ΔE_1 a K shell excitation; ΔE_2 a single-electron excitation; P, a plasmon

the energy is released in the form of photons (or X-rays) (Fig. 6.5b). The X-rays emitted are fingerprints of elements and thus are used for chemical microanalysis. The holes, created by the ionization process, in deep-core states may alternatively be filled by electrons from outer shells; energy released in this process may ionize another outer shell electron (Fig. 6.5c), resulting in the emission of Auger electrons. Accompanying these processes, second electrons can be emitted from the valence band. The difference between Fig. 6.5b, c is the emission of a photon in the former and a secondary electron in the latter.

Inelastic energy losses of fast electrons have been applied in TEM to characterize local chemical and electronic structures (see Egerton 1986, Colliex 1984 and Spence 1988b, for reviews). Signals of these inelastic excitations can be collected by an electron energy loss spectrometer (EELS). Solid-state effects are contained in the energy loss near-edge structures (ELNES) and extended energy loss fine structure (EXELFS) (see Teo and Joy 1981; Colliex et al. 1985, for reviews). The equivalent

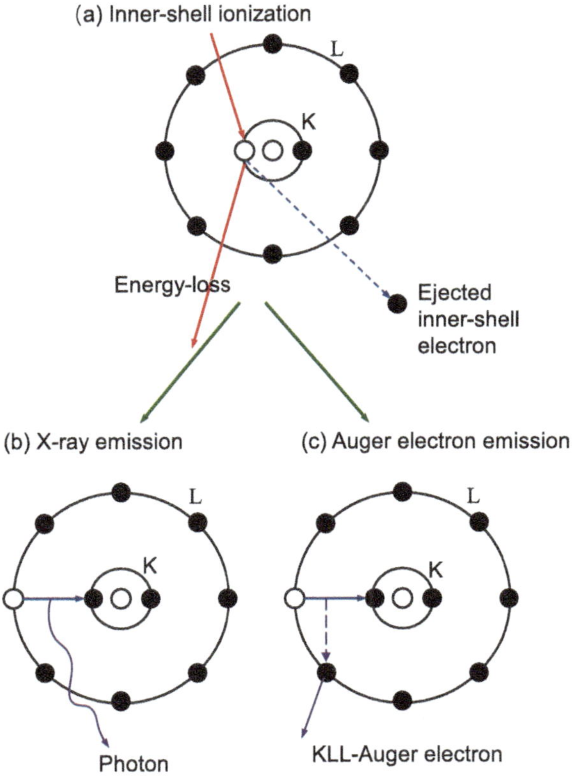

Fig. 6.5 Inelastic interactions of an incident electron with atom-bounded electrons, resulting in **a** inner shell ionization, **b** X-ray emission, and **c** Auger electron emission

analyses using X-ray absorption edges have been described by Stern (1982). Crystal orientation effects in EELS have been observed by Taftø and Krivanek (1982).

6.1.3 Bremsstrahlung

There are continuous energy loss processes in electron scattering besides quantum excitations introduced in the last section. Continuous energy loss spectra can be generated by an electron that penetrates the specimen and undergoes collisions with atoms in it. The electromagnetic radiation produced is known as Bremsstrahlung (German for braking radiation). The X-rays are usually generated. Bremsstrahlung is a classical topic discussed in textbooks on electromagnetic theory. The angular distribution of the energy radiation rate is the Poynting vector

$$\frac{d^2\Delta E}{d\Omega dt} = r\mathbf{r} \cdot [\mathbf{E} \times \mathbf{H}]$$

where **E** and **H** are the electric and magnetic field vectors, respectively, produced by the moving electron, the radiation solid angle $d\Omega = \sin\Theta' d\Theta' d\phi$, and electron energy loss ΔE. For simplicity we assume $d\mathbf{v}/dt$ is parallel to $\mathbf{v}$ for transmission electrons (Konopinski 1981); it yields

$$\frac{d^2\Delta E}{d\Omega dt} \approx \frac{e^2}{16\pi^2\varepsilon_0 c_0^3}\left(\frac{d\mathbf{v}}{dt}\right)^2 \frac{\sin^2\Theta'}{[1-(v/c_0)\cos\Theta']^5}$$

The total radiation power is

$$\frac{d\Delta E}{dt} \approx \frac{e^2\gamma^6}{6\pi\varepsilon_0 c_0^3}\left(\frac{d\mathbf{v}}{dt}\right)^2$$

When penetrating the specimen, the force that acts on the electron is $\mathbf{F} = e\nabla V$, where V is the crystal potential. Thus, an incident electron is accelerated and decelerated when approaching and leaving an atom due to electrostatic attraction and repulsion, respectively. Acceleration of the electron is $d\mathbf{v}/dt = \mathbf{F}/m_e = e\nabla\ V/m_e$. Thus the angular distribution of the radiated energy is proportional to $|d\mathbf{v}/dt|^2 \sin\Theta' = e^2/m_e^2\ |dV/\ dz|^2 \sin^2\Theta'$ under nonrelativistic approximation (Konopinski 1981). Therefore, Bremsstrahlung leads to continuous energy loss that increases with an increasing scattering angle. Radiation energy is emitted in the form of X-rays. Thus, Bremsstrahlung is responsible for the background observed in energy-dispersive X-ray spectroscopy (EDS). Total energy radiated by the electron is

$$\Delta E \approx \frac{e^2\gamma^6}{6\pi\varepsilon_0 c_0^3}\int_0^{dv} dt\left(\frac{d\mathbf{v}}{dt}\right)^2 \approx \frac{e^4\gamma^6}{6\pi m_e^2\varepsilon_0 c_0^3 v}\int_0^{d} dz\left(\frac{dV}{dz}\right)^2$$

and it is proportional to the sample thickness d. The distribution of ΔE as a function of radiation frequency ω gives rise to part of the background observed in EELS. A detailed analysis of the power spectrum of ΔE can be found in many textbooks (see, for example Konopinski 1981).

6.1.4 Electron Compton Scattering

The other continuous energy loss process is the energy transfer due to the collision of the incident electron with a crystal electron. This process is similar to the inelastic two-ball collision in classical physics, and it is known as electron Compton scattering or e-e scattering. Energy and momentum conservation are required in this process. Since large momentum transfer is involved in electron Compton scattering, this process may be responsible in part for the diffuse background observed in electron diffraction patterns of light elements. Electron Compton scattering leads to large losses (a broad peak from a few tens eV to several hundred eV) with a peak that moves to high-energy losses for an increasing scattering angle (Williams et al. 1984). The energy spread of the Compton peak results from Doppler broadening due to the orbital velocity or momentum distribution of atomic electrons.

In the electron energy loss spectrum, the signal background is primarily contributed to by electromagnetic radiation, e-e scattering, and multiple valence excitations. Which of these is the dominant factor is determined by the electronic structure of the material and the EELS acquisition geometry. As shown in the last section, the angular distribution of Bremsstrahlung is determined by $\sin^2 \Theta'$. Therefore, electrons that have suffered Bremsstrahlung are distributed at high angles; thus, the Bremsstrahlung effect is best seen if the EELS spectra are acquired from large-angle diffusely scattered electrons off Bragg beams. The same character is retained for electron Compton scattering.

6.2 Yoshioka's Equations for Inelastically Scattered Electrons

The original theory for describing the scattering of high-energy inelastic electrons in crystals was due to Kainuma (1955) and Yoshioka (1957). Based on quantum mechanical theory, a set of coupled Schrödinger-like equations, considering all transitions among the ground state and excited states of the crystal, have been derived. Electron multiple elastic and multiple inelastic scattering are governed by these equations. The essential part of inelastic-scattering theory is to solve these coupled equations for various scattering processes. We first introduce Yoshioka's equations, then apply them to investigate absorption effects of inelastic scattering in quantitative electron diffraction and imaging.

6.2.1 Basic Equations

To approach the inelastic scattering, we follow Yoshioka's method for deriving coupled equations. If the incident electron and the crystal are considered as an isolated system, so there is no energy exchange with the environment (Fig. 6.6), the Schrödinger equation of the system is

$$\begin{aligned}\left[-\frac{\hbar^2}{2m_0}\nabla^2 + H_c(\mathbf{r}_1, \ldots, \mathbf{r}_M) + \gamma \mathrm{H}(\mathbf{r}, \mathbf{r}_1, \ldots, \mathbf{r}_M)\right] \Phi_s(\mathbf{r}, \mathbf{r}_1, \ldots, \mathbf{r}_M) \\ = E\Phi_s(\mathbf{r}, \mathbf{r}_1, \ldots, \mathbf{r}_M)\end{aligned} \tag{6.1}$$

where $-\hbar^2/2m_0\nabla^2$ is the kinetic energy of the electron, H_c is the crystal Hamiltonian, and H describes the interaction between the incident electron and the crystal.

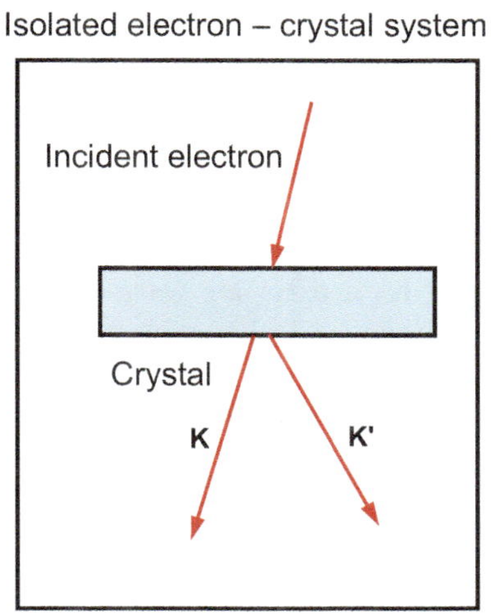

Fig. 6.6 A theoretical model of the inelastic interaction of an incident electron with the crystal

The $\Phi_s(\mathbf{r}, \mathbf{r}_1, \ldots, \mathbf{r}_M)$ is the wave function of the system, which depends on $\mathbf{r}$ the coordinates of the incident electron and on $\mathbf{r}_1, \ldots, r_M$ the coordinates of the electrons and ions of the crystal. The total energy of this isolated system is constant. For the case in which the system is not isolated from the environment, the thermal fluctuation effect has to be considered (see Chap. 17).

The eigen state of the system is expressed as a product of the eigen state of the crystal with that of the incident electron

$$\Phi_s(\mathbf{r}, \mathbf{r}_1, \ldots, \mathbf{r}_M) = \sum_n a_n(\mathbf{r}_1, \ldots, \mathbf{r}_M)\, \Psi_n(\mathbf{r}) \tag{6.2}$$

where a_n is the nth eigen state of the crystal of energy ε_n, and it is determined by

$$H_c a_n(\mathbf{r}_1, \ldots, \mathbf{r}_M) = \varepsilon_n a_n(\mathbf{r}_1, \ldots, \mathbf{r}_M) \tag{6.3}$$

The Ψ_0 in (6.2) describes the elastic scattered wave of energy E, and Ψ_n describes the inelastically scattered wave of energy $E_n = E - \varepsilon_n$ after exciting the nth crystal state, with $n = 1, 2, \ldots, m \ldots$ Substituting (6.2) and (6.3) into (6.1), multiplying both sides by a_n^*, and integrating over the coordinates $\mathbf{r}_1, \ldots, \mathbf{r}_M$ of crystal electrons, we obtain

$$\left(\nabla^2 + 4\pi^2 K_0^2\right) \Psi_0 = \frac{2m_e}{\hbar^2} \sum_m H_{0m}(\mathbf{r}) \Psi_m \tag{6.4a}$$

and

$$\left(\nabla^2 + 4\pi^2 K_n^2\right) \Psi_n = \frac{2m_e}{\hbar^2} \sum_m H_{nm}(\mathbf{r}) \Psi_m \tag{6.4b}$$

where

$$K_n^2 = \frac{2m_0}{h^2} E_n \tag{6.5a}$$

$$H_{nm} = \int a_n^* H a_m d\mathbf{r}_1, \ldots, d\mathbf{r}_M \tag{6.5b}$$

and $\mathbf{K}_0$ and $\mathbf{K}_n$ are wave vectors of elastically scattered electrons and nth state inelastically scattered electrons, respectively. These are Yoshioka's coupled equations for inelastic scattering. The summation of m is for all excited states, including the ground state of $m = 0$. The interaction Hamiltonian H is a real function, so that $H_{nm}^* = H_{mn}$. Diagonal elements of the H matrix are related to the crystal potential, with $H_{nn} \approx -eV$. Thus (6.5b) can be rewritten as

$$\left[\nabla^2 + 4\pi^2 \left(K_n^2 + U\right)\right] \Psi_n = \frac{2m_e}{\hbar^2} \sum_{m \neq n} H_{nm} \Psi_m \tag{6.6}$$

where U is the effective crystal potential. The H_{nm} are excitation matrix elements responsible for exciting the crystal from the m th to the nth state. The term on the right-hand side of (6.6) are sources for generating n th-state inelastically scattered electrons. The U is responsible for the elastic diffraction of the electron before and after inelastic excitation, resulting in the formation of Kikuchi patterns.

Using the Bloch theorem (Howie 1963), the crystal wave functions satisfies

$$a_n\left(\mathbf{r}_1 + \mathbf{R}_m, \mathbf{r}_2 + \mathbf{R}_m \ldots, \mathbf{r}_M + \mathbf{R}_m\right) = \exp\left(2\pi i \mathbf{q}_n \cdot \mathbf{R}_m\right) a_n\left(\mathbf{r}_1, \mathbf{r}_2, \ldots, \mathbf{r}_M\right) \tag{6.7}$$

where $\mathbf{q}_n$ is defined as the wave vector of the nth crystal state, and it is restricted within the first Brillouin zone (BZ). Equation (6.7) holds for any lattice vector $\mathbf{R}_m$; it can thus be easily shown from (6.5b) that when H is a function of interelectron distances $\mathbf{r} - \mathbf{r}_1, \mathbf{r} - \mathbf{r}_2, \ldots, \mathbf{r} - \mathbf{r}_\mathrm{M}$, only

$$H_{nm}(\mathbf{r}) = \exp\left(-2\pi i \mathbf{q}_{nm} \cdot \mathbf{r}\right) \sum_g H_g^{nm}\left(\mathbf{q}_{nm}\right) \exp(2\pi i \mathbf{g} \cdot \mathbf{r}) \tag{6.8a}$$

and the Fourier coefficients of $H_{nm}(\mathbf{r})$ are

$$H_g^{nm}\left(\mathbf{q}_{nm}\right) = \int d\mathbf{r} \exp\left[-2\pi i\left(\mathbf{g} - \mathbf{q}_{nm}\right) \cdot \mathbf{r}\right] H_{nm}(\mathbf{r}) \tag{6.8b}$$

The $H_{nm}(\mathbf{r})$ is thus characterized by a set of Fourier coefficients H_g^{nm} for the various reciprocal lattice vectors $\mathbf{g}$ and by a vector $\mathbf{q}_{nm} = \mathbf{q}_n - \mathbf{q}_m$ corresponding to the change in the wave vector of the electrons of the crystal due to inelastic transition, i.e., to the wave vector of some crystal excitation, such as a phonon, which is created in the transition $m \to n$. The determination of H_g^{nm} for each different inelastic-scattering process is given in the following sections.

6.2.2 *Incoherence and Coherence of Inelastically Scattered Electrons*

The intensity at the crystal exit face is proportional to the probability density distribution of the electron. Due to the fact that the system consists of two interacting subsystems, the crystal and the incident electron, the probability of the crystal being excited from a_m to a_n and the probability of the electron being scattered from $\Psi_m(\mathbf{r})$ to $\Psi_n(\mathbf{r})$ are coupled. This coupling relationship may be understood as the result of energy and momentum transfers. Therefore, the density distribution of the fast electron at the crystal exit face may be written (Fanidis et al. 1992) as

$$I_e(\mathbf{r}) = \int d\mathbf{r}_1 \ldots d\mathbf{r}_M \left|\Phi_s(\mathbf{r}, \mathbf{r}_1, \ldots, \mathbf{r}_M)\right|^2 \tag{6.9}$$

where integrations are over the coordinates of all crystal electrons. Using (6.2),

$$\begin{aligned} I_e(\mathbf{r}) &= \sum_n \sum_m \Psi_n^*(\mathbf{r})\Psi_m(\mathbf{r}) \int d\mathbf{r}_1 \ldots d\mathbf{r}_M a_n^*(\mathbf{r}_1, \ldots, \mathbf{r}_M)\, a_m(\mathbf{r}_1, \ldots, \mathbf{r}_M) \\ &= \sum_n \left|\Psi_n(\mathbf{r})\right|^2 \end{aligned} \tag{6.10}$$

where the orthonormal relation of the crystal eigen states was used. Similarly the diffraction pattern of the electrons is

$$I_e(\mathbf{u}) = \sum_n \left|\Psi_n(\mathbf{u})\right|^2 \tag{6.11}$$

Equations (6.10) and (6.11) have a clear physical meaning: For an isolated incident electron-crystal system, the intensity in the image or diffraction pattern is an incoherence superposition of electrons corresponding to different excited states. This is an extremely important principle in dealing with inelastically scattered electrons. Since there are many crystal excited states and the intensity of each state has to be calculated while considering its coupling (or interaction) with other states, the theoretical treatment of inelastic scattering becomes much more complex than elastic-scattering theory.

More importantly, electrons inelastically scattered by the same final state $|n\rangle$, characterized by energy loss ε_n and momentum transfer $\mathbf{q}_n$, are coherent (Fujimoto and Kainuma 1963). Elastically rescattered electrons have the same characteristics as elastic electrons, so that their interference forms lattice images (Shuman et al. 1986; Hashimoto et al. 1992). A detailed discussion regarding HRTEM images of inelastic electrons is given in Chap. 13.

6.2.3 Conservation of Intensity

We now start from (6.4b) to prove the conservation law of electron intensity (Kambe and Moliere 1970). First we use (6.4b) to calculate [Ψ_n^* (6.4b)−Ψ_n (6.4b)*], where (6.4b)* is the complex conjugate of (6.4b), which gives

$$\Psi_n^* \nabla^2 \Psi_n - \Psi_n \nabla^2 \Psi_n^* = \frac{2m_e}{\hbar^2} \left[\Psi_n^* \sum_m H_{nm} \Psi_m - \Psi_n \sum_m H_{nm}^* \Psi_m^* \right] \tag{6.12}$$

Summing over n in (6.12), using $H_{nm}^* = H_{mn}$ and integrating over $\mathbf{r}$, a simple relation is obtained

$$\sum_n \int_\Sigma d\mathbf{S} \cdot \left[\Psi_n^* \nabla \Psi_n - \Psi_n \nabla \Psi_n^* \right] = 0 \tag{6.13}$$

where the surface integral is on the surface Σ of the crystal. Using the standard definition of electron current density vector in quantum mechanics,

$$\mathbf{j}_n = \frac{\hbar}{2im_e} \left(\Psi_n^* \nabla \Psi_n - \Psi_n \nabla \Psi_n^* \right) \tag{6.14}$$

Equation (6.13) is readily written as

$$\sum_{n \neq 0} \int_\Sigma d\mathbf{S} \cdot \mathbf{j}_n = - \int_\Sigma d\mathbf{S} \cdot \mathbf{j}_0 \tag{6.15}$$

Equation (6.15) is nothing but the law of conservation of total electrons. The rate at which elastic electrons decrease equals the rate at which the total of inelastic electrons increases.

6.2.4 Absorption Phenomenon

In conventional diffraction contrast imaging, images are formed by selecting a single Bragg-reflected beam using a small objective aperture, so that electrons inelastically scattered out of the aperture angular range do not contribute to the image, and these are thus considered to have been absorbed. From the preceding discussion, therefore, electrons that contribute to inelastic scattering are effectively absorbed by the crystal as long as the elastic-scattering component Ψ_0 is concerned. This is why inelastic scattering is normally considered an absorption effect in images of Bragg-reflected electrons.

It is important to point out that the absorption effect used in electron diffraction is only a phenomenological treatment of electron loss due to inelastic scattering.

In practice, however, inelastically scattered electrons are not lost but distributed at angles not restricted by Bragg's law. Chapter 6 discusses the quantitative calculation of absorption functions for various inelastic-scattering processes. Diffraction of absorbed electrons (i.e., inelastically scattered electrons) is the topic of the next few chapters.

6.3 Effects of Inelastic Excitations on an Elastic Wave

In Part I, the absorption effect of electron scattering is usually characterized by an imaginary potential. Here we use Yoshioka's theory to derive the form of this potential. For simplification, we use the single inelastic-scattering approximation, in which the electron is scattered inelastically only once. Thus (6.6) is approximated (for $n \neq 0$) as

$$\left[\nabla^2 + 4\pi^2\left(K_n^2 + U\right)\right]\Psi_n \approx \frac{2m_e}{\hbar^2}H_{n0}\Psi_0 \tag{6.16}$$

If the channeling effect is ignored, the probability of creating inelastically scattered electrons is entirely determined by H_{n0}. Roughly speaking, elastic diffraction of the inelastically scattered electrons, either before or after inelastic excitation, does not change the total number but the angular distribution of the inelastically scattered electrons. Hence as far as the absorption function (or the total number of inelastic electrons) is concerned, the elastic diffraction effect of the inelastically scattered electrons can be ignored. Thus based on the first-order approximation, (6.16) is approximated as

$$\left(\nabla^2 + 4\pi^2 K_n^2\right)\Psi_n \approx \frac{2m_e}{\hbar^2}H_{n0}\Psi_0 \tag{6.17}$$

Equation (6.17) means that inelastic electrons are generated when incident electrons interact with the crystal, but inelastic electrons are not diffracted by the crystal. Green's function solution of this equation is (see Appendix D)

$$\Psi_n(\mathbf{r}) = -\frac{m_e}{2\pi\hbar^2}\int d\mathbf{r}'\frac{\exp\left(2\pi i K_n\left|\mathbf{r}-\mathbf{r}'\right|\right)}{|\mathbf{r}-\mathbf{r}'|}H_{n0}\left(\mathbf{r}'\right)\Psi_0\left(\mathbf{r}'\right) \tag{6.18}$$

Now substituting the function $\Psi_n(\mathbf{r})$ given by (6.18) into (6.6) for the elastically scattered wave ($n = 0$), we obtain

$$\left[\nabla^2 + 4\pi^2\left(K_0^2 + U\right)\right]\Psi_0 - \frac{2m_e}{\hbar^2}\int d\mathbf{r}' A\left(\mathbf{r},\mathbf{r}'\right)\Psi_0\left(\mathbf{r}'\right) = 0 \tag{6.19}$$

where

$$A\left(\mathbf{r},\mathbf{r}'\right) = -\frac{m_e}{2\pi\hbar^2}\sum_{m\neq 0} H_{0m}(\mathbf{r})H_{m0}\left(\mathbf{r}'\right)\frac{\exp\left(2\pi i K_m\left|\mathbf{r}-\mathbf{r}'\right|\right)}{\left|\mathbf{r}-\mathbf{r}'\right|} \tag{6.20}$$

Equations (6.19) and (6.20) were derived from the first-order approximation. Higher order terms may be obtained iteratively (Ohtsuki and Yanagawa 1966; Dederichs 1966). It can easily be proven that the periodic condition $A\left(\mathbf{r}+\mathbf{R}_m, \mathbf{r}'+\mathbf{R}_m\right) = A\left(\mathbf{r},\mathbf{r}'\right)$ is satisfied. The function $A\left(\mathbf{r},\mathbf{r}'\right)$ characterizes the effect of inelastic scattering on elastic electrons, and it is equivalent to introducing a nonlocal optical potential function. We now proceed to the more detailed derivation of (6.20).

6.3.1 Mixed Dynamic Form Factor

For the convenience of discussion in the next few sections, we first introduce the concept of the electron density operator. In general, for $n \neq m$, the interaction of an incident electron with a crystal can be expressed as

$$\begin{aligned} H_{nm}(\mathbf{r}) &= \langle n|\frac{1}{4\pi\varepsilon_0}\left\{\sum_J \frac{e^2}{|\mathbf{r}-\mathbf{r}_J|} - \sum_k \frac{Z_k e^2}{|\mathbf{r}-\mathbf{R}_k|}\right\}|m\rangle \\ &= \frac{e^2}{4\pi^2\varepsilon_0}\int \frac{d\tau}{\tau^2}\exp(2\pi i\tau\cdot\mathbf{r})\rho_{nm}(\tau) \end{aligned} \tag{6.21}$$

where $\mathbf{r}_J$ and $\mathbf{R}_k$ are positions of the Jth crystal electron and the kth nucleus, respectively, τ and τ' are reciprocal-space vectors, and

$$\rho_{nm}(\tau) \equiv < a_n \mid \sum_J \exp\left(-2\pi i\tau\cdot\mathbf{r}_J\right) \mid a_m > \tag{6.22}$$

The $\rho_{nm}(\tau)$ is the Fourier-transformed electron density operator in the Heisenberg representation, and it is a complex function associated directly with charge distribution in the crystal. The inverse Fourier transform of (6.21) gives

$$\begin{aligned} \rho_{nm}(\tau) &= \frac{4\pi^2\varepsilon_0}{e^2}\tau^2\int d\mathbf{r}H_{nm}(\mathbf{r})\exp(-2\pi i\tau\cdot\mathbf{r}) \\ &= \frac{4\pi^2\varepsilon_0}{e^2}\sum_g H_g^{nm}\left(\mathbf{q}_{nm}\right)\tau^2\delta\left(\tau+\mathbf{q}_{nm}-\mathbf{g}\right) \end{aligned} \tag{6.23}$$

where (6.8) was used. The mixed dynamic form factor is defined as

$$
\begin{aligned}
S\left(\tau, \tau'\right) &\equiv \sum_{n\neq 0} \rho_{n0}(\tau)\rho_{n0}\left(-\tau'\right) \\
&= \frac{16\pi^4\varepsilon_0^2}{e^4}\tau^2\tau'^2 \int d\mathbf{r} \int d\mathbf{r}' \left[\sum_{n\neq 0} H_{0n}(\mathbf{r})H_{n0}\left(\mathbf{r}'\right)\right] \exp\left(-2\pi i\tau\cdot\mathbf{r} + 2\pi i\tau'\cdot\mathbf{r}'\right) \\
&= \frac{16\pi^4\varepsilon_0^2}{e^4}\sum_g\sum_h \tau^2\tau'^2 \left[\sum_{n\neq 0} H_g^{0n}\left(\mathbf{q}_{0n}\right) H_h^{n0}\left(\mathbf{q}_{n0}\right)\right] \\
&\quad \times \delta\left(\tau + \mathbf{q}_{0n} - \mathbf{g}\right)\delta\left(\tau' - \mathbf{q}_{n0} + \mathbf{h}\right)
\end{aligned}
\tag{6.24}
$$

The mixed dynamic form factor is a key quantity for characterizing an inelastic-scattering process. The role played by $S\left(\tau, \tau'\right)$ in inelastic scattering is equivalent to that taken by the crystal potential V in elastic scattering. The $S\left(\tau, \tau'\right)$ and $V(\mathbf{r})$ are the only two structurally related quantities that determine the behavior of electron scattering in crystals. The calculation of the $S\left(\tau, \tau'\right)$ factor for each inelastic-scattering process is given in the following sections.

6.3.2 Absorption Potential-Reciprocal-Space Description

To introduce the form of the imaginary component of the crystal potential, we solve (6.19) using the Bloch wave method, as described in Chap. 2. Thus

$$
\left[K_0^2 - (\mathbf{k}+\mathbf{g})^2\right]C_g + \sum_h \left(U_{g-h} + U'_{gh}\right)C_h = 0 \tag{6.25}
$$

where

$$
U'_{gh} = -\frac{2m_e}{4\pi^2\hbar^2 V_c}\int d\mathbf{r}\int d\mathbf{r}' A\left(\mathbf{r}, \mathbf{r}'\right)\exp\left(-2\pi i\mathbf{k}_g\cdot\mathbf{r} + 2\pi i\mathbf{k}_h\cdot\mathbf{r}'\right) \tag{6.26}
$$

$\mathbf{k}_g = \mathbf{K}_0 + \mathbf{g} + v\hat{z}$, and V_c is the volume of crystal. The U'_{gh} is a $N \times N$ complex matrix. Nondiagonal elements of U'_{gh} are nonzero because the absorption potential is not a local function (see (6.20)), depending strongly on the history of electron inelastic scattering. In other words, U' cannot simply be written as a function of one variable $\mathbf{r}$. Equation (2.9) is the same as (6.25) if the potential matrix element U_{g-h} is replaced by $U_{g-h} + U'_{gh}$. We now use Green's function introduced in (4.16)–(4.18a) and (4.18b) to derive the expression for the imaginary potential in reciprocal space

$$
-\frac{\exp\left[2\pi i K_m\left|\mathbf{r}-\mathbf{r}'\right|\right]}{4\pi\left|\mathbf{r}-\mathbf{r}'\right|} = \int d\mathbf{u}\frac{\exp\left[2\pi i\mathbf{u}\cdot\left(\mathbf{r}-\mathbf{r}'\right)\right]}{4\pi^2\left(K_m^2 - u^2\right)} \tag{6.27}
$$

Substituting (6.20) and (6.27) into (6.26) and expressing the correction potential as a complex function (Yoshioka 1957)

$$U'_{gh} = U'^{(R)}_{gh} + iU'^{(I)}_{gh} \tag{6.28}$$

the real part is

$$U'^{(R)}_{gh} = -\frac{m_e^2}{4\pi^4\hbar^4 V_c}\sum_{m\neq 0}\int d\tau(\mathbf{u})\frac{E'_{0m}\left(\mathbf{k}_g - \mathbf{u}\right)E'_{m0}\left(\mathbf{u}-\mathbf{k}_h\right)}{\left(\mathrm{K}_m^2 - u^2\right)} \tag{6.29}$$

where the integration is over all the **u** space except a spherical shell defined by $|\mathbf{u}| = K_m$ (i.e., the integral takes the principal value with respect to the singularity at $u = \mathrm{K_m}$); the imaginary part is

$$U'^{(I)}_{gh} = -\frac{m_e^2}{4\pi^3\hbar^4 V_c}\sum_{m\neq 0}\frac{1}{K_m}\int d\sigma(\mathbf{u})E'_{0m}\left(\mathbf{k}_g - \mathbf{u}\right)E'_{m0}\left(\mathbf{u}-\mathbf{k}_h\right) \tag{6.30}$$

with

$$E'_{0m}(\mathbf{u}) = \int d\mathbf{r}H_{0m}(\mathbf{r})\exp(-2\pi i\mathbf{u}\cdot\mathbf{r}) \tag{6.31}$$

the integral $\sigma(\mathbf{u})$ is over the surface of $|\mathbf{u}| = K_m$. Thus U'_{gh} is not a Hermitian matrix. The imaginary part $U'^{(I)}_{gh}$ is responsible for absorption. The real part $U'^{(R)}_{gh}$ is the result of virtual inelastic scattering (see Sect. 6.3.6). The $U'^{(R)}_{gh}$ is usually very small (Gjønnes 1962a; Rez 1976), so that it is negligible in comparison with the crystal potential. The most important effect is the imaginary component $U'^{(I)}_{gh}$.

Since (6.31) can be further simplified by using (6.8),

$$E'_{0m}(\mathbf{u}) = \sum_{g'} H_{g'}^{0m}\left(\mathbf{q}_{0m}\right)\delta\left(\mathbf{u} + \mathbf{g}' - \mathbf{q}_{0m}\right) \tag{6.32}$$

the imaginary potential is rewritten as

$$U'^{(I)}_{gh} = -\frac{m_e^2}{4\pi^3\hbar^4 K_0 V_c}\sum_{g'}\sum_{m\neq 0}\left[H_{g'}^{0m}\left(\mathbf{q}_{0m}\right)H_{h-g-g'}^{m0}\left(\mathbf{q}_{m0}\right)\right]\int d\sigma(\mathbf{u})\delta\left(\mathbf{u}-\mathbf{k}_g-\mathbf{g}'+\mathbf{q}_{0m}\right) \tag{6.33}$$

where integrating $\int d\sigma(\mathbf{u})\delta\left(\mathbf{u}-\mathbf{k}_g-\mathbf{g}'+\mathbf{q}_{0m}\right)$ represents summation over the reflections $\mathbf{g}'$ falling on the $|\mathbf{u}| = K$ spherical surface and satisfying $\mathbf{u} = \mathbf{k}_g - \mathbf{q}_{0m} + \mathbf{g}'$. Using the dynamic form factor introduced in (6.24) and (6.33) becomes

$$U'^{(I)}_{gh} = -\frac{m_e^2 e^4}{64\pi^7\hbar^4\varepsilon_0^2 K_0 V_c}\int d\sigma(\mathbf{u})\frac{S\left(\mathbf{k}_g - \mathbf{u}, \mathbf{k}_h - \mathbf{u}\right)}{|\,\mathbf{k}_g - \mathbf{u}\,|^2|\,\mathbf{k}_h - \mathbf{u}\,|^2} \tag{6.34}$$

Therefore the calculation of imaginary potential is determined by the mixed dynamic form factor. Finally the matrix formalism of the Bloch wave theory has made treatment of the nonlocal absorption function rather easy and complete.

6.3.3 *Absorption Potential-Real-Space Description*

In multislice calculations, the crystal potential in real space is normally used. We now consider the absorption effect of inelastic scattering for a thin crystal slice (Coene and Van Dyck 1990). We first separate the plane wave component by defining $\Psi_m = \exp\left[2\pi i \mathbf{K}_m \cdot \mathbf{r}\right] \Phi_m$; under the small-angle scattering and single inelastic-scattering approximations, (6.6) becomes

$$4\pi i K_{0z} \frac{\partial \Phi_0}{\partial z} + 4\pi^2 U \Phi_0 \approx \frac{2m_e}{\hbar^2} \sum_{n=1} H_{0n} \Phi_n \exp\left[2\pi i \left(\mathbf{K}_n - \mathbf{K}_0\right) \cdot \mathbf{r}\right] \qquad (6.35)$$

and

$$4\pi i K_{nz} \frac{\partial \Phi_n}{\partial z} + 4\pi^2 U \Phi_0 \approx \frac{2m_e}{\hbar^2} H_{n0} \Phi_0 \exp\left[2\pi i \left(\mathbf{K}_0 - \mathbf{K}_n\right) \cdot \mathbf{r}\right] \qquad (6.36)$$

In light of the discussion leading to (6.17), the crystal potential U can be eliminated if we are interested only in the total number of inelastic electrons generated by H_{n0}. Thus the solution of (6.36) is approximated by

$$\Phi_n \approx \frac{1}{4\pi i \mathbf{K}_{nz}} \frac{2m_e}{\hbar^2} \int_0^z dz\, H_{n0} \Phi_0 \exp\left[2\pi i \left(\mathbf{K}_0 - \mathbf{K}_n\right) \cdot \mathbf{r}\right] \qquad (6.37)$$

Substituting (6.37) into (6.35), for a thin crystal slice, we obtain

$$4\pi i K_0 \frac{\partial \Phi_0}{\partial z} + 4\pi^2 U \Phi_0 + i \frac{m_e^2}{\pi \hbar^4 K_{0z}} \sum_{n=1} H_{0n} \int_0^{\Delta z} dz\, H_{n0} \Phi_0 \approx 0 \qquad (6.38)$$

Hence the imaginary component of the projected potential of the slice is

$$U'(\mathbf{b}) \approx \frac{m_e^2 \Delta z}{8\pi^3 \hbar^4 K_{0z}} \sum_{n=1} \left| \int_0^{\Delta z} dz\, H_{n0}(\mathbf{r}) \right|^2 \qquad (6.39)$$

In comparison to the discussion outlined in Sect. 6.3.2, the imaginary component of the crystal potential U' is approximated as a local function, so that (6.39) can be used only for a qualitative estimation of electron inelastic absorption. This approximation is necessary in the real-space approach, and it is the result of ignoring elastic rescattering of inelastic electrons in (6.37).

6.3.4 Interpretating the Imaginary Potential

The introduction of the imaginary potential into the Schrödinger equation is the basis for describing the attenuation effect (Slater 1937; Heidenreich 1962). To illustrate the physics involved in the mathematical treatment of absorption, we start from the Schrödinger equation, considering the imaginary potential

$$\left[\nabla^2 + 4\pi^2\left(K_0^2 + U + iU'\right)\right]\Psi_0 = 0 \tag{6.40}$$

Following the same method used in (6.12)–(6.15), from (6.40) we obtain

$$\Psi_0^*\nabla^2\Psi_0 - \Psi_0\nabla^2\Psi_0^* = -8\pi^2 iU'\Psi_0^*\Psi_0 \tag{6.41}$$

Using the probability current defined in (6.14) and (6.41) is rewritten as

$$\nabla\cdot\mathbf{j}_0 = -\frac{4\pi^2\hbar}{m_e}U'\left|\Psi_0\right|^2 \tag{6.42}$$

Hence the current density $\mathbf{j}_0$ is not conserved but diminished at a rate proportional to the imaginary potential U' and the density function $|\Psi_0|^2$. This loss corresponds to removing electrons from the energy state E_0 to other states; U' is the absorption sink.

The imaginary potential can be correlated with an absorption factor. For simplicity, we use the phase object approximation (see Sect. 3.1); thus the electron wave function after interacting with the crystal is

$$\Phi = \exp\left[i\sigma\left(V + iV'\right)z\right] = \exp[i\sigma Vz]\exp\left[-\sigma V'z\right] \tag{6.43}$$

where V is the crystal potential and V' is the absorption potential. The electron intensity is thus

$$I = |\Phi|^2 = \exp\left[-2\sigma V'z\right] \tag{6.44}$$

Therefore V' is related to the inelastic mean free path Λ by

$$2\sigma V' = \frac{1}{\Lambda} \tag{6.45}$$

The frequency $\omega^{(i)}$ at which the inelastic collision occurs is thus

$$\omega^{(i)} = \frac{v}{\Lambda} \tag{6.46}$$

Combining (6.45) and (6.46) yields

$$V' = \frac{\hbar}{2e}\omega^{(i)} \tag{6.47}$$

It is obvious that V' is now directly associated with the frequency of inelastic collisions. Further interpretation is straightforward using the following relations. Since $\delta t_i = 1/\omega^{(i)}$ is the time interval between successive collisions, an electron whose energy is E in the time interval Δt_i is subject to an energy uncertainty δE_i, given by the Heisenberg inequality

$$\delta E_i \delta t_i \geq \hbar \tag{6.48}$$

The minimum value of δE_i is

$$\delta E_i = \frac{h}{2eV'\delta t_i} = \hbar\omega^{(i)} = \frac{eV'}{2} \tag{6.49}$$

which shows that the imaginary potential V' is the minimum uncertainty in energy experienced by an electron moving in a medium where the inelastic mean free path is Λ. If the motion is along the z axis with momentum component p_z, the conjugate inequality is

$$\delta p_z \delta z \geq \hbar \tag{6.50}$$

Since $\delta p_z = h\delta K_z$, where δK_z is the uncertainty in the wave vector,

$$\delta K_z = \frac{1}{2\pi\,\delta z} = \frac{1}{2\pi\,\Lambda} = \frac{\sigma V'}{\pi} \tag{6.51}$$

since $\delta z_i = \Lambda$. Therefore, the $\sigma V'$ factor (also called the damping factor) is the minimum uncertainty of the wave number for an electron moving in a medium in which the inelastic mean free path for energy loss acts of excited state i is $\delta z_i = \Lambda$.

6.3.5 *Effect of Inelastic Absorption in Quantitative Electron Microscopy*

To show the importance of inelastic absorption in electron diffraction and imaging, calculations were made for a GaAs [110] crystal 80 Å thick with and without including the TDS absorption effect. For relatively lower image resolution, TDS absorption does not show a significant effect in the simulated amplitude and phase of the electron exit wave function. However differences appear when the image resolu-

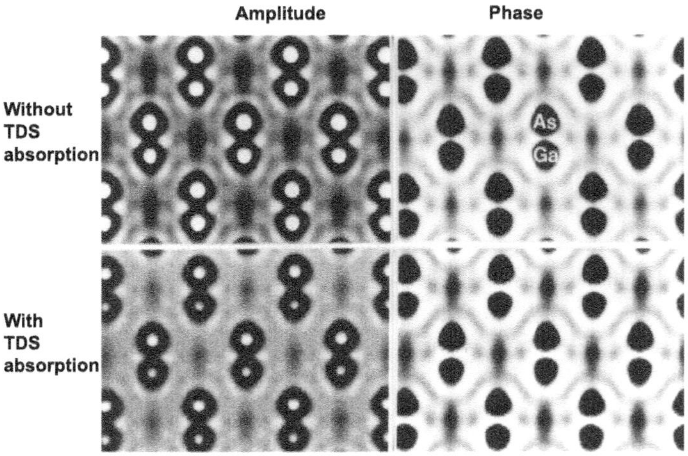

Fig. 6.7 The simulated electron amplitude and phase distributions at the exit face of a GaAs [110] crystal for diffracted beams limited by an objective aperture of radius 1.0 Å^{-1}, corresponding to a resolution limit of 1 Å, with and without including the TDS absorption effect; crystal thickness 80 Å, electron beam energy 200 keV, Cs = 1.2 mm, and $a_{Ga} = a_{As} = 0.063$ Å (Wang 1993b)

tion approaches 1 Å, as shown in Fig. 6.7: The TDS absorption becomes important in the amplitude image especially near the atomic nuclei, but the phase distribution is almost unaffected. As shown in Fig. 6.13, the TDS absorption function is a sharp function centered at each nuclear site, and its spatial variation becomes significant only within a distance shorter than about 1 Å around atomic equilibrium positions. Thus the TDS absorption effect is strong for waves distributed within about 1 Å around the atomic nucleus, and the spatial frequencies of these waves are distributed at an angular range larger than 1 Å^{-1}. Figure 6.7 clearly shows the importance of inelastic absorption in quantitative electron microscopy. It is thus necessary to calculate accurately the absorption potential in quantitative structure determination.

6.3.6 *Virtual Inelastic Scattering*

As shown in Sect. 6.3.4, the imaginary component of the crystal potential is the result of inelastic scattering, and it is responsible for the damping of elastically scattered electron intensity. This is the dominant effect of inelastic scattering on elastic electrons. As noticed in (6.29), a real component is also introduced by inelastic scattering, corresponding to the virtual inelastic-scattering process in inelastic excitation (Yoshioka 1957), as shown in Fig. 6.8.

In inelastic electron scattering, the transition of an electron from ground state 0 (or elastic state) to the nth-excited state is characterized by an excitation matrix element H_{n0}; the transition of the electron from the nth state to the ground state is represented by H_{0n}. In the former, the electron loses energy ε_n, but in the latter

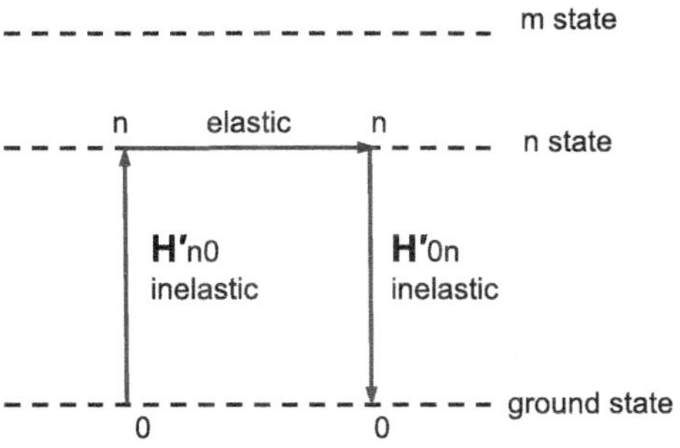

Fig. 6.8 An energy level diagram showing the virtual inelastic-scattering process in electron diffraction

the electron is expected to gain the same amount of energy ε_n. Both processes may have the same probability because $|H_{0n}| = |H_{n0}|$. Theoretically it is possible for an electron to transit first from $0 \to n$ and then return from $n \to 0$, so that the electron virtually has no energy loss at the exit face of the crystal. Thus the electron is still an elastic electron, and it is eventually not absorbed. But the entire process is different from that for a purely elastically scattered electron. Thus a small correction in the real component of the crystal potential is added. In practice, however, since the crystal is not isolated from the environment, so that the probability of the incident electron gaining energy ε_n is effectively equivalent to the entire electron-crystal system gaining energy ε_n from the heat reservoir. The probability of generating this energy fluctuation is determined by $\exp[-\varepsilon_n/k_BT]$, which is very small for core-shell excitations. This simply means that it is almost impossible for an incident electron to gain energy. This is probably why we never observe an incident electron gaining energy in scattering by crystals.

6.4 Inelastic-Scattering Process I: Phonon Excitation

6.4.1 Phonons in Crystals

Atomic vibrations are determined by interactions among atoms inside the crystal. The vibration of each atom consists of many modes having different frequencies and wave vectors. The vibration of a single atom can be represented by a simple harmonic oscillator, which can be precisely solved in quantum mechanics. Since energy levels of the oscillator can be quantized, these can be simply expressed as $\varepsilon_n = \hbar\omega\,(n_s + 1/2)$, where ω is the classical vibration angular frequency of the oscillator and n_s is an integer (or quantum number) that characterizes the n_sth-excited state of the oscillator. Actually n_s is the occupation number of phonons in the n_sth-excited state. The energy level diagram for phonon states is shown in Fig. 6.9. The sequence number of the state represents the number of phonons in the state. If the $n_s = 3$ is excited, for example, it means the simultaneous creation of three phonons

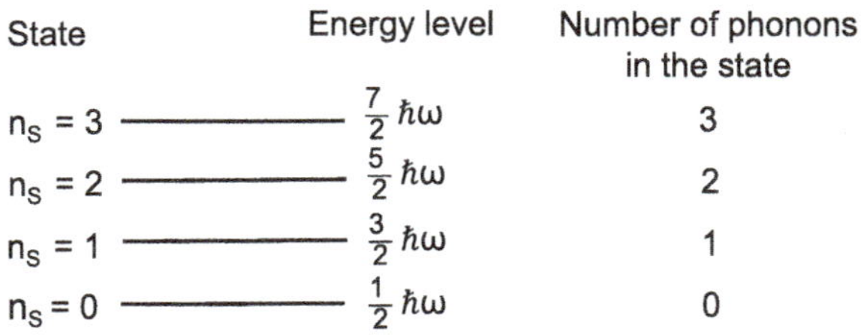

Fig. 6.9 An energy level diagram of a simple harmonic oscillator

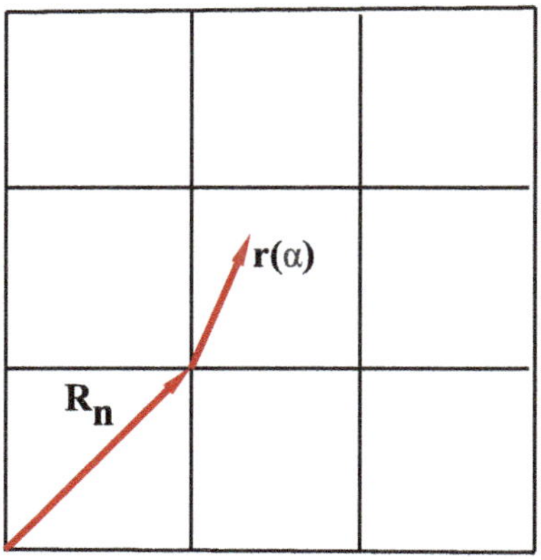

Fig. 6.10 A schematic diagram showing atom locations in a unit cell

in this scattering process, which is usually referred to as a mul-tiphonon-scattering process.

In crystals, three-dimensional interactions of atoms produce many phonon modes, each of which is characterized by a wave vector $\mathbf{q}$, polarization vector $\mathbf{e}$, frequency ω, and dispersion branch j. Based on harmonic oscillators and adiabatic approximations, in which all atoms are assumed to interact with harmonic forces and crystal electrons move as though the ions were fixed in their instantaneous positions, we consider phonon modes existing in a perfect crystal of an infinite number of unit cells. The equilibrium position vector of the nth unit cell relative to the origin is denoted by (see Fig. 6.10)

$$\mathbf{R}_n \equiv n_1\mathbf{a} + n_2\mathbf{b} + n_3\mathbf{c} \tag{6.52}$$

where n_1, n_2, and n_3 are any three integers. The instantaneous position of the α th atom in the nth unit cells is

$$\mathbf{r}\binom{n}{\alpha} \equiv \mathbf{R}_n + \mathbf{r}(\alpha) + \mathbf{u}\binom{n}{\alpha} \tag{6.53}$$

where $\mathbf{r}(\alpha)$ is the equilibrium position of the αth atom relative to the nth cell, and $\mathbf{u}\binom{n}{\alpha}$ is the time-dependent displacement vector of the atom. Lattice dynamics was initiated by Born (1942a, b), and it was introduced in many books (see Brüesch 1982, for example), so here we only outline some main results. The vibration displacement of an atom is governed by

$$M_\alpha \frac{\partial^2 u_\beta\binom{n}{\alpha}}{\partial t^2} = -\sum_{\alpha'}\sum_{n'}\sum_{\beta'} F_{\beta\beta'}\begin{pmatrix} nn' \\ \alpha\alpha' \end{pmatrix} u_\beta \begin{pmatrix} n' \\ \alpha' \end{pmatrix} \tag{6.54}$$

where $\beta', \beta = x, y$, or z; $F_{\beta\beta'}\begin{pmatrix} nn' \\ \alpha\alpha' \end{pmatrix}$ is defined as the harmonic force constant evaluated at the equilibrium atomic positions

$$F_{\beta\beta'}\begin{pmatrix} nn' \\ \alpha\alpha' \end{pmatrix} \equiv \left[\frac{\partial^2 E_p}{\partial u_\beta\binom{n}{\alpha}\partial u_{\beta'}\binom{n'}{\alpha'}}\right]_0$$

The E_p is the interaction potential energy of the crystal, and M_α is the mass of the αth atom. The solution of (6.54) can be written as the sum of harmonic oscillators of momentum $\mathbf{q}$, frequency $\omega_j(\mathbf{q})$, and polarization vector $\mathbf{e}\left(\alpha\middle|{}^{\mathbf{q}}_{j}\right)$

$$\mathbf{u}\begin{pmatrix} n \\ \alpha \end{pmatrix} = \frac{1}{(2N_0M_\alpha)^{1/2}}\sum_q\sum_j \frac{1}{\left[\omega_j(q)^{1/2}\right]}\mathbf{e}\left(\alpha\,\middle|{}^{q}_{j}\right)\exp\{2\pi i\mathbf{q}\cdot[\mathbf{R_n}+\mathbf{r}(\alpha)]\}$$
$$\times\left\{A\begin{pmatrix} q \\ j \end{pmatrix}\exp\left[-i\omega_j(q)t\right] + A^*\begin{pmatrix} -q \\ j \end{pmatrix}\exp\left[i\omega_j(\mathbf{q})t\right]\right\} \tag{6.56a}$$

or in normal coordinates

$$\mathbf{u}\begin{pmatrix} n \\ \alpha \end{pmatrix} = \frac{\hbar^{1/2}}{(2N_0M_\alpha)^{1/2}}\sum_{\mathbf{q}}\sum_j \frac{1}{\left[\omega_j(\mathbf{q})\right]^{1/2}}\mathbf{e}\left(\alpha\,\middle|{}^{\mathbf{q}}_{j}\right)$$
$$\times\exp\{2\pi i\mathbf{q}\cdot[\mathbf{R}_n+\mathbf{r}(\alpha)]\}\left[a^+\begin{pmatrix} -\mathbf{q} \\ j \end{pmatrix} + a\begin{pmatrix} \mathbf{q} \\ j \end{pmatrix}\right] \tag{6.56b}$$

where N_0 is the total number of primitive unit cells in the crystal; j indicates phonon branches; $a^+\binom{-\mathbf{q}}{j}$ and $a\binom{\mathbf{q}}{j}$ are defined as the creation and annihilation operators of a phonon with wave vector $\mathbf{q}$ and frequency $\omega_j(\mathbf{q})$, respectively; these have the following properties,

$$a^+\begin{pmatrix} -\mathbf{q} \\ j \end{pmatrix}\left|n_s\right> = \sqrt{n_s+1}\left|n_s+1\right> \tag{6.57a}$$

and

$$a\begin{pmatrix} \mathbf{q} \\ j \end{pmatrix}\left|n_s\right> = \sqrt{n_s}\left|n_s-1\right> \tag{6.57b}$$

If n_s is the occupation number of phonons with wave vector $\mathbf{q}$ and frequency ω_j in the phonon state $|n_s\rangle$. Therefore a phonon is characterized by ($\mathbf{q}$, ω_j, $\mathbf{e}$). The equation determining the phonon polarization vector is

$$\sum_{\alpha'}\sum_{\beta'}\left[D_{\beta\beta'}\begin{pmatrix}\mathbf{q}\\ \alpha\alpha'\end{pmatrix} - \delta_{\beta\beta'}\delta_{\alpha\alpha'}\omega^2(\mathbf{q})\right]e_{\beta'}\left(\alpha'\,|\,q\right) = 0 \tag{6.58}$$

If n_0 is the number of atoms in the primitive unit cell, and **D** is a $3n_0$-dimensional matrix defined by

$$D_{\beta\beta'}\binom{\mathbf{q}}{\alpha\alpha'} \equiv (M_\alpha M_{\alpha'})^{-1/2}\sum_{n'} F_{\beta\beta}\begin{pmatrix}nn'\\ \alpha\alpha'\end{pmatrix}\exp\left\{2\pi i\mathbf{q}\cdot\left[\mathbf{r}\begin{pmatrix}n'\\ \alpha'\end{pmatrix} - \mathbf{r}\begin{pmatrix}n\\ \alpha\end{pmatrix}\right]\right\} \tag{6.59a}$$

for a monatomic crystal,

$$D_{\beta\beta'}(\mathbf{q}) = -\frac{2}{M}\sum_{L\neq 0} F_{\beta\beta'}(L)\sin^2[\pi\mathbf{q}\cdot\mathbf{r}(L)] \tag{6.59b}$$

where an abbreviation $L = \alpha - \alpha'$ [i.e., $\mathbf{r}(L) = \mathbf{r}(\alpha) - \mathbf{r}\left(\alpha'\right)$] was used. This form shows explicitly that elements of the dynamic matrix **D** depend only on the difference $L = \alpha - \alpha'$, and they are therefore independent of α. The condition for the homogeneous system of (6.58) to have a nontrivial solution is that the determinant of the coefficients vanishes

$$\left| D_{\beta\beta'}\begin{pmatrix}\mathbf{q}\\ \alpha\alpha'\end{pmatrix} - \delta_{\beta\beta'}\delta_{\alpha\alpha'}\omega^2(\mathbf{q}) \right| = 0 \tag{6.60}$$

Equation (6.60) is $3n_0$ th degree in $\omega^2(\mathbf{q})$; the $3n_0$ solutions for each value of **q** are denoted by $\omega_j^2(\mathbf{q})$, where $j = 1, 2, \ldots, 3n_0$. The polarization vector determined according to (6.58) is normalized by

$$\sum_\alpha e^*\left(\alpha\big|{}^{\mathbf{q}}_{j}\right)e\left(\alpha\big|{}^{\mathbf{q}}_{j'}\right) = \delta_{jj'} \tag{6.61a}$$

and

$$\sum_j e^*_\beta\left(\alpha\big|{}^{q}_{j}\right)\cdot e_{\beta'}\left(\alpha\big|{}^{q}_{j}\right) = \delta_{\beta\beta'}\delta_{\alpha\alpha'} \tag{6.61b}$$

The lattice dynamics introduced here are used to calculate the dispersion relation of crystal vibrations in Chap. 12.

6.4.2 Perturbation Effect of Atomic Vibrations on Crystal Potential

The crystal potential is perturbed by atomic vibrations. This perturbation potential may not be periodic, so that diffuse scattering distributed between Bragg beams is generated. This is the electron-phonon interaction process, which is responsible for thermal diffuse scattering. The crystal potential is written as the sum of time-independent and time-dependent components; the former are determined by equilibrium positions of the atoms, and the latter by instantaneous displacements of the atoms

$$V(\mathbf{r}, t) = V_0(\mathbf{r}) + \Delta V(\mathbf{r}, t) \tag{6.62a}$$

where the instantaneous crystal potential is

$$V(\mathbf{r}, t) = \sum_n \sum_\alpha \left[V_\alpha \left(\mathbf{r} - \mathbf{R}_n - \mathbf{r}(\alpha) - \mathbf{u}\binom{n}{\alpha} \right) \right] \tag{6.62b}$$

The time average of the crystal potential is

$$\begin{aligned} V_0(\mathbf{r}) &= < \sum_n \sum_\alpha V_\alpha \left(\mathbf{r} - \mathbf{R}_n - \mathbf{r}(\alpha) - \mathbf{u}\binom{n}{\alpha} \right) > \\ &= \sum_n \sum_\alpha V_{0_\alpha} (\mathbf{r} - \mathbf{R}_n - \mathbf{r}(\alpha)) \end{aligned} \tag{6.62c}$$

The time-dependent crystal potential due to atomic vibrations is Takagi (1958a, b)

$$\Delta V(\mathbf{r}, t) = \sum_n \sum_\alpha \left\{ V_\alpha \left(\mathbf{r} - \mathbf{R}_n - \mathbf{r}(\alpha) - \mathbf{u}\binom{n}{\alpha} \right) - V_{0_\alpha} (\mathbf{r} - \mathbf{R}_n - \mathbf{r}(\alpha)) \right\} \tag{6.63}$$

The $V_{0\alpha}(\mathbf{r})$ is the time-averaged atomic potential, and $V_\alpha(\mathbf{r})$ is the instantaneous atomic potential

$$V_\alpha(\mathbf{r}) = \int d\tau f_\alpha^e(\tau) \exp[2\pi i \tau \cdot \mathbf{r}] \tag{6.64a}$$

$$V_0(\mathbf{r}) = < V(\mathbf{r}) > = \sum_n \sum_\alpha V_{0_\alpha} (\mathbf{r} - \mathbf{R}_n - \mathbf{r}(\alpha)) \tag{6.64b}$$

$$V_{0_\alpha}(\mathbf{r}) = \int d\tau f_\alpha^e(\tau) < \exp\left\{2\pi i\tau\cdot\left[\mathbf{r}-\mathbf{u}\binom{n}{\alpha}\right]\right\} > \\ \approx \int d\tau f_\alpha^e(\tau)\exp[-W_\alpha]\exp[2\pi i\tau\cdot\mathbf{r}] \tag{6.65}$$

$$W_\alpha = 2\pi^2 < \left|\tau\cdot\mathbf{u}\binom{n}{\alpha}\right|^2 > \tag{6.66}$$

A general relation of

$$< \exp x > \approx \exp\left(< x^2 > /2\right) \quad < x > = 0 \tag{6.67}$$

was used; $<>$ represents time average. The factor $\exp[-W_\alpha]$ is the DW factor; it is sensitive to variation in specimen temperature. The DW factor is always involved in elastic electron diffraction (see (1.16)), because the crystal potential in elastic scattering is the time-averaged potential. Proper consideration of the DW factor is important in quantitative electron microscopy.

Figure 6.11 shows electron-scattering factors of Si with and without inclusion of the DW factor. Differences are visible only at high scattering angles. In general, atomic vibration displacement is not a periodic function; thus the thermally perturbed potential $\Delta V(\mathbf{r})$ is a nonperiodic, time-dependent function, as shown in Fig. 6.12. The deviation potential ΔV, however, may not be small in comparison to V_0, so that the diffuse scattering produced by atomic thermal vibrations cannot be ignored in quantitative electron microscopy. Figure 6.13a shows the instantaneous potential V, thermal equilibrium potential V_0, and the difference $\Delta V = V - V_0$ of a silicon atom. If the instantaneous position of the atom is its equilibrium lattice site, ΔV is a symmetric sharp function. If the atom is vibrating to an instantaneous lattice position located at 0.01 nm from the equilibrium site, ΔV is no longer symmetric, as shown in Fig. 6.13b. Since all atoms are vibrating with some phase differences,

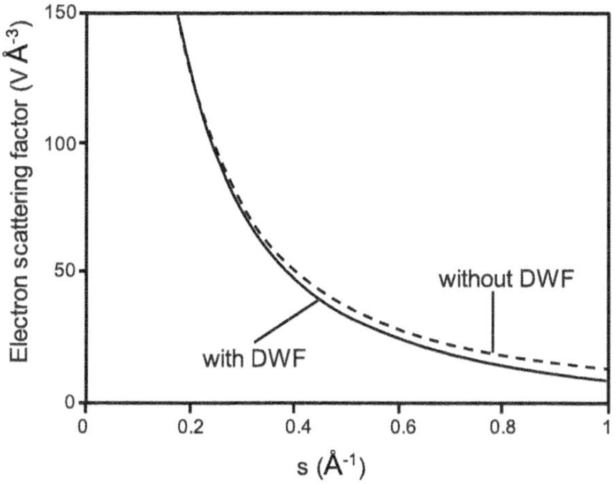

Fig. 6.11 The electron-scattering factor of silicon atom with (solid line) and without (dashed line) inclusion of the DW factor, which takes into account the effect of atomic thermal vibration on its scattering power. The mean vibration amplitude for silicon is taken as 0.007 nm; $s = \sin\theta/\lambda$ is the scattering vector

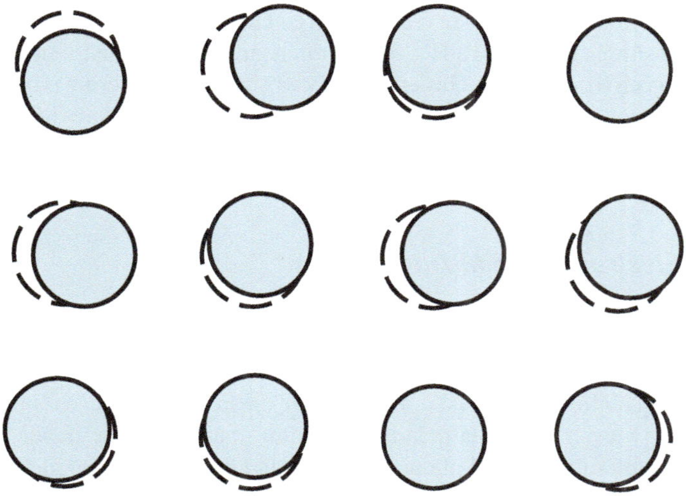

Fig. 6.12 The perturbation of a crystal structure by instantaneous atom displacements due to thermal vibration, resulting in a time-dependent correction in the crystal potential

Fig. 6.13 The potential of a silicon atom with (V_0) and without (V) including the DW factor. The V is the instantaneous atomic potential, and V_0 is the time-averaged atomic potential. The equilibrium position of the atom is $b = 0$; $\Delta V = V - V_0$ (dash-dotted line) is the deviation of the atomic potential from the time-averaged potential V_0 when the atomic displacement is **a** $u_\kappa = 0$ or **b** $u_\kappa = 0.01$ nm. The arrowhead in (**b**) indicates displacement of the atom

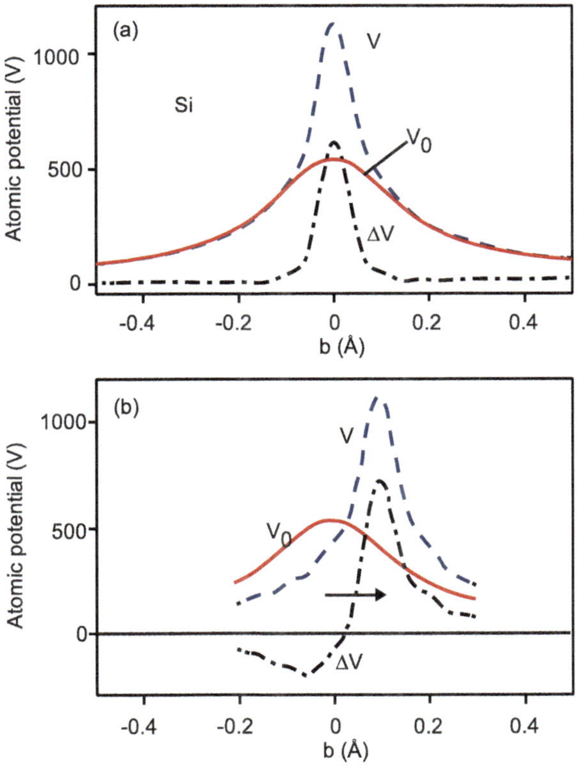

the ΔV function is not periodic. Therefore electrons scattered by phonons do not fall onto positions of Bragg reflections, but they are diffusely distributed in angular ranges between Bragg peaks. This is probably why phonon scattering is usually called thermal diffuse scattering. The analysis of diffuse scattering can provide information on interactions of crystal atoms.

6.4.3 Electron Phonon Interactions

Electron-phonon interaction is an inelastic process in high-energy electron scattering. The perturbation of the crystal potential by atomic vibrations eventually affects the scattering ability of the crystal. Phonons are created or annihilated in the process due to energy, even very small, and momentum transfers of the incident electron. Therefore scattering-induced phonon creation and annihilation are inelastic excitation processes that involve the transition of crystal vibration states.

The interaction Hamiltonian for creating a single phonon of momentum $\mathbf{q}$ and frequency $\omega_j(\mathbf{q})$ can be derived after expanding $\Delta V(\mathbf{r})$ in terms of atomic displacements (Whelan 1965a, b), for single-phonon scattering

$$\begin{aligned} H(\mathbf{r},\mathbf{q},\omega) &= < n_s(\mathbf{q},\omega)+1|[-e\Delta V(\mathbf{r})]|n_s(\mathbf{q},\omega) > \\ &= e < n_s(\mathbf{q},\omega)+1\left|\sum_n\sum_\alpha \left\{\mathbf{u}\binom{n}{\alpha}\cdot\nabla V_\alpha\left(\mathbf{r}-\mathbf{R}_n-\mathbf{r}(\alpha)\right)\right\}\right| n_s(\mathbf{q},\omega) > \\ &= e\sum_\alpha\sum_n A_\alpha\left(\omega_j(\mathbf{q})\right)\mathbf{e}\left(\alpha|{}^{\mathbf{q}}_{j}\right)\cdot\nabla\int d\mathbf{u}\, f_\alpha^e(\mathbf{u}) \\ &\quad\times \exp\{2\pi i\mathbf{u}\cdot[\mathbf{r}-\mathbf{R}(n)-\mathbf{r}(\alpha)]\}\exp\{2\pi i\mathbf{q}\cdot[\mathbf{R}_n+\mathbf{r}(\alpha)]\} \\ &= 2\pi \mathrm{ie}\sum_\alpha\sum_g A_\alpha\left(\omega_j(\mathbf{q})\right)\mathbf{e}\left(\alpha|{}^{\mathbf{q}}_{j}\right)\cdot(\mathbf{g}+\mathbf{q})f_\alpha^e(\mathbf{g}+\mathbf{q}) \\ &\quad\times \exp[2\pi i(\mathbf{g}+\mathbf{q})\cdot\mathbf{r}]\exp[-2\pi i\mathbf{g}\cdot\mathbf{r}(\alpha)] \end{aligned} \tag{6.68}$$

where $< n_s >$ is the average occupation number of the phonon state $|n_s(\mathbf{q},\omega)\rangle$

$$< n_s(\mathbf{q},\omega) > = \frac{\sum\limits_{n_s=0} n_s \exp\left[-(n_s+1/2)\hbar\omega_j/k_BT\right]}{\sum\limits_{n_s=0}\exp\left[-(n_s+1/2)\hbar\omega_j/k_BT\right]} = \frac{1}{\exp\left(\hbar\omega_j/k_BT\right)-1} \tag{6.69a}$$

and

$$A_\alpha\left(\omega_j(\mathbf{q})\right) \equiv \frac{\left\{\hbar\left[\langle n_s(\mathbf{q},\omega)\rangle+1\right]\right\}^{1/2}}{\left[2\omega_j(\mathbf{q})M_\alpha N_0\right]^{1/2}} \tag{6.69b}$$

is the atomic vibration amplitude in phonon mode $\omega_j(\mathbf{q})$, and k_B is the Boltzmann constant. The comparison of (6.68) with (6.8) gives

$$H_g(-\mathbf{q}) = 2\pi \mathrm{i}e \sum_{\alpha} A_\alpha \left(\omega_j(\mathbf{q})\right) \mathbf{e}\left(\alpha|_j^q\right) \cdot (\mathbf{g}+\mathbf{q}) f_\alpha^e(\mathbf{g}+\mathbf{q}) \exp[-2\pi i \mathbf{g} \cdot \mathbf{r}(\alpha)] \tag{6.70}$$

The transition matrix element introduced in (6.70) is used in the dynamic calculation of single-phonon excitation.

6.4.4 *Phonon Dispersion Surfaces*

Analogous to dispersion surfaces in electron diffraction, the relation $\omega_j(\mathbf{q})$ between phonon frequency and its wave vector is called the phonon dispersion relation. This relation determines allowed phonon modes in the crystal. The calculation of phonon dispersion surfaces is an essential part of lattice dynamics. To illustrate the basic physics involved in lattice vibrations, we consider a one-dimensional atom chain, as shown in Fig. 6.14. Two atoms separated by a are assumed in the primitive cell. Equations of motion, under the assumptions that (1) each atom interacts only with its nearest neighbor atoms and (2) harmonic force constants are identical between all pairs of nearest neighbor atoms, are

$$M_1 \frac{d^2 u_{2s+1}}{dt^2} = F\left(u_{2s+2} + u_{2s} - 2u_{2s+1}\right) \tag{6.71a}$$

$$M_2 \frac{d^2 u_{2s}}{dt^2} = F\left(u_{2s+1} + u_{2s-1} - 2u_{2s}\right) \tag{6.71b}$$

The solution is assumed to be

$$u_{2s+2} = Y_1 \exp[2\pi i(2s+1)qa - i\omega t] \tag{6.72a}$$

$$u_{2s} = Y_2 \exp[2\pi i 2sqa - i\omega t] \tag{6.72b}$$

On substituting (6.72) into (6.71), we have

$$-\omega^2 M_1 Y_1 = F Y_2[\exp(2\pi iqa) + \exp(-2\pi iqa)] - 2FY_1 \tag{6.73a}$$

$$-\omega^2 M_2 Y_2 = F Y_1[\exp(2\pi iqa) + \exp(-2\pi iqa)] - 2FY_2 \tag{6.73b}$$

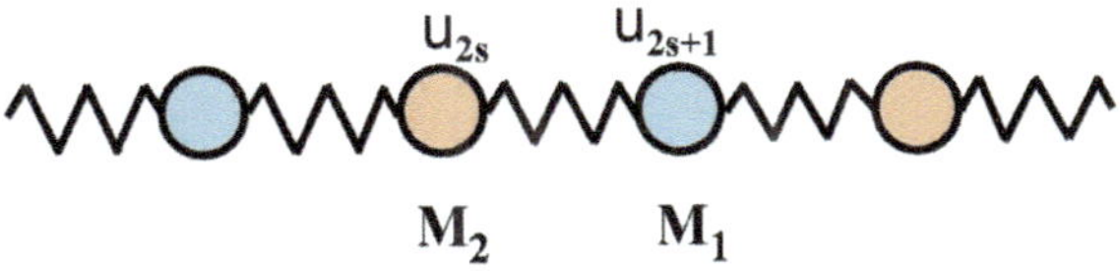

Fig. 6.14 The structural model of a diatomic linear lattice structure

The nontrivial solution of (6.73a) and (6.73b) exists only if the determinant of the coefficients of Y_1 and Y_2 vanishes; that is

$$\begin{vmatrix} 2F - M_1\omega^2 & -2F\cos 2\pi qa \\ -2F\cos 2\pi qa & 2F - M_2\omega^2 \end{vmatrix} = 0 \tag{6.74}$$

which gives

$$\omega_{\pm}^2 = F\left(\frac{1}{M_1} + \frac{1}{M_2}\right) \pm F\left[\left(\frac{1}{M_1} + \frac{1}{M_2}\right)^2 - \frac{4\sin^2 2\pi qa}{M_1 M_2}\right]^{1/2} \tag{6.75a}$$

For small q,

$$\begin{array}{ll} \omega_+^2 \approx 2F\left(\frac{1}{M_1} + \frac{1}{M_2}\right), & \frac{Y_2}{Y_1} \approx -\frac{M_1}{M_2} \\ \omega_-^2 \approx \frac{8\pi^2 F a^2}{M_1+M_2} q^2, & \frac{Y_2}{Y_1} \approx 1 \end{array} \tag{6.75c}$$

Phonon modes are thus classified as optical and acoustic branches. The optical branch $\omega_+$for which ω remains constant when $\mathbf{q}$ approaches zero (see Fig. 6.15), may be approximated by the Einstein model, in which $\omega = \omega_0$ independent of $\mathbf{q}$. In the optical branch, adjacent atoms tend to vibrate out of phase (Fig. 6.16). Acoustic branches, for which ω_-tends to zero when q approaches zero, may be approximated by the Debye model, in which $\omega = 2\pi c_w q$, where c_w is the phase velocity (Fig. 6.15). In this branch, paired neighbor atoms tend to vibrate in phase (Fig. 6.16).

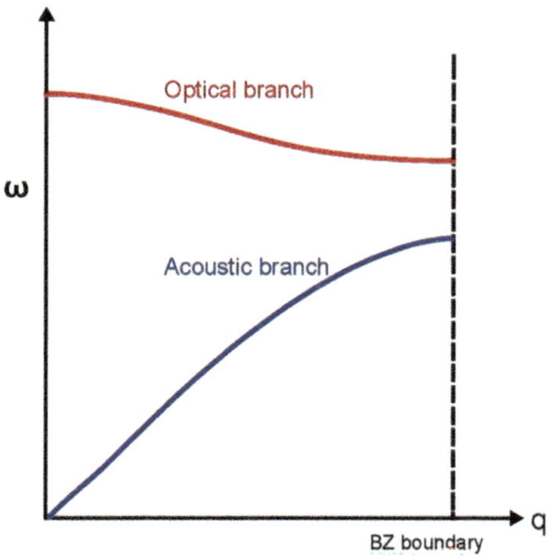

Fig. 6.15 Optical and acoustical phonon branches of the dispersion relation for a diatomic linear lattice with $M_1 > M_2$

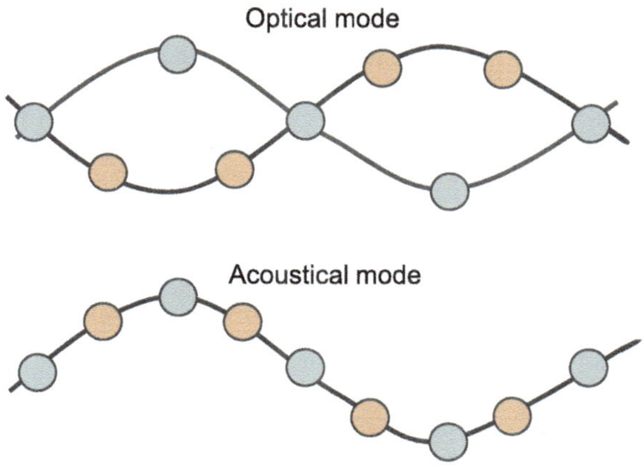

Fig. 6.16 Transverse optical and transverse acoustic waves in a diatomic linear lattice, illustrated by particle displacements for two modes at the same wavelength

6.4.5 Debye-Waller Factor

We now calculate the DW factor using the expression of atomic thermal displacement given by (6.56a) (Ohtsuki 1983). At thermal equilibrium, phonons are not created nor annihilated. Thus

$$\begin{aligned} W_\alpha &= 2\pi^2 < [\tau \cdot \mathbf{u}\begin{pmatrix} n \\ \alpha \end{pmatrix}]^2 >= 2\pi^2 < n_s | \left[\tau \cdot \mathbf{u}\begin{pmatrix} n \\ \alpha \end{pmatrix}\right]^2 | n_s > \\ &= \frac{\pi^2 \hbar}{N_0 M_\alpha} \sum_{\mathbf{q}} \sum_j \frac{1}{\omega_j(\mathbf{q})} \left[\tau \cdot \mathbf{e}\left(\alpha|{}^{\mathbf{q}}_{j}\right)\right]^2 \langle n_s | \left[a^+\begin{pmatrix} -\mathbf{q} \\ j \end{pmatrix} a\begin{pmatrix} \mathbf{q} \\ j \end{pmatrix} + a\begin{pmatrix} -\mathbf{q} \\ j \end{pmatrix} a^+\begin{pmatrix} \mathbf{q} \\ j \end{pmatrix}\right] | n_s \rangle \\ &= \frac{2\pi^2 \hbar}{N_0 M_\alpha} \sum_{\mathbf{q}} \sum_j \frac{1}{\omega_j(\mathbf{q})} \left[\tau \cdot \mathbf{e}\left(\alpha|{}^{\mathbf{q}}_{j}\right)\right]^2 (\langle n_s \rangle + 1/2) \end{aligned} \tag{6.76}$$

because

$$< n_s | \left[a^+\begin{pmatrix} -\mathbf{q} \\ j \end{pmatrix} a\begin{pmatrix} \mathbf{q}' \\ j' \end{pmatrix}\right] | n_s >= n_s \delta_{q,q'} \delta_{j,j'}$$

$$< n_s | \left[a^+\begin{pmatrix} -\mathbf{q} \\ j \end{pmatrix} a^+\begin{pmatrix} \mathbf{q} \\ j' \end{pmatrix}\right] | n_s >= 0$$

and

$$\langle n_s | \left[a^-\begin{pmatrix} -\mathbf{q} \\ j \end{pmatrix} a^-\begin{pmatrix} \mathbf{q}' \\ j' \end{pmatrix}\right] | n_s >= 0$$

Using (6.69a) for the occupation number of phonons, the DW factor can be rewritten as

$$W_\alpha = \frac{\pi^2 \hbar}{N_0 M_\alpha} \sum_{\mathbf{q}} \sum_j \frac{1}{\omega_j(\mathbf{q})} \left[\tau \cdot \mathbf{e}\left(\alpha|{}^{\mathbf{q}}_{j}\right)\right]^2 \coth\left[\frac{\hbar\omega_j(\mathbf{q})}{2k_B T}\right] \tag{6.77a}$$

Theoretically the DW factor can be calculated if phonon dispersion relations are known. In the high-temperature limiting case,

$$W_\alpha \approx \frac{2\pi^2 k_B T}{N_0 M_\alpha} \sum_{\mathbf{q}} \sum_j \frac{\left[\tau \cdot \mathbf{e}\left(\alpha \mid {}^{\mathbf{q}}_{j}\right)\right]^2}{\omega_j^2(\mathbf{q})} \tag{6.77b}$$

In the phonon spectra of three-dimensional-crystals, there are a total of $3n_0$ branches of phonon dispersion surfaces, among which are only three acoustic branches and the remaining $3\,(n_0 - 1)$ are optical branches. Thus there are no optical modes if there is only one atom per primitive cell. The acoustic branches are particularly important in TDS because of their large contribution to the mean square atomic vibration amplitude (or equivalently W_α factor). In the Debye model, the mean vibration amplitude of an atom is determined (Sears and Shelley 1991; see Sect. 7.6) by

$$\overline{a_\kappa^2} = \frac{3\hbar^2}{M_\kappa k_B T_D}\left[\frac{T^2}{T_D^2}\int_0^{T_D/T} dx \frac{x}{\exp(x) - 1} + \frac{1}{4}\right] \tag{6.78}$$

It is obvious that the mean square vibration amplitude is inversely proportional to the mass of the atom. Table 6.1 lists the mean vibration amplitude of several monatomic crystals at different temperatures. The calculation of DW factor using the Debye model is given in Sect. 7.6.

6.4.6 Mixed Dynamic Form Factor for Multiphonon Excitations

In phonon excitation, it is possible to generate one or more phonons during interaction with the same atom (atom A in Fig. 6.17a). Energy and momentum are transferred from the incident electron to the atom. The atom subsequently acts as a source for generating vibration waves in the crystal, analogous to the creation of water waves by a stone falling into a pond. This phonon excitation process is called the multiphonon-scattering process, in which the change in the electron wave vector is the sum of wave vectors of all phonons generated in a single scattering process. A detailed description of multiphonon excitation process is given in Sect. 7.5. Multiple-phonon excitation refers to a process in which phonons are created consecutively during the interaction with different atoms, such as A, B, and C atoms in Fig. 6.17b.

In electron scattering, the Hamiltonian for creating m phonons is

$$H_{m0}(\mathbf{r}) = \langle m|[-e\Delta V(\mathbf{r})]|0\rangle \tag{6.79}$$

Table 6.1 Atomic mean vibration amplitudes for some monoatomic cubic-structured crystals[a]

Element	Structure	Temperature	$\left[\overline{a_\kappa{}^2}\right]^{1/2}$ (Å)	$2\pi^2\overline{a_\kappa{}^2}$ (Å^2)
Al	fcc	600	0.152	0.4561
		290	0.099	0.1935
		86	0.064	0.0809
		10	0.057	0.0641
Au	fcc	900	0.168	0.5571
		293	0.084	0.1393
		94	0.048	0.0455
Cu	fcc	1273	0.179	0.6325
		900	0.154	0.4681
		293	0.085	0.1426
		100	0.053	0.0554
Li	bcc	293	0.209	0.8622
		90	0.157	0.4866
Pb	fcc	600	0.296	1.7295
		300	0.206	0.8377
		150	0.206	0.8377
C	Diamond	293	0.050	0.0493
Si	Diamond	293	0.075	0.1110

[α] A complete list of a_κ for more complex structures is given in the International Tables for Crystallography, Table 3.3.5.1A-C

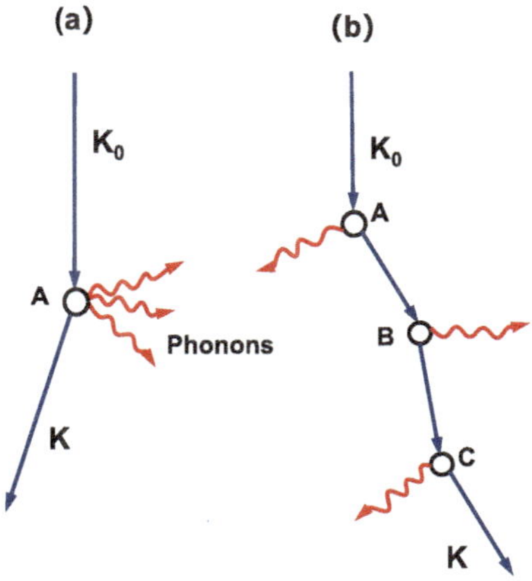

Fig. 6.17 Schematic diagrams showing the scattering processes of **a** multiphonon and **b** multiple phonons. The **K** is the wave vector of the incident electron after phonon excitation; *A*, *B*, and *C* are scattering sites

Using the expression for $\Delta V(\mathbf{r})$ given in (6.63), the Fourier transform of $H_{m0}(\mathbf{r})$ is

$$H_{m0}(\mathbf{Q}) = -\,e < m|\sum_n\sum_\alpha f_\alpha^e(\mathbf{Q})\exp\left[-2\pi i\mathbf{Q}\cdot\left(\mathbf{R}_n + \mathbf{r}(\alpha) + \mathbf{u}\binom{n}{\alpha}\right)\right]|0\rangle$$
$$= -\,e\sum_n\sum_\alpha f_\alpha^e(\mathbf{Q})\exp\left[-2\pi i\mathbf{Q}\cdot(\mathbf{R}_n + \mathbf{r}(\alpha))\right]$$
$$\times < m \mid \exp\left[-2\pi i\mathbf{Q}\cdot\mathbf{u}\binom{n}{\alpha}\right] \mid 0 > \tag{6.80}$$

We now consider the following relation:

$$\sum_{m\neq 0} H_{0m}(\mathbf{Q})H_{m0}(-\mathbf{Q}') = e^2\sum_n\sum_\alpha\sum_{n'}\sum_{\alpha'} f_\alpha^e(\mathbf{Q})f_{\alpha'}^e(-\mathbf{Q}')\exp\{-2\pi i\mathbf{Q}\cdot[\mathbf{R}_n+\mathbf{r}(\alpha)]\}$$
$$\times\exp\{2\pi i\mathbf{Q}'\cdot\mathbf{r}(\alpha')]\}\{\sum_{m\neq 0} < 0|\exp[-2\pi i\mathbf{Q}\cdot\mathbf{u}\tbinom{n}{\alpha})]|m >$$
$$\times < m\left|\exp\left[2\pi i\mathbf{Q}'\cdot\mathbf{u}\binom{n'}{\alpha}\right]\right|0 >\} \tag{6.81}$$

Note that the sum of n is over all unit cells, and the sum of m is over all phonon states. We now consider terms in the last { } bracket

$$\{\cdots\} = \sum_{m\neq 0}\left[< 0|\exp\left[-2\pi i\mathbf{Q}\cdot\mathbf{u}\binom{n}{\alpha}\right]|m >< m|\exp\left[2\pi i\mathbf{Q}'\cdot\mathbf{u}\binom{\alpha'}{n'}\right]|0 >\right]$$
$$= < 0|\exp\left[-2\pi i\left[\mathbf{Q}\cdot\mathbf{u}\binom{n}{\alpha} - \mathbf{Q}'\cdot\mathbf{u}\binom{n'}{\alpha'}\right]\right]|0 >$$
$$- < 0|\exp[-2\pi i\mathbf{Q}\cdot\mathbf{u}\binom{n}{\alpha})]|0 >< 0\left|\exp\left[-2\pi i\mathbf{Q}'\cdot\mathbf{u}\binom{\alpha'}{n'}\right]\right|0 > \tag{6.82}$$

where the completeness relation $\sum_m |m >< m| = 1$ was used. Using the relations (Ohtsuki 1983)

$$< \exp(x) >= \exp\left[< x^2 > /2\right] \tag{6.83}$$

and

$$\langle\exp(x)\exp(y)\rangle = \exp\left(\langle x^2\rangle/2 + \langle y^2\rangle/2 + \langle xy\rangle + \cdots\right) \tag{6.84}$$

for $\langle x\rangle = 0$ and $\langle y\rangle = 0$, (6.82) is simplified as

$$
\begin{aligned}
\{\cdots\} &= \exp\{-2\pi^2 < 0|\left[\mathbf{Q}\cdot\mathbf{u}\binom{n}{\alpha} - \mathbf{Q}'\cdot\mathbf{u}\binom{n'}{\alpha'}\right]^2|0>\} \\
&\quad - \exp\left\{-2\pi^2 < 0|\left[\mathbf{Q}\cdot\mathbf{u}\binom{n}{\alpha}\right]^2|0>\right\}\exp\left\{-2\pi^2 < 0|\left[\mathbf{Q}'\cdot\mathbf{u}\binom{n'}{\alpha'}\right]^2|0>\right\} \\
&= \exp\left\{-2\pi^2 < 0|[\mathbf{Q}\cdot\mathbf{u}\binom{n}{\alpha}]^2|0>\right\}\exp\left\{-2\pi^2 < 0|\left[\mathbf{Q}'\cdot\mathbf{u}\binom{n'}{\alpha'}\right]^2|0>\right\} \\
&\quad \times\left\{\exp\left[4\pi^2 < 0|\left[\mathbf{Q}\cdot\mathbf{u}\binom{n}{\alpha}\mathbf{Q}'\cdot\mathbf{u}\binom{n'}{\alpha'}\right]|0>\right] - 1\right\}
\end{aligned} \tag{6.85}
$$

Defining

$$
F_{n\alpha n'\alpha'}(\mathbf{Q},\mathbf{Q}') = 2\pi^2 < 0|[\mathbf{Q}\cdot\mathbf{u}\binom{n}{\alpha}][\mathbf{Q}'\cdot\mathbf{u}\binom{n'}{\alpha'}]|0> \tag{6.86}
$$

as the correlation function between the vibrations of the (n,α) and (n',α') atoms, substituting (6.85) into (6.81), and using (6.76), we have

$$
\begin{aligned}
&\sum_{m\neq 0} H_{0m}(\mathbf{Q})H_{m0}(-\mathbf{Q}') \\
&= e^2\sum_n\sum_\alpha\sum_{n'}\sum_{\alpha'} f_\alpha^e(\mathbf{Q})f_{\alpha'}^e(-\mathbf{Q}') \\
&\quad \times\exp\{-2\pi i\mathbf{Q}\cdot[\mathbf{R}_n + \mathbf{r}(\alpha)]\}\exp\{2\pi i\mathbf{Q}'\cdot[\mathbf{R}_{n'} + \mathbf{r}(\alpha')]\} \\
&\quad \times\exp[-W_\alpha(\mathbf{Q}) - W_{\alpha'}(\mathbf{Q}')]\{\exp[2F_{n\alpha n'\alpha'}(\mathbf{Q},\mathbf{Q}')] - 1\}
\end{aligned} \tag{6.87}
$$

The $|0>$ state is actually the elastic state of the system; the average occupation number of phonons in the elastic state at temperature T is $\langle n_s\rangle$, as defined in (6.69a). We now consider the quantities defined in (6.86) under thermal equilibrium conditions

$$
\begin{aligned}
F_{n\alpha n'\alpha'}(\mathbf{Q},\mathbf{Q}') &= 2\pi^2 < n_s|[\mathbf{Q}\cdot\mathbf{u}\binom{n}{\alpha}][\mathbf{Q}'\cdot\mathbf{u}\binom{n'}{\alpha'}]|n_s> \\
&= \frac{\pi^2\hbar}{N_0(M_\alpha M_{\alpha'})^{1/2}}\sum_{\mathbf{q}}\sum_j\sum_{\mathbf{q}'}\sum_{j'}\frac{1}{[\omega_j(\mathbf{q})\omega_{j'}(\mathbf{q}')]^{1/2}}\times \\
&\quad \times\left[\mathbf{Q}\cdot\mathbf{e}\left(\alpha|{}^{\mathbf{q}}_{j}\right)\mathbf{Q}'\cdot\mathbf{e}\left(\alpha'|{}^{\mathbf{q}'}_{j'}\right)\right]\exp\{2\pi i\mathbf{q}\cdot[\mathbf{R}_n + \mathbf{r}(\alpha)]\} \\
&\quad \times\exp\{-2\pi i\mathbf{q}'\cdot[\mathbf{R}_{n'} + \mathbf{r}(\alpha')]\} < n_s|\left[a^+\binom{-\mathbf{q}}{j} + a\binom{\mathbf{q}}{j}\right]\left[a^+\binom{-\mathbf{q}'}{j'} + a\binom{\mathbf{q}'}{j'}\right]|n_s> \\
&= \frac{2\pi^2\hbar}{N_0\sqrt{M_\alpha M_{\alpha'}}}\sum_{\mathbf{q}}\sum_j\frac{1}{[\omega_j(\mathbf{q})\omega_j(-\mathbf{q})]^{1/2}}\left[\mathbf{Q}\cdot\mathbf{e}\left(\alpha|{}^{\mathbf{q}}_{j}\right)\mathbf{Q}'\cdot\mathbf{e}\left(\alpha'|{}^{-\mathbf{q}}_{j}\right)\right] \\
&\quad \times(< n_s > +1/2)\exp\{2\pi i\mathbf{q}\cdot[\mathbf{R}(n) + \mathbf{r}(\alpha) - \mathbf{R}(n') - \mathbf{r}(\alpha')]\}
\end{aligned} \tag{6.88}
$$

since

$$< n_s \left| \left[a \begin{pmatrix} \mathbf{q} \\ j \end{pmatrix} a^+ \begin{pmatrix} -\mathbf{q}' \\ j' \end{pmatrix} + a^+ \begin{pmatrix} \mathbf{q} \\ j \end{pmatrix} a \begin{pmatrix} -\mathbf{q}' \\ j' \end{pmatrix} \right] \right| n_s >= (2 < n_s > +1)\, \delta_{\mathbf{q},\mathbf{q}'} \delta_{j,j'}$$

The mixed dynamic form factor is obtained by substituting (6.87) and (6.88) into (6.24). The calculation of $F_{n\alpha n'\alpha'}$ using the Debye model is given in Sect. 7.7.

6.4.7 Absorption Potential

The absorption potential for multiphonon excitation is obtained by substituting (6.87) into (6.30). The Fourier coefficients of the imaginary potential are

$$\begin{aligned} U_{gh}^{\prime(I)} &= -\frac{e^2 m_e^2}{4\pi^3 \hbar^4 V_c K_0} \int d\sigma(\mathbf{u}) \sum_n \sum_{n'} \sum_\alpha \sum_{\alpha'} f_\alpha^e (\mathbf{k}_g - \mathbf{u})\, f_{\alpha'}^e (\mathbf{u} - \mathbf{k}_h) \\ &\times \exp\left[-2\pi i \left(\mathbf{k}_g - \mathbf{u}\right) \cdot (\mathbf{R}(n) + \mathbf{r}(\alpha))\right] \\ &\times \exp\left[2\pi i \left(\mathbf{k}_h - \mathbf{u}\right) \cdot \left(\mathbf{R}\left(n'\right) + \mathbf{r}\left(\alpha'\right)\right)\right] \exp\left[-W_\alpha \left(\mathbf{k}_g - \mathbf{u}\right) - W_{\alpha'} \left(\mathbf{k}_h - \mathbf{u}\right)\right] \\ &\times \left\{\exp\left[2F_{n\alpha n'\alpha'}\left(\mathbf{k}_g - \mathbf{u}, \mathbf{k_h} - \mathbf{u}\right)\right] - 1\right\} \end{aligned} \quad (6.89)$$

Further simplified results can be obtained from this equation if some approximations are made. If there is no phase correlation between atoms distribution in different cells, so that $F_{n\alpha n'\alpha'}$ is independent of n and n' and it is denoted by $F_{\alpha\alpha'}\delta_{n,n'}$, we have

$$\begin{aligned} U_{gh}^{\prime(I)} &= -\frac{e^2 m_e^2}{4\pi^3 \hbar^4 \Omega K_0} \int d\sigma(\mathbf{u}) \sum_\alpha \sum_{\alpha'} f_\alpha^e (\mathbf{k}_g - \mathbf{u})\, f_{\alpha'}^e (\mathbf{u} - \mathbf{k}_h) \\ &\times \exp\left[-2\pi i \left(\mathbf{k}_g - \mathbf{u}\right) \cdot \mathbf{r}(\alpha)\right] \exp\left[2\pi i \left(\mathbf{k}_h - \mathbf{u}\right) \cdot \mathbf{r}\left(\alpha'\right)\right] \\ &\times \exp\left[-W_\alpha \left(\mathbf{k}_g - \mathbf{u}\right) - W_{\alpha'} \left(\mathbf{k}_h - \mathbf{u}\right)\right] \left\{\exp\left[2F_{\alpha\alpha'}\left(\mathbf{k}_g - \mathbf{u}, \mathbf{k}_h - \mathbf{u}\right)\right] - 1\right\} \end{aligned} \quad (6.90a)$$

If the atomic vibrations are described by the Einstein model, (6.90a) is simplified as

$$\begin{aligned} U_{gh}^{\prime(I)} &= -\frac{e^2 m_e^2}{4\pi^3 \hbar^4 \Omega K_0} \int d\sigma(\mathbf{u}) \sum_\alpha f_\alpha^e (\mathbf{k}_g - \mathbf{u})\, f_{\alpha'}^e (\mathbf{u} - \mathbf{k}_h) \exp[-2\pi i (\mathbf{g} - \mathbf{h}) \cdot \mathbf{r}(\alpha)] \\ &\times \exp\left[-W_\alpha \left(\mathbf{k}_g - \mathbf{u}\right) - W_\alpha \left(\mathbf{k}_h - \mathbf{u}\right)\right] \left\{\exp\left[2F_{\alpha\alpha}\left(\mathbf{k}_g - \mathbf{u}, \mathbf{k}_h - \mathbf{u}\right)\right] - 1\right\} \end{aligned} \quad (6.90b)$$

The imaginary component of the crystal potential produced by phonon excitations has been calculated by Yoshioka and Kainuma (1962), Whelan (1965b), Humphreys and Hirsch (1968), and Radi (1970) based on some simplified lattice vibration models. In general, since the mathematical expression of phonon excitation is rather lengthy and complex, the accurate calculation of $U_g'^{(I)}$ is difficult. A reasonable estimation can be obtained if the Einstein model is used (Bird and King 1990; Allen and Rossouw 1990). In (6.90b), the major error comes from extrapolating the atom-scattering factor at large scattering angles, because integration is over the entire reflection sphere. This problem has been considered by Bird and King (1990), who have developed a FORTRAN subroutine that calculates the absorption potential due to TDS. The mean free path and total absorption coefficient of phonon diffuse scattering are illustrated in Chap. 7 based on the semiclassical approach.

To summarize data provided by the numerical calculations of Bird and King (1990), Seale and Sheinin (1993) introduced a parametrization equation that allows the TDS imaginary atomic-scattering factor to be calculated simply. The absorption electron-scattering factor is expressed as

$$f^{e'}(s) = A(W)\exp\left[-b(W)s^{n(W)}\right] \tag{6.91a}$$

with

$$A(W) = A_A\,(W + B_A)^{n_A} + C_A,\quad b(W) = A_b W^{n_b} + C_b,\quad n(W) = A_n \exp\left(b_n\sqrt{W}\right) + C_n$$

where A_A, B_A, n_A, C_A, A_b, n_b, C_b, A_n, b_n, and C_n are fitting parameters, which have been tabulated for all the elements in Seale and Sheinin (1994).

The calculated absorption form factors for $Z_0 = 10, 20, 30$, and 40, respectively are shown in Fig. 6.18a. The mean vibration amplitude of the atom is taken as $\langle a^2\rangle^{1/2} = 0.128$ Å. The x s indicate data from the program by Bird and King (1990), and the continuous solid line is the result of parameterization fitting. Figure 6.18b shows the calculated absorption form factors for zinc with three different mean vibration amplitudes.

For the convenience of calculations, the absorption factor $U_g{}'$ may be written into a parametric fit of the form (Voss et al. 1980)

$$\frac{U_g'}{U_g} = -b_f|g| + c_f|g|^2 \tag{6.91b}$$

where b_f and c_f are fitting parameters. Figure 6.19 shows the calculated $U_g^{(I)}/U_g$ for several elements of different atomic numbers as a function of the reciprocal vector g.

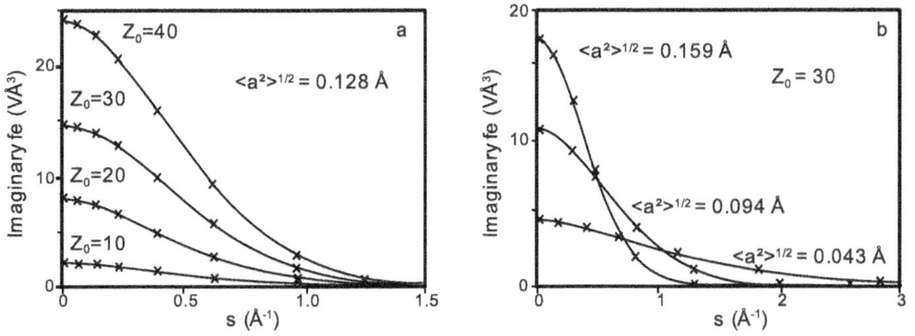

Fig. 6.18 The calculated neutral atom absorption factor due to thermal diffuse scattering. **a** Atomic numbers ranging from 10 – 40, with $\langle a^2\rangle^{1/2} = 0.128$ Å. **b** Three atomic mean vibration amplitudes for zinc with $Z_0 = 30$ (Seale and Sheinin 1993). Solid lines are calculated using the parameterization equation, and the crosses are given by the program provided by Bird and King (1990)

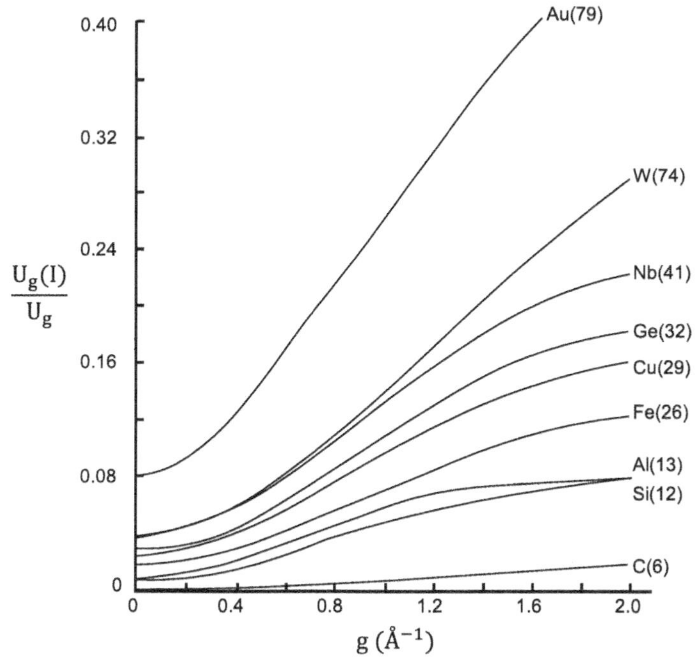

Fig. 6.19 Computed values of $U_g^{(I)}/U_g$ at 100 kV as a function of g for various elements at a temperature of 300 K (Humphreys and Hirsch 1968). The increase of the ratio at large g shows the drastic increase of TDS absorption at large scattering angles

6.5 Inelastic-Scattering Process II: Valence Excitation

6.5.1 Dielectric Response Theory of Valence Excitations

When an electron passes through a metal foil, the most noticeable energy loss is due to plasmon oscillations in the sea of conduction electrons. For an ideal case in which electrons can move freely in the sea, the system can be treated as an electron gas. The collective excitation of electrons in valence states produces numerous low-energy excited states, resulting in the energy loss of electrons. Figure 6.20 shows a valence spectrum recorded at 100 keV from a thin $SrTiO_3$ crystal. The first three peaks are referred to as the valence loss spectrum. The peak appearing at about 47 eV is the Ti $M_{2,3}$ ionization edge.

Many sophisticated quantum mechanical many-body theories have been developed to consider electron interactions in metals (Bohm 1953; Pine 1953; Nozieres and Pine 1958; Ferrell, 1956 and 1957). However theories become very complex if the solid is a nonmetal material, since the electron gas model is no longer valid. The valence band of a crystalline material is a collection of many electron energy states, and it is usually characterized by a density of states (DOS) function. The valence electron wave functions are directly associated with chemical bonding. The energy loss of an electron results from the interband transition of valence electrons to the conducting band. It is difficult, in general, to treat the excitation of each individual state of the valence band. Thus statistical properties of valence electrons are usually characterized by a dielectric function $\varepsilon(\omega, \mathbf{q})$, which depends on the frequency ω of the plasmon oscillation and the wave vector $\mathbf{q}$ of the disturbance (Ritchie 1957). The fast-incident electron is represented as a point charge with a well-defined path, since we are concerned with processes in which fractional changes of energy and momentum of the electron are small. For an infinitely large isotropic medium (Fig. 6.21),

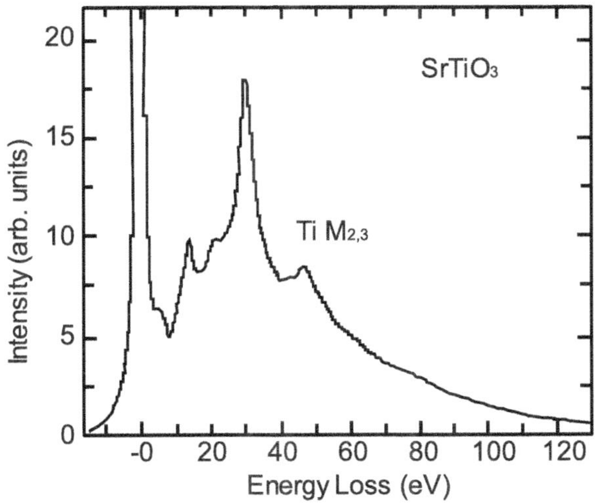

Fig. 6.20 An EELS valence loss spectrum of a thin $SrTiO_3$ crystal recorded at 100 keV

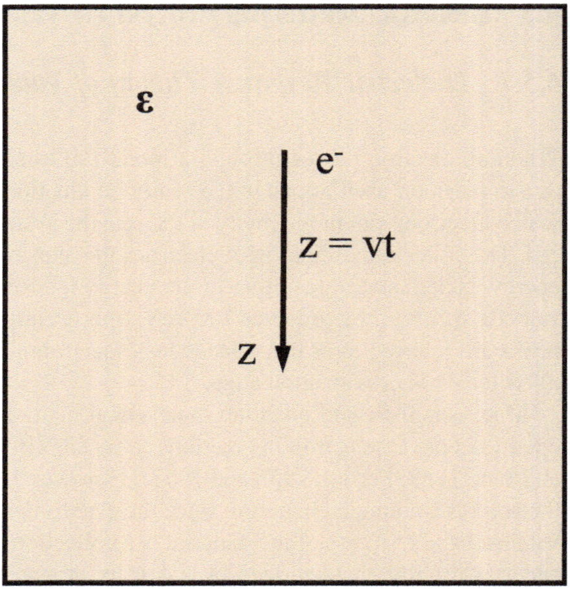

Fig. 6.21 A model of valence excitation in a homogeneous medium by a fast electron

the induced field generated by the moving electron is then determined by the Poisson equation.

$$\varepsilon(\omega, \mathbf{q})\nabla^2 V_e(\mathbf{r}, t) = \frac{e}{\varepsilon_0}\delta(x)\delta(y)\delta(z - vt) \tag{6.92}$$

where the electron is assumed to travel along the z axis. Transforming (6.92) into reciprocal space and frequency space, we have

$$V_e(\mathbf{q}, \omega) = -\frac{e}{4\pi^2\varepsilon_0\varepsilon(\omega, \mathbf{q})}\frac{\delta\left(2\pi q_z v - \omega\right)}{q^2} \tag{6.93a}$$

where

$$V_e(\mathbf{q}, \omega) = \frac{1}{2\pi}\int d\omega \int d\mathbf{q}\exp[-2\pi i\mathbf{q}\cdot\mathbf{r} + i\omega t]V_e(\mathbf{r}, t) \tag{6.93b}$$

The energy loss per unit path in the medium is given by

$$\begin{aligned} -\frac{dE}{dz} &= -e\frac{dV_e(\mathbf{r}, t)}{dz}\bigg|_{x=0,y=0,z=vt} \\ &= \frac{e^2}{4\pi^3\varepsilon_0 v^2}\int dq_x \int dq_y \int_0^\infty d\omega \frac{\omega}{q_x^2 + q_y^2 + (\omega/2\pi v)^2}\,\mathrm{Im}\left\{-\frac{1}{\varepsilon(\omega, \mathbf{q})}\right\} \end{aligned} \tag{6.94}$$

Thus the excitation probability of the valence states is

$$\frac{d^2P_v}{dzd\omega} = \frac{e^2}{4\pi^3\hbar\varepsilon_0 v^2}\int_{-\infty}^{\infty} dq_x \int_{-\infty}^{\infty} dq_y \frac{1}{q_x^2+q_y^2+(\omega/2\pi v)^2}\,\mathrm{Im}\left[-\frac{1}{\varepsilon(\omega,\mathbf{q})}\right] \quad (6.95)$$

where integrating $\mathbf{q}_t$ involves summing over electrons scattered to different angles. The angular distribution of the valence loss electron is a Lorentzian function $1/\left[\theta^2+\theta_E^2\right]$, with $\vartheta_E = \Delta E/2E$. For a homogeneous medium, where ε is independent of $\mathbf{q}$,

$$\begin{aligned}\frac{d^2P_v}{dzd\omega} &= \frac{e^2}{4\pi^2\varepsilon_0\hbar v^2}\,\mathrm{Im}\left[-\frac{1}{\varepsilon}\right]\ln\left[1+(2\pi q_c v/\omega)^2\right] \\ &\approx \frac{e^2}{2\pi^2\varepsilon_0\hbar v^2}\,\mathrm{Im}\left[-\frac{1}{\varepsilon}\right]\ln\left[(2\pi q_c v/\omega)\right]\end{aligned} \quad (6.96)$$

where q_c is the cutoff value of the wave vector and $2\pi q_c v/\omega \gg 1$. It is apparent that valence excitation is a delocalized scattering process, and $\mathrm{Im}[-1/\varepsilon(\omega,\mathbf{q})]$ is called the energy loss function. Equation (6.96) can be derived directly from quantum mechanics (Wang 1996).

6.5.2 *Mean Free Path and Absorption Potential*

The mean free path of valence excitation is related to the excitation probability by

$$\frac{1}{\Lambda} = \int_0^{\infty} d\omega \frac{d^2P_v}{dzd\omega} \quad (6.97)$$

For metals at 100 kV, the mean free path ranges from 50 – 100 nm. For oxides, such as MgO, Λ can be as large as 250 nm. A comprehensive description of dielectric response theory of valence excitations has been given by Raether (1980). Figure 6.22 shows the calculated and observed electron mean free paths of aluminum and carbon as a function of electron microscope accelerating voltage. In general, Λ is determined by the structure of the valence band, and it does not have a simple relation with the atomic number. According to (6.45), the mean free path is related to the absorption potential by $2\sigma V' = 1/\Lambda$; thus the imaginary component of the potential for valence excitation is

$$V' = \frac{e^2}{4\pi^2\varepsilon_0\sigma\hbar v^2}\int_0^{\infty} d\omega\,\mathrm{Im}\left[-\frac{1}{\varepsilon}\right]\ln\left[1+(2\pi q_c v/\omega)^2\right] \quad (6.98)$$

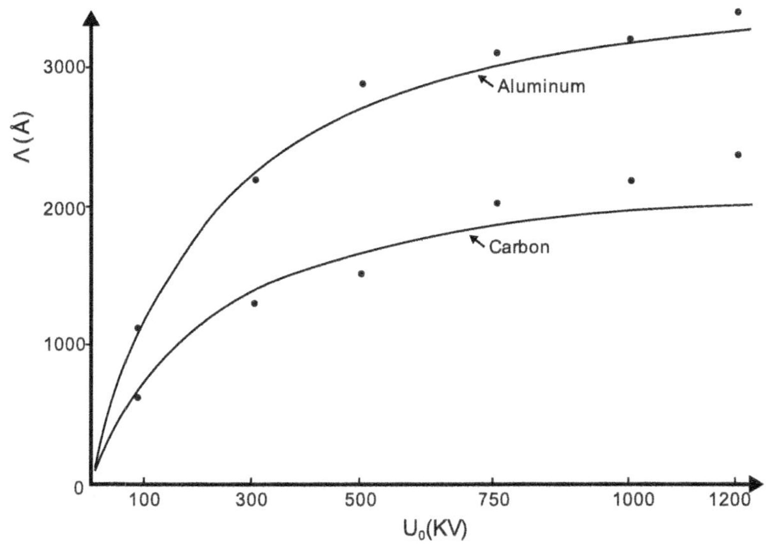

Fig. 6.22 The mean free path length for volume plasmon excitations in aluminum and carbon as a function of the incident electron energy. The solid line indicates calculated values, and points represent measured values (Raether 1980)

For a homogeneous medium, the absorption potential V' is independent of $\mathbf{r}$. This is the result of delocalized scattering. Therefore for a uniformly thin foil, the absorption effect introduced by volume plasmon excitation is equivalent to introducing a mean absorption factor $\exp\left(-2\sigma V'd\right) = \exp(-d/\Lambda)$ in the final calculated image or diffraction intensity. This absorption factor does not affect the contrast of the image, but it decreases the intensity of the entire elastically scattered electrons. For those cases involving interfaces of two phases, conclusions are different, as shown in the next section.

Since the absorption potential of valence excitation is constant (6.98), the Fourier transform of V' is zero for all Bragg reflections except the (000) transmitted beam. This result is consistent with that derived from more sophisticated theory (Radi 1968, 1970).

6.5.3 *Interface and Surface Excitations*

The dielectric response theory introduced in Sect. 6.5.1 can also be applied to calculate the excitation of surfaces (see Bah et al. 1992, for an introduction). For a general case, an interface is formed by two media with dielectric constants ε_1 and ε_2 (Fig. 6.23). The surface excitation can be considered as a one-electron excitation corresponding to the point probe case in STEM. The incident electron is assumed

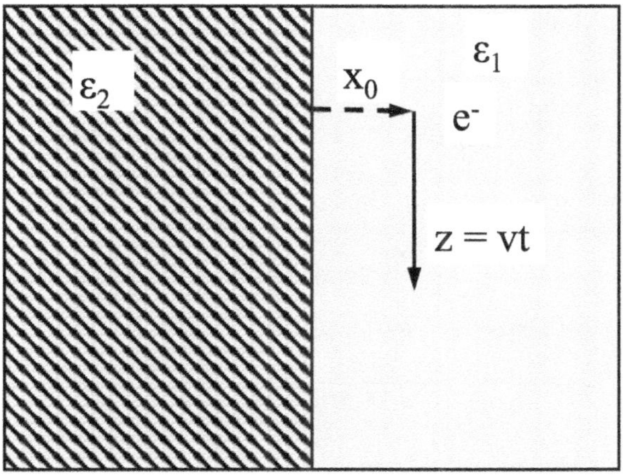

Fig. 6.23 A model of the dielectric response theory of interface excitation for two homogeneous media. The incident electron is traveling along the z axis at a distance x_0 from the interface

to travel at a distance x_0 in a medium ε_1 parallel to the interface. Thus the field generated in space is the solution of (6.99a).

$$\varepsilon(\omega, \mathbf{q})\nabla^2 V_e(\mathbf{r}, t) = \frac{e}{\varepsilon_0}\delta\left(x - x_0\right)\delta(y)\delta(z - vt) \tag{6.99a}$$

The interface is assumed to be infinitely long to simplify the mathematical operation. We now consider the solution to (6.99a) under the nonrelativistic approximation (Howie 1983). A more sophisticated treatment that considers the relativistic effect was given by Garcia-Molina et al. (1985).

The solution of (6.99a) can be separated into two components

$$V_e(\mathbf{r}, t) = V^{(s)}(\mathbf{r}, t) + V^{(\mathrm{id})}(\mathbf{r}, t) \tag{6.99b}$$

where $V^{(s)}(\mathbf{r}, t)$ is the potential field generated by the electron itself and $V^{(\mathrm{id})}(\mathbf{r}, t)$ is the potential field generated by induced charges distributed at the interface. The interaction of electrons with $V^{(\mathrm{id})}(\mathbf{r}, t)$ results in energy loss. Using the Fourier transform of V_e,

$$V_e(\mathbf{r}, t) = \int d\omega \int dq_y \int dq_z \exp\left[2\pi i\left(q_y y + q_z z\right) - i\omega t\right] V_e\left(x, \mathbf{q}_n, \omega\right) \tag{6.100a}$$

or

$$V_e\left(x, \mathbf{q}_n, \omega\right) = \frac{1}{2\pi}\int dt \int dy \int dz \exp\left[-2\pi i\left(q_z y + q_z z\right) + i\omega t\right] V_e(\mathbf{r}, t) \tag{6.100b}$$

Equation (6.99a) becomes

$$\varepsilon(\omega, \mathbf{q})\left[\frac{d^2}{dx^2} - 4\pi^2 q_n^2\right] V_e(x, \mathbf{q}_n, \omega) = \frac{e}{\varepsilon_0}\delta(x - x_0)\,\delta(\omega - 2\pi q_z v) \qquad (6.101)$$

where $\mathbf{q}_n = (q_y, q_z)$ and $q_n = (q_y^2 + q_z^2)^{1/2}$. The solution of (6.101) is

$$V_e(x, \mathbf{q}_n, \omega) = a(\mathbf{q}_n, \omega)\exp(2\pi q_n x) \quad x < 0 \qquad (6.102a)$$

$$V_e(x, \mathbf{q}_n, \omega) = b(\mathbf{q}_n, \omega)\exp(-2\pi q_n x) + \frac{e}{4\pi\varepsilon_0\varepsilon_1 q_n}\exp(-2\pi q_n |x - x_0|)\,\delta(\omega - 2\pi q_z v) \quad x > 0 \qquad (6.102b)$$

Matching the boundary condition at $x = 0$, coefficients $a(\mathbf{q}_n, \omega)$ and $b(\mathbf{q}_n, \omega)$ are determined as

$$a(\mathbf{q}_n, \omega) = \left(\frac{2\varepsilon_1}{\varepsilon_1 + \varepsilon_2}\right)\left(\frac{e}{4\pi\varepsilon_0\varepsilon_1 q_n}\right)\exp(-2\pi q_n x_0)\,\delta(\omega - 2\pi q_z v) \qquad (6.103a)$$

and

$$b(\mathbf{q}_n, \omega) = \left(\frac{\varepsilon_1 - \varepsilon_2}{\varepsilon_1(\varepsilon_1 + \varepsilon_2)}\right)\left(\frac{e}{4\pi\varepsilon_0\varepsilon_1 q_n}\right)\exp(-2\pi q_n x_0)\,\delta(\omega - 2\pi q_z v) \qquad (6.103b)$$

The potential energy of the incident electron in its induced field is

$$E_p = -eV^{(\mathrm{id})}(\mathbf{r}, t) = -\frac{e^2}{8\pi^2\varepsilon_0 v}\int_{-\infty}^{\infty} d\omega \int_{-\infty}^{\infty} dq_y \frac{\exp\left\{-2\pi\left[q_y^2 + (\omega/2\pi v)^2\right]^{1/2} |x + x_0|\right\}}{\left[q_y^2 + (\omega/2\pi v)^2\right]^{1/2}} \frac{\varepsilon_1 - \varepsilon_2}{\varepsilon_1(\varepsilon_1 + \varepsilon_2)} \times \exp\left[-2\pi i q_y y + i\left(\frac{\omega}{v}z - \omega t\right)\right] \qquad (6.104)$$

The energy loss of the electron results from the deceleration effect of its induced charges, so that the rate at which the electron energy is lost per unit traveling distance is related to the differential excitation probability by

$$-\frac{dE}{dz} = \left.\frac{dE_p}{dz}\right|_{x=x_0, y=0, z=vt} = \int_0^{\infty} d\omega\hbar\omega\frac{d^2 P_S}{d\omega dz} \qquad (6.105)$$

Substituting (6.104) into (6.105), we obtain

$$\frac{d^2 P_s}{d\omega dz} = \frac{e^2}{2\pi^2 \hbar v^2 \varepsilon_0} \int_0^\infty dq_y \frac{\exp\left\{-4\pi \left[q_y^2 + (\omega/2\pi v)^2\right]^{1/2} |x_0|\right\}}{\left[q_y^2 + (\omega/2\pi v)^2\right]^{1/2}} \operatorname{Im}\left[-\frac{\varepsilon_1 - \varepsilon_2}{\varepsilon_1 (\varepsilon_1 + \varepsilon_2)}\right] \tag{6.106}$$

where a relation of $\varepsilon(-\omega, \mathbf{q}) = \varepsilon^*(\omega, \mathbf{q})$ was used. If the dielectric function $\varepsilon(\omega, \mathbf{q})$ is assumed to be independent of wave vector $\mathbf{q}$, the excitation probability of the interface is

$$\frac{d^2 P_s}{d\omega dz} = \frac{e^2}{2\pi^2 \hbar v^2 \varepsilon_0} \operatorname{Im}\left[-\frac{\varepsilon_1 - \varepsilon_2}{\varepsilon_1 (\varepsilon_1 + \varepsilon_2)}\right] K_0 (2\omega x_0/v) \tag{6.107a}$$

where the modified Bessel function is

$$K_0 (2\omega x_0/v) = \int_0^\infty du \frac{\exp\left\{-2\left[u^2 + (\omega/v)^2\right]^{1/2} |x_0|\right\}}{\left[u^2 + (\omega/v)^2\right]^{1/2}} \tag{6.107b}$$

The localization of interface excitation is determined by the spatial variation of function $K_0 (2\omega x_0/\pi v)$. It is worthwhile to point out that (6.106) has to be used if we wish to include the dispersion relation of $\varepsilon(\omega, \mathbf{q})$.

The total excitation probability of the valence band is the sum of the volume and surface plasmons. From (6.96) and (6.107a), we have

$$\begin{aligned}\frac{d^2 P}{dz d\omega} =& \frac{e^2}{2\pi^2 \varepsilon_0 \hbar v^2} \left\{ \operatorname{Im}\left[-\frac{1}{\varepsilon_1}\right] \left[\ln (2\pi q_c v/\omega) - K_0 (2\omega x_0/v)\right] \right. \\ &\left. + \operatorname{Im}\left[-\frac{2}{\varepsilon_1 + \varepsilon_2}\right] K_0 (2\omega x_0/v) \right\}\end{aligned} \tag{6.108}$$

where the first term is the excitation probability of the volume plasmon as a function of the incident electron impact distance from the interface, and the second term is the excitation probability of the interface plasmon.

From the mathematical expression, the Bessell function $K_0 (2\omega x_0/v)$ tends to infinity at $x_0 = 0$. This unphysical result came from two sources. One source is the assumption of $\varepsilon(\omega)$ being independent of q, i.e., no dispersion. The other source is the unlimited momentum transfer. For valence excitation, there is a cutoff q_c for q, so that the upper limit of q integration in (6.107b) should be replaced by q_c.

For simplification, the surface profile imaging of GaAs (100) can be taken as an example. Figure 6.24 shows the calculated excitation probabilities of volume plasmon and interface plasmon (or surface plasmon in this case). The variations of dP_v/dz and dP_i/dz are quite slow when electrons move close to the interface from the GaAs side ($x < 0$). The only sharp change occurs when the electron is about 0.3 nm away from the interface. In the vacuum side ($x > 0$), the interaction of the electron beam with

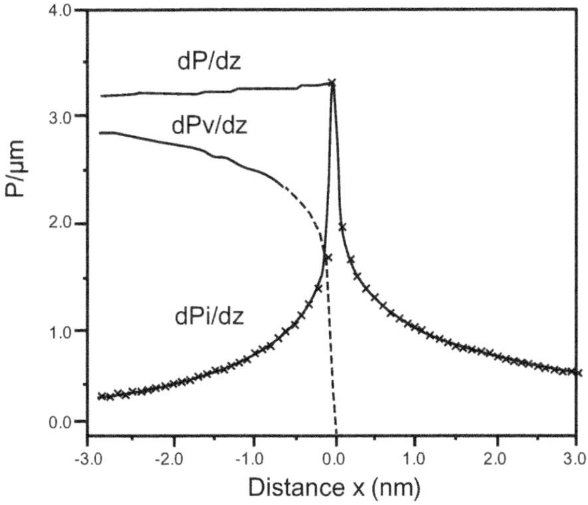

Fig. 6.24 Calculated GaAs volume plasmon (dP_v/dz) and interface plasmon (i.e., GaAs with vacuum) (dP_i/dz) excitation probabilities per unit distance as a function of the impact parameter x. The $dP/dz = dP_v/dz + dP_i/dz$ is the total excitation probability of plasmons, with $x < 0$ inside the GaAs crystal and $x > 0$ in the vacuum. The variation of dP_i/dz close to the interface ($x = 0$) indicates the localization effect of surface excitations; electron energy is 200 keV

the interface is a long-range interaction, resulting in some residual surface plasmon excitation even at $x > 3$ nm. This is the result of nonlocalized inelastic excitation. In the vacuum side, only surface excitation occurs. Although variations of dP_v/dz and dP_i/dz are quite significant near the interface, the total excitation probability $dP/dz = dP_i/dz + dP_v/dz$ is almost a constant inside GaAs.

Figure 6.25a, b show the calculated valence loss spectra (*dashed line*) and the experimentally observed spectra (*solid line*) of a MgO cube. Volume excitation is dominant if the electron probe directly penetrates the cube (Fig. 6.25a); the surface excitation is dominant if the electron probe is moved away from the cube (Fig. 6.25b). Calculated spectra not only show major features of the spectra, but they also exhibit the same absolute scattering probability in reference to observed data.

The classical dielectric response theory is an easy method for describing valence excitations in high-energy electron scattering. For metals under the free electron approximation, quantum mechanical charge density fluctuation theory (Ashley et al. 1974; Ashley and Ferrell 1976) has shown results equivalent to those obtained using the dielectric response theory (Ritchie 1981; Ritchie and Howie 1988). In the excitations of small metal particles (Fujimoto and Komaki 1968; Wang and Cowley 1987a and 1987b), both theories have shown identical results (Ferrell and Echnenique 1985; Kohl 1983). The final result, however, is easy to obtain based on the dielectric response theory.

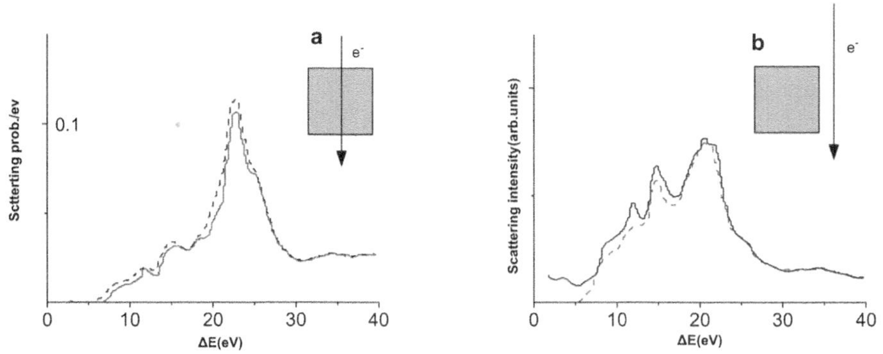

Fig. 6.25 **a** Volume and **b** surface excitation spectra acquired at 100 kV from an MgO cube of about 120 nm thickness. Dashed lines are simulated spectra based on dielectric response theory (Walls and Howie 1989)

6.5.4 *The Mixed Dynamic Form Factor and Generalized Dielectric Function*

The mixed dynamic form factor defined in (6.24) is a key quantity for describing imaging theories of inelastically scattered electrons (Rose 1976a, b; Kohl 1983). We now introduce the relationship between $S\left(\mathbf{u}, \mathbf{u}'\right)$ and the matrix elements of inelastic excitation. From (6.23), the Fourier transform of the excitation matrix elements is directly written in the form

$$H_{nm}(\mathbf{u}) = \frac{e^2}{4\pi^2\varepsilon_0}\frac{\rho_{nm}(\mathbf{u})}{u^2} \tag{6.109a}$$

thus

$$\sum_{m\neq 0} H_{0m}(\mathbf{u})H_{m0}\left(-\mathbf{u}'\right) = \frac{e^4}{16\pi^4\varepsilon_0^2}\frac{1}{u^2u'^2}S\left(\mathbf{u}, \mathbf{u}'\right) \tag{6.109b}$$

More importantly, the mixed dynamic form factor is directly related to the generalized dielectric function $\varepsilon_{\tau\tau}{}'(\omega)$ (Kohl and Rose 1985) by

$$S\left(\tau, \tau'\right) = \int_{-\infty}^{\infty} d\omega \frac{2\pi i\hbar\varepsilon_0}{e^2\left[1 - \exp\left(-\hbar\omega/k_BT\right)\right]}\left(\frac{\tau^2}{\varepsilon_{\tau\tau'}(\omega)} - \frac{\tau'^2}{\varepsilon^*_{\tau'\tau}(\omega)}\right) \tag{6.110}$$

If we take the special case of $\tau = \tau'$, (6.110) is simplified as

$$S(\tau, \tau) = \int_{-\infty}^{\infty} d\omega \frac{4\pi\hbar\varepsilon_0}{e^2\left[1 - \exp\left(-\hbar\omega/k_BT\right)\right]}\tau^2\,\mathrm{Im}\left[-\frac{1}{\varepsilon_{\tau\tau}(\omega)}\right] \tag{6.111}$$

where $\mathrm{Im}\left[-1/\varepsilon_{\tau\tau}(\omega)\right]$ is just the energy loss function (Raether 1980).

For valence excitation, the condition $|\hbar\omega| \gg k_B T$ is always satisfied. Thus the integration of ω from $-\infty$ to 0 vanishes in (6.111), since $\left[1 - \exp\left(-\hbar\omega/k_B T\right)\right]^{-1} \approx 0$. Under this approximation, (6.110) becomes

$$S\left(\tau, \tau'\right) = \frac{2\pi i \hbar \varepsilon_0}{e^2} \int_0^{\infty} d\omega \left[\frac{\tau^2}{\varepsilon_{\tau\tau'}(\omega)} - \frac{\tau'^2}{\varepsilon^*_{\tau'\tau}(\omega)}\right] \tag{6.112}$$

Substituting (6.112) into (6.109a) and (6.109b), the function used in the diffraction theory of valence excitation is expressed as

$$\sum_{n \neq 0} H_{0n}(\tau) H_{n0}\left(-\tau'\right) = \frac{i\hbar e^2}{8\pi^3 \varepsilon_0} \int_0^{\infty} d\omega \left[\frac{1}{\tau'^2 \varepsilon_{\tau\tau'}(\omega)} - \frac{1}{\tau^2 \varepsilon^* \tau' \tau(\omega)}\right] \tag{6.113}$$

The introduction of the generalized dielectric function has greatly simplified calculating the mixed dynamic form factor in valence excitation.

6.6 Inelastic-Scattering Process III: Atomic Inner Shell Excitation

Single-electron excitation, or atomic inner shell excitation, is an important inelastic-scattering process in electron scattering. Figure 6.26 shows an EELS spectrum of a thin $SrTiO_3$ crystal. The Ti $L_{2,3}$, OK, and Ti L_1 edges are shown. The energy loss of the incident electrons is the result of ionizing core-shell electrons. The transition probability of electrons to different valence states is determined by the density of states of the valence band. This is observed in the oxygen K edge in Fig. 6.26. The most important application of atomic inner shell excitation is in high spatial resolution chemical analysis. We now discuss the diffraction behavior of electrons which have excited atomic inner shell electrons.

6.6.1 Excitation Matrix

In high-energy electron scattering, atomic core-shell electrons can be excited to transit from the ground state $| \varphi_0(\mathbf{r} - \mathbf{X}_S) >$ to the state $| \varphi_m(\mathbf{r} - \mathbf{X}_S) >$, discrete (bound state), or continuum (free electron) state at excitation energy ε_n measured from state $| \varphi_0 >$, where $\mathbf{X}_S = \mathbf{R}_m + \mathbf{r}_\alpha$ denotes the position of the atom to be excited. We now use the approach of Maslen and Rossouw (1984) to derive the transition matrix of atomic inner shell excitation. The model is essentially the tight-binding model proposed by Humphreys and Whelan (1969), which allows no overlay of

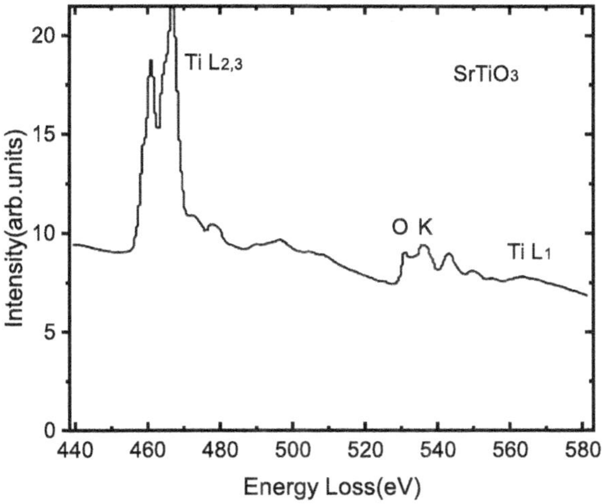

Fig. 6.26 The EELS spectrum of a thin $SrTiO_3$ crystal showing the ionization edges of Ti and O inner shell electrons

atomic wave functions on different lattice sites. We assume a simple lattice of N noninteracting, one-electron atoms, distributed one atom per cell, each at a position $\mathbf{X}_s$. The electron spin effect is ignored. The electronic ground state wave function $| a_0 >$ of the crystal may be described by a Slater (1937) determinant of atomic orbitals,

$$| a_0 >= \frac{1}{(N!)^{1/2}} \det \left[\varphi_0 \left(\mathbf{r} - \mathbf{X}_1\right), \varphi_0 \left(\mathbf{r} - \mathbf{X}_2\right), \ldots, \varphi_0 \left(\mathbf{r} - \mathbf{X}_s\right), \ldots, \varphi_0 \left(\mathbf{r} - \mathbf{X}_N\right)\right] \quad (6.114)$$

Let

$$|D_m >= \frac{1}{(N!)^{1/2}} \det \left[\varphi_0 \left(\mathbf{r} - \mathbf{X}_1\right), \varphi_0 \left(\mathbf{r} - \mathbf{X}_2\right), \ldots, \varphi_n \left(\mathbf{r} - \mathbf{X}_S\right), \ldots, \varphi_0 \left(\mathbf{r} - \mathbf{X}_N\right)\right] \quad (6.115)$$

denotes the determinant obtained from $| a_0 >$ on replacing $\varphi_0 \left(\mathbf{r} - \mathbf{X}_S\right)$ by $\varphi_n \left(\mathbf{r} - \mathbf{X}_S\right)$, where the $s = (m + a_1)$ th atom is assumed to be excited from the ground state to the nth-excited state. The normalized crystal wave function, with wave vector $\mathbf{q}$ corresponding to ionization or excitation of some crystal atoms (Maslen and Rossouw 1984) is

$$|a_n\rangle = \frac{1}{(N!)^{1/2}} \sum_{m=1}^{N_0} \exp\left(2\pi i \mathbf{q} \cdot \mathbf{R}_m\right) |\mathrm{D}_m\rangle \quad (6.116)$$

Using (6.115) and (6.116), the transition matrix element is

$$H_{n0}(\mathbf{r}) = \frac{e^2}{4\pi\varepsilon_0}\sum_m\sum_{\alpha_1}\exp\left(-2\pi i\mathbf{q}\cdot\mathbf{R}_m\right)$$
$$\int d\mathbf{r}'\frac{\varphi_n^*\left(\mathbf{r}'-\mathbf{R}_m-\mathbf{r}_{\alpha_1}\right)\varphi_0\left(\mathbf{r}'-\mathbf{R}_m-\mathbf{r}_{\alpha_1}\right)}{|\mathbf{r}-\mathbf{r}'|} \tag{6.117}$$

the sum of α_1 is over the same type of atoms within the unit cell that will be excited. Taking a Fourier transform of (6.117), using the identity in Eq. (E.2) in Appendix E, we obtain

$$H_{n0}(\mathbf{u}) = \frac{e^2}{4\pi\varepsilon_0}\sum_g\sum_{\alpha_1}\exp\left[-2\pi i(\mathbf{g}-\mathbf{q})\cdot\mathbf{r}_{\alpha_1}\right]\frac{\rho_{n0}^{\prime(\alpha_1)}(\mathbf{g}-\mathbf{q})}{|\mathbf{g}-\mathbf{q}|^2}\delta(\mathbf{u}-\mathbf{g}+\mathbf{q}) \tag{6.118}$$

where the density operator for single-electron excitation is

$$\rho_{n0}^{\prime(\alpha_1)}(\tau) \equiv < \varphi_n(\mathbf{r})|\exp(-2\pi i\tau\cdot\mathbf{r})|\varphi_0(\mathbf{r}) > \tag{6.119}$$

Thus the Fourier coefficients of the transition matrix elements are

$$H_g^{n0}(\mathbf{q}) = \frac{e^2}{4\pi^2\varepsilon_0}\sum_{\alpha_1}\exp\left[-2\pi i(\mathbf{g}-\mathbf{q})\cdot\mathbf{r}_{\alpha_1}\right]\frac{\rho_{n0}^{\prime(\alpha_1)}(\mathbf{g}-\mathbf{q})}{|\mathbf{g}-\mathbf{q}|^2} \tag{6.120}$$

The calculation of $\rho_{n0}^{\prime(\alpha_1)}$ is the core part of single-electron excitation (Inokuti 1971; Whelan 1965a). An appropriate representation of the atomic wave function has to be considered for different purposes (Saldin and Rez 1987; Maslen 1987; Weickenmeier and Kohl 1989).

6.6.2 Absorption Potential

Substituting (6.120) into (6.33) and assuming there is only one excited state, the imaginary component of the crystal potential contributed by single-electron excitation is

$$U_{gh}^{\prime(I)} = -\frac{m_e^2 e^4}{64\pi^7\hbar^4\varepsilon_0^2 K_0 V_c}\sum_{g'}\sum_{\alpha_1}\sum_{\alpha_2}\exp\left[-2\pi i\left(\mathbf{g}'+\mathbf{q}\right)\cdot\mathbf{r}_{\alpha_1}\right]$$
$$\times\exp\left[-2\pi\left(\mathbf{h}-\mathbf{g}-\mathbf{g}'-\mathbf{q}\right)\cdot\mathbf{r}_{\alpha_2}\right]$$
$$\times\frac{\rho_{0n}^{\prime(\alpha_1)}\left(\mathbf{g}'+\mathbf{q}\right)\rho_{n0}^{\prime(\alpha_2)}\left(\mathbf{h}-\mathbf{g}-\mathbf{g}'-\mathbf{q}\right)}{|\mathbf{g}'+\mathbf{q}|^2\,|\mathbf{h}-\mathbf{g}-\mathbf{g}'-\mathbf{q}|^2}\int d\sigma(\mathbf{u})\delta\left(\mathbf{u}-\mathbf{k}_g-\mathbf{g}'-\mathbf{q}\right) \tag{6.121}$$

For single-electron excitation, if the scattering from each atom site is treated as incoherent, the terms with $\alpha_1 \neq \alpha_2$ vanish; thus (6.121) becomes

$$U_{gh}^{\prime(I)} = -\frac{m_e^2 e^4}{64\pi^7 \hbar^4 \varepsilon_0^2 K_0 V_c} \sum_{\alpha_1} \exp\left[2\pi i(\mathbf{g}-\mathbf{h}) \cdot \mathbf{r}_{\alpha_1}\right]$$
$$\times \sum_{g'} \frac{\rho_{0n}^{\prime(\alpha')}\left(\mathbf{g}'+\mathbf{q}\right) \rho_{n0}^{\prime(\alpha_1)}\left(\mathbf{h}-\mathbf{g}-\mathbf{g}'-\mathbf{q}\right)}{|\mathbf{g}'+\mathbf{q}|^2 \mid \mathbf{h}-\mathbf{g}-\mathbf{g}'-\mathbf{q}|^2}$$
$$\times \int d\sigma(\mathbf{u})\delta\left(\mathbf{u}-\mathbf{k}_g-\mathbf{g}'-\mathbf{q}\right) \tag{6.122}$$

The effect of atomic thermal vibration may be introduced if the $\exp\left[2\pi i(\mathbf{g}-\mathbf{h}) \cdot \mathbf{r}_{\alpha_1}\right]$ term is averaged over time, resulting in the appearance of the DW factor

$$U_{gh}^{\prime(I)} = -\frac{m_e^2 e^4}{64\pi^7 \hbar^4 \varepsilon_0^2 K_0 V_c} \sum_{\alpha_1} \exp\left[2\pi i(\mathbf{g}-\mathbf{h}) \cdot \mathbf{r}_{\alpha_1} - W_{\alpha_1}(\mathbf{g}-\mathbf{h})\right]$$
$$\times \sum_{g'} \frac{\rho_{0n}^{\prime(\alpha_1)}\left(\mathbf{g}'+\mathbf{q}\right) \rho_{n0}^{\prime(\alpha_1)}(\mathbf{h}-\mathbf{g}-\mathbf{g}'-\mathbf{q})}{|\mathbf{g}'+\mathbf{q}|^2|\mathbf{h}-\mathbf{g}-\mathbf{g}'-\mathbf{q}|^2} \int d\sigma(\mathbf{u})\delta(\mathbf{u}-\mathbf{k}_g-\mathbf{g}'-\mathbf{q}) \tag{6.123}$$

The accuracy of the calculated imaginary potential strongly depends on the model chosen for the atomic inner shell electron wave functions. Calculations have been made using the Thomas-Fermi atomic model (Yoshioka 1957) and atomic wave functions (Whelan 1965a; Allen and Rossouw 1990). The K (for $Z_0 < 15$) and L (for $Z_0 < 30$) shell excitations of light elements can be reasonably modeled using the hydrogenlike atomic wave function (for a review, see Egerton 1986). For the M shells of heavier elements, many-body effects as well as inner-shell-screening effects are particularly important, making the calculation rather difficult. Calculations by Allen and Rossouw (1990) have shown that the absorption potential introduced by K shell ionization is much smaller than that introduced by TDS.

Based on the real-space approach, the absorption potential V' given by (6.39) has been calculated by Coene and Van Dyck (1990) for single-electron inner shell excitations, and their results have been given as

$$V'(\mathbf{b}) = \left(1 + E/m_0 c_0^2\right)^2 \pi \cdot \lambda \sum_{p=1}^{4} \frac{c_p}{d_p} \exp\left(-\pi^2 b^2/d_p\right) \tag{6.124}$$

where parameters c_p and d_p, which are independent of crystal structure and accelerating voltage, have been tabulated in Coene and Van Dyck (1990). This provides a quick method for estimating the absorption potential. Figure 6.27 shows the calculated absorption potential V' for oxygen due to single-electron excitation. For easy

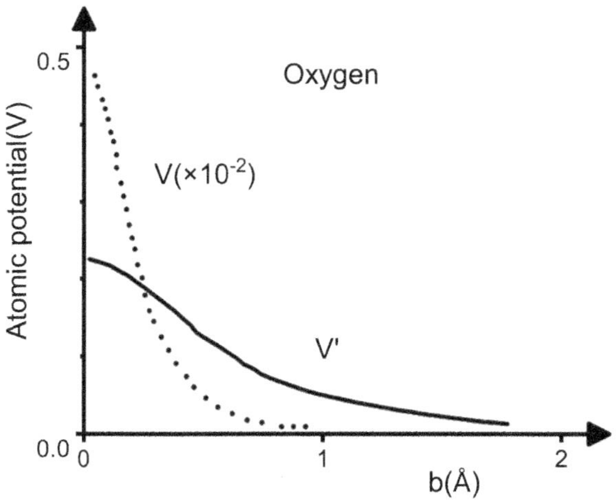

Fig. 6.27 The calculated oxygen absorption potential V' (*solid line*) due to single-electron excitations. The elastic-scattering potential V (*dashed line*) is reduced by a factor of 100 to be compared with V' (Coene and Van Dyck 1990)

comparison, the elastic-scattering factor is reduced by a factor of 100. It is apparent that the full width at half maximum (FWHM) of V' is much wider than that of V.

6.7 Diffraction And Channeling Effects in X-Ray and Auger Electron Emissions

In electron beam microanalysis, signals from atomic inner shell ionizations are collected to determine specimen chemical composition using techniques of electron energy loss spectroscopy and EDS. The rate at which inelastic signals are generated depends strongly on the diffracting conditions of the incident electron, because the channeling current is determined by Bloch waves that have been excited. Intensities of ionization edges in EELS depend on the crystal orientation (Disko et al. 1982). The X-ray emission rate from inner shell electron transitions is sensitive to the channeling status of electrons along atomic planes (for a review, see Spence 1988b), leading to a new technique of ALCHEMI, which is used to define distributions of minor impurities in the unit cell. This section discusses the diffraction and channeling effects of incident electrons on the generation of inelastic signals.

6.7.1 Localization in Atomic Inner Shell Excitation

In atomic inner shell excitations, the localization effect refers to the maximum impact distance (in classical term) of an incident electron from the nucleus to initiate effective excitation of core-shell electrons. Electronic transition is considered to occur

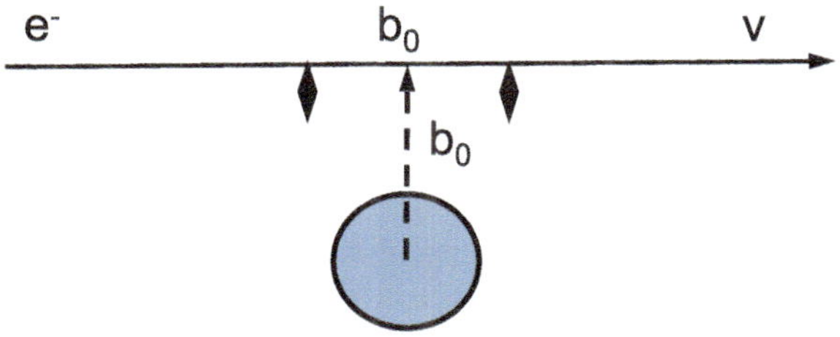

Fig. 6.28 A model for evaluating the localization distance of an inelastic excitation process

only when the impact parameter is smaller than the localization distance. A simple approximated criterion for localization of electronic excitation is given in terms of the Bohr adiabatic impact parameter (Howie 1979). At distance b_0 from the track of a fast electron, the pulse of an electric field caused by its passage at velocity v has a duration time $\Delta t \approx b_0/v$ (Fig. 6.28). Hence we expect electronic excitations ΔE to be caused only by fast electrons with impact parameters $b_0 \leq \hbar v/\Delta E$, since $\Delta t \Delta E \approx \hbar$. Crystallographic effects involving planes of spacing d_g thus do not occur unless $b_0 < d_g$, so the criterion for their existence is

$$\Delta E \geq \frac{\hbar v}{d_g} \tag{6.125a}$$

and

$$b_0 \approx \frac{\hbar v}{\Delta E} \tag{6.125b}$$

For 100 keV electrons and an interplanar distance of 0.2 nm, the criterion implies that crystallographic effects on localization are observed at small angles for losses ΔE in excess of 1 or 2 keV. Localization effects may be observed at smaller losses ΔE but only if a detector confined to scattering angles comparable with or greater than the Bragg angle θ_g is used (Craven et al. 1978).

From (6.125b), the localization parameter is inversely proportional to the energy transferred in the inelastic excitation process. Therefore for deeply bounded inner shell electrons, X-rays are not emitted unless the impact distance of the incident electron approaches the nucleus. In this case, the excitation probability of inner shell electrons is approximately described by a Dirac delta function. The validity of the delta function approximation relies on the energy loss of the electron.

6.7.2 Delocalization in Electron Impact Ionization in Crystals

In general, significant errors can be introduced if the atomic inner shell excitation is treated according to the perfectly localized inelastic excitation approximation, in which the excitation probability function in real space is mathematically expressed as a delta function centered at the atomic nucleus. This approximation means that excitation occurs only when the incident electron directly hits the atomic nucleus. The channeling phenomenon of the incident beam in Sect. 2.3 is an additional factor that introduces crystallographic and dynamic diffraction effects in generating X-rays. This effect has been used to locate impurity atoms in crystalline materials, i.e., the ALCHEMI technique (see Spence 1992, for a review). The delocalization effect in atomic ionization in crystal environments has been considered by several authors (Cherns et al. 1973; Rossouw 1981; Allen 1993; Allen and Rossouw 1993). In this section, we intend to derive a full dynamic description of electron inelastic scattering in crystals with consideration to possible delocalization and channeling effects. Errors introduced by the delta function approximation in X-ray excitation will be examined.

We pointed out in Sect. 6.3.4 that $\nabla \cdot \mathbf{j}_0$ is the rate per unit volume at which the number of elastic electrons decreases. From (6.15), the rate at which inelastic electrons are generated per unit volume is

$$\frac{dn_{ie}}{d\mathbf{r}} = -\nabla \cdot \mathbf{j}_0 = i\frac{\hbar}{2m_e}\left[\psi_0^* \nabla^2 \psi_0 - \psi_0 \nabla^2 \psi_0^*\right] \tag{6.126}$$

We now use the Bloch wave solution of ψ_0 to calculate the yield of the inelastic signal

$$\psi_0(\mathbf{r}) = \sum_i \alpha_i \sum_g C_g^{(i)} \exp\left[2\pi i(\mathbf{K}+\mathbf{g})\cdot\mathbf{r} + 2\pi i v_i z\right] \tag{6.127}$$

Substituting (6.127) into (6.126) and integrating over the crystal volume, the total number of inelastic electrons to be generated is

$$\begin{aligned} n_{ie} &= i V_c \frac{2\pi^2\hbar}{m_e} \sum_i \sum_j \alpha_j^* \alpha_i \sum_g \sum_h C_h^{(j)*} C_g^{(i)} \left[-(\mathbf{K}+v_i\hat{z}+\mathbf{g})^2 + \left(\mathbf{K}+v_j^*\hat{z}+\mathbf{h}\right)^2\right] \\ &\times \int d\mathbf{b} \exp[2\pi i(\mathbf{g}-\mathbf{h})\cdot\mathbf{b}] \int_0^d dz \exp\left[2\pi i\left(v_i - v_j\right)z\right] \\ &= V_c \frac{\pi\hbar}{m_e} \sum_i \sum_j \alpha_j^* \alpha_i \sum_g C_g^{(j)*} C_g^{(i)} \left[-(\mathbf{K}+v_i\hat{z}+\mathbf{g})^2 + \left(\mathbf{K}+v_j^*\hat{z}+\mathbf{g}\right)^2\right] \\ &\times \frac{\exp\left[2\pi i\left(v_i - v_j\right)d\right] - 1}{\left(v_i - v_j\right)} \end{aligned} \tag{6.128}$$

where v_j may be complex due to the introduction of an imaginary potential into the Schrödinger equation; **g** and **h** are assumed to be in the ZOLZ. We now consider equations satisfied by Bloch wave coefficients,

$$\left[K_0^2 - (\mathbf{K}+v_i\hat{z}+\mathbf{g})^2\right] C_g^{(i)} + \sum_h \left[U_{g-h} + U'_{gh}\right] C_h^{(i)} = 0 \tag{6.129}$$

Taking a complex conjugate of (6.129) for the jth Bloch wave, we have

$$\left[K_0^2 - \left(\mathbf{K} + v_j^* \hat{z} + \mathbf{g}\right)^2\right] C_g^{(j)*} + \sum_h \left[U_{g-h}^* + U_{gh}^{\prime *}\right] C_h^{(j)*} = 0 \tag{6.130}$$

Subtracting (6.130) multiplied by $C_g^{(i)}$ from (6.129) multiplied by $C_g^{(j)*}$, we obtain

$$\begin{aligned} &C_g^{(j)*} C_g^{(i)} [-(\mathbf{K} + v_j \hat{z} + \mathbf{g})^2 + \left(\mathbf{K} + v_j^* \hat{z} + \mathbf{g}\right)^2] \\ &\quad = -C_g^{(j)*} \sum_h \left(U_{g-h} + U_{gh}^{\prime}\right) C_h^{(i)} + C_g^{(i)} \sum_h \left(U_{g-h}^* + U_{gh}^{\prime *}\right) C_h^{(j)*} \end{aligned} \tag{6.131}$$

Substituting this relation into (6.128) and using $U_{g-h} = U_{h-g}^*$, we have

$$n_{ie} = V_c \frac{\pi \hbar}{m_e} \sum_i \sum_j \alpha_j^* \alpha_i \sum_g \sum_h C_g^{(j)*} C_h^{(i)} \left(U_{hg}^{\prime *} - U_{gh}^{\prime}\right) \frac{\exp\left[2\pi i \left(v_i - v_j\right) d\right] - 1}{\left(v_i - v_j\right)} \tag{6.132}$$

We now examine the imaginary potential U' given in Sect. 6.6.2. To simplify the final expression, we introduce an approximation by assuming that the sum over $\mathbf{g}'$ in (6.123) is dominated by a single term, with $\mathbf{g}' = -\mathbf{g}$. This approximation is based on evaluating the integral of the delta function in (6.123); thus

$$\begin{aligned} U_{gh}^{\prime(I)} \approx &-\frac{m_e^2 e^4}{64\pi^7 \hbar^4 \varepsilon_0^2 K_0 \Omega} \sum_{\alpha_1} \exp\left[2\pi i (\mathbf{g} - \mathbf{h}) \cdot \mathbf{r}_{\alpha_1}\right. \\ &\left. - W_{\alpha_1}(\mathbf{g} - \mathbf{h})\right] \frac{\rho' 0_n^{(\alpha_1)}(-\mathbf{g} + \mathbf{q}) \rho_{n0}^{\prime(\alpha_1)}(\mathbf{h} - \mathbf{q})}{|\mathbf{g} - \mathbf{q}|^2 |\mathbf{h} - \mathbf{q}|^2} \end{aligned} \tag{6.133}$$

which satisfies

$$U_{hg}^{\prime(I)*} = U_{gh}^{\prime(I)}$$

By neglecting the real component of $U_{gh}^{\prime}$, substituting (6.133) into (6.132), the rate at which the inelastic transition is excited is

$$n_{ie} = \frac{m_e e^4 N_0}{16\pi^6 \hbar^3 \varepsilon_0^2 K_0} \sum_i \sum_j \sum_g \sum_h \text{(B.W.) (Sites) (Kin.) (Thickness)} \tag{6.134a}$$

where Bloch wave components are

$$\text{(B.W.)} = \alpha_j^* \alpha_i C_g^{(j)*} C_h^{(i)} \tag{6.134b}$$

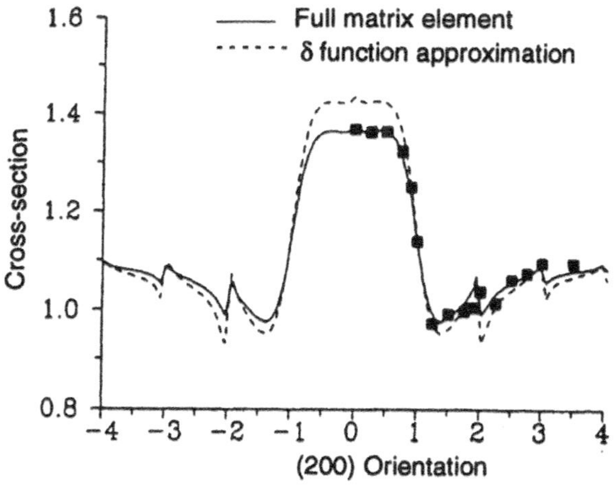

Fig. 6.29 A comparison of the measured X-ray emission rate of Al under the (200) systematic row orientation with those calculated based on hydrogenlike (*solid line*) and delta function (*dashed line*) models. The squared points are the experimentally measured X-ray emission signal; $T = 300$ K, $E_0 = 300$ keV, and $d = 348$ nm. The vertical axis is scaled for arbitrary units (Allen and Rossouw 1993)

The atomic sites term is

$$(\text{Sites}) = \sum_{\alpha_1} \exp\left[2\pi i(\mathbf{g}-\mathbf{h})\cdot\mathbf{r}_{\alpha_1} - W_{\alpha_1}(\mathbf{g}-\mathbf{h})\right] \tag{6.134c}$$

The thickness-dependent term is

$$(\text{Thickness}) = \frac{\exp\left[2\pi i\left(\upsilon_i - \upsilon_j\right)d\right] - 1}{\left(\upsilon_i - \upsilon_j\right)} \tag{6.134d}$$

and the kinematic scattering term

$$(\text{Kin.}) = \frac{\rho_{0n}^{\prime(\alpha_1)}(-\mathbf{g}+\mathbf{q})\rho_{n0}^{\prime(\alpha_1)}(\mathbf{h}-\mathbf{q})}{|\mathbf{g}-\mathbf{q}|^2|\mathbf{h}-\mathbf{q}|^2} \tag{6.134e}$$

Equations (6.134a)–(6.134e) is consistent with the result of Allen and Rossouw (1993).

Excitations of light elements tend to be less localized because shell electrons are distributed at larger distances from the nucleus due to weaker attraction. Figure 6.29 shows the theoretically calculated and experimentally observed X-ray emission rate of aluminum as a function of the crystal orientation. The zero point corresponds to the (200) Bragg condition. For aluminum, the inelastic-scattering matrix is calculated from either the hydrogenlike model or the delta function model (i.e., the perfectly localized scattering model). Significant differences are shown in Fig. 6.29, and it is apparent that the calculation based on the hydrogenlike model fits the experimental observation best. Figure 6.29 also shows the dependence of X-ray emission on the diffracting (or channeling) condition.

6.8 Minimum Momentum Transfer in Inelastic Scattering

Two fundamental quantities, momentum transfer and energy loss, are involved in any inelastic-scattering process. For the electron-crystal system, the laws of conservation of energy and conservation of momentum have to be satisfied; a relationship between electron momentum transfer and energy loss is thus expected. For simplification, we discuss a case in which no diffraction is allowed and the incident electron is inelastically scattered through an angle θ' as the result of inelastic interaction with the crystal.

6.8.1 Conservation of Energy

The total energy of an incident electron is written as

$$w = E_0 + m_0 c_0^2 = \gamma m_0 c_0^2 \tag{6.135}$$

The momentum of the electron is

$$p = \gamma m_0 v = \hbar K_0 \tag{6.136}$$

Combining (6.135) and (6.136), we obtain

$$w = \left[\left(m_0 c_0^2 \right)^2 + p^2 c_0^2 \right]^{1/2} = \left[\left(m_0 c_0^2 \right)^2 + \hbar^2 K_0^2 c_0^2 \right]^{1/2} \tag{6.137}$$

The conservation of energy requires

$$w - \Delta E = w' = \left[\left(m_0 c_0^2 \right)^2 + \hbar^2 K^2 c_0^2 \right]^{1/2} \tag{6.138}$$

where w' is the total energy of the electron after inelastic scattering. Substituting (6.137) into (6.138), an equation relating the change in magnitude of the electron wave number to the energy loss is derived

$$K^2 = K_0^2 - \frac{2\gamma m_0 \Delta E}{\hbar^2} + \frac{\Delta E^2}{\hbar^2 c_0^2} \tag{6.139}$$

This relation is independent of the scattering angle.

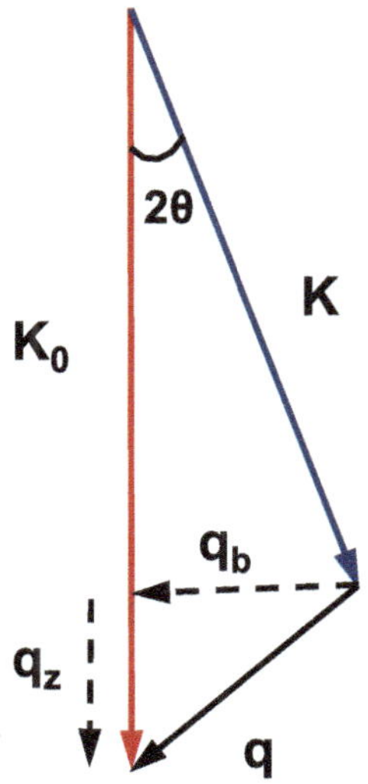

Fig. 6.30 An electron-scattering diagram showing the momentum transfer in inelastic scattering

6.8.2 Conservation of Momentum

The conservation of momentum is shown in Fig. 6.30 using the vector triangle diagram, which requires

$$\mathbf{K}_0 - \mathbf{K} = \mathbf{q} \tag{6.140}$$

Applying the cosine rule,

$$q^2 = K_0^2 + K^2 - 2K_0K\cos\theta' \tag{6.141}$$

where $\theta' = 2\theta$, the angle between $\mathbf{K}_0$ and $\mathbf{K}$. The minimum momentum transfer occurs at $\theta' = 0$, so that $q_{\min} = K_0 - K$. For small energy loss, the last term in (6.139) is dropped; thus

$$(K - K_0)(K + K_0) \approx -\frac{2\gamma m_0 \Delta E}{\hbar^2}$$

or

$$q_{\min} = K_0 - K \approx \frac{\gamma m_0 \Delta E}{\hbar^2 K_0} = K_0\theta_E \tag{6.142}$$

where $\theta_E = \Delta E/\gamma m_0 v^2$ is the characteristic angle of inelastic scattering for a particular energy loss ΔE. In general, (6.141) can be rewritten as

$$q^2 = (K_0 - K)^2 + 2K_0K\left(1 - \cos\theta'\right) = (q_{\min})^2 + 4K_0K\sin^2\left(\theta'/2\right) \tag{6.143}$$

For small-angle scattering,

$$q^2 \approx K_0^2 \left(\theta'^2 + \theta_E^2\right) \tag{6.144}$$

which means that the momentum transfer in the direction parallel to the beam direction is approximately $q_z \approx K_0\theta_E = q_{\text{min}}$. This is independent of the electron-scattering angle; the momentum transfer in the direction perpendicular to the incident beam direction (i.e., in the diffraction plane) is $q_b = K_0\theta'$. These relations are used in future discussions.

6.9 Summary

Chapter 6 describes the inelastic-scattering processes involved in electron diffraction. The main point of Chap. 6 is to examine the effects of inelastic excitations on the diffraction behavior of elastically scattered electrons. Yoshioka's coupled equations were introduced, which are the first principles equations for the remaining chapters in this book. A rigorous theoretical basis is given to describe the absorption effect on elastic electron diffraction. Contributions to the imaginary potential from various inelastic-scattering processes were examined in detail. Finally we illustrated the physical basis for understanding electron diffraction effects on the emission of X-ray and Auger electrons.

Chapter 7
Semiclassical Theory of Thermal Diffuse Scattering

Thermal diffuse scattering is an important inelastic-scattering process in electron diffraction. Since the energy loss from phonon-scattered electrons is much smaller than 0.1 eV, which is rather small in comparison to either the energy resolution of the current available energy filter or the energy spread of the electron source, it is probably impossible to separate the intensity of elastically scattered electrons from that of the TDS electrons by using an energy filter. Therefore the contribution of TDS electrons has to be considered in quantitative data analysis in electron microscopy. In this chapter reviews some theoretical approaches for describing the diffraction of TDS electrons based on the semiclassical model. We first use the two-beam approximation to illustrate the basic principle for dealing with TDS. Then the theory is extended to the many-beam case. The coherent effect in TDS is examined and a coherent length is defined. Finally we discuss the contribution of Huang scattering in electron diffraction pattern. Several other dynamic theories of TDS are given in Chaps. 8–11.

7.1 Frozen Lattice Model

We have described the creation and annihilation of phonons in electron diffraction. The mathematical treatment is rather lengthy. Since atomic vibration is a time-dependent process, its perturbation on crystal potential is time-dependent. The question here is whether or not we need to use the time-dependent Schrödinger equation to describe the diffraction behavior of TDS electrons. We must first examine the following factors before a clear answer can be given.

In high-energy electron diffraction, the traveling velocity of the electron is larger than one-half of the speed of light ($v = 0.53c_0$ for 100−keV electrons). For a thin crystal of thickness $d = 100$ nm, the time that the electron interacts with the crystal atoms is on the order of $\Delta t \approx 6 \times 10^{-15}$ s. In general, the vibration period T of crystal atoms is on the order of 10^{-13} s (Sinha 1973; Mitra and Massa 1982). The

Z. L. Wang, *Elastic and Inelastic Scattering in Electron Diffraction and Imaging*, Springer Series in Solid-State Sciences 205, https://doi.org/10.1007/978-3-031-90819-4_7

large-amplitude, low-frequency acoustic waves that produce most of the atomic displacements and are responsible for TDS are many times slower. Thus the condition of $\Delta t \ll T$ is satisfied. In reality, the electron wave packet is unlocalized in its traveling direction due to a small spread in the time of emission (Loan et al. 1991). The length of the wave packet can be estimated with Heisenberg's uncertainty principle $\Delta E \Delta \tau \geq \hbar$. Assuming a 0.25 eV coherent energy spread for a field emission tip in the electron microscope, the emission time is $\sim 2.5 \times 10^{-15}$ s; the wave packet length is $\sim 3 \times 10^{2}$ nm; and the interaction time is $\sim 2 \times 10^{-2}$ vibration period T. Therefore whether the electron is treated classically or quantum mechanically, the electron-crystal interaction time is much shorter than the atom vibration period, and the atoms are seen as stationary by the incident electron. Thus the scattering of an incident electron is basically a quasielastic-scattering (without energy loss) process by distorted frozen crystal lattices.

For a field emission tip, the beam current is on the order of 10^{12} electrons per second, so that the average time between successive electrons passing through the specimen is about 100 atomic vibration periods. This delay is sufficiently large so that atomic displacements seen by successive electrons are essentially uncorrelated. The physical process of accumulating millions of electrons, each of which has been scattered by an independent lattice configuration, can be considered as a Monte Carlo integration. This treatment includes Bragg scattering and quasielastic thermal diffuse (or phonon) scattering, but not valence or atomic inner shell inelastic excitations. In addition, the average distance between two successive incident electrons is on the order of 10^{5} nm; this means that we can not only completely neglect the interaction between incident electrons, but also the scattering of the electron is completely independent of that of the previous electron or the next electron. In other words, single-electron-scattering theory holds.

Theoretical calculation has found that the lifetime of a phonon is on the order of 10^{-13} s (Björkman et al. 1967; Woll and Kohn 1962). The limited lifetime of phonons is due to anharmonic effects in atom vibrations. This is much larger than the interaction time Δt of the electron with the crystal. Thus there is no phonon decay during the scattering of an incident electron. The average time interval between two incident electrons is $\sim 10^{-12}$ s, which is much longer than the lifetime of phonons. Therefore the crystal can be considered in its ground state for each incident electron, and there is no need to consider phonons created by the previous incident electron.

In practice, it usually takes more than 0.05 s to record a diffraction pattern or image. During this period of time, the crystal lattice has experienced millions of different vibration configurations. Thus the acquired electron diffraction pattern is an incoherent summation of independent scatterings for millions of electrons. This is equivalent to time averaged over the intensity distribution produced by one-electron scattering.

Therefore an electron diffraction pattern or image is considered as the sum of scattering intensities for many instantaneous pictures of the displaced atoms. In other words, TDS is actually a statistically averaged, quasielastic scattering of electrons by crystals whose atoms are arranged in different configurations for every incident electron. Figure 7.1 shows two schematic crystal configurations at time $t = 0$ and

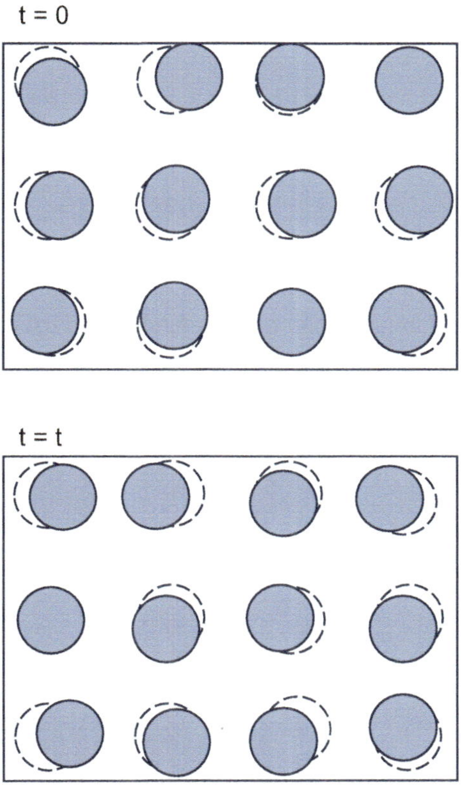

Fig. 7.1 A diagram of frozen lattice configurations of a crystal at time **a** $t = 0$ and **b** $t = t$. This is the basic picture of classical TDS theory

$t = t$. This is the so-called frozen lattice model used in classical TDS theory. The main task involved in the theory is to perform the time average of electron diffraction intensities for a vast number of different thermal vibration configurations. This statistical average can easily be performed under the kinematic scattering approximation (James 1965; Cochran 1963; Maradudin et al. 1971). But the calculation becomes huge if dynamic scattering effects are important.

7.2 Two-Beam TDS Theory

The frozen lattice model was first introduced by Hall and Hirsch (1965) for calculating the absorption coefficients of TDS. For simplification, a two-beam Bragg reflection approximation is made, and the Einstein model, in which all atoms are vibrating independently without any phase correlation, is assumed. We now consider a Bloch wave

$$B_i = \frac{1}{\sqrt{2}} \left\{ \exp\left(2\pi i \mathbf{K}_0 \cdot \mathbf{r}\right) - (-1)^i \exp\left[2\pi i \left(\mathbf{K}_0 + \mathbf{g}\right) \cdot \mathbf{r}\right] \right\} \tag{7.1}$$

with $i = 1$ and 2. The intensity under the kinematic scattering approximation for this Bloch wave to be scattered to a plane wave $\exp(2\pi i \mathbf{K} \cdot \mathbf{r})$ by the atoms at $\mathbf{r}_k{}'$ is

$$I = \frac{1}{2} \left| \sum_{\kappa} \left\{ f^e(\tau) \exp\left(2\pi i \tau \cdot \mathbf{r}'_\kappa\right) - (-1)^i f^e(\tau - \mathbf{g}) \exp\left[2\pi i (\tau - \mathbf{g}) \cdot \mathbf{r}_\kappa{}'\right] \right\} \right|^2 \tag{7.2}$$

where $\tau = \mathbf{K} - \mathbf{K}_0$ is a vector in reciprocal space that has the same meaning as $\mathbf{u}$. For simplification, we assume that the crystal is monatomic. Thermal vibration introduces a small displacement to the position of each atom $\mathbf{r}_\kappa{}' = \mathbf{r}_\kappa + \mathbf{u}_\kappa$. Substituting this relation into (7.2), taking a time average over the intensity distribution, we have

$$\begin{aligned}
I &= \frac{1}{2} \sum_{\kappa} \sum_{\kappa'} < \{ \left[f^e(\tau)\right]^2 \exp\left[2\pi i \tau \cdot (\mathbf{r}_\kappa + \mathbf{u}_\kappa - \mathbf{r}_{\kappa'} - \mathbf{u}_{\kappa'})\right] \\
&+ \left[f^e(\tau - \mathbf{g})\right]^2 \exp\left[2\pi i (\tau - \mathbf{g}) \cdot (\mathbf{r}_\kappa + \mathbf{u}_\kappa - \mathbf{r}_{\kappa'} - \mathbf{u}_{\kappa'})\right] \\
&- (-1)^i 2 f^e(\tau) f^e(\tau - \mathbf{g}) \cos\left[2\pi \tau \cdot (\mathbf{r}_\kappa + \mathbf{u}_\kappa - \mathbf{r}_{\kappa'} - \mathbf{u}_{\kappa'}) + 2\pi \mathbf{g} \cdot (\mathbf{r}_{\kappa'} + \mathbf{u}_{\kappa'})\right] \} > \\
&= \frac{1}{2} \left[f^e(\tau)\right]^2 \{ N + \sum_{\kappa \neq \kappa'} \sum \exp\left[2\pi i \tau \cdot (\mathbf{r}_\kappa - \mathbf{r}_{\kappa'})\right] \exp[-2W(\tau)] \} \\
&+ \frac{1}{2} \left[f^e(\tau - \mathbf{g})\right]^2 \{ N + \sum_{\kappa \neq \kappa'} \sum \exp\left[2\pi i (\tau - \mathbf{g}) \cdot (\mathbf{r}_\kappa - \mathbf{r}_{\kappa'})\right] \exp[-2W(\tau - \mathbf{g})] \} \\
&- (-1)^i f^e(\tau) f^e(\tau - \mathbf{g}) \sum_{\kappa} \sum_{\kappa'} < \cos\left[2\pi \tau \cdot (\mathbf{r}_\kappa + \mathbf{u}_\kappa - \mathbf{r}_{\kappa'} - \mathbf{u}_{\kappa'}) \right. \\
&\left. + 2\pi \mathbf{g} \cdot (\mathbf{r}_{\kappa'} + \mathbf{u}_{\kappa'})\right] >
\end{aligned} \tag{7.3}$$

where $W(\tau)$ is the DW factor. Since for simple fcc and bcc cubic structures

$$\begin{aligned}
&\sum_{\kappa} \sum_{\kappa'} < \cos\left[2\pi \tau \cdot (\mathbf{r}_\kappa + \mathbf{u}_\kappa - \mathbf{r}_{\kappa'} - \mathbf{u}_{\kappa'}) + 2\pi \mathbf{g} \cdot (\mathbf{r}_{\kappa'} + \mathbf{u}_{\kappa'})\right] > \\
&= N \exp[-W(\mathbf{g})] + \sum_{\kappa \neq \kappa'} \sum < \cos\left[2\pi \tau \cdot (\mathbf{r}_\kappa + \mathbf{u}_\kappa) + 2\pi (\mathbf{g} - \tau) \cdot (\mathbf{r}_{\kappa'} + \mathbf{u}_{\kappa'})\right] > \\
&= N \exp[-W(\mathbf{g})] + \sum_{\kappa \neq \kappa'} \sum \mathrm{Re} \left\{ \exp\left[2\pi i \tau \cdot \mathbf{r}_\kappa + 2\pi i (\mathbf{g} - \tau) \cdot \mathbf{r}_{\kappa'}\right] \right. \\
&\left. \times < \exp\left[2\pi \tau \cdot \mathbf{u}_\kappa + 2\pi (\mathbf{g} - \tau) \cdot \mathbf{u}_{\kappa'}\right] > \right\} \\
&= N \exp[-W(\mathbf{g})] + \sum_{\kappa \neq \kappa'} \sum \cos\left[2\pi \tau \cdot \mathbf{r}_\kappa + 2\pi (\mathbf{g} - \tau) \cdot \mathbf{r}_{\kappa'}\right] \\
&\times \exp[-W(\tau) - W(\mathbf{g} - \tau)]
\end{aligned} \tag{7.4}$$

Equation (7.3) becomes

$$
\begin{aligned}
I =&\frac{1}{2}\left[f^e(\tau)\right]^2\{N+\sum_{\kappa\neq\kappa'}\sum\exp\left[2\pi i\tau\cdot(\mathbf{r}_\kappa-\mathbf{r}_{\kappa'})\right]\exp[-2\,W(\tau)]\}\\
&+\frac{1}{2}\left[f^e(\tau-\mathbf{g})\right]^2\{N+\sum_{\kappa\neq\kappa'}\sum\exp\left[2\pi i(\tau-\mathbf{g})\cdot(\mathbf{r}_\kappa-\mathbf{r}_{\kappa'})\right]\exp[-2\,W(\tau-\mathbf{g})]\}\\
&-(-1)^i f^e(\tau)f^e(\tau-\mathbf{g})\{N\exp[-W(\mathbf{g})]\\
&+\sum_{\kappa\neq\kappa'}\sum\cos\left[2\pi\tau\cdot\mathbf{r}_\kappa+2\pi(\mathbf{g}-\tau)\cdot\mathbf{r}_{\kappa'}\right]\exp[-W(\tau)-W(\mathbf{g}-\tau)]\}
\end{aligned}
\quad (7.5)
$$

Part of the intensity can be considered Bragg scattering, and the remaining is TDS. Bragg scattering is obtained by calculating the intensity scattered by the crystal with atoms whose scattering amplitude is $\{f^e\exp[-W(\tau)]\}$, which is

$$
\begin{aligned}
I_B =&\frac{1}{2}\left[f^e(\tau)\right]^2\sum_\kappa\sum_{\kappa'}\exp\left[2\pi i\tau\cdot(\mathbf{r}_\kappa-\mathbf{r}_{\kappa'})\right]\exp[-2\,W(\tau)]\\
&+\frac{1}{2}\left[f^e(\tau-\mathbf{g})\right]^2\sum_\kappa\sum_{\kappa'}\exp\left[2\pi i(\tau-\mathbf{g})\cdot(\mathbf{r}_\kappa-\mathbf{r}_{k'})\right]\exp[-2\,W(\tau-\mathbf{g})]\\
&-(-1)^i f^e(\tau)f^e(\tau-\mathbf{g})\exp[-W(\tau)-W(\mathbf{g}-\tau)]\\
&\times\sum_\kappa\sum_{\kappa'}\cos\left[2\pi\tau\cdot\mathbf{r}_\kappa+2\pi(\mathbf{g}-\tau)\cdot\mathbf{r}_{\kappa'}\right]
\end{aligned}
\quad (7.6)
$$

Subtracting (7.6) from (7.5), the TDS intensity is

$$
\begin{aligned}
I_{\mathrm{TDS}} =&\frac{1}{2}N\big(f^e(\tau)^2\{1-\exp[-2\,W(\tau)]\}+\left[f^e(\tau-\mathbf{g})\right]^2\{1-\exp[-2\,W(\tau-\mathbf{g})]\}\\
&-(-1)^i 2f^e(\tau)f^e(\tau-\mathbf{g})\{\exp[-W(\mathbf{g})]-\exp[-W(\tau)-W(\mathbf{g}-\tau)]\}\big)
\end{aligned}
\quad (7.7)
$$

Equation (7.7) was derived from the Einstein model without considering phonon dispersion. Several points can be made from (7.7). For a large-angle TDS, so that $\tau \gg g$, the third term is positive for Bloch Wave $1\,(i = 1)$, but it is negative for Bloch Wave $2(i = 2)$. The contribution to TDS from Bloch Wave 1 is more than that from Bloch Wave 2. This accounts for the larger anomalous absorption effect of Bloch Wave 1, whose intensity is concentrated near atomic columns. This means that the probability to be diffusely scattered by atoms is significantly larger than that of Bloch Wave 2, whose intensity is concentrated between atomic columns. Equation (7.7) can be easily extrapolated to cases in which more than one type of atom is involved

$$\begin{aligned} I_{\text{TDS}} =& \frac{1}{2} N_0 \sum_{\alpha} \Big(\left[f_\alpha^e(\tau)\right]^2 \{1 - \exp[-2W_\alpha(\tau)]\} \\ &+ \left[f_\alpha^e(\tau - \mathbf{g})\right]^2 \left\{1 - \exp\left[-2W_\alpha(\tau - \mathbf{g})\right]\right\} \\ &- (-1)^i 2 f_\alpha^e(\tau) f_\alpha^e(\tau - \mathbf{g}) \exp(-2\pi i \mathbf{g} \cdot \mathbf{r}_\alpha) \\ &\left\{\exp\left[-W_\alpha(\mathbf{g})\right] - \exp\left[W_\alpha(\tau) - W_\alpha(\mathbf{g} - \tau)\right]\right\} \Big) \end{aligned} \tag{7.8}$$

where N_0 is the number of unit cells in the crystal. Based on analogous procedures, the distribution of TDS electrons, considering phase correlations of atomic thermal vibrations, has been derived under the single-phonon and double-phonon approximations (Hall 1965; see Sect. 7.5).

The first term in (7.8) is the angular distribution of the thermal diffusely scattered electron from the (000) beam. Figure 7.2a shows the calculated elastic-scattering intensity $(f^e)^2 \exp(-2W)$ and TDS intensity $(f^e)^2 [1 - \exp(-2W)]$ of silicon as a function of the scattering vector **s**. The elastic scattering is dominant if the scattering vector $s = \sin\theta/\lambda$ is smaller than 1 ; the TDS becomes dominant if $s > 1.25$. For crystalline materials, almost all of the elastic-scattering intensity is accumulated at the Bragg beams, as determined by the Bragg law. Positions of the $\{\ell 00\}$ systematic reflections, for example, are indicated by vertical solid lines. Excitations of Bragg beams are determined by diffracting conditions. Thermal diffusely scattered electrons, however, are diffusely distributed between Bragg peaks. Therefore it is possible to tilt the crystal to orientations far from the low-index zone axis so that there are no strong Bragg beams; thus the TDS dominates at high angles.

In electron diffraction, the intensity distributed between $s \to s + ds$ is $\left[2\pi s\, (f^e)^2 \exp(-2\,W) ds\right]$ for elastic scattering and $\{2\pi s\, [f^e]^2 [1 - \exp(-2\,W)] ds\}$. for TDS. Figure 7.2b gives a plot of these functions for silicon. In electron diffraction, scattering angles much larger than low-index Bragg angles are referred to as high angles. Figure 7.2b shows the dominant factor of TDS at high angles.

In scanning electron microscopy, electron backscattered diffraction patterns are formed by diffraction of a large angle (> 90°) Rutherford and thermal diffuse scattered electrons when they leave the crystal. Large-angle backscattering is considered to be primarily generated by TDS. By including many-beam effects just
using the Bloch wave approach, the theoretical scheme just outlined has been extended and applied to calculate electron backscattering diffraction patterns (Hirsch and Humphreys 1970; Spencer et al. 1972; Marthinsen and Høier 1986).

Some structural information can be obtained by analyzing diffusely scattered electron diffraction patterns. The third term in (7.3) is the result of

$$f_3(\tau) = f^e(\tau) f^e(\tau - \mathbf{g}) \sum_{\kappa} \sum_{\kappa'} < \cos\left[2\pi\tau \cdot (\mathbf{r}_\kappa + \mathbf{u}_\kappa) + 2\pi(\mathbf{g} - \tau) \cdot (\mathbf{r}_{\kappa'} + \mathbf{u}_{\kappa'})\right] >$$

The Fourier transform of f_3 can provide the position as well as correlation of the fluctuation responsible for diffuse scattering (Gjønnes and Høier 1971). The real-

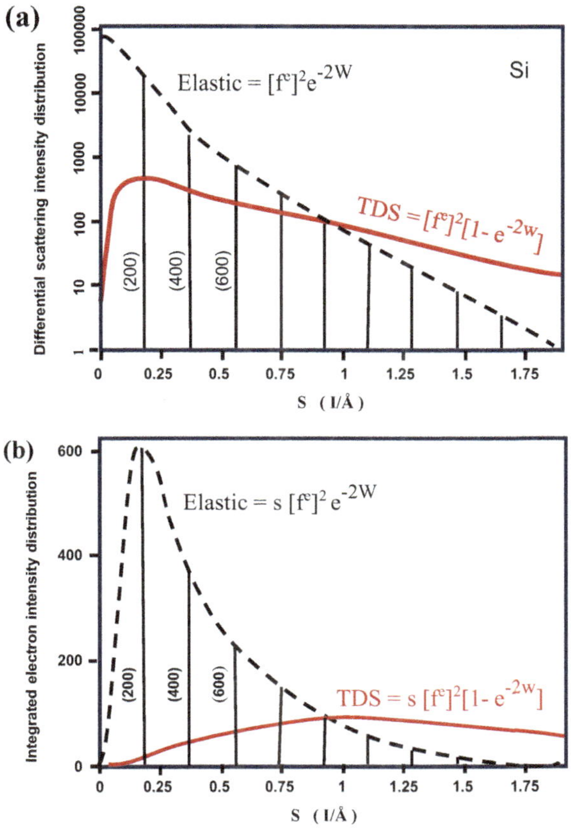

Fig. 7.2 Calculated **a** radial and **b** projected radial intensity distribution functions for the elastically and thermal diffusely scattered electrons by silicon crystal if only the (000)-transmitted beam is considered *see text*. For simplicity, only positions of the $\{\ell 00\}$ row reflections are indicated by vertical solid lines in the figure. The mean the vibration amplitude of Si is taken as $a_{\mathrm{Si}} = 0.007$ nm

space form of f_3 is called the distribution correlation function. Detailed analysis of this function has explained fine details observed in diffuse-scattering patterns of a thin MgO crystal (Gjønnes and Watanabe 1966).

7.3 Total Absorption Coefficient

As illustrated in Chap. 6, inelastic scattering introduces an absorption effect in the elastically scattered electrons. To calculate the absorption coefficient, we integrate diffuse scattering over the surface of the Ewald sphere. Thus the rate at which elastic electron intensity decreases is determined by

$$\frac{dI_e}{dz} = -\left(\mu_0 + \mu_g\right) I_e \tag{7.9a}$$

where absorption coefficients are determined by

Table 7.1 Calculated TDS absorption coefficients for (111) systematic reflections of Al at 40kV[a]

g	$\mu\left(\times 10^{-3}\text{Å}^{-1}\right)$
000	0.667
111	0.591
222	0.440
333	0.288
444	0.165
555	0.081

[a] Doyle (1970)

$$\mu_0 = \frac{1}{K_0^2 \Omega_0} \sum_\alpha \int d\sigma(\tau) \left(\left[f_\alpha^e(\tau) \right]^2 \{1 - \exp[-2W_\alpha(\tau)]\} \right) \tag{7.9b}$$

and

$$\begin{aligned} \mu_g = & \frac{1}{K_0^2 \Omega_0} \sum_\alpha \int d\sigma(\tau) f_\alpha^e(\tau) f_\alpha^e(\tau - \mathbf{g}) \exp(-2\pi i \mathbf{g} \cdot \mathbf{r}_\alpha) \\ & \times \left\{ \exp\left[-W_\alpha(\mathbf{g})\right] - \exp\left[-W_\alpha(\tau) - W_\alpha(\mathbf{g} - \tau)\right] \right\} \end{aligned} \tag{7.9c}$$

If there is no Bragg reflection, the mean free path of TDS can be estimated by

$$\Lambda \approx \frac{1}{\mu_0} \tag{7.10}$$

Calculated absorption coefficients based on the Einstein model and multiphonon model (Sect. 7.5) are compared in Table 7.2.

The mean free path of TDS electrons has been calculated as a function of incident electron energy (Howie and Stern 1972). At 100keV, $\Lambda \approx 100$ nm for aluminum and copper. Their results also show that the mean free paths for TDS and valence loss electrons are of the same order. Numerical calculations according to (7.9) have been performed by Doyle (1969) for aluminum, who listed absorption coefficients for low-index Bragg reflections, as shown in Table 7.1.

Figure 7.3 shows calculated mean free paths Λ_{TDS} for TDS as a function of temperature T and incident electron energy E_0 for Si and GaAs. The calculation was based on the Einstein model. Note the decrease in Λ_{TDS} from Si to GaAs by a factor of 6 and the large change between room temperature and absolute zero for GaAs, due to its low Debye temperature ($\Theta_D = 271$ K for GaAs and $\Theta_D = 645$ K for Si). At 300 keV and 300 K, Λ_{TDS} is 1195 and 217 nm for Si and GaAs, respectively.

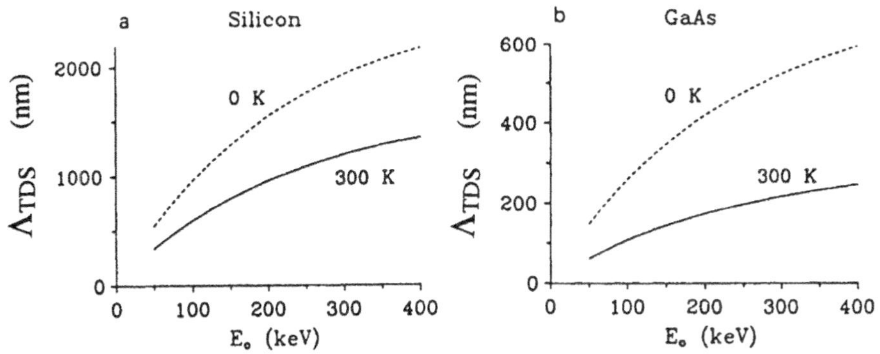

Fig. 7.3 Calculated TDS mean free paths for **a** Si and **b** GaAs at T = 0 K *(dashed line)* and 300 K *(solid line)* (Rossouw et al. 1990)

7.4 Many-Beam TDS Theory

The two-beam TDS theory (Sect. 7.3) based on the Einstein model can be extended to cases involving many Bloch waves (Rossouw and Bursill 1985). The electron wave functions before and after inelastic scattering can be written, respectively, as

$$\Psi = \sum_j \alpha_j \sum_h C_h^{(j)} \exp\left[2\pi i \left(\mathbf{k}_j + \mathbf{h}\right) \cdot \mathbf{r}\right] \tag{7.11a}$$

and

$$\Psi' = \sum_i \alpha_i \sum_g C_g^{(i)} \exp\left[2\pi i \left(\tau + \mathbf{k}_i + \mathbf{g}\right) \cdot \mathbf{r}\right] \tag{7.11b}$$

The intensity scattered to $\tau = \mathbf{K} - \mathbf{K}_0$ in reciprocal space is

$$I = < \left| \int d\mathbf{r}\Psi'^*(\mathbf{r}) \sum_\kappa V_\kappa \left(\mathbf{r} - \mathbf{r}'_\kappa\right) \Psi(\mathbf{r}) \right|^2 > \tag{7.12}$$

Substituting (7.11) into (7.12),

$$\begin{aligned} I = {} & \mathrm{SUM} \sum_\kappa \sum_{\kappa'} f_\kappa^{e*}\left(\mathbf{Q}_{ijgh}\right) f_{\kappa'}^{e}\left(\mathbf{Q}_{i'j'g'h'}\right) \\ & \times < \exp\left[-2\pi i \mathbf{Q}_{ijgh} \cdot \left(\mathbf{r}_\kappa + \mathbf{u}_k\right) + 2\pi i \mathbf{Q}_{i'j'g'h'} \cdot \left(\mathbf{r}_{\kappa'} + \mathbf{u}_{\kappa'}\right)\right] > \end{aligned} \tag{7.13}$$

where a SUM operator is defined to be

$$\mathrm{SUM} = \sum_{ij}\sum_{gh}\sum_{i'j'}\sum_{gh} \alpha_i^*\left(\mathbf{K}'\right)\alpha_j(\mathbf{K})\alpha_{i'}\left(\mathbf{K}'\right)\alpha_{j'}^*(\mathbf{K})$$
$$\times\, C_g^{(i)^*}\left(\mathbf{K}'\right)C_h^{(j)}(\mathbf{K})C_{g'}^{(i')}\left(\mathbf{K}'\right)C_{h'}^{(j')*}(\mathbf{K}) \qquad (7.14a)$$

with

$$\mathbf{Q}_{ijgh} = \tau + \mathbf{k}_i - \mathbf{k}_j + \mathbf{g} - \mathbf{h} = \tau + \left(v_i - v_j\right)\hat{z} + \mathbf{g} - \mathbf{h} \qquad (7.14b)$$

and

$$\mathbf{Q}_{i'j'g'h'} = \tau + \mathbf{k}_{i'} - \mathbf{k}_{j'} + \mathbf{g}' - \mathbf{h}' = \tau + \left(v_{i'} - v_{j'}\right)\hat{z} + \mathbf{g}' - \mathbf{h}' \qquad (7.14c)$$

We now split the sum over atoms into sums with $\kappa = \kappa'$ and $\kappa \neq \kappa'$ and call them T_1 and T_2, respectively. After taking the time average based on the Einstein model, we have

$$T_1 = \sum_{\kappa} f_\kappa^{e*}\left(\mathbf{Q}_{ijgh}\right)f_\kappa^{e}\left(\mathbf{Q}_{i'j'g'h'}\right)\exp\left[-2\pi i\left(\mathbf{Q}_{ijgh} - \mathbf{Q}_{ij'g'h'}\right)\cdot\mathbf{r}_k\right]$$
$$\times \exp\left[-W_\kappa\left(\mathbf{Q}_{ijgh} - \mathbf{Q}_{i'j'g'h'}\right)\right] \qquad (7.15a)$$
$$T_2 = \sum\sum_{\kappa\neq\kappa} f_\kappa^{e*}\left(\mathbf{Q}_{ijgh}\right)f_{\kappa'}^{e}\left(\mathbf{Q}_{i'j'g'h'}\right)\exp\left[-2\pi i\mathbf{Q}_{ijgh}\cdot\mathbf{r}_\kappa + 2\pi i\mathbf{Q}_{i'j'g'h'}\cdot\mathbf{r}_{\kappa'}\right]$$
$$\times \exp\left[-W_\kappa\left(\mathbf{Q}_{ijgh}\right) - W_{k'}\left(\mathbf{Q}_{i'j'g'h'}\right)\right] \qquad (7.15b)$$

Total scattering is then

$$\mathrm{I} = \mathrm{SUM}\left(T_1 + T_2\right) \qquad (7.16)$$

The Bragg contribution is obtained by deriving the intensity when the atomic-scattering factor f_κ^e is replaced by $\left[f_\kappa^e \exp\left(-W_\kappa\right)\right]$. Using arguments analogous to those for obtaining (7.16), we find

$$I_B = \mathrm{SUM}\left(B_1 + B_2\right) \qquad (7.17)$$

where

$$B_1 = \sum_{\kappa} f_\kappa^{e*}\left(\mathbf{Q}_{ijgh}\right) f_\kappa^{e}\left(\mathbf{Q}_{i'j'g'h'}\right)\exp\left[-2\pi i\left(\mathbf{Q}_{ijgh} - \mathbf{Q}_{i'j'g'h'}\right)\cdot\mathbf{r}_k\right]$$
$$\times \exp\left[-W_\kappa\left(\mathbf{Q}_{ijgh}\right) - W_\kappa\left(\mathbf{Q}_{i'j'g'h'}\right)\right] \qquad (7.18a)$$

and

$$B_1 = T_2 \qquad (7.18b)$$

Thus the TDS intensity is

$$I_{\text{TDS}} = I - I_B = \text{SUM} \sum_{\kappa} f_{\kappa}^{e*}(\mathbf{Q}_{ijgh})\, f_{\kappa}^{e}(\mathbf{Q}_{i'j'g'h'}) \exp\left[-2\pi i\left(\mathbf{Q}_{ijgh} - \mathbf{Q}_{i'j'g'h'}\right)\cdot \mathbf{r}_k\right]$$
$$\times \left\{\exp\left[-W_{\kappa}\left(\mathbf{Q}_{ijgh} - \mathbf{Q}_{i'j'g'h'}\right)\right] - \exp\left[-W_{\kappa}\left(\mathbf{Q}_{ijgh}\right) - W_{\kappa}\left(\mathbf{Q}_{i'j'g'h'}\right)\right]\right\} \tag{7.19}$$

The preceding studies were carried out for a three-dimensional periodic crystal that considered the dependence of crystal potential on z. For simplification, if we are interested only in the scattering confined to the ZOLZ, so that crystal potential can be assumed to be independent of z, integrating z can be directly performed in (7.12) at the beginning of the calculation. The final result is simplified as

$$I_{\text{TDS}} = N_0 \sum_{gh} \sum_{g'h'} \sum_{\alpha} [\text{Kin.}][\text{Site}] \sum_{ij} \sum_{i'j'} [\text{B.W.}][\text{Thickness}] \tag{7.20a}$$

The kinematic scattering term is

$$[\text{Kin.}] = f_{\alpha}^{e^*}(\tau + \mathbf{g} - \mathbf{h}) f_{\alpha}^{e}\left(\tau + \mathbf{g}' - \mathbf{h}'\right)\left\{\exp\left[-W_{\alpha}\left(\mathbf{g} - \mathbf{h} - \mathbf{g}' + \mathbf{h}'\right)\right]\right.$$
$$\left. - \exp\left[-W_{\alpha}(\tau + \mathbf{g} - \mathbf{h}) - W_{\alpha}\left(\tau + \mathbf{g}' - \mathbf{h}'\right)\right]\right\} \tag{7.20b}$$

The term associated with atom sites is

$$[\text{Site}] = \exp\left[-2\pi i\left(\mathbf{g} - \mathbf{h} - \mathbf{g}' + \mathbf{h}'\right)\cdot \mathbf{r}_{\alpha}\right] \tag{7.20c}$$

The Bloch wave term is

$$[\text{B.W.}] = \alpha_i^*\left(\mathbf{K}'\right)\alpha_j(\mathbf{K})\alpha_{i'}\left(\mathbf{K}'\right)\alpha_{j'*}(\mathbf{K})$$
$$\times C_g^{(i)*}\left(\mathbf{K}'\right) C_h^{(j)}(\mathbf{K}) C_{g'}^{(i')}\left(\mathbf{K}'\right) C_h^{(j')*}(\mathbf{K}) \tag{7.20d}$$

The term related to specimen thickness d is

$$[\text{Thickness}] = \frac{\left\{\exp\left[-2\pi\left(\nu_i - \nu_j\right)d\right] - 1\right\}}{\left(\nu_i - \nu_j\right)} \frac{\left\{\exp\left[2\pi i\left(\nu_{i'} - \nu_{j'}\right)d\right] - 1\right\}}{\left(\nu_{i'} - \nu_{j'}\right)} \tag{7.20e}$$

As shown in Sect. 8.3, the thickness term is different from the one derived directly from Yoshioka's coupled equations (8.13a) and (8.13b), because summation over the momenta transferred in the z direction was not introduced here. In combination to the theory introduced in Sect. 6.7.2, (7.19) has been modified to calculate the backscattered electron diffraction patterns (Rossouw et al. 1994).

7.5 Multiphonon Excitations

In the Einstein model, the vibration frequency of an atom is independent of the phonon wave vector, so that TDS contributes a featureless background to the diffraction pattern. In the Debye model, however, vibrations of atoms have phase correlations. Thus the phonon dispersion surface may produce some new features in the diffraction pattern. In TDS, single phonon or multiphonons can be created during a single scattering of the fast electron by the crystal. The kinematical scattering theory for this case is given in Sect. 7.7. In this section, we consider the two-beam phonon excitation theory based on the Debye model (Hall 1965). For simplification, we consider a monatomic crystal of cubic structure. Our discussion can be directly conducted from (7.3) for the two-beam case

$$
\begin{aligned}
I = &\frac{1}{2}\sum_{\kappa}\sum_{\kappa'}\left[f^e(\tau)\right]^2 \exp\left[2\pi i\tau\cdot(\mathbf{r}_\kappa-\mathbf{r}_{\kappa'})\right] < \exp\left[2\pi i\tau\cdot(\mathbf{u}_\kappa-\mathbf{u}_{\kappa'})\right] > \\
&+\frac{1}{2}\sum_{\kappa}\sum_{\kappa'}\left[f^e(\tau-\mathbf{g})\right]^2 \exp\left[2\pi i(\tau-\mathbf{g})\cdot(\mathbf{r}_\kappa-\mathbf{r}_{\kappa'})\right] \\
&\times < \exp\left[2\pi i(\tau-\mathbf{g})\cdot(\mathbf{u}_\kappa-\mathbf{u}_{\kappa'})\right] > \\
&-(-1)^i\sum_{\kappa}\sum_{\kappa'} f^e(\tau)f^e(\tau-\mathbf{g})\,\mathrm{Re}\left\{\exp\left[2\pi i\tau\cdot(\mathbf{r}_\kappa-\mathbf{r}_{\kappa'})+2\pi i\mathbf{g}\cdot\mathbf{r}_{\kappa'}\right]\right. \\
&\left.\times < \exp\left[2\pi i\tau\cdot(\mathbf{u}_\kappa-\mathbf{u}_{\kappa'})+2\pi i\mathbf{g}\cdot\mathbf{u}_{\kappa'}\right] >\right\}
\end{aligned}
\tag{7.20f}
$$

The time-dependent displacement of an atom can be expressed as

$$
\mathbf{u}_\kappa = \sum_q\sum_p A_{qp}\mathbf{e}_p\cos\left[\omega_p(\mathbf{q})t - 2\pi\mathbf{q}\cdot\mathbf{r}_\kappa - \Theta_{qp}\right]
\tag{7.21}
$$

where A_{qp} is the amplitude of the phonon with wave vector $\mathbf{q}$ and polarization vector $\mathbf{e}_p$. The phonon has frequency $\omega_p(\mathbf{q})$ and phase shift Θ_{qp}; the latter is added to account for incoherence in different phonon modes. We now consider the following terms:

$$
\begin{aligned}
&< \exp\left[2\pi i\tau\cdot(\mathbf{u}_\kappa-\mathbf{u}_{\kappa'})\right] >= \exp\{-2\pi^2 < [\tau\cdot(\mathbf{u}_\kappa-\mathbf{u}_{\kappa'})]^2 >\} \\
&= \exp\{-4\pi^2\sum_q\sum_p < A_{qp}^2 > (\mathbf{e}_p\cdot\tau)^2\left(1-\cos\left[2\pi\mathbf{q}\cdot(\mathbf{r}_\kappa-\mathbf{r}_{\kappa'})\right]\right)\} \\
&= \exp\{-\sum_q\sum_p G_{qp}(\tau)\left(1-\cos\left[2\pi\mathbf{q}\cdot(\mathbf{r}_\kappa-\mathbf{r}_{\kappa'})\right]\right)\} \\
&= \exp[-2\,W(\tau)]\exp\{\sum_q\sum_p G_{qp}(\tau)\cos\left[2\pi\mathbf{q}\cdot(\mathbf{r}_\kappa-\mathbf{r}_{\kappa'})\right]\}
\end{aligned}
\tag{7.22}
$$

where the mean square amplitude of the phonon is defined by

$$< A_{qp}^2 >= \frac{\hbar(< n_s > +1/2)}{N_0 M \omega_p} \tag{7.23a}$$

and

$$G_{qp}(\tau) = 4\pi^2 < A_{qp}^2 > (\mathbf{e}_p \cdot \tau)^2 \tag{7.23b}$$

Similarly

$$< \exp\left[2\pi i(\tau - \mathbf{g}) \cdot (\mathbf{u}_\kappa - \mathbf{u}_{\kappa'})\right] >=$$
$$\exp[-2\,W(\tau - \mathbf{g})] \exp\left\{\sum_q \sum_p G_{qp}(\tau - \mathbf{g}) \cos\left[2\pi \mathbf{q} \cdot (\mathbf{r}_\kappa - \mathbf{r}_{\kappa'})\right]\right\} \tag{7.24}$$

The third term in (7.20) is

$$\begin{aligned} &< \exp\left[2\pi\tau \cdot (\mathbf{u}_\kappa - \mathbf{u}_{\kappa'}) + 2\pi \mathbf{g} \cdot \mathbf{u}_{\kappa'}\right] > \\ &=< \exp\left[2\pi\tau \cdot \mathbf{u}_\kappa - 2\pi(\tau - \mathbf{g}) \cdot \mathbf{u}_{\kappa'}\right] > \\ &= \exp[-W(\tau) - W(\tau - \mathbf{g})] \\ &\times \exp\{\sum_q \sum_p \left[G_{qp}(\tau) G_{qp}(\tau - \mathbf{g})\right]^{1/2} \cos\left[2\pi \mathbf{q} \cdot (\mathbf{r}_\kappa - \mathbf{r}_{\kappa'})\right]\} \end{aligned} \tag{7.25}$$

Substituting (7.22), (7.24), and (7.25) into (7.20), the total scattering intensity is

$$\begin{aligned} I = &\frac{1}{2} \sum_\kappa \sum_{\kappa'} \left[f^e(\tau)\right]^2 \exp\left[2\pi i \tau \cdot (\mathbf{r}_\kappa - \mathbf{r}_{\kappa'})\right] \\ &\times \exp[-2\,W(\tau)] \exp\{\sum_q \sum_p G_{qp}(\tau) \cos\left[2\pi \mathbf{q} \cdot (\mathbf{r}_\kappa - \mathbf{r}_{\kappa'})\right]\} \\ &+ \frac{1}{2} \sum_\kappa \sum_{\kappa'} \left[f^e(\tau - \mathbf{g})\right]^2 \exp\left[2\pi i (\tau - \mathbf{g}) \cdot (\mathbf{r}_\kappa - \mathbf{r}_{\kappa'})\right] \exp[-2\,W(\tau - \mathbf{g})] \\ &\times \exp\{\sum_q \sum_p G_{qp}(\tau - \mathbf{g}) \cos\left[2\pi \mathbf{q} \cdot (\mathbf{r}_\kappa - \mathbf{r}_{\kappa'})\right]\} \\ &- (-1)^i \sum_\kappa \sum_{\kappa'} f^e(\tau) f^e(\tau - \mathbf{g}) \cos\left[2\pi\tau \cdot (\mathbf{r}_\kappa - \mathbf{r}_{\kappa'}) + 2\pi \mathbf{g} \cdot \mathbf{r}_{\kappa'}\right] \\ &\times \exp[-W(\tau) - W(\tau - \mathbf{g})] \\ &\times \exp\{\sum_q \sum_p \left[G_{qp}(\tau) G_{qp}(\tau - \mathbf{g})\right]^{1/2} \cos\left[2\pi \mathbf{q} \cdot (\mathbf{r}_\kappa - \mathbf{r}_{\kappa'})\right]\} \end{aligned} \tag{7.26}$$

Equation (7.26) can be written as a series in the power of G_{qp} according to

$$\exp x = \sum_{L=0} \frac{x^L}{L!}$$

The $L = 0$ term corresponds to the Bragg scattering

$$\begin{aligned}I_B =& \frac{1}{2}\left[f^e(\tau)\right]^2 \exp[-2W(\tau)]P_s(\tau) + \frac{1}{2}\left[f^e(\tau-\mathbf{g})\right]^2 \exp[-2W(\tau-\mathbf{g})]P_s(\tau-\mathbf{g}) \\ & - (-1)^i \sum_{\kappa}\sum_{\kappa'} f^e(\tau) f^e(\tau-\mathbf{g}) \cos\left[2\pi\tau\cdot(\mathbf{r}_\kappa - \mathbf{r}_{\kappa'}) + 2\pi\mathbf{g}\cdot\mathbf{r}_{\kappa'}\right] \\ & \times \exp[-W(\tau) - W(\tau-\mathbf{g})]\end{aligned} \tag{7.27a}$$

where

$$P_s(\tau) = \sum_{\kappa}\sum_{\kappa'} \exp\left[2\pi i\tau\cdot(\mathbf{r}_\kappa - \mathbf{r}_{\kappa'})\right] \tag{7.27b}$$

The $L = 1$ term corresponds to single-phonon excitation

$$\begin{aligned}(I_{\mathrm{TDS}})_1 =& \frac{1}{4}\left[f^e(\tau)\right]^2 \exp[-2W(\tau)] \sum_q\sum_p G_{qp}(\tau)\left[P_s(\tau+\mathbf{q}) + P_s(\tau-\mathbf{q})\right] \\ & + \frac{1}{4}\left[f^e(\tau-\mathbf{g})\right]^2 \exp[-2W(\tau-\mathbf{g})] \\ & \times \left\{\sum_q\sum_p G_{qp}(\tau-\mathbf{g})\left[P_s(\tau-\mathbf{g}+\mathbf{q}) + P_s(\tau-\mathbf{g}-\mathbf{q})\right]\right\} \\ & - (-1)^i f^e(\tau) f^e(\tau-\mathbf{g}) \exp[-W(\tau) - W(\tau-\mathbf{g})] \\ & \times \sum_q\sum_p \{\left[G_{qp}(\tau) G_{qp}(\tau-\mathbf{g})\right]^{1/2} \\ & \times \sum_{\kappa}\sum_{\kappa'} \cos\left[2\pi\tau\cdot(\mathbf{r}_\kappa - \mathbf{r}_{\kappa'}) + 2\pi\mathbf{g}\cdot\mathbf{r}_{\kappa'}\right] \cos\left[2\pi\mathbf{q}\cdot(\mathbf{r}_\kappa - \mathbf{r}_{\kappa'})\right]\}\end{aligned} \tag{7.28a}$$

The $L = 2$ term corresponds to double-phonon excitation

$$\begin{aligned}(I_{\mathrm{TDS}})_2 =& \frac{1}{4}\sum_{\kappa}\sum_{\kappa'} \left[f^e(\tau)\right]^2 \exp\left[2\pi i\tau\cdot(\mathbf{r}_\kappa - \mathbf{r}_{\kappa'})\right] \\ & \times \exp[-2W(\tau)]\{\sum_q\sum_p G_{qp}(\tau)\cos\left[2\pi\mathbf{q}\cdot(\mathbf{r}_\kappa - \mathbf{r}_{\kappa'})\right]\}^2 \\ & + \frac{1}{4}\sum_{\kappa}\sum_{\kappa'} \left[f^e(\tau-\mathbf{g})\right]^2 \exp\left[2\pi i(\tau-\mathbf{g})\cdot(\mathbf{r}_\kappa - \mathbf{r}_{\kappa'})\right] \\ & \times \exp[-2W(\tau-\mathbf{g})]\{\sum_q\sum_p G_{qp}(\tau-\mathbf{g})\cos\left[2\pi\mathbf{q}\cdot(\mathbf{r}_\kappa - \mathbf{r}_{\kappa'})\right]\}^2 \\ & - \frac{1}{2}(-1)^i \sum_{\kappa}\sum_{\kappa'} f^e(\tau) f^e(\tau-\mathbf{g}) \cos\left[2\pi\tau\cdot(\mathbf{r}_\kappa - \mathbf{r}_{\kappa'}) + 2\pi\mathbf{g}\cdot\mathbf{r}_{\kappa'}\right]\end{aligned}$$

$$\times \exp[-W(\tau) - W(\tau - \mathbf{g})]\{\sum_q \sum_p [G_{qp}(\tau)G_{qp}(\tau - \mathbf{g})]^{1/2}$$

$$\times \cos\left[2\pi\mathbf{q} \cdot (\mathbf{r}_k - \mathbf{r}_{\kappa'})\right]\}^2 \qquad (7.28b)$$

Similarly the Lth-order expansion of (7.26) in G_{qp} corresponds to the L-phonon excitation (James 1965; Borie 1970; Paskin 1958)

$$\begin{aligned}
(I_{\text{TDS}})_L =& \frac{1}{2\,L!} \sum_\kappa \sum_{\kappa'} [f^e(\tau)]^2 \exp[2\pi i \tau \cdot (\mathbf{r}_\kappa - \mathbf{r}_{\kappa'})] \exp[-2\,W(\tau)] \\
& \times \{\sum_q \sum_p G_{qp}(\tau) \cos\left[2\pi\mathbf{q} \cdot (\mathbf{r}_\kappa - \mathbf{r}_{\kappa'})\right]\}^L \\
& + \frac{1}{2\,L!} \sum_\kappa \sum_{\kappa'} [f^e(\tau - \mathbf{g})]^2 \exp\left[2\pi i (\tau - \mathbf{g}) \cdot (\mathbf{r}_\kappa - \mathbf{r}_{\kappa'})\right] \exp[-2\,W(\tau - \mathbf{g})] \\
& \times \{\sum_q \sum_p G_{qp}(\tau - \mathbf{g}) \cos\left[2\pi\mathbf{q} \cdot (\mathbf{r}_\kappa - \mathbf{r}_{\kappa'})\right]\}^L \\
& - \frac{(-1)^i}{L!} \sum_\kappa \sum_{\kappa'} f^e(\tau) f^e(\tau - \mathbf{g}) \cos\left[2\pi\tau \cdot (\mathbf{r}_\kappa - \mathbf{r}_{\kappa'}) + 2\pi\mathbf{g} \cdot \mathbf{r}_{\kappa'}\right] \\
& \times \exp[-W(\tau) - W(\tau - \mathbf{g})] \\
& \times \{\sum_q \sum_p [G_{qp}(\tau)G_{qp}(\tau - \mathbf{g})]^{1/2} \cos\left[2\pi\mathbf{q} \cdot (\mathbf{r}_\kappa - \mathbf{r}_{\kappa'})\right]\}^L
\end{aligned} \qquad (7.29)$$

To simplify (7.29), we make an approximation by assuming that the three polarization vectors e_p are mutually perpendicular and the phonon frequency is independent of polarization direction, so that

$$\sum_q \sum_p G_{qp}(\tau) \cos\left[2\pi\mathbf{q} \cdot (\mathbf{r}_\kappa - \mathbf{r}_{\kappa'})\right] = \tau^2 \mathrm{E}_{\kappa\kappa'} \qquad (7.30a)$$

$$\sum_q \sum_p [G_{qp}(\tau)G_{qp}(\tau - \mathbf{g})]^{1/2} \cos\left[2\pi\mathbf{q} \cdot (\mathbf{r}_\kappa - \mathbf{r}_{\kappa'})\right] = \tau \cdot (\tau - \mathbf{g}) \mathrm{E}_{\kappa\kappa'} \qquad (7.30b)$$

where

$$\mathrm{E}_{\kappa\kappa'} = 4\pi^2 \sum_q \frac{\hbar(< n_s > +1/2)}{N_0 M \omega} \cos\left[2\pi\mathbf{q} \cdot (\mathbf{r}_\kappa - \mathbf{r}_{\kappa'})\right] \qquad (7.30c)$$

If the sum over $\mathbf{q}$ is replaced by an integral, (7.30c) becomes

$$\mathrm{E}_{\kappa\kappa'} = 4\pi^2 \Omega \frac{\hbar}{M} \int_{\mathrm{BZ}} d\mathbf{q} \left[\frac{1}{\exp(\hbar\omega/k_B T) - 1} + \frac{1}{2}\right] \frac{\cos\left[2\pi\mathbf{q} \cdot (\mathbf{r}_k - \mathbf{r}_{k'})\right]}{\omega} \qquad (7.31)$$

It is important to note that the average occupation number of phonons $< n_s >$ depends on the phonon frequency ω. Finally (7.29) becomes

$$
\begin{aligned}
(I_{\mathrm{TDS}})_L =& \frac{1}{2\,L!}\sum_{\kappa}\sum_{\kappa'}\left[f^e(\tau)\right]^2 \exp\left[2\pi i\tau\cdot(\mathbf{r}_\kappa-\mathbf{r}_{\kappa'})\right]\exp[-2\,W(\tau)]\left\{\tau^2E_{\kappa\kappa'}\right\}^L \\
&+\frac{1}{2\,L!}\sum_{\kappa}\sum_{\kappa'}\left[f^e(\tau-\mathbf{g})\right]^2\exp\left[2\pi i(\tau-\mathbf{g})\cdot(\mathbf{r}_\kappa-\mathbf{r}_{\kappa'})\right] \\
&\times\exp[-2\,W(\tau-\mathbf{g})]\left\{|\tau-\mathbf{g}|^2E_{\kappa\kappa'}\right\}^L \\
&-\frac{(-1)^i}{L!}\sum_{\kappa}\sum_{\kappa'}f^e(\tau)f^e(\tau-\mathbf{g})\cos\left[2\pi\tau\cdot(\mathbf{r}_\kappa-\mathbf{r}_{\kappa'})+2\pi\mathbf{g}\cdot\mathbf{r}_{\kappa'}\right] \\
&\times\exp[-W(\tau)-W(\tau-\mathbf{g})]\left\{\tau\cdot(\tau-\mathbf{g})E_{\kappa\kappa'}\right\}^L
\end{aligned}
\tag{7.32}
$$

The total intensity of TDS is a sum of (7.32) for $L \geq 1$

$$
\begin{aligned}
I_{\mathrm{TDS_L}} =& \frac{1}{2}\left[f^e(\tau)\right]^2\exp[-2\,W(\tau)]J_1(\tau)+\frac{1}{2}\left[f^e(\tau-\mathbf{g})\right]^2\exp[-2\,W(\tau-\mathbf{g})]J_1(\tau-\mathbf{g}) \\
&-(-1)^i f^e(\tau)f^e(\tau-\mathbf{g})\exp[-W(\tau)-W(\tau-\mathbf{g})]J_2(\tau,\mathbf{g})
\end{aligned}
\tag{7.33a}
$$

where

$$
J_1(\tau)=\sum_{\kappa}\sum_{\kappa'}\cos\left[2\pi\tau\cdot(\mathbf{r}_\kappa-\mathbf{r}_{\kappa'})\right]\left[\exp\left(\tau^2E_{\kappa\kappa'}\right)-1\right]
\tag{7.33b}
$$

and

$$
J_2(\tau,\mathbf{g})=\sum_{\kappa}\sum_{\kappa'}\cos\left[2\pi\tau\cdot(\mathbf{r}_\kappa-\mathbf{r}_{\kappa'})+2\pi\mathbf{g}\cdot\mathbf{r}_{\kappa'}\right]\left\{\exp\left[\tau\cdot(\tau-\mathbf{g})E_{\kappa\kappa'}\right]-1\right\}
\tag{7.33c}
$$

Finally the absorption coefficient is calculated by Hall (1965)

$$
\mu_g=\frac{1}{K^2N_0\Omega_0}\int d\sigma(\tau)I_{\mathrm{TDS}}
\tag{7.34}
$$

where integration is over the surface of the Ewald sphere.

Figure 7.4 shows the calculated TDS intensity distributions for the single phonon, multiphonons, and Einstein models of gold. Three important observations can be made. First the single phonon and multiphonon curves are almost the same at low scattering angles, indicating the dominant effect of single-phonon scattering at small angles because $u = 2\sin\theta/\lambda$. Second the contribution of multiphonons is more significant at high scattering angles, suggesting that the large momentum transfer, which is indispensable for scattering electrons to high angles (in comparison to Bragg angles

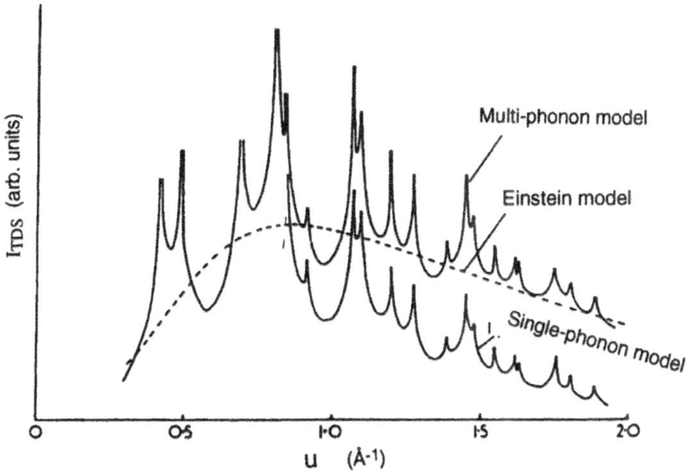

Fig. 7.4 Calculated diffusely scattered electron intensity distributions as given by the Einstein model, single-phonon and multiphonon models of gold at 900 °C. The single-phonon and multiphonon curves are almost the same at small scattering angles; the difference increases with increasing u. The curve for the multiphonon model follows that of the Einstein model at large u (Hall 1965). The scattering with u>1 is usually considered as high-angle scattering

Table 7.2 Calculated μ_0 and μ_g for Thermal Scattering in the Range $\tau = 0$ to 30 nm^{-1} for Gold with g = (200)[a]

	100 K			300 K			900 K		
Model	μ_0 (μm^{-1})	μ_g (μm^{-1})	Λ (μm^{-1})	μ_0 (μm^{-1})	μ_g (μm^{-1})	Λ (μm^{-1})	μ_0 (μm^{-1})	μ_g (μm^{-1})	Λ (μm^{-1})
Einstein	14.9	13.7	67	34.7	30.4	29	64.7	53.5	15
Multiphonon	15.8	14.5	63	35.3	31.6	28	67.1	55.4	15
Single-phonon (Random orientation)	13.6	12.5	74	24.7	21.7	40	33.0	26.0	30

[a] Hall (1965)

for low-index reflections) is the resultant momenta of several phonons. Finally the Einstein model can reasonably predict the intensity profile, disregarding fine peaks due to phonon dispersions, of TDS electrons at both low and high angles. Therefore the absorption effect of TDS in the high-angle range can be fairly accurately simulated by the Einstein model. This greatly simplifies mathematical calculations in some cases.

A set of calculated absorption coefficients and the corresponding TDS mean free path of gold is shown in Table 7.2. Differences between values calculated with the multiphonon and single-phonon models become larger with increasing temperature. More importantly, the mean free path of TDS in gold is about 30 nm at 300 K; this is rather short.

7.6 Evaluating The Debye-Waller Factor

The Debye-Waller factor is an important quantity in electron scattering. In this section, we use the Debye model to approximate the DW factor. The formal expression of the DW factor was defined in (6.77a)

$$W_\alpha = \frac{2\pi^2\hbar}{N_0 M_\alpha}\sum_{\mathbf{q}}\sum_j \frac{1}{\omega_j(\mathbf{q})}\left[\tau\cdot\mathbf{e}\left(\alpha\middle|{}^{\mathbf{q}}_{j}\right)\right]^2\left\{\frac{1}{\exp\left[\hbar\omega_j(\mathbf{q})/k_BT\right]-1}+\frac{1}{2}\right\} \tag{7.35}$$

We now follow the method of Warren (1990) to simplify the calculation of (7.35). We assume that all vibration waves can be considered as pure longitudinal and pure transverse. The velocities of all longitudinal waves are replaced by an average longitudinal velocity and velocities of all transverse waves by an average transverse velocity. Each average velocity is considered to be a constant independent of the wave vector **q**.

To perform the summation in (7.35), the BZ is replaced by a sphere of radius q_m, whose volume is equal to that of the BZ$(4/3)\pi q_m^3 = V_{\mathrm{BZ}}$. The density of points (or states) in the sphere is N_0/V_{BZ}, and the summation of **q** is replaced by integration throughout the sphere. For each type of wave, longitudinal or transverse, the polarization vector **e** is equally distributed at all orientations relative to τ. For all waves whose vector **q** terminates in the hollow sphere, we can use the average

$$<\left[\tau\cdot\mathbf{e}\left(\alpha\middle|{}^{\mathbf{q}}_{j}\right)\right]^2> = \tau^2 < \cos^2(\tau,\mathbf{e}) > = \tau^2/3$$

With these approximations, (7.35) becomes

$$W_\alpha = \frac{2\pi^2\hbar\tau^2}{3N_0M_\alpha}\sum_j\int_0^{q_m} dq4\pi q^2\frac{1}{\omega_j(q)}\left\{\frac{1}{\exp\left[\hbar\omega_j(q)/k_BT\right]-1}+\frac{1}{2}\right\}\frac{N_0}{V_{\mathrm{BZ}}} \tag{7.36}$$

In terms of a constant velocity v_j for each branch of the phonon, $\omega_j = 2\pi v_j q$. The integral variable q is replaced by ω, and (7.36) becomes

$$W_\alpha = \frac{2\pi^2\hbar\tau^2}{M_\alpha}\sum_j\frac{1}{{\omega_{jm}}^3}\int_0^{\omega_{jm}} d\omega\ \omega\left\{\frac{1}{\exp\left[\hbar\omega/k_BT\right]-1}+\frac{1}{2}\right\} \tag{7.37}$$

where $\omega_{jm} = 2\pi\nu_j q_m$. Let $x = \hbar\omega/k_B T$ and $x_j = \hbar\omega_{jm}/k_B T$; then (7.37) becomes

$$W_\alpha = \frac{2\pi^2 k_B T \tau^2}{M_\alpha} \sum_j \frac{1}{\omega_{jm}^2} \left[\frac{1}{x_j} \int_0^{x_j} dx \frac{x}{\exp(x) - 1} + \frac{x_j}{4} \right] \tag{7.38}$$

The sum of j includes one longitudinal (ω_L) and two transverse (ω_T) waves. If we use an average Debye temperature defined as $T_D = \hbar\omega_D/k_B$, where ω_D is determined by $3/\omega_D^2 = 1/\omega_L^2 + 2/\omega_T^2$, (7.38) becomes

$$W_\alpha = \frac{6\pi^2\hbar^2\tau^2}{M_\alpha k_B T_D} \left[\left(\frac{T}{T_D}\right)^2 \int_0^{T_D/T} dx \frac{x}{\exp(x) - 1} + \frac{1}{4} \right] \tag{7.39}$$

The mean square vibration amplitude of the atom is

$$\overline{a_\alpha^2} = \frac{W_\alpha}{2\pi^2\tau^2} = \frac{3\hbar^2}{M_\alpha k_B T_D} \left[\left(\frac{T}{T_D}\right)^2 \int_0^{T_D/T} dx \frac{x}{\exp(x) - 1} + \frac{1}{4} \right] \tag{7.40}$$

This is just (6.78). The calculations of (7.39) and (7.40) are ready to be performed if the Debye temperature is known. The approximations made in this section for calculating the DW factor are called the Warren approximation.

7.7 Coherent Length in Thermal Diffuse Scattering

Vibrations of crystal atoms may preserve certain phase relationships as characterized by the phonon dispersion surface and polarization vectors. Thus instantaneous atom displacements follow patterns determined by crystal structure and interatomic forces, so that diffuse scattering generated from different atom sites may retain some phase correlation because it is coherent. However the coherence between TDS waves generated from different atom sites may damp with increasing interatomic distance. Thus incoherent scattering is generated if two atoms are located far enough away from one another. This section determines the coherent length (or volume) in TDS.

For simplicity, we use kinematic scattering theory to explore this coherence. The dynamical theory is given in Sect. 11.5. Following the discussion in Sect. 7.2, the TDS intensity distribution is the difference between total scattering and Bragg scattering

$$
\begin{aligned}
I_{\mathrm{TDS}} &= <|\sum_{\kappa} f^e(\tau)\exp[2\pi i\tau\cdot(\mathbf{r}_k+\mathbf{u}_k)]|^2> \\
&\quad -|<\sum_{\kappa} f^e(\tau)\exp[2\pi i\tau\cdot(\mathbf{r}_\kappa+\mathbf{u}_k)]>|^2 \\
&= \sum_{\kappa}\sum_{\kappa'}\left[f^e(\tau)\right]^2\exp[2\pi\tau\cdot(\mathbf{r}_\kappa-\mathbf{r}_{k'})]\{<\exp[2\pi i\tau\cdot(\mathbf{u}_\kappa-\mathbf{u}_{\kappa'})]> \\
&\quad -<\exp(2\pi i\tau\cdot\mathbf{u}_k)><\exp(2\pi i\tau\cdot\mathbf{u}_{\kappa'})>\} \\
&= \sum_{\kappa}\sum_{\kappa'}\left[f^e(\tau)\right]^2\exp[2\pi\tau\cdot(\mathbf{r}_\kappa-\mathbf{r}_{k'})]\exp(-2W) \\
&\quad \times\left\{\exp\left[4\pi^2<(\tau\cdot\mathbf{u}_\kappa)\left(\tau\cdot\mathbf{u}'_\kappa\right)>\right]-1\right\} \\
&= \sum_{\kappa}\left[f^e(\tau)\right]^2[1-\exp(-2W)] \\
&\quad +\sum_{\kappa}\sum_{\kappa'\neq\kappa}\left[f^e(\tau)\right]^2\exp[2\pi i\tau\cdot(\mathbf{r}_\kappa-\mathbf{r}_{\kappa'})]\exp(-2W) \\
&\quad \times\left\{\exp\left[4\pi^2<(\tau\cdot\mathbf{u}_\kappa)\left(\tau\cdot\mathbf{u}'_\kappa\right)>\right]-1\right\}
\end{aligned}
\tag{7.41}
$$

The first term is the incoherent-scattering result, and the second term is the coherent scattering. The coherence between the thermal diffuse scattering generated from atom sites κ and κ' is determined by $\left\{\exp\left[4\pi^2<(\tau\cdot\mathbf{u}_\kappa)(\tau\cdot\mathbf{u}_{\kappa'})>\right]-1\right\}$. Evaluating this function gives a coherence length (or coherent volume).

Under the harmonic oscillators approximation, from (7.21) and (7.23) we have

$$
Y_{\kappa\kappa'}=4\pi^2<(\tau\cdot\mathbf{u}_\kappa)(\tau\cdot\mathbf{u}_{\kappa'})>=\sum_{q}\sum_{i}G_{qp}\cos(2\pi\mathbf{q}\cdot\Delta\mathbf{r}_{\kappa\kappa'})=\tau^2E_{\kappa\kappa'} \tag{7.42}
$$

where $E_{\kappa\kappa'}$ was given by (7.30c) and $\Delta\mathbf{r}_{\kappa\kappa'}=(\mathbf{r}_{0\kappa}-\mathbf{r}_{0\kappa'})$ is a vector of interatomic distance. We now use the same approximations introduced in Sect. 7.6 to calculate (7.42). The BZ is replaced by a sphere of radius q_m and the orientation average

$$
<\left[\tau\cdot\mathbf{e}\left(\alpha|^{\mathbf{q}}_j\right)\right]^2>=\frac{\tau^2}{2}\int_0^\pi d\theta\sin\theta\cos^2\theta=\tau^2/3
$$

Equation (7.42) is approximated as

$$
\begin{aligned}
Y_{\kappa\kappa'} &\approx \frac{4\pi^2\hbar\tau^2}{3N_0\sqrt{M_\kappa M_{\kappa'}}}\sum_j\int_0^{q_m}dq\,2\pi q^2\int_0^\pi d\theta\sin\theta\frac{1}{\omega_j(q)}\left\{\langle n_s\rangle+\frac{1}{2}\right\} \\
&\quad \cos(2\pi q\,\Delta r_{\kappa\kappa'}\cos\theta)\,\frac{N_0}{V_{\mathrm{BZ}}}
\end{aligned}
$$

$$= \frac{\hbar\tau^2}{q_m^3\sqrt{M_\kappa M_{\kappa'}}}\sum_j \frac{1}{\nu_j}\int_0^{q_m} dq \frac{\sin(2\pi q \Delta r_{\kappa\kappa'})}{\Delta r_{\kappa\kappa'}}\left\{\frac{1}{\exp(2\pi\hbar\nu_j q/k_B T)-1}+\frac{1}{2}\right\} \tag{7.43}$$

At a high-temperature limiting case (7.43) is approximated as

$$Y_{\kappa\kappa'} \approx \frac{k_B T\tau^2}{q_m^2\sqrt{M_\kappa M_{\kappa'}}}\sum_j \frac{1}{\nu_j^2}\frac{\mathrm{Si}(\Theta_{\kappa\kappa'})}{\Theta_{\kappa\kappa'}} \tag{7.44}$$

where

$$\mathrm{Si}(\Theta) = \int_0^\theta du \frac{\sin u}{u}, \quad \Theta_{\kappa\kappa'} = 2\pi q_m |\Delta \mathrm{r}_{\kappa\kappa'}|$$

Comparing (7.44) to the high-temperature limiting case for the DW factor [from (7.37)],

$$W_\kappa \approx \frac{k_B T\tau^2}{2q_m^2 M_\kappa}\sum_j \frac{1}{\nu_j^2} \tag{7.45}$$

Equation (7.44) is rewritten as

$$Y_{\kappa\kappa'} \approx 2W_\kappa\left(\frac{M_\kappa}{M_{\kappa'}}\right)^{1/2}\frac{\mathrm{Si}(\Theta_{\kappa\kappa'})}{\Theta_{\kappa\kappa'}} = 4\pi^2\overline{a_{\kappa^2}}\left(\frac{M_\kappa}{M_{\kappa'}}\right)^{1/2}\tau^2\frac{\mathrm{Si}(\Theta_{\kappa\kappa'})}{\Theta_{\kappa\kappa'}} \tag{7.46}$$

Substituting (7.46) and (7.42) into (7.41),

$$I_{\mathrm{TDS}} = \sum_\kappa [f^e(\tau)]^2\{1-\exp[-2W]\}$$
$$+\sum_\kappa\sum_{\kappa'\neq\kappa}[f^e(\tau)]^2\exp[2\pi i\tau\cdot(\mathbf{r}_\kappa-\mathbf{r}_{\kappa'})]\exp[-2W]\left\{\exp\left[2W\frac{\mathrm{Si}(\Theta_{\kappa\kappa'})}{\Theta_{\kappa\kappa'}}\right]-1\right\} \tag{7.47}$$

Therefore the thermal diffuse scattering generated from atom sites κ and κ' is considered to be incoherent if the two atoms are separated enough so that

$$4\pi^2\overline{a^2}\tau^2\frac{\mathrm{Si}(\Theta_{\kappa\kappa'})}{\Theta_{\kappa\kappa'}} \ll 1 \tag{7.48}$$

or

$$\frac{\mathrm{Si}(\Theta_{\kappa\kappa'})}{\Theta_{\kappa\kappa'}} \ll \frac{1}{4\pi^2\overline{a^2}\tau^2} \tag{7.49}$$

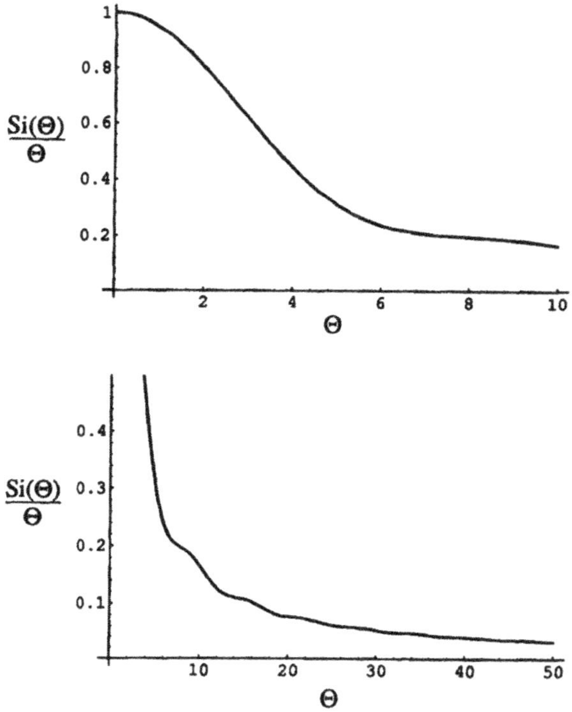

Fig. 7.5 Plots of $\mathrm{Si}(\Theta)/\Theta$ as a function of Θ

where $\overline{a^2}$ is the mean square atomic vibration amplitude. The interatomic distance larger than which (7.48) holds is defined as the coherent length. For diffraction pattern calculations, the coherent volume is thus defined as a sphere with a diameter equal to the coherent length. For image simulations, however, the coherent volume is a cigar-shaped atomic column whose size is the coherent length (see Sect. 11.5.5). In this case, only the longitudinal coherence matters. From (7.49) the coherent length depends on the scattering angle ($\tau = 2\sin\theta/\lambda$).

Two conclusions can be drawn from (7.49). Since $\mathrm{Si}\,(\Theta_{\kappa\kappa'})\,/\Theta_{\kappa\kappa'}$ drops quickly with increasing interatomic distance, as shown in Fig. 7.5, the coherent length for a low-angle TDS is much shorter than that for a high-angle TDS. The coherent length of light elements is larger than that of heavy elements because $\overline{a^2}$ decreases with increasing atomic mass. The coherence of the TDS has a strong impact on the applicability of the incoherent-scattering model for simulating high-angle annular dark-field (HAADF) STEM images (see Sect. 11.5.5).

The mean square vibration amplitude $\overline{a^2}$ is on the order of 0.01Å^2 for most materials; (7.49) is approximated as

$$\frac{\mathrm{Si}\,(\Theta_{\kappa\kappa'})}{\Theta_{\kappa\kappa'}} \ll \frac{2.5}{\tau^2} \tag{7.50}$$

where τ is measured in Å^{-1}. In HAADF-STEM imaging, τ is about 2Å^{-1} for 100-kV electrons; (7.50) is approximately satisfied for $\Theta_{\kappa\kappa'} > 15$. Since the radius of the first BZ is typically about 0.2Å^{-1}, the diffuse scattering from two atom sites is considered to be completely incoherent if their interatomic distance $|\Delta\mathbf{r}_{\kappa\kappa'}| > 10$ Å, which is equivalent to 4–5 atomic columns apart. This conclusion has a great impact on the incoherent-imaging theory of Z-contrast imaging introduced in Chap. 11.

For a low-angle TDS,

$$\tau \approx 0.5\text{Å}^{-1}, \quad \frac{\mathrm{Si}\left(\Theta_{\kappa\kappa'}\right)}{\Theta_{\kappa\kappa'}} \ll 10$$

This condition is satisfied by any interatomic distance. Thus scattering is purely incoherent, and the second term in (7.41) can be omitted. Therefore, the Einstein model holds for low-angle TDS.

In the dynamical scattering theory to be shown in Sect. 11.5.5, an evaluation has to be performed for a function defined as

$$\begin{aligned} X_{\kappa\kappa'} &= 4\pi^2 < (\tau\cdot\mathbf{u}_\kappa)\left(\tau'\cdot\mathbf{u}_{\kappa'}\right) > \\ &= \frac{4\pi^2\hbar}{N_0\sqrt{M_\kappa M_{\kappa'}}}\sum_q\sum_i \frac{(< n_s > +1/2)}{\omega_i(\mathbf{q})}\left(\tau\cdot\mathbf{e}_{q,i}\right)\left(\tau'\cdot\mathbf{e}_{q,i}\right)\cos\left(2\pi\mathbf{q}\cdot\Delta\mathbf{r}_{\kappa\kappa'}\right) \end{aligned} \tag{7.51}$$

Since the orientation average $< \left(\tau\cdot\mathbf{e}_{q,i}\right)\left(\tau'\cdot\mathbf{e}_{q,i}\right) >= \tau\cdot\tau'/3$, (7.51) can be evaluated by the same procedures used for $Y_{\kappa\kappa'}$. For a high-temperature limiting case, the result is

$$X_{\kappa\kappa'} \approx 4\pi^2\overline{a_\kappa^2}\left(\frac{M_\kappa}{M_{\kappa'}}\right)^{1/2}\tau\cdot\tau'\frac{\mathrm{Si}\left(\Theta_{\kappa\kappa'}\right)}{\Theta_{\kappa\kappa'}} \tag{7.52}$$

7.8 Diffuse Scattering of Imperfect Crystals

7.8.1 *Huang Scattering*

In X-ray diffraction, diffuse scattering produced by point defects was first observed by Huang (1947). Thus this type of diffuse scattering produced by lattice imperfection is called Huang scattering. It has features analogous with TDS, such as weakening of Bragg reflections and fine-intensity maxima between Bragg beams. The TDS is usually mixed with the diffuse scattering from defects. Importantly, the TDS is sensitive to variations in specimen temperature but the Huang scattering is not. These characteristics can be used to distinguish TDS from Huang scattering.

There are possibly five types of crystal imperfections that may produce diffuse scattering: (1) lattice distortions due to point defects produced by vacancies or interstitial atoms (Trinkaus 1972; Gjønnes 1993); (2) atom displacements near dislocation cores and lines; (3) structural relaxation arising from atoms of different sizes occupying a single sublattice (Warren et al. 1951; Butler et al. 1992); (4) atomic displacement and relaxation at surfaces and interfaces; and (5) short-range ordering due to atom substitutions or vacancies. The kinematic diffuse-scattering theories of imperfect crystals have comprehensively been described by Cowley (1981). We now use the two-beam Bloch wave description to discuss the diffuse scattering from imperfect crystals.

7.8.2 *Diffuse Scattering Produced by Point Defects*

Using the same procedures presented in Sect. 7.2, diffuse scattering of an imperfect crystal can also be treated. Lattice distortion in each unit cell due to thermal vibration can be mathematically described by a time-dependent function. In Huang scattering, on the other hand, atom displacements are time-independent but vary from cell to cell. Therefore the time average is replaced by structural average on different cell configurations. Lattice distortion in Huang scattering depends on the elasticity of the material (Robertson 1991). Thus anisotropy static atom displacement due to the presence of neighbor atoms, which is equivalent to the polarization direction of phonons, may produce various fine details and streaks in the diffraction pattern (Trinkaus 1972).

For simplification, we discuss a case in which static displacement $\mathbf{v}_\kappa$ as well as thermal vibration displacement $\mathbf{u}_\kappa$ are introduced to each atom site (Hall et al. 1966), so that $\mathbf{r}'_\kappa = \mathbf{r}_\kappa + \mathbf{v}_\kappa + \mathbf{u}_\kappa$. Based on the same assumptions made in Sect. 7.2, the diffraction intensity is

$$\begin{aligned}
\mathrm{I} =& \frac{1}{2}\sum_{\kappa}\sum_{\kappa'}\left[f^e(\tau)\right]^2 < \exp\left[2\pi i\tau\cdot\left(\mathbf{r}_\kappa{}' - \mathbf{r}_{\kappa'}{}'\right)\right] > \\
&+ \frac{1}{2}\sum_{\kappa}\sum_{\kappa'}\left[f^e(\tau - \mathbf{g})\right]^2 < \exp\left[2\pi i(\tau - \mathbf{g})\cdot\left(\mathbf{r}_\kappa{}' - \mathbf{r}_{\kappa'}{}'\right)\right] > \\
&- (-1)^i f^e(\tau) f^e(\tau - \mathbf{g})\{< \exp\left[2\pi i(\tau - \mathbf{g})\cdot\mathbf{r}_\kappa{}' - 2\pi i\tau\cdot\mathbf{r}_{\kappa'}{}'\right] > \\
&+ < \exp\left[2\pi i(\tau - \mathbf{g})\cdot\mathbf{r}_{\kappa'}{}' - 2\pi i\tau\cdot\mathbf{r}'_\kappa\right] >\}
\end{aligned} \tag{7.53}$$

where <> indicates the time average of lattice configurations. We now consider the first term,

$$
\begin{aligned}
I_1 &= \frac{1}{2}\sum_{\kappa}\sum_{\kappa'}\left[f^e(\tau)\right]^2 < \exp\left[2\pi i\tau\cdot(\mathbf{r}'_\kappa - \mathbf{r}'_{\kappa'})\right] > \\
&= \frac{1}{2}\sum_{\kappa}\sum_{\kappa'}\left[f^e(\tau)\right]^2 \exp\left[2\pi i\tau\cdot(\mathbf{r}_\kappa - \mathbf{r}_{\kappa'})\right] \\
&\times \exp\left[2\pi i\tau\cdot(\mathbf{v}_\kappa - \mathbf{v}_{\kappa'})\right] < \exp\left[2\pi i\tau\cdot(\mathbf{u}_\kappa - \mathbf{u}_{\kappa'})\right] > \\
&= \frac{1}{2}\left[f^e(\tau)\right]^2]\{N + \sum_{\kappa\neq\kappa'}\sum \exp\left[2\pi i\tau\cdot(\mathbf{r}_\kappa - \mathbf{r}_{\kappa'})\right] \\
&\times \exp\left[2\pi i\tau\cdot(\mathbf{v}_\kappa - \mathbf{v}_{\kappa'})\right] < \exp\left[2\pi i\tau\cdot(\mathbf{u}_\kappa - \mathbf{u}_{\kappa'})\right] >\} \\
&= \frac{1}{2}\left[f^e(\tau)\right]^2 \{N + \sum_{\kappa\neq\kappa'}\sum \exp\left[2\pi i\tau\cdot(\mathbf{r}_\kappa - \mathbf{r}_{\kappa'})\right] \\
&\times \exp\left[2\pi i\tau\cdot(\mathbf{v}_\kappa - \mathbf{v}_{\kappa'})\right]\exp[-2W(\tau)]\}
\end{aligned} \tag{7.54}
$$

where the Einstein model was used to describe TDS. We also consider the following random-walk model to calculate the lattice average of $\exp\left[2\pi i\tau\cdot(\mathbf{v}_\kappa - \mathbf{v}_{\kappa'})\right]$ (Cochran and Kartha 1956).

The fundamental assumption is the random distribution of defects, which results in displacement components $\Delta\nu_1, \Delta\nu_2, \ldots, \Delta\nu_L$, parallel to τ and varying in a random manner. A quantity

$$
v = \sum_{m=1}^{L} \Delta\nu_m
$$

is assumed to have a Gaussian distribution, where L is the number of displacement components. Thus for a fixed value of $\mathbf{r}_\kappa - \mathbf{r}_{\kappa'}$, there are N values of $\nu_{\kappa T} - \nu_{\kappa' T}$, and these N points have a Gaussian distribution about zero with variance $\overline{(\nu_{\kappa T} - \nu_{\kappa' T})^2}$. This distribution of points can be replaced by a continuous distribution

$$
G_p(v) = \frac{1}{\left[2\pi\overline{(\nu_{\kappa T} - \nu_{\kappa' T})^2}\right]^{1/2}} \exp\left[-\frac{\nu^2}{2\,\overline{(\nu_{\kappa T} - \nu_{\kappa' T})^2}}\right] \tag{7.55}
$$

The configuration average is equivalent to

$$
\begin{aligned}
&\overline{\exp\left[2\pi i\tau\,(\nu_{\kappa T} - \nu_{\kappa' T})\right]} = \frac{\int_{-\infty}^{\infty} d\nu \exp(2\pi i\tau\nu)G_p(\nu)}{\int_{-\infty}^{\infty} d\nu G_p(\nu)} \\
&= \exp[-2\pi^2\tau^2\overline{(\nu_{\kappa T} - \nu_{\kappa' T})^2}] = \exp\{-2\pi^2\tau^2[\overline{(\nu_{\kappa T})^2} + \overline{(\nu_{\kappa' T})^2} - 2\overline{(\nu_{\kappa T}\nu_{\kappa' T})}]\}
\end{aligned}
$$

The term $\overline{(\nu_{\kappa T})^2}$ may equal $\overline{(\nu_{\kappa' T})^2}$ if atomic sites are equivalent in simple cubic structure. Thus (7.54) becomes

$$I_1 = \frac{1}{2}\left[f^e(\tau)\right]^2 \left[N + \sum_{\kappa \neq \kappa'}\sum \exp\left[2\pi i \tau \cdot (\mathbf{r}_\kappa - \mathbf{r}_{\kappa'})\right] \exp[-2\,W(\tau)] \right.$$

$$\left. \times \exp\left\{-2\pi^2\tau^2\left[\overline{(\nu_{\kappa\tau})^2} + \overline{(\nu_{\kappa'\tau})^2}\right] + 4\pi^2\tau^2\overline{(\nu_{\kappa\tau}\nu_{\kappa'\tau})}\right\}\right] \quad (7.56)$$

If displacements have a Gaussian distribution, then $\overline{(\nu_{\kappa\tau}\nu_{\kappa'\tau})} = 0$ for $\kappa \neq \kappa'$. Assuming this to be the case, after subtracting Bragg scattering, the diffusely scattered intensity from the (000) beam is

$$I_{1D} = \frac{N}{2}\left[f^e(\tau)\right]^2 \left\{1 - \exp\left[-2\,W(\tau) - 2\,W^{(d)}(\tau)\right]\right\} \quad (7.57)$$

where $W^{(d)}(\tau) = 2\pi^2\tau^2\overline{(\nu)^2}$ is equivalent to the DW factor for TDS, but it is temperature-independent. Finally the absorption coefficient is

$$\mu_0 = \frac{1}{K_0^2\Omega_0}\int d\sigma(\tau)\left[f^e(\tau)\right]^2 \left\{1 - \exp\left[-2\,W(\tau) - 2\,W^{(d)}(\tau)\right]\right\} \quad (7.58)$$

By the same token, we obtain

$$\mu_g = \frac{1}{K_0^2\Omega_0}\int d\sigma(\tau) f^e(\tau) f^e(\tau - \mathbf{g})\left\{\exp\left[-W(\mathbf{g}) - W^{(d)}(\mathbf{g})\right]\right.$$
$$\left. - \exp\left[-W(\tau) - W(\mathbf{g}-\tau) - W^{(d)}(\tau) - W^{(d)}(\mathbf{g}-\tau)\right]\right\} \quad (7.59)$$

We now use the method introduced by Hall et al. (1966) to calculate the average mean square displacement produced by point defects. A sphere of radius R is chosen to be centered at each point defect; displacements of the N' atoms inside the sphere are affected only by the strain field of the point defect at the center, so the effect due to other defects can be neglected. This assumption is valid only if the defect concentration is low. Thus the displacement around a misfitting sphere is

$$\mathbf{v} = \frac{c'}{r^3}\mathbf{r} \quad (7.60)$$

where c' is a constant depending on the degree of misfit. It is easy to show

$$\overline{(\nu_\tau)^2} = \frac{4\pi^2 c'^2}{3\,N'\Omega_0}\left(\frac{1}{r_{\min}} - \frac{1}{R}\right) \quad (7.61)$$

where $r_{\min}$ is the inner cutoff radius, which is of the order of the interatomic distance or the diameter of the impurity atom; R can be chosen so that N' is the number of matrix atoms for each impurity atom. For a normal concentration $1/R$ can be neglected in (7.61). The value of $\overline{(\nu_T)^2}$ is thus critically affected by the selection of c', which depends on the degree of misfit between the impurity and matrix atoms and the elastic constants of the impurity and matrix (Mott and Nabarro 1940).

The contrast produced by static displacements associated with misfitting dopant atoms has been observed in silicon doped with boron (Perovic et al. 1991). It was found that the contrast of the dopant layer is brighter when the image was formed with the diffusely scattered electrons. This type of contrast is called Huang-scattering contrast (Wang 1994).

7.9 Summary

This chapter reviews the classical thermal diffuse-scattering theory. The physical basis of the frozen lattice model is shown. The absorption coefficients of TDS are calculated. Diffuse scattering due to multiphonon processes is considered based on the two-beam theory, and then the theory is extended to the many-beam case. The coherent effect of TDS significantly affects the interpretation of Z-contrast images formed by thermal diffusely scattered electrons (see Chap. 11). Huang scattering is also a source of diffuse scattering in addition to TDS. This is important when considering the contrast mechanism of high-angle annular dark-field STEM images (see Chap. 11).

Chapter 8
Dynamic Inelastic Electron Scattering I: Bloch Wave Theory

Creating, propagating, and scattering inelastically scattered electrons are fully incorporated into Yoshioka's equations. The key part of the theoretical work is to solve these coupled equations for the process of interest. In this chapter we first discuss the Bloch wave approach for single inelastic-scattering processes. Then the theory is extended to cases where double inelastic scattering is appreciable. The equations derived can be applied to inelastic-scattering processes provided the appropriate transition matrix elements are known. Full dynamic scattering effects are considered. Finally the Bloch wave solution for imperfect crystals is given.

8.1 Solution of Yoshioka's Equations

In Sect. 6.2 we derived Yoshioka's coupled equation for electron inelastic scattering. The diffraction of inelastically scattered electrons is determined by the solution of

$$\left[\nabla^2 + 4\pi^2\left(U + K_n^2\right)\right]\Psi_n = \frac{2m_e}{\hbar^2}\sum_{m\neq n} H_{nm}\Psi_m \tag{8.1}$$

which is a set of coupled equations for $n = 0, 1, 2\ldots$, and $m = 0, 1, 2\ldots$. In the Bloch wave theory, the solution to (8.1) is written as a superposition of Bloch waves with z-dependent superposition coefficients (Howie 1963; Humphreys and Whelan 1969)

$$\Psi_n(\mathbf{r}) = \sum_i \phi_n^{(i)}(z) B_i\left(\mathbf{k}_n^{(i)}, \mathbf{r}\right) \tag{8.2}$$

where

Z. L. Wang, *Elastic and Inelastic Scattering in Electron Diffraction and Imaging*, Springer Series in Solid-State Sciences 205, https://doi.org/10.1007/978-3-031-90819-4_8

$$B_i\left(\mathbf{k}_n^{(i)}, \mathbf{r}\right) = \sum_g C_g^{(i)}(\mathbf{K}_n) \exp\left[2\pi i\left(\mathbf{k}_n^{(i)} + \mathbf{g}\right)\cdot\mathbf{r}\right] \tag{8.3a}$$

which is an eigen state of

$$\left[\nabla^2 + 4\pi^2\left(U + K_n^2\right)\right] B_i = 0 \tag{8.3b}$$

The z axis is chosen as the thickness direction of the thin slab in the Laue case. Substituting (8.2) and (8.3) into (8.1),

$$\sum_i \left\{ \frac{d^2\phi_n^{(i)}(z)}{dz^2} B_i\left(\mathbf{k}_n^{(i)}, \mathbf{r}\right) + 2\frac{d\phi_n^{(i)}(z)}{dz}\frac{\partial B_i\left(\mathbf{k}_n^{(i)}, \mathbf{r}\right)}{\partial z} \right\} = \frac{2m_e}{\hbar^2}\sum_{m\neq n} H_{nm}(\mathbf{r}) \sum_i \phi_m^{(i)}(z) B_i\left(\mathbf{k}_m^{(i)}, \mathbf{r}\right) \tag{8.4}$$

neglecting the $\left[d^2\phi_n^{(i)}(z)/dz^2\right] B_i$ term under the small-angle scattering approximation, multiplying both sides by $B_j^*\left(\mathbf{k}_n^{(j)}, \mathbf{r}\right)$, and integrating over x and y, for the ZOLZ reflections with $g_z = 0$, a set of equations that determines the z-dependent scattering amplitude $\phi_n^{(i)}(z)$ is obtained (Ishida 1970)

$$\frac{d\phi_n^{(i)}(z)}{dz} = \sum_{m\neq n}\sum_j S_{nm}^{(i)(j)} \exp\left[2\pi i \delta_{nm}^{(i)(j)} z\right] \delta\left(\mathbf{k}_{mb}^{(j)} - \mathbf{k}_{nb}^{(i)} - \mathbf{q}_{mb} + \mathbf{q}_{nb}\right) \phi_m^{(j)}(z) \tag{8.5a}$$

with

$$S_{nm}^{(i)(j)} = \frac{-im_e}{2\pi\hbar^2\left[k_n^{(i)}\right]_z} \sum_g\sum_h C_g^{(i)*}(\mathbf{K}_n) H_{g-h}^{nm}(\mathbf{q}_n - \mathbf{q}_m) C_h^{(j)}(\mathbf{K}_m) \tag{8.5b}$$

and

$$\delta_{nm}^{(i)(j)} = \left[k_m^{(j)}\right]_z - \left[k_n^{(i)}\right]_z - (q_m)_z + (q_n)_z \tag{8.5c}$$

The

$$\delta\left(\mathbf{k}_{mb}^{(j)} - \mathbf{k}_{nb}^{(i)} - \mathbf{q}_{mb} + \mathbf{q}_{nb}\right)$$

function in (8.5a) represents the conservation of momentum in the plane perpendicular to the incident beam direction

$$\mathbf{k}_{mb}^{(j)} - \mathbf{k}_{nb}^{(i)} = \mathbf{q}_{mb} - \mathbf{q}_{nb} \tag{8.5d}$$

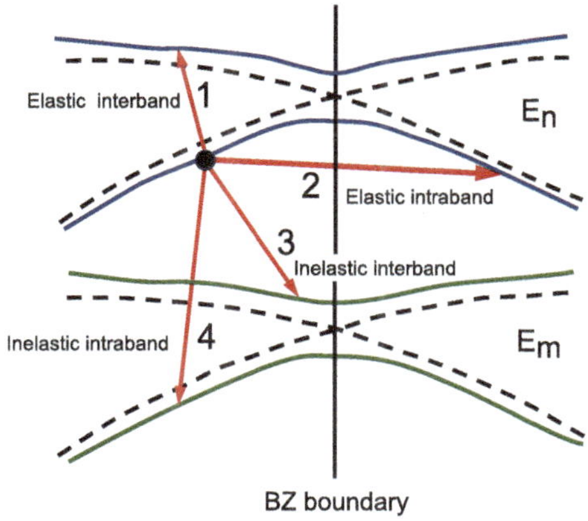

Fig. 8.1 Scattering transitions involving two dispersion surfaces (each having two branches) for energies E_n and E_m

This relation holds for each Bloch wave. The momentum in z direction may not be conserved because of the finite thickness of the crystal. Using (8.5), Ishida (Ishida 1970) gave a full description of the absorption behavior of inelastic waves in different excited states and dispersion branches.

Figure 8.1 shows two dispersion surfaces (each of two branches) for two different energies E_n and E_m under the two-beam approximation. Transitions from a wave vector $\mathbf{K}_n$ state on the first surface to other states on either the first or the second surface and possibly on the other branch of the surface can then be classified as elastic or inelastic, interband or intraband transition. Thus interband and intraband scattering refer to the transition between different Bloch waves and between wave points on the same branch of the dispersion surface, respectively. For reasons of clarity, the two surfaces have been drawn rather far apart in K space.

We now consider a case of small-angle scattering. For valence excitations, the momentum transfer is so small that the change in Bragg conditions is negligible. In this case, the interaction potential has a long range (i.e., delocalized) and its Fourier components are almost zero for large $\mathbf{g}$s, leaving H_0^{nm} as the only remaining term. Equation (8.5b) now becomes

$$S_{nm}^{(i)(j)} \approx -\frac{im_e}{2\pi\hbar^2 K_{0z}} H_0^{nm} \delta_{ij} \tag{8.6}$$

Therefore interband transitions of the fast electron are forbidden under the small-angle inelastic-scattering approximation, and transition probabilities for intraband scattering are the same for all branches of the dispersion surface. We conclude that intraband transition cannot play a serious role in the mechanism of diffraction contrast imaging. In other words, valence loss electrons preserve the contrast of elastically scattered electrons (Howie 1963). This conclusion holds only for low-resolution diffraction contrast imaging. For atomic resolution phase contrast imaging, chromatic aberration effects must be considered (see Chap. 13).

For phonon scattering, the energy loss of the electron is so small that the two sets of dispersion surfaces are practically coincident. The rapid variation of interaction potential gives rise to large-angle scattering, so that the interband transition becomes dominant (Fujimoto and Howie 1966). The interband-transited electrons may not preserve the contrast of elastically scattered electrons.

8.2 Iterative Method

Equation (8.5a) can be solved iteratively. The zero-order solution of (8.5a) is

$$\phi_m^{(i)}(z)\big|_0 = C_0^{(i)*}(\mathbf{K}_0)\,\delta_{m,0}$$

which satisfies the boundary condition at $z = 0$ for the incident plane wave $\exp(2\pi i \mathbf{K}_0 \cdot \mathbf{r})$. Thus the first-order solution is obtained by substituting the zero-order solution into (8.5a) (Rez et al. 1977; Maslen and Rossouw 1984); the result is

$$\phi_n^{(i)}(z)\big|_1 = \sum_j C_0^{(j)*}(\mathbf{K}_0) \int_0^z dz_1\, S_{n0}^{(i)(j)} \exp\left[2\pi i \delta_{n0}^{(i)(j)} z_1\right] \tag{8.7}$$

where $S_{n0}^{(i)(j)}$ is a function of z_1. This is the basic equation for calculating the diffraction intensity of single-inelastically scattered electrons. Substituting the first-order solution (8.7) into (8.5a), the second-order term due to double inelastic electron scattering is obtained

$$\begin{aligned}\phi_n^{(i)}(z)\big|_2 = &\sum_{m\neq n}\sum_{jj'} C_0^{(j')*}(\mathbf{K}_0) \int_0^z dz_2 S_{nm}^{(i)(j)} \exp\left[2\pi i \delta_{nm}^{(i)(j)} z_2\right] \\ &\times \int_0^{z_2} dz_1 S_{m0}^{(j)(j')} \exp[2\pi i \delta_{m0}^{(j)(j')} z_1].\end{aligned} \tag{8.8}$$

This is the equation to be used to calculate the diffraction intensity of the double inelastically scattered electrons (see Sect. 8.5). The wave vector of the Bloch wave is defined as

$$\mathbf{k}_m^{(j)} = \mathbf{K}_m + \nu_m^{(j)} \hat{z} \tag{8.9a}$$

with

$$\mathbf{K}_m = \mathbf{K}_0 - \mathbf{q}_m \tag{8.9b}$$

and $\mathbf{q}_m$ is the change of wave vector of electrons in the crystal due to transition to the m th-excited state.

Substituting (8.3) into (8.2), the intensity of electrons inelastically scattered to $(\mathbf{g} - \mathbf{q}_b)$ is the modulus square of the amplitude for the plane wave $\exp\left[2\pi i\left(\mathbf{K}_0 + \mathbf{g} - \mathbf{q}_b\right) \cdot \mathbf{r}\right]$, which is

$$I_n\left(\mathbf{g} - \mathbf{q}_b\right) = \sum_{q_z}\sum_{ii'} \phi_n^{(i)}(d)\left[\phi_n^{(i')}(d)\right]^* C_g^{(i)}(\mathbf{K}) C_g^{(i')*}(\mathbf{K}) \exp\left\{2\pi i\left[\nu_n^{(i)} - \nu_n^{(i')}\right] d\right\} \quad (8.10)$$

where $\mathbf{K} = \mathbf{K}_0 - \mathbf{q}$; the sum of q_z is over all discrete states in the z direction, with $q_z = p/d$, where p is an integer (Young and Rez 1975). The key to (8.10) is finding the quantity

$$\phi_n^{(i)}(d)\left[\phi_n^{(i')}(d)\right]^*.$$

We now discuss first- and second-order scattering results.

8.3 Diffraction of Single Inelastically Scattered Electrons

The first-order iterative solution (8.7) of (8.5a) describes the diffraction of single-inelastically scattered electrons, which means that the electron can be elastically scattered many times but inelastically scattered only once. Substituting (8.7) into (8.10) and assuming that $S_{n0}^{(i)(j)}$ is independent of z (this is exact for the ZOLZ reflections), we have

$$\begin{aligned} I_1\left(\mathbf{g} - \mathbf{q}_b\right) = &\sum_{ij}\sum_{i'j'} C_0^{(j)*}\left(\mathbf{K}_0\right) C_0^{(j')}\left(\mathbf{K}_0\right) C_g^{(i)}(\mathbf{K}) C_g^{(i)*}(\mathbf{K}) \\ &\times S_{n0}^{(i)(j)} S_{n0}^{(i')(j')*} \exp\left\{2\pi i\left[\nu_n^{(i)} - \nu_n^{(i')}\right] d\right\} \\ &\times \int_0^d dz_1 \exp\left\{2\pi i\left[k_{0z}^{(j)} - k_{nz}^{(i)}\right] z_1\right\} \int_0^d dz_2 \exp\left\{-2\pi i\left[k_{0z}^{(j')} - k_{nz}^{(i')}\right] z_2\right\} \\ &\times \sum_{q_z} \exp\left[2\pi i q_z\left(z_1 - z_2\right)\right] \end{aligned} \quad (8.11)$$

By using identity (Young and Rez 1975)

$$\sum_{q_z} \exp\left[2\pi i q_z\left(z_1 - z_2\right)\right] = \sum_p \exp\left[i\frac{2\pi p}{d}\left(z - z'\right)\right] = \delta\left(z - z'\right) \quad (8.12)$$

and substituting the $S_{n0}^{(i)(j)}$ function for its definition, (8.11) is finalized as Rez et al. (1977) and Maslen and Rossouw (1984)

$$I_1(\mathbf{g}-\mathbf{q}_b) = \frac{m_e^2}{4\pi^2\hbar^4 K_{0z}^2} \sum_{ii'} \sum_{jj'} \sum_{g_1 h_1} \sum_{g_2 h_2} [\text{B.W.}]_1 [\text{Kin.}]_1 [\text{Thickness}]_1 \tag{8.13a}$$

where the dynamic scattering of the inelastic electrons before and after the inelastic event is contained in the Bloch wave part

$$[\text{B.W.}]_1 = c_{g_1}^{(i)*}(\mathbf{K})c_{h_1}^{(j)}(\mathbf{K}_0)c_{g_2}^{(i')}(\mathbf{K})c_{h_2}^{(j')*}(\mathbf{K}_0)c_0^{(j)*}(\mathbf{K}_0)c_0^{(j')}(\mathbf{K}_0)c_g^{(i)}(\mathbf{K})c_g^{(i')*}(\mathbf{K}) \tag{8.13b}$$

$[\text{B.W.}]_1$ is responsible for the formation of Kikuchi patterns. The kinematic inelastic electron-scattering part is

$$[\text{Kin.}]_1 = H_{g_1-h_1}^{n0}(\mathbf{q}_b)\, H_{g_2-h_2}^{n0*}(\mathbf{q}_b) \tag{8.13c}$$

which is responsible for angular spreading from inelastic electron scattering. The thickness-dependent part is

$$[\text{Thickness}]_1 = \frac{\sin\left\{2\pi\left[\nu^{(j)}-\nu^{(j')}\right]d\right\} - \sin\left\{2\pi\left[\nu^{(i)}-\nu^{(i')}\right]d\right\}}{2\pi\left[\nu^{(j)}-\nu^{(j')}-\nu^{(i)}+\nu^{(i')}\right]} \tag{8.13d}$$

An equivalent result was also obtained based on the first Born approximation of inelastic scattering of Bloch waves (Saldin and Rez 1987). To see the physical meaning, we take the small thickness approximation of (8.13a)–(8.13d) so that dynamic scattering effects vanish

$$I_1(\mathbf{g}-\mathbf{q}_b) \approx d\frac{m_e^2}{4\pi^2\hbar^4 K_{0z}^2}\left|H_g^{(1)}(\mathbf{q}_b)\right|^2 \tag{8.14}$$

where the substitution of $H_g^{(1)} = H_g^{n0}$ was made. The symbol change is for the convenience of Sect. 8.5, in which double inelastic scattering is discussed. The total scattered intensity is proportional to specimen thickness d. The integrated intensity of the entire single-inelastically scattered electrons is

$$I_{1t} = \Omega \sum_g \int_{\text{BZ}} d\mathbf{q}_b I_1(\mathbf{g}-\mathbf{q}_b) = d\frac{\Omega m_e^2}{4\pi^2\hbar^4 K_{0z}^2} \sum_g \int_{\text{BZ}} d\mathbf{q}_b \left|H_g^{(1)}(\mathbf{q}_b)\right|^2 \tag{8.15}$$

Scattering at $(\mathbf{g} - \mathbf{q}_b)$ is determined by scattering matrix elements H_g^{n0}, which was derived for different inelastic-scattering processes (see Chap. 6); the scattering matrix elements are listed below.

For atomic inner shell excitations, the kinematic scattering term (Rossouw 1985; see Sect. 6.6.1) is

$$[\text{Kin.}]_a = \frac{e^4}{16\pi^4\varepsilon_0^2}\sum_\alpha \exp\left[-2\pi i\left(\mathbf{Q}_1 - \mathbf{Q}_2\right)\cdot\mathbf{r}_\alpha\right]\frac{\rho_{n0}^{\prime(\alpha_1)}\left(\mathbf{Q}_1\right)}{Q_1^2}\frac{\rho_{on}^{\prime(\alpha_1)}\left(-\mathbf{Q}_2\right)}{Q_2^2} \tag{8.16}$$

where $\mathbf{Q}_1 = \mathbf{g}_1 - \mathbf{h}_1 - \mathbf{q}$ and $\mathbf{Q}_2 = \mathbf{g}_2 - \mathbf{h}_2 - \mathbf{q}$. The sum over α is added to take the contributions of all atoms within the unit cell.

For TDS, using the Einstein model,

$$\begin{aligned}[\text{Kin.}]_{\text{TDS}} =& \sum_\alpha f_\alpha^e\left(\mathbf{Q}_1\right) f_\alpha^{e^*}\left(\mathbf{Q}_2\right)\exp\left[-2\pi i\left(\mathbf{Q}_1 - \mathbf{Q}_2\right)\cdot\mathbf{r}_\alpha\right]\\ &\times\left\{\exp\left[-W_\alpha\left(\mathbf{Q}_1 - \mathbf{Q}_2\right)\right] - \exp\left[-W_\alpha\left(\mathbf{Q}_1\right) - W_\alpha\left(\mathbf{Q}_2\right)\right]\right\}\end{aligned} \tag{8.17}$$

For single-phonon excitation, from (6.56a) and (6.56b),

$$H_g = 2\pi \text{ie}\sum_\alpha A_\alpha\left(\omega_j(\mathbf{q})\right)\mathbf{e}\left(\alpha\middle|{}^{\mathbf{q}}_{j}\right)\cdot(\mathbf{g} - q) f_\alpha^e(\mathbf{g} - \mathbf{q})\exp[-2\pi i\mathbf{g}\cdot\mathbf{r}(\alpha)] \tag{8.18}$$

Thus

$$\begin{aligned}[\text{Kin.}]_s =& 4\pi^2 e^2\sum_\alpha\sum_{\alpha'} A_\alpha(\omega_j(\mathbf{q}))A_{\alpha'}(\omega_j(\mathbf{q}))\mathbf{e}\left(\alpha\middle|{}^{\mathbf{q}}_{j}\right)\\ &\cdot\mathbf{Q}_1\mathbf{e}^*\left(\alpha'\middle|{}^{\mathbf{q}}_{j}\right)\cdot\mathbf{Q}_2 f_\alpha^{\mathbf{e}}\left(\mathbf{Q}_1\right) f_{\alpha'}^{\mathbf{e}^*}\left(\mathbf{Q}_2\right)\\ &\times\exp\left[-2\pi i(\mathbf{g}_1 - \mathbf{h}_1)\cdot\mathbf{r}(\alpha)\right]\exp\left[2\pi i(\mathbf{g}_2 - \mathbf{h}_2)\cdot\mathbf{r}(\alpha')\right]\end{aligned} \tag{8.19}$$

By substituting the corresponding kinematic scattering term, (8.13) can be applied to calculate the diffraction pattern of any inelastic excitation process. However a separate calculation has to be conducted for each wave vector $\mathbf{q}$, making the amount of calculation rather large.

As shown in the [Kin.] term, atom locations in the unit cell affect the final calculated result. For localized inelastic scattering, each atomic site is equivalent to a point emitter in the crystal for generating an inelastic wave. The angular distribution of the emitted inelastic wave is determined by the inelastic-scattering cross section of the atom, but the subsequent diffraction behavior of the wave is determined by the elastic rescattering of the projected local crystal structure, similar to interior point source diffraction within a unit cell.

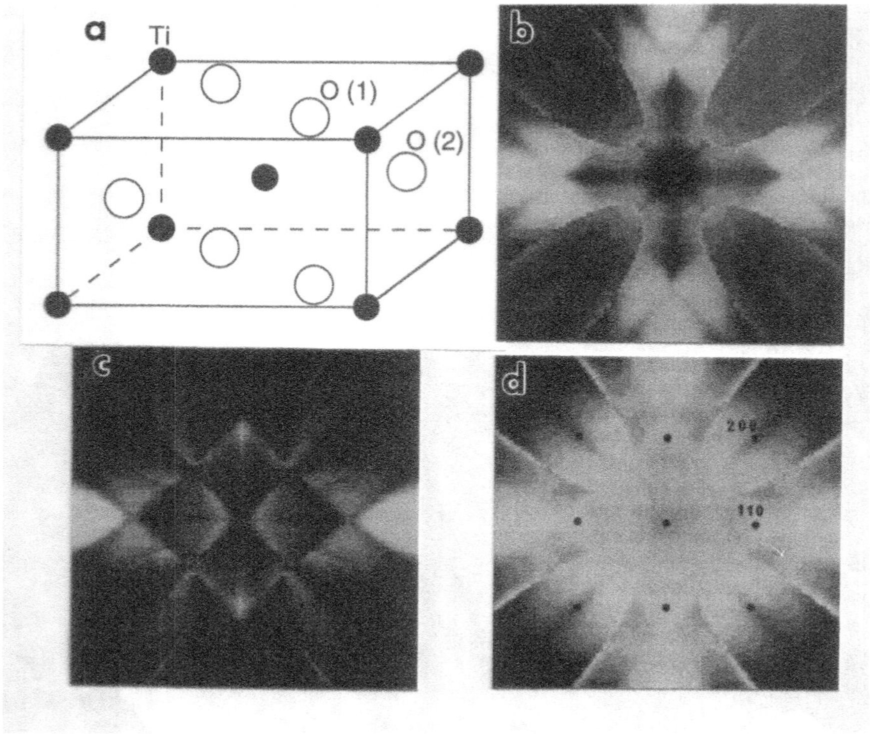

Fig. 8.2 **a** The atomic structure of TiO_2 showing nonequivalent oxygen sites. **b** A nine-beam simulated [001] angular distribution map to TDS electrons generated from Ti sites (maximum 0.33, minimum 0.014). **c** A TDS from two oxygen sites with the same projection symmetry (maximum 0.05, minimum 0.014). **d** A summation of TDS from all oxygen sites (maximum 0.071, minimum 0.028), with a superimposed diffraction pattern. The Einstein model was assumed; isotropic mean displacements were taken as 0.067 Å for O and 0.075 Å for Ti. (Courtesy of Drs. Rossouw and Bursill (1985))

Figure 8.2b–d show a set of nine-beam, computed TDS electron diffraction patterns based on the Einstein model. For Ti, scattering occurs near the center, and streaks along $\langle 110 \rangle$ are introduced (Fig. 8.2b). The TDS waves generated from the two oxygen sites with the same projection symmetry create a pattern distinct from that of Ti. The summed O-site TDS intensity shows excess lobes directed inward from the $\{200\}$ beams to the origin. Fine features observed in the patterns are the result of dynamic elastic rescattering.

8.4 Theory of Kikuchi Patterns

The theory of Kikuchi patterns was originated by Kainuma (1955), and it has been applied to interpret the fine details observed in Kikuchi patterns (Fujimoto and Kainuma 1963; Kainuma and Kogiso 1968; Okamoto et al. 1971). The theory proposed by Kainuma (1955) involves equations analogous to the Yoshioka equations, which are solved in a proper Bloch wave representation to find the angular distribution of inelastically scattered electrons at the exit face of the crystal.

As illustrated in Sect. 6.1, Kikuchi patterns are formed by inelastically scattered electrons with scattering angles in competition with the Bragg angles. Since valence loss is a process with very small momentum transfer, it is unlikely to make a significant contribution to the formation of Kikuchi lines if the crystal thickness is so small that the multiple valence loss-scattering effect is negligible. In general, the probability of atomic inner shell excitation is quite small in comparison to that of TDS. The mean free path of TDS is comparable to that of valence loss. Therefore it is generally believed that Kikuchi patterns are formed primarily by TDS electrons. In this section, we use (8.13) to discuss the physics involved in the formation of Kikuchi patterns and the possible structural information that can be retrieved from data analysis. Equation (8.13a) contains diffraction effects of the electron before and after inelastic excitation. Since the Kikuchi pattern is formed by the diffraction of the electron after inelastic scattering, the diffraction effect of the electron before inelastic excitation can thus be neglected, which means setting $C_g^{(j)}(\mathbf{K}_0) = \delta_{g0}$. For simplification, TDS is characterized by the Einstein model. Thus (8.13) becomes

$$
\begin{aligned}
I_1(\mathbf{g}-\mathbf{q}_b) \approx\ & \frac{m_e^2}{4\pi^2\hbar^4 K_{0z}^2} \sum_{ii'} \sum_{\mathbf{g}_1\mathbf{g}_2} C_{\mathbf{g}_1}^{(i)*}(\mathbf{K}) C_{\mathbf{g}_2}^{(i')}(\mathbf{K}) C_{\mathbf{g}}^{(i)}(\mathbf{K}) C_{\mathbf{g}}^{(i')*}(\mathbf{K}) \\
& \times \frac{\sin\left\{2\pi[\nu^{(i)}-\nu^{(i')}]d\right\}}{2\pi[\nu^{(i)}-\nu^{(i')}]} \sum_\alpha \exp\left[-2\pi i(\mathbf{g}_1-\mathbf{g}_2)\cdot\mathbf{r}_\alpha\right] f_\alpha^e(\mathbf{g}_1-\mathbf{q}) f_\alpha^{e^*}(\mathbf{g}_2-\mathbf{q}) \\
& \times \left\{\exp\left[-W_\alpha(\mathbf{g}_1-\mathbf{g}_2)\right] - \exp\left[-W_\alpha(\mathbf{g}_1-\mathbf{q}) - W_\alpha(\mathbf{g}_2-\mathbf{q})\right]\right\}
\end{aligned} \tag{8.20}
$$

Under the two-beam approximation (see Sect. 2.3), (8.20) becomes (Bird and Wright 1989)

$$
\begin{aligned}
I_1(\mathbf{g}-\mathbf{q}_b) \sim\ & d\left\{1 - \frac{1}{2(1+w^2)}\left[1 - \frac{\sin(2\pi\Delta\nu d)}{2\pi\Delta\nu d}\right]\right\} \sum_\alpha T_\alpha(0,0,\mathbf{q}) \\
& + d\left\{\frac{1}{2(1+w^2)}\left[1 - \frac{\sin(2\pi\Delta\nu d)}{2\pi\Delta\nu d}\right]\right\} \sum_\alpha T_\alpha(\mathbf{g},\mathbf{g},\mathbf{q}) \\
& + d\left\{\frac{w}{(1+w^2)}\left[1 - \frac{\sin(2\pi\Delta\nu d)}{2\pi\Delta\nu d}\right]\right\} \sum_\alpha \cos\left(2\pi\mathbf{g}\cdot\mathbf{r}_\alpha + \vartheta'\right) T_\alpha(0,\mathbf{g},\mathbf{q}) \\
& + d\left[\frac{1}{(1+w^2)^{1/2}} \frac{\sin^2(\pi\Delta\nu d)}{\pi\Delta\nu d}\right] \sum_\alpha \sin\left(2\pi\mathbf{g}\cdot\mathbf{r}_\alpha + \vartheta'\right) T_\alpha(0,\mathbf{g},\mathbf{q})
\end{aligned} \tag{8.21a}
$$

where

$$T_\alpha(\mathbf{g}_1, \mathbf{g}_2, \mathbf{q}) = f_\alpha^e(\mathbf{g}_1 - \mathbf{q})\, f_\alpha^{e*}(\mathbf{g}_2 - \mathbf{q}) \times \left\{\exp\left[-W_\alpha(\mathbf{g}_1 - \mathbf{g}_2)\right] - \exp\left[-W_\alpha(\mathbf{g}_1 - \mathbf{q}) - W_\alpha(\mathbf{g}_2 - \mathbf{q})\right]\right\} \quad (8.21b)$$

$$\Delta\nu = \frac{\left[(KS_g)^2 + |U_g|^2\right]^{1/2}}{K_0} \quad (8.21c)$$

$$w = \frac{KS_g}{|U_g|} \quad (8.21d)$$

and ϑ' is the phase factor of U_g.

As it stands, (8.21a) refers to only one pair of linked Kikuchi lines. The w dependence of the scattering intensity is the most important factor that determines the intensity of Kikuchi patterns with the change of excitation error S_g. If $T_\alpha(\mathbf{g}_1, \mathbf{g}_2, \mathbf{q}) \approx T_\alpha(0, 0, \mathbf{q})$, then the sum of the first and second terms in (8.21a) is approximately constant for crystal thickness d. Thus these two terms are approximately orientation-independent, and they contribute a uniform background in the diffraction pattern. The second and fourth terms are symmetrical about $w = 0$, and they give either a peak (bright line) or a trough (dark line), depending on the sign of the term. The third term is antisymmetric about the Bragg position, and it gives a bright-on dark-line profile. Terms containing phase ($2\pi\mathbf{g} \cdot \mathbf{r}_\alpha + \vartheta'$) represent the variation of Kikuchi patterns with the change of structure phase factor. These characteristics may be applied to measure the polarity of noncentrosymmetric crystals (Bird and Wright 1989). Examples illustrated here qualitatively fit the experimental observations in Sect. 6.1.

Equation (8.21a) clearly shows that the Kikuchi pattern tends to be weak in intensity and featureless if the kinematic scattering term $T_\alpha(\mathbf{g}_1, \mathbf{g}_2, \mathbf{q})$ drops quickly with an increase in the electron-scattering angle. This again shows that TDS must be the dominant process in forming Kikuchi patterns.

8.5 Diffraction of Double Inelastically Scattered Electrons

Following similar procedures, (8.8) and (8.10) are applied to calculate the diffraction intensity of double inelastically scattered electrons (Wang and Li 1994) in which two-step inelastic excitations are involved, as in Fig. 8.3. The transition of the crystal state from the ground state 0 (i.e., the elastic state) to an intermediate state m is excited by an inelastic interaction process

$$H_g^{(1)} = H_g^{m0}$$

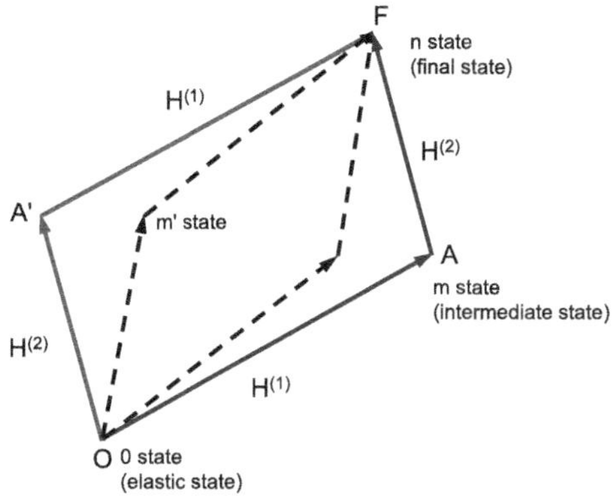

Fig. 8.3 A schematic diagram showing double inelastic-scattering processes. The intermediate states are m and m'. For each intermediate state m, there are two possible routes for double inelastic scattering: $O \rightarrow A$ due to $H^{(1)}$ followed by $A \rightarrow F$ due to $H^{(2)}$, and $O \rightarrow A'$ due to $H^{(2)}$ followed by $A' \rightarrow F$ due to $H^{(1)}$. The two routes are designed to distinguish the sequence of inelastic excitations

such as phonon scattering. The transition of the crystal from the intermediate state m to the final state n is excited by another inelastic interaction process, such as single-electron excitation, with

$$H_g^{(2)} = H_g^{nm}$$

The initial state and the final state are known, but the intermediate state m can be any state with a varying wave vector. In practice, it is believed that excitations $0 \rightarrow m \rightarrow n$ and $0 \rightarrow m' \rightarrow n$ are two incoherent processes if $m \neq m'$ (see Sect. 6.2.2). This incoherence is assumed in the following mathematical operations. Contributions of all intermediate states are contained in the sum over m in (8.8), and the major task here is to perform the summation over m before numerical calculations.

Now let us first consider a quantity in (8.10) for the second-order inelastic scattering, which is defined as

$$\begin{aligned}
R_{ii'} &= \sum_{q_{nz}} \Psi_n^{(i)}(d)\Big|_2 \; \Psi_n^{(i')*}(d)\Big|_2 \\
&= \sum_{q_{nz}} \sum_{m \neq n} \sum_{jj'} \sum_{ff''} C_0^{(j')*}(\mathbf{K}_0)\, C_0^{(f'')}(\mathbf{K}_0) \int_0^d dz_2\, S_{nm}^{(i)(j)}(z_2) \exp\left[2\pi i \delta_{nm}^{(i)(j)} z_2\right] \\
&\times \int_0^{z_2} dz_1\, S_{m0}^{(j)(j')}(z_1) \exp\left[2\pi i \delta_{m0}^{(j)(j')} z_1\right] \int_0^d dz_2'\, S_{nm}^{(i')(f)*}(z_2') \exp\left[-2\pi i \delta_{nm}^{(i')(f)} z_2'\right] \\
&\times \int_0^{z_2'} dz_1'\, S_{m0}^{(f)(f'')*}(z_1') \exp\left[-2\pi i \delta_{m0}^{(f)(f'')} z_1'\right]
\end{aligned} \qquad (8.22)$$

If the first inelastic electron-scattering process is phonon excitation, for example, the sum of m is over all intermediate states of different wave vectors $\mathbf{q}_{\mathrm{m}} = (\mathbf{q}_{mb}, q_{mz})$; i.e.,

$$\sum_{m \neq n} = \sum_{\mathbf{q}_{mb}} \sum_{q_{mz}}$$

Two assumptions are made to simplify (8.22) (Rez et al. 1977; Maslen and Rossouw 1984). First, the $S_{nm}^{(i)(j)}$ function is approximated to be independent of z, so it can be removed from the z integral. Second we assume that the $S_{nm}^{(i)(j)}$ function varies slowly with a change in q_{n2}, so it can also be removed from the summation over q_{nz}. Thus (8.22) is approximated as

$$\begin{aligned} R_{ii'} \approx & \sum_{\mathbf{q}_{mb}} \sum_{jj'} \sum_{ff''} C_0^{(j')*}(\mathbf{K}_0)\, C_0^{(f'')}(\mathbf{K}_0)\, S_{nm}^{(i)(j)} S_{m0}^{(j)(j')} S_{nm}^{(i')(f)*} S_{m0}^{(f)(f'')*} \\ & \times \sum_{q_{mz}} \sum_{q_{nz}} \int_0^d dz_2 \exp\left[2\pi i \delta_{nm}^{(i)(j)} z_2\right] \int_0^d dz_2' \exp\left[-2\pi i \delta_{nm}^{(i')(f)} z_2'\right] \\ & \times \int_0^{z_2} dz_1 \exp\left[2\pi i \delta_{m0}^{(j)(j')} z_1\right] \int_0^{z_2'} dz_1' \exp\left[-2\pi i \delta_{m0}^{(f)(f'')} z_1'\right] \end{aligned} \tag{8.23}$$

Since d is small, the allowed values of q_z are discrete, with an average spacing of $1/d$. Thus

$$q_{nz}^{(i)} = q_{nz}^{(i')} = \frac{n}{d}, \quad q_{mz}^{(j)} = q_{mz}^{(f)} = \frac{m}{d}$$

independent of the branches of the dispersion surface (Rez et al. 1977), where m and n are integers. Using the definition of $\delta_{nm}^{(i)(j)}$ (8.5c and (8.12), (8.23) can be approximated as

$$\begin{aligned} R_{ii'} \approx & \sum_{jj'} \sum_{\mathrm{ff''}} C_0^{(j')*}(\mathbf{K}_0)\, C_0^{(f'')}(\mathbf{K}_0) \sum_{\mathbf{q}_{mb}} S_{nm}^{(i)(j)} S_{m0}^{(j)(j')} S_{nm}^{(i')(f)*} S_{m0}^{(f)(f'')*} \\ & \times \int_0^d dz_2 \exp\left\{2\pi i \left[\nu^{(j)} - \nu^{(i)} - \nu^{(f)} + \nu^{(i')}\right] z_2\right\} \\ & \times \int_0^{z_2} dz_1 \exp\left\{2\pi i \left[\nu^{(j')} - \nu^{(j)} - \nu^{(f'')} + \nu^{(f)}\right] z_1\right\} \end{aligned} \tag{8.24}$$

where an approximation of $v_m^{(i)} \approx v^{(i)}$ was made because the energy loss of the high-energy electrons is much smaller than the kinetic energy. Furthermore, the sum of $\mathbf{q}_{mb}$ can be changed to an integration in the first BZ (Wang 1991); thus

$$\begin{aligned}
&\sum_{\mathbf{q}_{mb}} S_{nm}^{(i)(j)} S_{m0}^{(j)(j')} S_{nm}^{(i')(f)*} S_{m0}^{(f)(f'')*} \\
&= \Omega \int_{BZ} d\mathbf{q}_{mb} S_{nm}^{(i)(j)} S_{m0}^{(j)(j')} S_{nm}^{(i')(f)*} S_{m0}^{(f)(f'')*} \\
&= \Omega \left(\frac{m_0}{2\pi\hbar^2 K_{0z}} \right)^4 \sum_{g_1 h_1} \sum_{g_1' h_1'} \sum_{g_2 h_2} \sum_{g_2' h_2'} C_{g_1}^{(i)**}(\mathbf{K}) C_{h_1'}^{(j')}(\mathbf{K}_0) C_{g_2}^{(i')}(\mathbf{K}) C_{h_2'}^{(f'')*}(\mathbf{K}_0) \\
&\times \int_{BZ} d\mathbf{q}_{mb} \left[C_{h_1}^{(j)}(\mathbf{K}_m) C_{g_1'}^{(j)*}(\mathbf{K}_m) C_{h_2}^{(f)*}(\mathbf{K}_m) C_{g_2'}^{(f)}(\mathbf{K}_m) \right. \\
&\left. \times H_{g_1 - h_1}^{(2)}(\mathbf{q}_b - \mathbf{q}_{mb}) H_{g_1' - h_1'}^{(1)}(\mathbf{q}_{mb}) H_{g_2 - h_2}^{(2)*}(\mathbf{q}_b - \mathbf{q}_{mb}) H_{g_2' - h_2'}^{(1)*}(\mathbf{q}_{mb}) \right]
\end{aligned} \tag{8.25}$$

with $\mathbf{K}_m = \mathbf{K} - \mathbf{q}_m$. Finally using (8.25), substituting (8.24) into (8.10), the intensity at $(\mathbf{g} - \mathbf{q}_b)$ due to double inelastic electron scattering is determined by Wang and Li (1994)

$$\begin{aligned}
I_2(\mathbf{g} - \mathbf{q}_b) = &\frac{\Omega m_e^4}{16\pi^4 \hbar^8 K_{0z}^4} \sum_{ii'} \sum_{jj'} \sum_{ff''} \sum_{g_1 h_1} \sum_{g_1' h_1'} \sum_{g_2 h_2} \sum_{g_2' h_2'} \\
&\times \int_{BZ} d\mathbf{q}_{mb} [\text{B.W.}]_2 [\text{Inelastic}]_2 [\text{Thickness}]
\end{aligned} \tag{8.26a}$$

where

$$\begin{aligned}
[\text{B.W.}]_2 = &C_{g_1}^{(i)*}(\mathbf{K}) C_{g_1'}^{(j)*}(\mathbf{K}_m) C_{h_1}^{(j)}(\mathbf{K}_m) C_{h_1'}^{(j')}(\mathbf{K}_0) \\
&\times C_{g_2}^{(i')}(\mathbf{K}) C_{g_2'}^{(f)}(\mathbf{K}_m) C_{h_2}^{(f)*}(\mathbf{K}_m) \\
&\times C_{h_2'}^{(f'')^*}(\mathbf{K}_0) C_g^{(i)}(\mathbf{K}) C_g^{(i')*}(\mathbf{K}) C_0^{(j')^*}(\mathbf{K}_0) C_0^{(f'')}(\mathbf{K}_0)
\end{aligned} \tag{8.26b}$$

with $\mathbf{K} = \mathbf{K_0} - \mathbf{q}_b$.

$$[\text{Thickness}]_2 = \left(\frac{\cos\left\{2\pi\left[\nu^{(j)} - \nu^{(f)}\right]d\right\} - \cos\left\{2\pi\left[\nu^{(i)} - \nu^{(i')}\right]d\right\}}{2\pi\left[\nu^{(j')} - \nu^{(i)} + \nu^{(i')} - \nu^{(f)}\right]} \right.$$
$$\left. - \frac{\cos\left\{2\pi\left[\nu^{(j')} - \nu^{(f'')}\right]d\right\} - \cos\left\{2\pi\left[\nu^{(i)} - \nu^{(i')}\right]d\right\}}{2\pi\left[\nu^{(i')} - \nu^{(i)} + \nu^{(j')} - \nu^{(f'')}\right]} \right)$$
$$\times \left\{2\pi\left[\nu^{(j')} - \nu^{(j)} - \nu^{(f'')} + \nu^{(f)}\right]\right\}^{-1} \tag{8.26c}$$

and

$$[\text{Inelastic}]_2 = H^{(2)}_{g_1-h_1}(\mathbf{q}_b - \mathbf{q}_{mb})\, H^{(1)}_{g_1-h'_1}(\mathbf{q}_{mb}) \left[H^{(2)}_{g_2-h_2}(\mathbf{q}_b - \mathbf{q}_{mb})\, H^{(1)}_{g'_2-h'_2}(\mathbf{q}_{mb})\right]^*$$
$$+ H^{(1)}_{g_1-h_1}(\mathbf{q}_b - \mathbf{q}_{mb})\, H^{(2)}_{g'_1-h'_1}(\mathbf{q}_{mb}) \left[H^{(1)}_{g_2-h_2}(\mathbf{q}_b - \mathbf{q}_{mb})\right.$$
$$\left. \times H^{(2)}_{g'_2-h'_2}(\mathbf{q}_{mb})\right]^* \left\{1 - \delta\left[H^{(2)}, H^{(1)}\right]\right\} \tag{8.26d}$$

where the second term is added to include the contribution of double inelastic processes in which the $H^{(2)}$ excitation occurs before the $H^{(1)}$ excitation (see Fig. 8.3). The second term disappears if the double inelastic electron scattering is a double excitation of the same state; this is characterized by a delta-like function defined as

$$\left\{1 - \delta\left[H^{(2)}, H^{(1)}\right]\right\} = \begin{cases} 0 \text{ for } H^{(2)} = H^{(1)} \\ 1 \text{ for } H^{(2)} \neq H^{(1)} \end{cases} \tag{8.27}$$

These equations could be further approximated by neglecting the $\mathbf{q}_m$-dependence of the $\mathbf{C}$ matrix, so that the C_g coefficients inside the $\mathbf{q}_{mb}$ integration are removed.

Kinematic scattering results can be obtained from (8.26a)–(8.26d) under the small thickness approximation

$$I_2(\mathbf{g} - \mathbf{q}_b) \approx \frac{d^2}{2} \frac{\Omega m_e^4}{16\pi^4 \hbar^8 K_{0z}^4}$$
$$\times \int_{\text{BZ}} d\mathbf{q}_{mb} \left[\left|\sum_h H^{(2)}_{g-h}(\mathbf{q}_b - \mathbf{q}_{mb})\, H^{(1)}_h(\mathbf{q}_{mb})\right|^2\right]$$
$$\times \left\{2 - \delta\left[H^{(2)}, H^{(1)}\right]\right\} \tag{8.28}$$

The physical meaning of (8.28) is described as follows. (1) The electron is elastically scattered to a Bragg spot $\mathbf{h}$; (2) it is scattered inelastically through an angle $(-\mathbf{q}_{mb})$ due to a momentum transfer of $\hbar\mathbf{q}_{mb}$ to the crystal (i.e., the first inelastic interaction $H_h^{(1)}$); (3) the electron undergoes the Umpklapp scattering, in which it is scattered through another angle $[\mathbf{g} - \mathbf{h} - (\mathbf{q}_b - \mathbf{q}_{mb})]$ due to the second inelastic interaction

$H^{(2)}_{g-h}$. The resultant wave vector change is $\mathbf{Q} = \mathbf{g} - \mathbf{q}_b$. The convolution operation is the result of electron angular redistribution due to inelastic scattering. The final intensity at $(\mathbf{g} - \mathbf{q}_b)$ is an incoherent summation of all intensities contributed by the scattering of different intermediate states. Total scattered electron intensity is proportional to d^2.

If the dynamic scattering effects are ignored, we have approximately

$$I_2(\mathbf{g} - \mathbf{q}_b) = \frac{d^2}{2} \frac{\Omega m_e^4}{16\pi^4 \hbar^8 K_{0z}^4} \times \int_{BZ} d\mathbf{q}_{mb} \left[\sum_h \left| H^{(2)}_{g-h}(\mathbf{q}_b - \mathbf{q}_{mb}) \right|^2 \left| H^{(1)}_h(\mathbf{q}_{mb}) \right|^2 \right] \left\{ 2 - \delta \left[H^{(2)}, H^{(1)} \right] \right\} \quad (8.29)$$

To examine the consistency of (8.29) with the result expected from the Poisson distribution law, we assume that the two inelastic processes are generated by the same type of event; i.e., $H_g^{(2)} = H_g^{(1)}$. Thus the total intensity of the electrons after double scattering by $H^{(1)}$ is the summation of the intensities distributed in the diffraction plane; i.e.,

$$\begin{aligned} I_{2t} &= \Omega \sum_g \int_{BZ} d\mathbf{q}_b I_2(\mathbf{g} - \mathbf{q}_b) \\ &= \frac{d^2}{2} \frac{\Omega^2 m_e^4}{16\pi^4 \hbar^8 K_{0z}^4} \int_{BZ} d\mathbf{q}_b \int_{BZ} d\mathbf{q}_{mb} \left[\sum_{gh} \mid H^{(1)}_{g-h}(\mathbf{q}_b - \mathbf{q}_{mb})\, H^{(1)}_h(\mathbf{q}_{mb})^2 \right] \\ &= \frac{d^2}{2} \frac{\Omega^2 m_e^4}{16\pi^4 \hbar^8 K_{0z}^4} \left[\sum_g \int_{BZ} d\mathbf{q}_b \left| H^{(1)}_g(\mathbf{q}_b) \right|^2 \right]^2 \end{aligned} \quad (8.30)$$

Comparing (8.30) with 8.15), a relationship of $I_{2t} = (1/2!)\,(I_{1t})^2$ is obtained just as expected from the Poisson distribution law (Whelan 1976; see Sect. 14.1).

The theory proposed in this section has fully considered the diffraction of the electron before and after inelastic excitations, and this theory can be applied to any two inelastic-scattering processes. In phonon excitation, matrix elements $H_g(\mathbf{q})$ depend on the assumption of phonon models, and they can easily be obtained for single-phonon excitation (Whelan 1965b). Separated calculations are needed for double-, triple-, or high-order multiphonon scattering. In practice, single-phonon excitation is a valid model for low-angle diffuse scattering. For high-angle diffuse scattering, the intensity contributed by multiphonon excitations is comparable with that made by single-phonon excitation (Sect. 7.5; Hall 1965). It is not possible to include the total excitations of multiphonons in a single formula based on the Bloch wave theory. It is thus desirable to propose an alternative theory that covers all phonon excitations in a single formula. This question is addressed in Chap. 10.

8.6 Coherent Double Inelastic Scattering Under Delta Function Localization Approximation

We now consider a special case of double inelastic scattering, TDS-atomic inner shell scattering. For simplification, we use the result for small crystal thicknesses and replace $H^{(1)}$ by $H^{(P)}$, where P means phonon scattering. From (8.29), the angular distribution of double inelastically scattered electrons is given by

$$I_2\left(\mathbf{g}-\mathbf{q}_b\right) \sim d^2 \int_{\mathrm{BZ}} d\mathbf{q}_{mb} \left| \sum_h H_{g-h}^{(2)}\left(\mathbf{q}_b-\mathbf{q}_{mb}\right) H_h^{(P)}\left(\mathbf{q}_{mb}\right) \right|^2 \tag{8.31}$$

where superscriptions m and n of H are omitted because momentum transfer $\mathbf{q}_m$ represents the different excited states m. This assumption holds if we consider only contributions made by inelastic-scattering processes of fixed energy loss but different momentum transfers (i.e., energy-filtered electron diffraction pattern). This is the case for TDS-atomic inner shell, double inelastic scattering, in which the energy loss of TDS is almost zero, and the energy loss of atomic inner shell excitation is the threshold of the corresponding ionization edge. Fourier coefficients can be written as a sum over different atomic sites

$$H_g(\mathbf{q}) = \sum_\alpha H_\alpha(\mathbf{g}-\mathbf{q}) \exp\left[-2\pi i(\mathbf{g}-\mathbf{q}) \cdot \mathbf{r}_\alpha\right] \tag{8.32}$$

Thus (8.31) becomes

$$\begin{aligned} I_2\left(\mathbf{g}-\mathbf{q}_b\right) \sim d^2 \int_{\mathrm{BZ}} d\mathbf{q}_{\mathrm{mb}} \,|\, \sum_h \sum_\alpha \sum_{\alpha'} H_{\alpha'}^{(2)}\left(\mathbf{g}-\mathbf{h}-\mathbf{q}_b+\mathbf{q}_{mb}\right) H_\alpha^{(P)}\left(\mathbf{h}-\mathbf{q}_{mb}\right) \\ \times \exp\left[-2\pi i\left(\mathbf{g}-\mathbf{h}-\mathbf{q}_b+\mathbf{q}_{mb}\right) \cdot \mathbf{r}_{\alpha'}\right] \exp\left[-2\pi i\left(\mathbf{h}-\mathbf{q}_{mb}\right) \cdot \mathbf{r}_\alpha\right]\big|^2 \end{aligned} \tag{8.33}$$

If the inelastic scattering is a perfectly localized scattering process, i.e., inelastic excitation occurs only when the electron hits the atomic nuclei (the delta function approximation), then this can be expressed mathematically as

$$H_\alpha'\left(\mathbf{r}-\mathbf{R}_n-\mathbf{r}_\alpha\right) \approx \delta\left(\mathbf{r}-\mathbf{R}_n-\mathbf{r}_\alpha\right) \tag{8.34a}$$

or in reciprocal space

$$H_\alpha(\mathbf{g}-\mathbf{q}) \approx H_\alpha(0) \tag{8.34b}$$

independent of $\mathbf{g}-\mathbf{q}$. Using (8.34b) for both Processes 1 and 2, (8.33) is simplified as

$$\begin{aligned} I_2(\mathbf{g}-\mathbf{q}_b) \sim & d^2 \int_{BZ} d\mathbf{q}_{mb} \mid \sum_\alpha \sum_{\alpha'} H_{\alpha'}^{(2)} H_\alpha^{(P)} \\ & \times \exp\left[-2\pi i\,(\mathbf{g}-\mathbf{q}_b+\mathbf{q}_{mb})\cdot\mathbf{r}_{\alpha'} + 2\pi i \mathbf{q}_{mb}\cdot\mathbf{r}_\alpha\right] \\ & \times \sum_h \exp\left[-2\pi i \mathbf{h}\cdot(\mathbf{r}_\alpha - \mathbf{r}_{\alpha'})\right] \mid^2 \\ & \sim d^2 \mid \sum_\alpha H_\alpha^{(2)} H_\alpha^{(P)} \exp\left[-2\pi i\,(\mathbf{g}-\mathbf{q}_b)\cdot\mathbf{r}_\alpha\right] \mid^2 \end{aligned} \tag{8.35}$$

where the identity

$$\sum_h \exp\left[-2\pi i \mathbf{h}\cdot(\mathbf{r}_\alpha - \mathbf{r}_{\alpha'})\right] = \delta_{\alpha\alpha'}$$

was used. Equation (8.35) gives an important conclusion: For thin crystals under kinematic scattering and the perfect localization (or delta function) approximations, the inelastic scattering does not occur unless the incident electron directly strikes the nucleus; the two inelastic-scattering processes tend to occur simultaneously during scattering by the same atom. The probability of the two inelastic-scattering processes occurring separately during scattering by different atoms is negligible.

Recently experimental techniques were introduced for the acquisition of EELS spectra of high-angle thermal diffuse scattered electrons in a TEM (Wang and Fisher 1993). This type of spectrum, called TDS-EELS, can be acquired in image mode using an annular dark-field objective aperture or in diffraction mode using an annular condenser aperture. The TDS-EELS is temperature-sensitive, and it may contain information on specific atomic Debye-Waller factors. The TDS-EELS is attributed to the double inelastic-scattering process: Electrons are scattered to high angles due to TDS, which does not introduce energy loss; then an energy loss $\Delta E = \varepsilon_n$ and an angular redistribution are introduced due to single-electron excitation of atomic inner shells. Double inelastic scattering can occur in two different ways. (1) In "coherent" double inelastic scattering (Fig. 8.4a), phonon (or TDS) and core-shell excitations occur simultaneously during scattering by the same atom A, so that the large momentum transfer (due to TDS) and large energy loss (due to core-shell excitation) are carried out in the scattering of the same atom. Electrons scattered this way carry thermal vibration and electronic structure information about the same atom. As previously discussed, this result holds under kinematic and delta function localization approximations. (2) In incoherent double inelastic scattering (Fig. 8.4b), TDS and core-shell excitations occur separately during scattering by different atoms. If, for example, phonon scattering occurs first, the electron is scattered to an intermediate

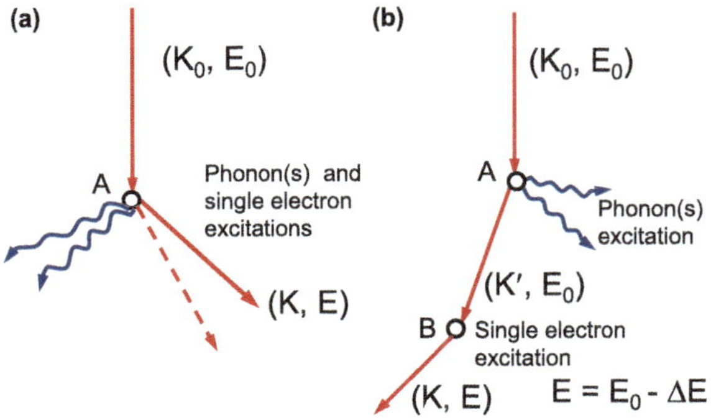

Fig. 8.4 Models of **a** coherent and **b** incoherent double inelastic-scattering processes (see text)

state $(\mathbf{K}', E_0)$ due to its interaction with atom A. Then the electron is rescattered by atom B to the final state $(\mathbf{K}, E)$ because of core-shell excitation. Information carried by electrons scattered in this way is less localized. Experimental results appear not to support the incoherent double inelastic-scattering model but the coherent double inelastic model (Wang and Fisher 1993).

Equation (8.35) simply provides the theoretical basis of the coherent double inelastic-scattering model. However it must be pointed out that this model was derived from approximations of small thickness, delta function localization, and no backscattering. Therefore its application in practice may be restricted due to finite localization of inelastic scattering.

The analysis of core-loss signals observed in TDS-EELS is based on an assumption that the diffuse-scattering background in electron diffraction patterns is produced purely by thermal diffuse scattering (without any energy loss) (Wang and Fisher 1993). As pointed out by Eaglesham and Berger (1994), electron Compton scattering (with energy loss from 100–2000 eV; see Sect. 6.1.3) may be responsible in part for the diffuse background observed in electron diffraction patterns, especially for specimens composed of light elements. Compton scattering, however, does not affect the analysis of ionization edges observed in TDS-EELS, because electrons double inelastically scattered by Compton atomic inner shell processes suffer a total energy loss of $(\Delta E_c + \varepsilon_n)$, where ΔE_c is the energy loss of electron Compton scattering, ranging from 100–2000 eV, and ε_n is the threshold energy (or edge energy) of the core loss. Signals observed near the ionization edge ε_n can come only from the TDS-atomic inner shell scattering, because TDS does not introduce significant energy loss. An anologous discussion also applies to Bremsstrahlung. The only possible effect of Compton scattering on TDS-EELS is the anomalous increase of the spectrum background due to the broad energy loss profile, so that the background may not be simulated by a function of the form $A\Delta E^r$. This has been observed experimentally (Wang and Fisher 1993).

8.7 Diffraction Contrast Images of Inelastically Scattered Electrons

In practice, defects are usually present in crystals. In Sects. 2.10.2 and 4.5, we discussed the diffraction theory of imperfect crystals for elastically scattered electrons. We now illustrate the principles of approaching images of inelastically scattered electrons of imperfect crystals. Images of stacking faults and dislocations formed by phonon- and plasmon-scattered electrons were first calculated by Melander (1975), Melander and Sandstrom (1975a, b). An equivalent approach was proposed by Rez (1976, 1983) based on Yoshioka equations.

8.7.1 Images of Stacking Faults

A stacking fault is usually created by a relative displacement **R** of the crystal at the top with respect to that at the bottom at depth d_1. Electron scattering from 0 to $d_1 - \Delta z'$, and from $d_1 + \Delta z'$ to d is governed by scattering equations of the perfect crystal, where $2\Delta z'$ is a small thickness region within which the stacking fault is contained. Inelastic scattering can take place either before or after the stacking fault. The effect of the fault is to introduce a phase shift $2\pi \mathbf{g} \cdot \mathbf{R}$ in the Bloch wave coefficients. Using the theorem proved in Sect. 4.4, the Bloch wave coefficient of the elastic wave after the stacking fault is expressed into a matrix equation

$$\phi_0\left(d_1 + \Delta z'\right) = \mathbf{P}_f\left(d_1\right)\phi_0(0) \tag{8.36}$$

where the scattering operator is

$$\left[\mathbf{P}_f(z)\right]_{gh} = \sum_i C_g^{(i)} \exp\left(2\pi i \nu_i z\right) C_h^{(i)*} \exp[2\pi i (\mathbf{g} - \mathbf{h}) \cdot \mathbf{R}] \tag{8.37}$$

A phase factor $\exp[2\pi i (\mathbf{g} - \mathbf{h}) \cdot \mathbf{R}]$ was added to incorporate the phase shift of Bloch wave coefficients due to lattice translation.

Under the single inelastic scattering approximation, (8.5) is expressed in matrix form if condition

$$\mathbf{k}_{0b}^{(j)} - \mathbf{k}_{nb}^{(i)} = \mathbf{q}_b$$

is satisfied

$$\frac{d\phi_n(z)}{dz} = \exp\left(-2\pi i q_z z\right) \mathbf{S}_{n0}\phi_0(z) \tag{8.38a}$$

with

$$\left(\mathbf{S}_{nm}\right)_{ij} = S_{nm}^{(i)(j)} \exp\left[2\pi i \left(k_{mz}^{(j)} - k_{nz}^{(i)}\right) z\right] \tag{8.38b}$$

We now consider the following two cases. First if inelastic scattering occurs after the stacking fault, the Bloch wave amplitude of the inelastic wave is

$$\phi_n^{(1)} = \int_{d_1}^{d} dz \exp(-2\pi i q_z z)\, \mathbf{S}_{n0}(z)\mathbf{P}_f\,(d_1)\, \phi_0(0) \tag{8.39}$$

Second if inelastic scattering occurs before electrons reach the stacking fault, the inelastic Bloch wave amplitude is

$$\phi_n^{(2)} = \mathbf{P}_f\,(d - d_1) \int_{0}^{d_1} dz \exp(-2\pi i q_z z)\, \mathbf{S}_{n0}(z)\phi_0(0) \tag{8.40}$$

The intensity scattered to $\mathbf{g} - \mathbf{q}$ is

$$I_n(\mathbf{g} - \mathbf{q}) \propto \sum_{q_z} \left| \sum_{i'} C_g^{(i')} \left\{ \left[\phi_n^{(1)}\right]_{i'} + \left[\phi_n^{(2)}\right]_{i'} \right\} \exp(2\pi i \nu_{i'} d) \right|^2 \tag{8.41}$$

where the sum over q_z includes contributions of inelastic states of different momenta in the z direction but the same momentum in the **b** plane. Equation (8.41) was used extensively by Rez (1976) to calculate stacking fault images of inelastically scattered electrons.

8.7.2 Solution of Yoshioka's Equations for Imperfect Crystals

In this section we calculate images formed by inelastically scattered electrons of crystals containing defects and dislocations. We consider the solution of Yoshioka's equation for a crystal with a continuous lattice distortion described by $\mathbf{R}(\mathbf{r})$. First we introduce the lattice distortion in the Bloch wave expression (8.3a)

$$B_i'\left(\mathbf{k}_n^{(i)}, \mathbf{r}\right) = \sum_g C_g^{(i)} \exp\left[2\pi i \mathbf{k}_n^{(i)} \cdot \mathbf{r} + 2\pi i \mathbf{g} \cdot (\mathbf{r} - \mathbf{R})\right] \tag{8.42}$$

Second inelastic transition matrix elements and crystal potential are modified as

$$H_{nm}(\mathbf{r}) = \exp(-2\pi i \mathbf{q}_{nm} \cdot \mathbf{r}) \sum_g H_g^{nm}(\mathbf{q}_{nm}) \exp[2\pi i \mathbf{g} \cdot (\mathbf{r} - \mathbf{R})] \tag{8.43a}$$

and

$$U(\mathbf{r}) = \sum_g U_g \exp(2\pi i\mathbf{g}\cdot\mathbf{r} - 2\pi i\mathbf{g}\cdot\mathbf{R}) \tag{8.43b}$$

The solution to (8.1) is written as

$$\Psi_n(\mathbf{r}) = \sum_i \phi_n^{(i)}(z) B_i'\left(\mathbf{k}_n^{(i)}, \mathbf{r}\right) \tag{8.44}$$

Substituting (8.44) into (8.1), and considering the small-angle scattering approximation, we have

$$\begin{aligned}
&\sum_i \left\{ 2\frac{d\phi_n^{(i)}(z)}{dz}\frac{\partial B_i'\left(\mathbf{k}_n^{(i)}, \mathbf{r}\right)}{\partial z} + \phi_n^{(i)}(z)\left[\nabla^2 \ +4\pi^2\left(U + K_n^2\right)\right] B_i'\left(\mathbf{k}_n^{(i)}, \mathbf{r}\right)\right\} \\
&\approx \frac{2m_e}{\hbar^2}\sum_{m\neq n} H_{nm}(\mathbf{r}) \sum_i \phi_m^{(i)}(z) B_i'\left(\mathbf{k}_m^{(i)}, \mathbf{r}\right)
\end{aligned} \tag{8.45}$$

It is important to note that the term $\left[\nabla^2 + 4\pi^2\left(U + K_n^2\right)\right] B_i'\left(\mathbf{k}_n^{(i)}, \mathbf{r}\right)$ is no longer equal to zero because of the nonperiodic lattice distortion in U. We now use the modified Bloch wave expression in (8.42) to examine this term. If the lattice distortion $\mathbf{R}$ is a slow variation function at atomic scale, neglecting the term containing $\nabla^2\mathbf{R}$, we have approximately

$$\begin{aligned}
&\left[\nabla^2 + 4\pi^2\left(U + K_n^2\right)\right] B_i'\left(\mathbf{k}_n^{(i)}, \mathbf{r}\right) \\
&\approx 4\pi^2 \sum_g \left[K_n^2 C_g^{(i)} - \left[\mathbf{k}_n^{(i)} + \mathbf{g} - \nabla(\mathbf{g}\cdot\mathbf{R})\right]^2 C_g^{(i)} + \sum_h U_{g-h} C_h^{(i)} \right] \\
&\times \exp\left[2\pi i\mathbf{k}_n^{(i)}\cdot\mathbf{r} + 2\pi i\mathbf{g}\cdot(\mathbf{r} - \mathbf{R})\right]
\end{aligned} \tag{8.46}$$

Using the Bloch wave solution in (2.10a) for the C coefficients,

$$\sum_h U_{g-h} C_h^{(i)} = -\left\{K_n^2 - \left[\mathbf{k}_n^{(i)} + \mathbf{g}\right]^2\right\} C_g^{(i)}$$

Equation (8.46) is approximated as

$$\begin{aligned}
&\left[\nabla^2 + 4\pi^2\left(U + K_n^2\right)\right] B_i'\left(\mathbf{k}_n^{(i)}, \mathbf{r}\right) \\
&\approx 8\pi^2 \sum_g \left\{\left[\mathbf{k}_n^{(i)} + \mathbf{g}\right]\cdot\nabla(\mathbf{g}\cdot\mathbf{R}) C_g^{(i)}\right\} \exp\left[2\pi i\mathbf{k}_n^{(i)}\cdot\mathbf{r} + 2\pi i\mathbf{g}\cdot(\mathbf{r} - \mathbf{R})\right]
\end{aligned} \tag{8.47}$$

Substituting (8.47) into (8.45), multiplying both sides by $B'_j\left(\mathbf{k}_n^{(j)}, \mathbf{r}\right)$, integrating over $\mathbf{b}$, and applying some algebra, we have

$$\begin{aligned}\frac{d\phi_n^{(i)}(z)}{dz} =& 2\pi i \sum_g \sum_j C_g^{(i)*} C_g^{(j)} \frac{\left[k_{nz}^{(j)} + g_z\right]}{k_{nz}^{(i)}} \mathbf{g} \cdot \frac{d\mathbf{R}(z)}{dz} \exp\left[2\pi i \left(\nu_j - \nu_i\right) z\right] \phi_n^{(j)}(z) \\ &+ \sum_{m \neq n} \sum_j S_{nm}^{(i)(j)} \exp\left[2\pi i \delta_{nm}^{(i)(j)} z\right] \delta\left(\mathbf{k}_{mb}^{(j)} - \mathbf{k}_{nb}^{(i)} - \mathbf{q}_{mb} + \mathbf{q}_{nb}\right) \phi_m^{(j)}(z)\end{aligned} \quad (8.48)$$

where the first term corresponds to the elastic diffuse scattering of inelastic electrons due to the presence of defects and dislocations [i.e., (4.42)] and the second term is the source for creating inelastic electrons as well as the dynamic diffraction effect [i.e., (8.5a)]. Electron multiple elastic, multiple inelastic, and diffuse scattering from the lattice distortion are comprehensively included in (8.48).

8.7.3 *Diffraction Contrast Imaging of Single Inelastically Scattered Electrons*

Under the single inelastic-scattering approximation, the solution of (8.48) determines the contrast of the image formed by the inelastically scattered electrons. Following the same procedure as demonstrated for a stacking fault, (8.48) can be written in matrix form for the inelastic wave with $n = 1, 2 \ldots$ (Rez 1983)

$$\frac{d\phi_n(z)}{dz} = \mathbf{B}(z)\phi_n(z) + \exp\left(-2\pi i q_z z\right) \mathbf{S}_{n0}\phi_0(z) \quad (8.49)$$

and for the elastic wave

$$\frac{d\phi_0(z)}{dz} = \mathbf{B}(z)\phi_0(z) \quad (8.50)$$

where $\mathbf{B}$ matrix elements are

$$\begin{aligned}[\mathbf{B}(z)]_{ij} &= 2\pi i \sum_g C_g^{(i)*} C_g^{(j)} \frac{\left(k_{nz}^{(j)} + g_z\right)}{k_{nz}^{(i)}} \mathbf{g} \cdot \frac{d\mathbf{R}(z)}{dz} \exp\left[2\pi i \left(\nu_j - \nu_i\right) z\right] \\ &\approx 2\pi i \sum_g C_g^{(i)*} C_g^{(j)} \mathbf{g} \cdot \frac{d\mathbf{R}(z)}{dz} \exp\left[2\pi i \left(\nu_j - \nu_i\right) z\right]\end{aligned} \quad (8.51a)$$

and $\mathbf{S}$ matrix elements are

$$(\mathbf{S}_{n0})_{ij} = S_{n0}^{(i)(j)} \exp\left[2\pi i \delta_{n0}^{(i)(j)} z\right] \tag{8.51b}$$

To find the solution of (8.49), a matrix $\mathbf{E}'(z)$ is defined as the solution of

$$\frac{d\mathbf{E}'(z)}{dz} = -\mathbf{E}'(z)\mathbf{B}(z) \tag{8.52}$$

We now consider the following mathematical operation, using of (8.49) and (8.52)

$$\begin{aligned} \frac{d\left[\mathbf{E}'(z)\phi_n(z)\right]}{dz} &= \frac{d\mathbf{E}'(z)}{dz}\phi_n(z) + \mathbf{E}'(z)\frac{d\phi_n(z)}{dz} \\ &= \exp\left(-2\pi i q_z z\right)\mathbf{E}'(z)\mathbf{S}_{n0}\phi_0(z) \end{aligned} \tag{8.53}$$

Integrating over z and using the boundary condition $[\phi_n(0)]_i = 0$ for $n \neq 0$, (8.53) becomes

$$\phi_n(d) = \mathbf{E}'(d)^{-1} \int_0^d dz \exp\left(-2\pi i q_z z\right)\mathbf{E}'(z)\mathbf{S}_{n0}\phi_0(z) \tag{8.54}$$

This is a formal solution of (8.49) provided the $\mathbf{E}'(z)$ matrix that satisfies (8.52) can be found. The selection of a boundary value for $\mathbf{E}'(0)$ may not affect the final results. The image intensity is thus

$$I_n(\mathbf{g} - \mathbf{q}) \propto \sum_{q_z} \left| \sum_{i'} C_g^{(i')} \left[\phi_n(d)\right]_{i'} \exp\left(2\pi i \nu_{i'} d\right)\right|^2 \tag{8.55}$$

The dislocation image, formed by the inelastically scattered electron, can be calculated from (8.55). The column approximation has to be made in calculating the low-resolution diffraction contrast images. For high-resolution images of inelastic electrons, the calculation has to be performed using the theory introduced in Chap. 13.

8.8 Summary

In this chapter, we illustrated the formal dynamic Bloch wave theory of the single and double inelastically scattered electrons. The theory is best suited for calculating diffraction patterns of inelastically scattered electrons for ZOLZ reflections. Contrast features observed in Kikuchi patterns are qualitatively interpreted based on the two-beam theory, and they are found to be due primarily to TDS for thin specimens. The theory is extended to calculate images formed by electrons that are inelastically scattered by crystals containing defects. A formal equation is introduced, and a method for solving the equation is illustrated.

Chapter 9
Reciprocity in Electron Diffraction and Imaging

In electron imaging, TEM and STEM are considered to be equivalent, so that images obtained in both optical configurations have many analogous characteristics, and sometimes they can be interpreted using the same theory. This analogy is based on the reciprocity theorem, which is stated as:

The amplitude of the disturbance at a point P due to radiation from a point Q, which has traversed any system involving elastic-scattering processes only, is the same as the amplitude of the disturbance that would be observed at Q if the point source were placed at P.

This theorem was proposed by Helmholtz (1886) in optics and first used by Laue (1935) in X-ray diffraction. The theorem has been applied in electron diffraction: Kikuchi pattern theory (Kainuma 1955), the correlation between STEM and TEM (Cowley 1969), symmetry of diffracted intensity (Pogany and Turner 1968), the theory of electron-channeling patterns (Fujimoto et al. 1972), and symmetry of CBED patterns (Buxton et al. 1976). This chapter introduces the theorem for both elastically and inelastically scattered electrons. This chapter also serves as a basis for Green's function theory in Chap. 10 and HAADF-STEM image in Chap. ??.

9.1 Reciprocity Theorem for Elastically Scattered Electrons

The reciprocity theorem has been proved using different theoretical approaches: Hermitian operators in quantum mechanics (Bilhorn et al. 1964), Born series (Pogany and Turner 1968), scattering matrix theory (Qin and Goodman 1989), and the space integral method (Gunning and Goodman 1992). Here we choose the Born series method for the convenience of discussion in the next section. If a point source is assumed to locate at point P in space (above the specimen, as shown in Fig. 9.1), the wave function in space is determined by

Z. L. Wang, *Elastic and Inelastic Scattering in Electron Diffraction and Imaging*, Springer Series in Solid-State Sciences 205, https://doi.org/10.1007/978-3-031-90819-4_9

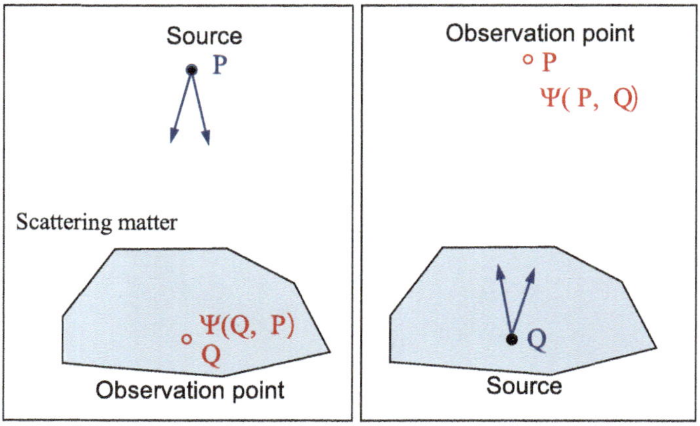

Fig. 9.1 An illustration of the reciprocity theorem. **a** The $\Psi(Q, P)$ is the wave observed at Q if a point source is placed at P; **b** $\Psi(P, Q)$ is the wave observed at P if a point source is placed at Q. The reciprocity theorem shows that $\Psi(Q, P) = \Psi(P, Q)$, i.e., the observation point and source can be exchanged without affecting the value of the observed wave function

$$\left(\nabla^2 + 4\pi^2 K^2\right) \Psi_0 \left(\mathbf{r}, \mathbf{r}_P\right) = -4\pi^2 U(\mathbf{r}) \Psi_0 \left(\mathbf{r}, \mathbf{r}_P\right) + \delta \left(\mathbf{r} - \mathbf{r}_P\right) \tag{9.1}$$

The solution of this equation can be written (see Appendix D) as

$$\Psi_0 \left(\mathbf{r}, \mathbf{r}_P\right) = G_0 \left(\mathbf{r}, \mathbf{r}_P\right) - 4\pi^2 \int d\mathbf{r}' G_0 \left(\mathbf{r}, \mathbf{r}'\right) U \left(\mathbf{r}'\right) \Psi_0 \left(\mathbf{r}', \mathbf{r}_P\right) \tag{9.2}$$

where Green's function is

$$G_0 \left(\mathbf{r}, \mathbf{r}'\right) = -\frac{\exp\left(2\pi i K_0 \left|\mathbf{r} - \mathbf{r}'\right|\right)}{4\pi \left|\mathbf{r} - \mathbf{r}'\right|} \tag{9.3}$$

From Sect. 4.2, we know that the solution of (9.2) at Q point inside (or below) the specimen may be expressed in a Born series

$$\begin{aligned}
\Psi_0 \left(\mathbf{r}_Q, \mathbf{r}_P\right) =& G_0 \left(\mathbf{r}_Q, \mathbf{r}_P\right) - 4\pi^2 \int d\mathbf{r}_1 G_0 \left(\mathbf{r}_Q, \mathbf{r}_1\right) U \left(\mathbf{r}_1\right) G_0 \left(\mathbf{r}_1, \mathbf{r}_P\right) \\
&+ \left(-4\pi^2\right)^2 \int d\mathbf{r}_1 \int d\mathbf{r}_2 G_0 \left(\mathbf{r}_Q, \mathbf{r}_1\right) G_0 \left(\mathbf{r}_1, \mathbf{r}_2\right) U \left(\mathbf{r}_1\right) U \left(\mathbf{r}_2\right) G_0 \left(\mathbf{r}_2, \mathbf{r}_P\right) \\
&+ \left(-4\pi^2\right)^3 \int d\mathbf{r}_1 \int d\mathbf{r}_2 \int d\mathbf{r}_3 G_0 \left(\mathbf{r}_Q, \mathbf{r}_1\right) G_0 \left(\mathbf{r}_1, \mathbf{r}_2\right) \\
&\times G_0 \left(\mathbf{r}_2, \mathbf{r}_3\right) U \left(\mathbf{r}_1\right) U \left(\mathbf{r}_2\right) U \left(\mathbf{r}_3\right) G_0 \left(\mathbf{r}_3, \mathbf{r}_P\right) + \ldots
\end{aligned} \tag{9.4}$$

Interchanging $\mathbf{r}_Q$ and $\mathbf{r}_P$ does not change the value of any order of the Born series; thus

$$\Psi_0\left(\mathbf{r}_Q, \mathbf{r}_P\right) = \Psi_0\left(\mathbf{r}_P, \mathbf{r}_Q\right) \tag{9.5}$$

Equation (9.5) is just the reciprocity theorem. Two points need to be emphasized here. First no restriction was made on the form of the crystal potential, which can be a complex function if the anomalous absorption effect is included. Second reciprocity holds for each order of the Born series. This is because the reciprocity theorem results from a time reversal in quantum mechanics. Therefore the reciprocity theorem is exact for elastically scattered electrons regardless of the shape of the crystal, the form of the crystal potential, scattering geometry, and electron energy.

9.2 Equivalence of TEM and STEM

Figure 9.2 compares TEM and STEM. In TEM, the specimen is illuminated by a plane wave (i.e., the source is at infinity), and the magnification effect of the objective lens takes place after the incident beam is diffracted by the specimen. In STEM, a fine electron probe is formed with a condenser lens by converging the electrons emitted from a fine tip. The beam is scanned across the specimen surface by deflection coils. The image is formed by collecting electrons scattered to particular angular ranges in the diffraction pattern when the electron probe is scanned across the specimen. We now prove mathematically the equivalence of images formed in TEM and STEM.

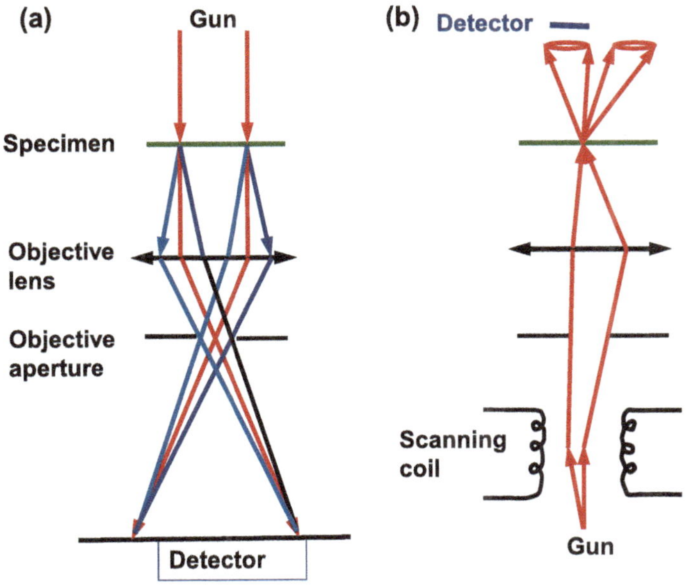

Fig. 9.2 An imaging ray diagrams in **a** TEM and **b** STEM

Based on Abbe's imaging theory (Sect. 1.7), the electron wave at the image plane is a convolution of the electron wave function at the exit face of the crystal with the transfer function of the objective lens

$$\Psi_{\mathrm{TEM}}(\mathbf{r},\mathbf{r}_P)=\Psi_0(\mathbf{r},\mathbf{r}_P)\otimes T_{\mathrm{obj}}(\mathbf{r}) \tag{9.6}$$

where

$$T_{\mathrm{obj}}(\mathbf{r})=\mathbf{FT}^{-1}\left[T_{\mathrm{obj}}(\mathbf{u})\right] \tag{9.7a}$$

is the transfer function of the objective lens, with

$$T_{\mathrm{obj}}(\mathbf{u})=A_{\mathrm{Ob}}(\mathbf{u})\exp\left(\frac{\pi i}{2}C_s\lambda^3u^4+\pi i\lambda u^2\Delta f\right) \tag{9.7b}$$

The C_s is the lens spherical aberration coefficient, Δf is the lens defocus, and A_{Ob} defines the shape of the objective aperture. Substituting (9.4) into (9.6), we have

$$\begin{aligned}\Psi_{\mathrm{TEM}(\mathbf{r}_Q,\mathbf{r}_P)}=&\left[G_0\left(\mathbf{r}_Q,\mathbf{r}_P\right)\otimes T_{\mathrm{obj}}\left(\mathbf{r}_Q\right)\right]\\&-4\pi^2\int d\mathbf{r}_1\left[G_0\left(\mathbf{r}_Q,\mathbf{r}_1\right)\otimes T_{\mathrm{obj}}\left(\mathbf{r}_Q\right)\right]U(\mathbf{r}_1)\,G_0(\mathbf{r}_1,\mathbf{r}_P)\\&+\left(-4\pi^2\right)^2\int d\mathbf{r}_1\int d\mathbf{r}_2\left[G_0\left(\mathbf{r}_Q,\mathbf{r}_1\right)\otimes T_{\mathrm{obj}}\left(\mathbf{r}_Q\right)\right]\\&\times G_0(\mathbf{r}_1,\mathbf{r}_2)\,U(\mathbf{r}_1)\,U(\mathbf{r}_2)\,G_0(\mathbf{r}_2,\mathbf{r}_P)\\&+\left(-4\pi^2\right)^3\int d\mathbf{r}_1\int d\mathbf{r}_2\int d\mathbf{r}_3\left[G_0\left(\mathbf{r}_Q,\mathbf{r}_1\right)\otimes T_{\mathrm{obj}}\left(\mathbf{r}_Q\right)\right]\\&\times G_0(\mathbf{r}_1,\mathbf{r}_2)\,G_0(\mathbf{r}_2,\mathbf{r}_3)\,U(\mathbf{r}_1)\,U(\mathbf{r}_2)\,U(\mathbf{r}_3)\,G_0(\mathbf{r}_3,\mathbf{r}_P)+\ldots\end{aligned} \tag{9.8}$$

In STEM, the condenser lens is equivalent to the objective lens in TEM, but it is placed before the electron beam interacting with the specimen. An electron probe, centered at $\mathbf{r}_Q$, is formed by the condenser lens, and it is expressed as $T_{\mathrm{con}}\left(\mathbf{r}-\mathbf{r}_Q\right)$. Since the wave generated from a point source located at $\mathbf{r}'$ is $G_0\left(\mathbf{r},\mathbf{r}'\right)$, the probe function is

$$\Psi_{oi}\left(\mathbf{r},\mathbf{r}_Q\right)=\int d\mathbf{r}'G_0\left(\mathbf{r},\mathbf{r}'\right)T_{\mathrm{con}}\left(\mathbf{r}'-\mathbf{r}_Q\right)=G_0\left(\mathbf{r},\mathbf{r}_Q\right)\otimes T_{\mathrm{con}}\left(\mathbf{r}_Q\right) \tag{9.9}$$

The wave function at the exit face of the specimen is the solution of the Schrödinger equation for an incident wave of the form given by (9.9), which is

$$\Psi_0\left(\mathbf{r},\mathbf{r}_Q\right)=\Psi_{oi}\left(\mathbf{r},\mathbf{r}_Q\right)-4\pi^2\int d\mathbf{r}'G_0\left(\mathbf{r},\mathbf{r}'\right)U\left(\mathbf{r}'\right)\Psi_0\left(\mathbf{r}',\mathbf{r}_Q\right) \tag{9.10}$$

The solution of (9.10) is expressed in the Born series as

$$\begin{aligned}
&\Psi_{\text{STEM}}\left(\mathbf{r}_P,\mathbf{r}_Q\right)=\left[G_0\left(\mathbf{r}_P,\mathbf{r}_Q\right)\otimes T_{\text{con}}\left(\mathbf{r}_Q\right)\right]\\
&-4\pi^2\int d\mathbf{r}_1 G_0\left(\mathbf{r}_P,\mathbf{r}_1\right)U\left(\mathbf{r}_1\right)\left[G_0\left(\mathbf{r}_1,\mathbf{r}_Q\right)\otimes T_{\text{con}}\ \left(\mathbf{r}_Q\right)\right.\\
&+\left(-4\pi^2\right)^2\int d\mathbf{r}_1\int d\mathbf{r}_2 G_0\left(\mathbf{r}_P,\mathbf{r}_1\right)G_0\left(\mathbf{r}_1,\mathbf{r}_2\right)U\left(\mathbf{r}_1\right)U\left(\mathbf{r}_2\right)\left[G_0\left(\mathbf{r}_2,\mathbf{r}_Q\right)\otimes T_{\text{con}}\left(\mathbf{r}_Q\right)\right]\\
&+\left(-4\pi^2\right)^3\int d\mathbf{r}_1\int d\mathbf{r}_2\int d\mathbf{r}_3 G_0\left(\mathbf{r}_P,\mathbf{r}_1\right)G_0\left(\mathbf{r}_1,\mathbf{r}_2\right),G_0\left(\mathbf{r}_2,\mathbf{r}_3\right)U\left(\mathbf{r}_1 U\left(\mathbf{r}_2\right)U\mathbf{r}_3\right)\\
&\times\left[G_0\left(\mathbf{r}_3,\mathbf{r}_Q\right)\otimes T_{\text{con}}\ \left(\mathbf{r}_Q\right)\right]+\ldots\\
&=\left[G_0\left(\mathbf{r}_P,\mathbf{r}_Q\right)\otimes T_{\text{con}}\left(\mathbf{r}_Q\right)\right]-4\pi^2\int d\mathbf{r}_1\left[G_0\left(\mathbf{r}_Q,\mathbf{r}_1\right)\otimes T_{\text{con}}\left(\mathbf{r}_Q\right)\right]U\left(\mathbf{r}_1\right)G_0\left(\mathbf{r}_1,\mathbf{r}_P\right)\\
&+\left(-4\pi^2\right)^2\int d\mathbf{r}_1\int d\mathbf{r}_2\left[G_0\left(\mathbf{r}_Q,\mathbf{r}_1\right)\otimes T_{\text{con}}\ \left(\mathbf{r}_Q\right)\right]G_0\left(\mathbf{r}_1,\mathbf{r}_2\right)U\left(\mathbf{r}_1\right)U\left(\mathbf{r}_2\right)G_0\left(\mathbf{r}_2,\mathbf{r}_P\right)\\
&+\left(-4\pi^2\right)^3\int d\mathbf{r}_1\int d\mathbf{r}_2\int d\mathbf{r}_3\left[G_0\left(\mathbf{r}_Q,\mathbf{r}_1\right)\otimes T_{\text{con}}\left(\mathbf{r}_Q\right)\right]\\
&\times G_0\left(\mathbf{r}_1,\mathbf{r}_2\right)G_0\left(\mathbf{r}_2,\mathbf{r}_3\right)U\left(\mathbf{r}_1\right)U\left(\mathbf{r}_2\right)U\left(\mathbf{r}_3\right)G_0\left(\mathbf{r}_3,\mathbf{r}_P\right)+\ldots
\end{aligned}\tag{9.11}$$

The last step was derived by using the reciprocity of $G_0\left(\mathbf{r},\mathbf{r}_1\right)=G_0(\mathbf{r}_1,\mathbf{r})$ and interchanging integral variables. If the transfer function of the objective lens in TEM is equal to the condenser lens probe-forming function in STEM (i.e., $T_{\text{obj}}=T_{\text{con}}$), then comparing (9.11) and (9.8) shows that

$$\Psi_{\text{TEM}}\left(\mathbf{r}_Q,\mathbf{r}_P\right)=\Psi_{\text{STEM}}\left(\mathbf{r}_P,\mathbf{r}_Q\right)\tag{9.12}$$

Therefore STEM is a time-reversed imaging process of TEM, and equivalent results are obtained if only elastic scattering is considered. This is the theoretical basis allowing theories developed for TEM imaging to interpret images of STEM (Crewe and Wall 1970; Cowley 1976). In practice, the shape of the image detector in STEM could make some difference. In STEM, the detected image intensity when the incident electron probe is scanned to $\mathbf{r}_Q$ is

$$\begin{aligned}
I_{\text{STEM}}\ \left(\mathbf{r}_Q\right)&=\int d\mathbf{u}D(\mathbf{u})\left|\Psi_{\text{STEM}}\ \left(\mathbf{u},\mathbf{r}_Q\right)\right|^2\\
&=\int d\mathbf{u}D(\mathbf{u})\left|\int d\mathbf{r}_P\exp\left(2\pi i\mathbf{u}\cdot\mathbf{r}_P\right)\Psi_{\text{STEM}}\ \left(\mathbf{r}_P,\mathbf{r}_Q\right)\right|^2
\end{aligned}\tag{9.13}$$

where D is the detector function and $\left|\Psi_{\text{STEM}}\ \left(\mathbf{u},\mathbf{r}_Q\right)\right|^2$ is the electron diffraction pattern when the incident electron probe is located at $\mathbf{r}_Q$. For a special case of point bright-field detector, so that $D(\mathbf{u})=\delta(\mathbf{u})$, (9.13) becomes

$$\begin{aligned}
I_{\text{STEM}}\left(\mathbf{r}_Q\right)&=\left|\int d\mathbf{r}_P|\Psi_{\text{STEM}}\left(\mathbf{r}_P,\mathbf{r}_Q\right)\right|^2=\left|\int d\mathbf{r}_P|\Psi_{\text{TEM}}\left(\mathbf{r}_Q,\mathbf{r}_P\right)\right|^2\\
&=I_{\text{TEM}}\left(\mathbf{r}_P\right)
\end{aligned}\tag{9.14}$$

where (9.12) was used. Thus precisely speaking, TEM bright-field image is identical to the bright-field STEM image if an on-axis point detector is used (Cowley 1969).

The equivalence of TEM and STEM, however, may not hold for inelastically scattered electrons if the electron energy loss is larger than a few eV. This is because the contrast transfer effect of the condenser and objective lenses in the STEM and TEM cases occurs before and after, respectively, the electrons interact with the crystal, so that the chromatic aberration effect present in TEM is absent in STEM.

The reciprocity of $\Psi_0\left(\mathbf{r}_P, \mathbf{r}_Q\right)$ and $\Psi_0\left(\mathbf{r}_Q, \mathbf{r}_P\right)$ is the direct result of time reversal in quantum mechanics (Gunning and Goodman 1992), and it is exact with no approximation. An alternative proof of the reciprocity theorem follows.

If V is time-independent, the wave scattered to $\mathbf{r}$ by a point source at $\mathbf{r}_P$ is determined by

$$\left[-\frac{\hbar^2}{2m_0}\nabla^2 - e\gamma V(\mathbf{r}) - E\right]\Psi_0\left(\mathbf{r}, \mathbf{r}_P\right) = \delta\left(\mathbf{r} - \mathbf{r}_P\right) \tag{9.15a}$$

and an identical source at $\mathbf{r}_Q$ is described by

$$\left[-\frac{\hbar^2}{2m_0}\nabla^2 - e\gamma V(\mathbf{r}) - E\right]\Psi_0\left(\mathbf{r}, \mathbf{r_Q}\right) = \delta\left(\mathbf{r} - \mathbf{r}_Q\right) \tag{9.15b}$$

If we integrate the result of multiplying (9.15a) by $\Psi\left(\mathbf{r}_{\mathbf{r}}\mathbf{r}_Q\right)$ and subtracting (9.15b) multiplied by $\Psi\left(\mathbf{r}, \mathbf{r_P}\right)$ over all space,

$$\begin{aligned}
&\Psi_0\left(\mathbf{r}_Q, \mathbf{r}_P\right) - \Psi_0\left(\mathbf{r}_P, \mathbf{r}_Q\right) \\
&= -\frac{\hbar^2}{2m_0}\int d\mathbf{r}\left[\Psi_0\left(\mathbf{r}, \mathbf{r}_Q\right)\nabla^2\Psi_0\left(\mathbf{r}, \mathbf{r}_P\right) - \Psi_0\left(\mathbf{r}, \mathbf{r}_P\right)\nabla^2\Psi_0\left(\mathbf{r}, \mathbf{r}_Q\right)\right] \\
&= -\frac{\hbar^2}{2m_0}\int d\mathbf{r}\nabla\cdot\left[\Psi_0\left(\mathbf{r}, \mathbf{r}_Q\right)\nabla\Psi_0\left(\mathbf{r}, \mathbf{r}_P\right) - \Psi_0\left(\mathbf{r}, \mathbf{r}_P\right)\nabla\Psi_0\left(\mathbf{r}, \mathbf{r}_Q\right)\right] \\
&= -\frac{\hbar^2}{2m_0}\int_S d\mathbf{S}\cdot\left[\Psi_0\left(\mathbf{r}, \mathbf{r}_Q\right)\nabla\Psi_0\left(\mathbf{r}, \mathbf{r}_P\right) - \Psi_0\left(\mathbf{r}, \mathbf{r}_P\right)\nabla\Psi_0\left(\mathbf{r}, \mathbf{r}_Q\right)\right]
\end{aligned} \tag{9.16}$$

This integral equals zero if the homogeneous boundary condition of the form

$$c_1\Psi_0 + c_2\frac{\partial\Psi_0}{\partial n} = 0$$

is satisfied at the boundary surface, where n is the normal direction of the boundary. Thus the final result is

$$\Psi_0\left(\mathbf{r}_Q, \mathbf{r}_P\right) = \Psi_0\left(\mathbf{r}_P, \mathbf{r}_Q\right) \tag{9.17}$$

and the reciprocity theorem is obtained.

9.3 Reciprocity Theorem for Inelastically Scattered Electrons

For inelastically scattered electrons, time reversal requires the scattering medium to be in its excited state. Thus images recorded in TEM and STEM using the inelastically scattered electrons may not be equivalent because the state reciprocal to the final state in TEM is given by $\Sigma_n \Psi_n$, which consists of a coherent superposition of the excited object waves, and this apparently is not equal to the initial state in STEM, where the object is in its ground state (Kohl and Rose 1985). However a suitable reciprocity theorem can be derived under some restrictions (Pogany and Turner 1968). To determine possible approximations, we start with Yoshioka's equations

$$\left(\nabla^2 + 4\pi^2 K_0^2\right) \Psi_0 = \sum_m H'_{0m} \Psi_m + \delta\left(\mathbf{r} - \mathbf{r}_P\right) \tag{9.18a}$$

and

$$\left(\nabla^2 + 4\pi^2 K_n^2\right) \Psi_n = \sum_m H'_{nm} \Psi_m \quad n = 1, 2, \ldots \tag{9.18b}$$

where $H'_{nm} = \left(2m_0/\hbar^2\right) H_{nm}$ is a transition matrix introduced for the convenience of short notation; a point electron source is introduced at $\mathbf{r}_p$ that represents the emission source of the elastic electron. Green's function solutions of (9.18a) and (9.18b) are expressed as

$$\Psi_0\left(\mathbf{r}, \mathbf{r}_P\right) = G_0\left(\mathbf{r}, \mathbf{r}_P\right) + \int d\mathbf{r}' G_0\left(\mathbf{r}, \mathbf{r}'\right) \sum_m H'_{0m}\left(\mathbf{r}'\right) \Psi_m\left(\mathbf{r}', \mathbf{r}_P\right) \tag{9.19a}$$

and

$$\Psi_n\left(\mathbf{r}, \mathbf{r}_P\right) = \int d\mathbf{r}' G_n\left(\mathbf{r}, \mathbf{r}'\right) \sum_m H'_{nm}\left(\mathbf{r}'\right) \Psi_m\left(\mathbf{r}', \mathbf{r}_P\right) \tag{9.19b}$$

with

$$G_n\left(\mathbf{r}, \mathbf{r}'\right) = -\frac{\exp\left(2\pi i K_n \left|\mathbf{r} - \mathbf{r}'\right|\right)}{4\pi \left|\mathbf{r} - \mathbf{r}'\right|} \tag{9.19c}$$

We now introduce an approximation of $K_n \approx K_0$, which means that the energy loss of electrons is negligible. This occurs in phonon scattering and small energy loss valence excitations. Thus Green's functions for elastic scattering and inelastic scattering are equal, $G_0 = G_n$. The purpose of introducing this approximation is to allow the interchange of the G_0 and G_n functions within the integrals without affecting the final result. We first look for solutions of (9.19a) and (9.19b) under this approximation. The zeroth-order solutions are

$$\Psi_0\left(\mathbf{r}, \mathbf{r}_P\right)\big|_0 = G_0\left(\mathbf{r}, \mathbf{r}_P\right) \tag{9.20a}$$

and

$$\Psi_n(\mathbf{r},\mathbf{r}_P)|_0 = 0 \tag{9.20b}$$

First-order solutions are obtained by substituting zeroth-order solution into (9.19a) and (9.19b), giving

$$\Psi_0(\mathbf{r},\mathbf{r}_P)|_1 = \int d\mathbf{r}_1 G_0(\mathbf{r},\mathbf{r}_1) < a_0(\mathbf{r}_c)\left|H'(\mathbf{r}_1,\mathbf{r}_c)\right| a_0(\mathbf{r}_c) > G_0(\mathbf{r}_1,\mathbf{r}_P) \tag{9.21a}$$

and

$$\Psi_n(\mathbf{r},\mathbf{r}_P)|_1 = \int d\mathbf{r}_1 G_0(\mathbf{r},\mathbf{r}_1) < a_n(\mathbf{r}_c)\left|H'(\mathbf{r}_1,\mathbf{r}_c)\right| a_0(\mathbf{r}_c) > G_0(\mathbf{r}_1,\mathbf{r}_P) \tag{9.21b}$$

where $\mathbf{r}_c$ denotes crystal electron positions, and $H' = 2m_0/\hbar^2 H$. Second-order terms are

$$\Psi_0(\mathbf{r},\mathbf{r}_P)|_2 = \int d\mathbf{r}_1 \int d\mathbf{r}_2 G_0(\mathbf{r},\mathbf{r}_1) \sum_m H'_{0m}(\mathbf{r}_1) G(\mathbf{r}_1,\mathbf{r}_2) H'_{m0}(\mathbf{r}_2) G_0(\mathbf{r}_2,\mathbf{r}_P) \tag{9.22a}$$

and

$$\Psi_n(\mathbf{r},\mathbf{r}_P)|_2 = \int d\mathbf{r}_1 \int d\mathbf{r}_2 G_0(\mathbf{r},\mathbf{r}_1) \sum_m H'_{nm}(\mathbf{r}_1) G(\mathbf{r}_1,\mathbf{r}_2) H'_{m0}(\mathbf{r}_2) G_0(\mathbf{r}_2,\mathbf{r}_P) \tag{9.22b}$$

We now consider the following relation:

$$\begin{aligned}\sum_m H'_{nm}(\mathbf{r}_1) H'_{mL}(\mathbf{r}_2) &= \sum_m < a_n(\mathbf{r}_c) \,|H'(\mathbf{r}_1,\mathbf{r}_c)|\, a_m(\mathbf{r}_c) >< a_m(\mathbf{r}'_c) \,|H'(\mathbf{r}_1,\mathbf{r}'_c)|\, a_L(\mathbf{r}'_c) > \\ &=< a_n(\mathbf{r}_c)\left|H'(\mathbf{r}_1,\mathbf{r}_c) H'(\mathbf{r}_2,\mathbf{r}_c)\right| a_L(\mathbf{r}_c) >\end{aligned} \tag{9.23}$$

since

$$\sum_m \left|a_m(\mathbf{r}_c) >< a_m(\mathbf{r}'_c)\right| = \delta(\mathbf{r}_c - \mathbf{r}'_c)$$

Thus (9.22a) and (9.22b) are rewritten as

$$\begin{aligned}\Psi_0(\mathbf{r},\mathbf{r}_P)|_2 = &\int d\mathbf{r}_1 \int d\mathbf{r}_2 G_0(\mathbf{r},\mathbf{r}_1) G_0(\mathbf{r}_1,\mathbf{r}_2) \\ &\times < a_0(\mathbf{r}_c)\left|H'(\mathbf{r}_1,\mathbf{r}_c) H'(\mathbf{r}_2,\mathbf{r}_c)\right| a_0(\mathbf{r}_c) > G_0(\mathbf{r}_2,\mathbf{r}_P)\end{aligned} \tag{9.24a}$$

and

$$\begin{aligned}\Psi_n(\mathbf{r},\mathbf{r}_P)|_2 = &\int d\mathbf{r}_1 \int d\mathbf{r}_2 G_0(\mathbf{r},\mathbf{r}_1) G_0(\mathbf{r}_1,\mathbf{r}_2) \\ &< a_n(\mathbf{r}_c)\left|H'(\mathbf{r}_1,\mathbf{r}_c) H'(\mathbf{r}_2,\mathbf{r}_c)\right| a_0(\mathbf{r}_c) > G_0(\mathbf{r}_2,\mathbf{r}_P)\end{aligned} \tag{9.24b}$$

An analogous relation can be derived for the Jth-order solution

$$\Psi_0(\mathbf{r},\mathbf{r}_P)|_J = \int d\mathbf{r}_1 \int d\mathbf{r}_2 \cdots \int d\mathbf{r}_J G_0(\mathbf{r},\mathbf{r}_1) G_0(\mathbf{r}_1,\mathbf{r}_2) \cdots G_0(\mathbf{r}_{J-1},\mathbf{r}_J) \times < a_0(\mathbf{r}_c) \left| H'(\mathbf{r}_1,\mathbf{r}_c) H'(\mathbf{r}_2,\mathbf{r}_c) \cdots H'(\mathbf{r}_J,\mathbf{r}_c) \right| a_0(\mathbf{r}_c) > G_0(\mathbf{r}_J,\mathbf{r}_P) \tag{9.25a}$$

$$\Psi_n(\mathbf{r},\mathbf{r}_P)|_J = \int d\mathbf{r}_1 \int d\mathbf{r}_2 \cdots \int d\mathbf{r}_J G_0(\mathbf{r},\mathbf{r}_1) G_0(\mathbf{r}_1,\mathbf{r}_2) \cdots G_0(\mathbf{r}_{J-1},\mathbf{r}_J) \times < a_n(\mathbf{r}_c) \left| H'(\mathbf{r}_1,\mathbf{r}_c) H'(\mathbf{r}_2,\mathbf{r}_c) \cdots H'(\mathbf{r}_J,\mathbf{r}_c) \right| a_0(\mathbf{r}_c) > G_0(\mathbf{r}_J,\mathbf{r}_P) \tag{9.25b}$$

By exchanging $\mathbf{r}$ with $\mathbf{r}_P$, $\mathbf{r}_1$ with $\mathbf{r_J}$, the following relations hold for every order of the expansion:

$$\Psi_0(\mathbf{r},\mathbf{r}_P)|_J = \Psi_0(\mathbf{r}_P,\mathbf{r})|_J \tag{9.26a}$$

$$\Psi_n(\mathbf{r},\mathbf{r}_P)|_J = \Psi_n(\mathbf{r}_P,\mathbf{r})|_J \tag{9.26b}$$

since $G_0\left(\mathbf{r},\mathbf{r}'\right) = G_0\left(\mathbf{r}',\mathbf{r}\right)$. Therefore under the approximation of $K_n = K_0$, we have proved the reciprocity theorem of inelastically scattered electrons for every order of the Born series. Finally we have

$$\Psi_0\left(\mathbf{r}_Q,\mathbf{r}_P\right) = \Psi_0\left(\mathbf{r}_P,\mathbf{r}_Q\right) \tag{9.27a}$$

and

$$\Psi_n\left(\mathbf{r}_Q,\mathbf{r}_P\right) = \Psi_n\left(\mathbf{r}_P,\mathbf{r}_Q\right) \tag{9.27b}$$

As shown in Sect. 6.2.2, the recorded intensity in the image is an incoherent superposition of the electrons in all states,

$$I\left(\mathbf{r}_Q,\mathbf{r}_P\right) = \sum_{n=0} \left|\Psi_n\left(\mathbf{r}_Q,\mathbf{r}_P\right)\right|^2 = \sum_{n=0} \left|\Psi_n\left(\mathbf{r}_P,\mathbf{r}_Q\right)\right|^2 = I\left(\mathbf{r}_P,\mathbf{r}_Q\right) \tag{9.28}$$

This is the intensity reciprocity theorem proposed by Pogany and Turner (1968). Therefore the intensity reciprocity theorem holds for inelastically scattered electrons as long as the electron energy loss is small. Therefore the reciprocity theorem is exact for thermal diffusely scattered electrons regardless of the shape of the crystal, the form of the crystal potential, the scattering geometry, or the electron energy.

9.4 Summary

In this chapter, we proved the reciprocity theorem of elastically and inelastically scattered electrons. The wave function reciprocity theorem is exact for elastically scattered electrons regardless of the shape of the crystal or the form of the crystal potential. The equivalence of TEM and STEM has been illustrated with the presence of lenses. It was shown that the STEM image acquired with an on-axis point detector is exactly the same as the TEM image provided with the identical optical transfer functions. The intensity reciprocity theorem holds for inelastically scattered electrons if electron energy loss is small. These results are used in Chaps. 10–13 to describe the diffraction and imaging of inelastically scattered electrons.

Chapter 10
Dynamic Inelastic Electron Scattering II: Green's Function Theory

10.1 Generalized Reciprocity Theorem

Elastic scattering of electrons from a frozen lattice configuration is treated as time-independent. The first theoretical objective in treating dynamic inelastic electron scattering is to average statistically over the quasielastic electron diffraction patterns produced by various crystal configurations before numerical calculations. Green's function theory is probably the method best suited for this purpose. For simplicity, we start with the equation of single inelastic electron excitation with consideration for the time-dependent (or configuration-dependent) thermal vibration perturbation on the crystal potential (ΔV)

$$\left(-\frac{\hbar^2}{2m_0}\nabla^2 - e\gamma V_0 - e\gamma\Delta V - E_n\right)\Psi_n = -H_{n0}\Psi_0 \tag{10.1}$$

and

$$\left(-\frac{\hbar^2}{2m_0}\nabla^2 - e\gamma V_0 - e\gamma\Delta V - E_0\right)\Psi_0 = 0 \tag{10.2}$$

Shifting the ΔV term to the right-hand side, (10.1) and (10.2) can be converted into integral equations with the use of Green's function (Kainuma 1955; Gjønnes 1966)

$$\Psi_n(\mathbf{r}) = \int d\mathbf{r}_1 \hat{G}_n(\mathbf{r}, \mathbf{r}_1)\left[e\gamma\Delta V(\mathbf{r}_1)\Psi_n(\mathbf{r}_1) - H_{n0}(\mathbf{q}_1, \mathbf{r}_1)\Psi_0(\mathbf{r}_1)\right] \tag{10.3}$$

and

$$\Psi_0(\mathbf{r}) = \Psi_0(\mathbf{K}_0, \mathbf{r}) + \int d\mathbf{r}_1 \hat{G}(\mathbf{r}, \mathbf{r}_1)\left[e\gamma\Delta V(\mathbf{r}_1)\Psi_0(\mathbf{r}_1)\right] \tag{10.4}$$

Z. L. Wang, *Elastic and Inelastic Scattering in Electron Diffraction and Imaging*, Springer Series in Solid-State Sciences 205, https://doi.org/10.1007/978-3-031-90819-4_10

where $\hat{G}_n$ and $\hat{G}$ are Green's functions satisfying

$$\left(-\frac{\hbar^2}{2m_0}\nabla^2 - e\gamma V_0 - E_n\right)\hat{G}_n(\mathbf{r}, \mathbf{r}_1) = \delta(\mathbf{r} - \mathbf{r}_1) \tag{10.5}$$

and

$$\left(-\frac{\hbar^2}{2m_0}\nabla^2 - e\gamma V_0 - E_0\right)\hat{G}(\mathbf{r}, \mathbf{r}_1) = \delta(\mathbf{r} - \mathbf{r}_1) \tag{10.6}$$

and $\Psi_0(\mathbf{K}_0, \mathbf{r})$ is the elastic wave function of the incident electron of wave vector $\mathbf{K}_0$. Equations (10.3) and (10.4) are the generalized result of the reciprocity theorem.

Equation (10.3) can be solved iteratively in a power series of ΔV and H_{n0}

$$\begin{aligned}\Psi_n(\mathbf{r}) = &-\int d\mathbf{r}_1 \hat{G}_n(\mathbf{r}, \mathbf{r}_1) H_{n0}(\mathbf{q}_1, \mathbf{r}_1)\Psi_0(\mathbf{K}_0, \mathbf{r}_1)\\ &- e\gamma \int d\mathbf{r}_1 \int d\mathbf{r}_1' \hat{G}_n(\mathbf{r}, \mathbf{r}_1)\Delta V(\mathbf{r}_1)\hat{G}_n(\mathbf{r}_1, \mathbf{r}_1') H_{n0}(\mathbf{q}_1, \mathbf{r}_1')\Psi_0(\mathbf{K}_0, \mathbf{r}_1')\\ &- e\gamma \int d\mathbf{r}_1 \int d\mathbf{r}_1' \hat{G}_n(\mathbf{r}, \mathbf{r}_1) H_{n0}(\mathbf{q}_1, \mathbf{r}_1)\hat{G}(\mathbf{r}_1, \mathbf{r}_1')\Delta V(\mathbf{r}_1')\Psi_0(\mathbf{K}_0, \mathbf{r}_1')\end{aligned} \tag{10.7}$$

where the first term is the atomic inner shell scattering due to the inelastic interaction term H_{n0} (see Sect. 6.6); the second term is the atomic inner shell TDS, double inelastic scattering; and the third term is the TDS-atomic inner shell, double inelastic scattering. The only difference between the second and the third terms is which excitation process occurs first.

The physical meaning of (10.7) is clear. The third term, for example, is interpreted as following: The elastic wave $[\Psi_0(\mathbf{K}_0, \mathbf{r}_1')]$ is diffusely scattered at point $\mathbf{r}_1'$ due to thermal vibrations of crystal atoms $[\Delta V(\mathbf{r}_1')]$. Then it is elastically rescattered $\left[\left(\hat{G}(\mathbf{r}_1, \mathbf{r}_1')\right)\right]$ from $\mathbf{r}_1'$ to $\mathbf{r}_1$ before the occurrence of the inelastic event $[H_{n0}(\mathbf{q}_1, \mathbf{r}_1)]$ at $\mathbf{r}_1$. Finally the wave is again elastically rescattered $\left[\hat{G}_n(\mathbf{r}, \mathbf{r}_1)\right]$ before exiting the crystal surface. The subscript n indicates the change of electron energy after single-electron excitation; integrations of $\mathbf{r}_1$ and $\mathbf{r}_1'$ sum over inelastic waves generated at different points inside the crystal.

Solving (10.4) iteratively, the elastically scattered electron wave is given by

$$\Psi_0(\mathbf{r}) = \Psi_0(\mathbf{K}_0, \mathbf{r}) + e\gamma \int d\mathbf{r}_1 \hat{G}(\mathbf{r}, \mathbf{r}_1)\Delta V(\mathbf{r}_1)\Psi_0(\mathbf{K}_0, \mathbf{r}_1) + \cdots \tag{10.8}$$

where the first term is the purely elastically scattered wave and the second term is the first-order TDS. Green's function theory has mathematically expressed the inelastic scattering in a simple way that can be adequately applied to TDS.

10.2 Fourier Transform of Green's Function

The solution of the inelastic wave is expressed by the integral form of Green's function. As shown in the next section, the two-dimensional Fourier transformation of Green's function is required for calculating diffraction patterns of inelastically scattered electrons

$$\hat{G}\left(\mathbf{u}, z, \mathbf{r}_1\right) = \int d\mathbf{b} \exp[-2\pi i \mathbf{u} \cdot \mathbf{b}] \hat{G}\left(\mathbf{r}, \mathbf{r}_1\right) \tag{10.9}$$

where $\hat{G}\left(\mathbf{r}, \mathbf{r}_1\right)$ is the solution to (10.4), and it represents the wave function at $\mathbf{r}$ due to the presence of a point source at $\mathbf{r}_1$. In this section, the calculation of (10.9) is derived analytically for the convenience of the next section.

Considering a case in which the observation point $\mathbf{r}$ is outside the crystal and $\mathbf{r}_1$ is the point source of electrons inside the crystal, the reciprocity relation $\hat{G}\left(\mathbf{r}, \mathbf{r}_1\right) = \widehat{G}\left(\mathbf{r}_1, \mathbf{r}\right)$ simply indicates that the same result would be obtained if the point source were moved to $\mathbf{r}$ outside the crystal and the observation point $\mathbf{r}_1$ were located inside the crystal. Therefore $\hat{G}\left(\tau_b, z, \mathbf{r}_1\right)$ can be rewritten as

$$\hat{G}\left(\tau_b, z, \mathbf{r}_1\right) = \int d\mathbf{b} \exp\left(-2\pi i \tau_b \cdot \mathbf{b}\right) \hat{G}\left(\mathbf{r}_1, \mathbf{r}\right) \tag{10.10}$$

where $\hat{G}\left(\mathbf{r}_1, \mathbf{r}\right)$ is the wave function inside the crystal ($\mathbf{r}_1$) due to a point source outside the crystal ($\mathbf{r}$), and it can be expressed according to the scattering theory in quantum mechanics as

$$\hat{G}\left(\mathbf{r}_1, \mathbf{r}\right) = G_s\left(\mathbf{r}_1, \mathbf{r}\right) + \hat{G}_c\left(\mathbf{r}_1, \mathbf{r}\right) \tag{10.11a}$$

where $G_s\left(\mathbf{r}_1, \mathbf{r}\right)$ is the wave emitted from the point source

$$G_s(\mathbf{r}_1, \mathbf{r}) = \frac{m_0}{2\pi\hbar^2} \frac{\exp\left(2\pi i K_n \left|\mathbf{r} - \mathbf{r}_1\right|\right)}{\left|\mathbf{r} - \mathbf{r}_1\right|} \tag{10.11b}$$

and $\hat{G}_c\left(\mathbf{r}_1, \mathbf{r}\right)$ is the wave scattered by the crystal. For the convenience of the following calculations, $G_s\left(\mathbf{r}_1, \mathbf{r}\right)$ is written in a Fourier integral form according to Sect. 4.2

$$G_s\left(\mathbf{r}_1, \mathbf{r}\right) = -\frac{m_0}{2\pi^2\hbar^2} \int d\mathbf{u} \frac{\exp\left[2\pi i \mathbf{u} \cdot \left(\mathbf{r} - \mathbf{r}_1\right)\right]}{\left(K_n^2 - u^2\right)} \tag{10.12}$$

The two-dimensional Fourier transform of $G_s(\mathbf{r}_1, \mathbf{r})$ is

$$
\begin{aligned}
G_s(\tau_b, z, \mathbf{r}_1) &= \int d\mathbf{b} \exp(-2\pi i \tau_b \cdot \mathbf{b})\, G_s(\mathbf{r}_1, \mathbf{r}) \\
&= -\frac{m_0}{2\pi^2\hbar^2} \exp(-2\pi i \tau_b \cdot \mathbf{b}_1) \\
&\times \int d\mathbf{b} \int d\mathbf{u} \frac{\exp[2\pi i (\mathbf{u}_b - \tau_b) \cdot (\mathbf{b} - \mathbf{b}_1)]}{\left(K_n^2 - u^2\right)} \exp\left[2\pi i u_z (z - z_1)\right] \\
&= -\frac{m_0}{2\pi^2\hbar^2} \exp(-2\pi i \tau_b \cdot \mathbf{b}_1) \\
&\times \int d\mathbf{b}' \int d\mathbf{u} \frac{\exp\left[2\pi i (\mathbf{u}_b - \tau_b) \cdot \mathbf{b}'\right]}{\left(K_n^2 - u^2\right)} \exp\left[2\pi i u_z (z - z_1)\right] \\
&= -\frac{m_0}{2\pi^2\hbar^2} \exp(-2\pi i \tau_b \cdot \mathbf{b}_1) \int du_z \frac{\exp\left[2\pi i u_z (z - z_1)\right]}{K_n^2 - \tau_b^2 - u_z^2} \\
&= -\frac{i m_0}{\pi\hbar^2} \frac{\exp(-2\pi i \tau_b \cdot \mathbf{b}_1)}{\left(K_n^2 - \tau_b^2\right)^{1/2}} \exp\left[2\pi i \left(K_n^2 - \tau_b^2\right)^{1/2} (z - z_1)\right] \\
&= -\frac{i m_0}{\pi\hbar^2} \frac{\exp(-2\pi i \tau_b \cdot \mathbf{b}_1)}{K_n\,|\cos\varphi_0|} \exp[2\pi i K_n (\cos\varphi_0)(z - z_1)]
\end{aligned}
\quad (10.13)
$$

where the substitution of $\tau_b = K_n \sin\varphi_0$ was made. We now return to calculating (10.9) using the relation given by (10.13)

$$
\begin{aligned}
\hat{G}(\tau_b, z, \mathbf{r}_1) &= -\frac{i m_0}{\pi\hbar^2} \frac{\exp(-2\pi i \tau_b \cdot \mathbf{b}_1)}{K_n\,|cos\varphi_0|} \exp[2\pi i K_n (\cos\varphi_0)(z - z_1)] \\
&+ \int d\mathbf{b} \exp(-2\pi i \tau_b \cdot \mathbf{b})\, \hat{G}_c(\mathbf{r}_1, \mathbf{r}) \\
&= a_z \exp(-2\pi i \tau \cdot \mathbf{r}_1) + \int d\mathbf{b} \exp(-2\pi i \tau_b \cdot \mathbf{b})\, \hat{G}_c(\mathbf{r}_1, \mathbf{r})
\end{aligned}
\quad (10.14)
$$

where $\tau_z = K_n \cos\varphi_0$ and

$$
a_z = \left[-\frac{i m_0}{\pi\hbar^2} \frac{\exp(2\pi i \tau_z z)}{\tau_z}\right]
$$

In (10.14), if the first term is taken as an incident plane wave of amplitude a_z and wave vector $(-\tau)$, the second term will be the wave scattered by the crystal due to the incident plane wave. Therefore $\hat{G}(\tau_b, z, \mathbf{r}_1)$ can be mathematically expressed (Gorodnichev et al. 1989; Dudarev et al. 1993a) as

$$
\hat{G}(\tau_b, z, \mathbf{r}_1) = a_z \Psi_0(-\tau, \mathbf{r}_1) \quad (10.15)
$$

where $\Psi_0(-\tau, \mathbf{r}_1)$ is the solution of the Schrödinger equation for an incident plane wave of wave vector $(-\tau)$ (with $\tau_z = K_n \cos\varphi_0$). The negative sign of the wave vector means that the electron strikes the crystal from the bottom. The elastic-scattering wave $\Psi_0(-\tau, \mathbf{r}_1)$ can be obtained by using conventional dynamic theories introduced in Chaps. 2–5. If the observation point is located at $z = \infty$, the variable z can be omitted in (10.15); thus

$$\hat{G}(\tau_b, \mathbf{r}_1) = a_z \Psi_0(-\tau, \mathbf{r}_1) \tag{10.16}$$

This important result is used in the following discussions.

10.3 First-Order TDS

We now consider an application of Green's function theory for calculating the diffraction pattern of first-order TDS electrons. Taking a Fourier transform of the second term in (10.8), the modulus square of the result is the first-order TDS intensity scattered to $\mathbf{u}_b$ in reciprocal space

$$\begin{aligned} I_1(\mathbf{u}_b) =& e^2\gamma^2 \int d\mathbf{r}_1 \int d\mathbf{r}_2 \hat{G}(\mathbf{u}_b, \mathbf{r}_1)\hat{G}^*(\mathbf{u}_b, \mathbf{r}_2) \\ & \times < \Delta V(\mathbf{r}_1)\Delta V(\mathbf{r}_2) > \Psi_0(\mathbf{K}_0, \mathbf{r}_1)\Psi_0^*(\mathbf{K}_0, \mathbf{r}_2) \end{aligned} \tag{10.17}$$

The most important advantage of Green's function method is that the thermal average can be performed before numerical calculations (1991).

In general, the electron angular distribution is calculated by assuming that the observation point is far from the crystal (i.e., $z = \infty$) in corresponding to the geometry of the Fraunhofer diffraction. Thus the relation shown in (10.16) holds, and (10.17) becomes

$$\begin{aligned} I_1(\mathbf{u}) = D_1 \sum_{q_z} \int d\mathbf{r}_1 \int d\mathbf{r}_2 \Psi_0(-\mathbf{K}_0 - \mathbf{u}_b, \mathbf{r}_1)\Psi_0^*(-\mathbf{K}_0 - \mathbf{u}_b, \mathbf{r}_2) \\ \times < \Delta V(\mathbf{r}_1)\Delta V(\mathbf{r}_2) > \Psi_0(\mathbf{K}_0, \mathbf{r}_1)\Psi_0^*(\mathbf{K}_0, \mathbf{r}_2) \end{aligned} \tag{10.18}$$

with $D_1 = e^2\gamma^2 m_0 \left[2\pi^2\hbar^2 E_n \cos^2\varphi_0\right]^{-1}$. Using $\Delta V(\mathbf{r}_1)$ given by (6.63), $\langle \Delta V(\mathbf{r}_1) \cdot \Delta V(\mathbf{r}_2)\rangle$ can be written (Afanas'ev and Kagan 1968) as

$$< \Delta V(\mathbf{r}_1)\Delta V(\mathbf{r}_2) >= \int d\mathbf{Q} \int d\mathbf{Q}' S_{\text{TDS}}(\mathbf{Q}, \mathbf{Q}') \exp\left(2\pi i \mathbf{Q}\cdot\mathbf{r}_1 - 2\pi i \mathbf{Q}'\cdot\mathbf{r}_2\right) \tag{10.19}$$

where

$$S_{\mathrm{TDS}}(\mathbf{Q},\mathbf{Q}')=\sum_{\kappa}\sum_{\kappa'} f_{\kappa}^{e}(\mathbf{Q})f_{\kappa'}^{e}(-\mathbf{Q}')\exp\left[-2\pi i\mathbf{Q}\cdot\mathbf{r}_{\kappa}+2\pi i\mathbf{Q}'\cdot\mathbf{r}'_{\kappa}\right]$$
$$\times\exp\left[-W_{\kappa}(\mathbf{Q})-W_{\kappa'}(\mathbf{Q}')\right]\left\{\exp\left[4\pi^{2}\langle(\mathbf{Q}\cdot\mathbf{u}_{\kappa})(\mathbf{Q}'\cdot\mathbf{u}_{\kappa'})\rangle\right]-1\right\} \tag{10.20}$$

The calculation of $W_{\kappa}(\mathbf{Q})$ was given in Sects. 6.4.5 and 7.5.

For a three-dimensional-periodic crystal structure, summation over κ can be separated into a summation over centers of unit cells $\mathbf{R}_n$ and a summation over atoms within a cell $\mathbf{r}_k=\mathbf{R}_n+\mathbf{r}_\alpha$. The quantity $\langle(\mathbf{Q}\cdot\mathbf{u}_k)\left(\mathbf{Q}'\cdot\mathbf{u}_{\kappa'}\right)\rangle$ (Afanas'ev and Kagan 1968) is

$$<(\mathbf{Q}\cdot\mathbf{u}_k)\left(\mathbf{Q}'\cdot\mathbf{u}_{k'}\right)>=\Omega\int_{\mathrm{BZ}}d\tau\sum_{j}\frac{\hbar\left(\langle n_s\rangle+1/2\right)}{\omega_j(\tau)\left(M_\alpha M_{\alpha'}\right)^{1/2}}$$
$$\times\left[\mathbf{Q}\cdot\mathbf{e}\left(\alpha\left|{}^{\tau}_{j}\right.\right)\right]\left[\mathbf{Q}'\cdot\mathbf{e}\left(\alpha'\left|{}^{-\tau}_{j}\right.\right)\right]\exp[2\pi i\tau\cdot(\mathbf{r}_{\kappa}-\mathbf{r}_{k'})] \tag{10.21}$$

Numerical calculations using (10.10) and (10.11) are feasible if the knowledge about phonon dispersion relationships is available. The analytical calculation of (10.21) was given in Sect. 7.7 using the Warren approximation.

In convergent beam electron diffraction, fine details introduced by phonon dispersions are usually obliterated because of the convolution from the beam convergence. Thus details introduced by phonon dispersion cannot be resolved in the diffraction pattern. Therefore good agreement has been obtained between calculated results from the Einstein model and experimentally observed results. However it must be pointed out that this agreement is the result of beam convergence (Wang 1992a, b).

Under the random vibrational phase approximation, (10.20) can be reduced to

$$S_{\mathrm{TDS}}(\mathbf{Q},\mathbf{Q}')=\sum_{\kappa} f_{\kappa}^{e}(\mathbf{Q})f_{\kappa}^{e}(-\mathbf{Q}')\exp\left[-2\pi i(\mathbf{Q}-\mathbf{Q}')\cdot\mathbf{r}_{\kappa}\right]$$
$$\times\left\{\exp\left[-W_{\kappa}(\mathbf{Q}-\mathbf{Q}')\right]-\exp\left[-W_{\kappa}(\mathbf{Q})-W_{\kappa}(\mathbf{Q}')\right]\right\} \tag{10.22}$$

Equation (10.18) can, in principle, be applied to the scattering object of arbitrary shape as long as a solution of the elastic-scattering wave can be found. The theory just presented can be applied to calculate the RHEED of phonon-scattered electrons.

For a thin parallel-sided slab crystal in the Laue case, (10.18) can be converted into the Bloch wave representation as

$$I_1(\mathbf{u}_b)\approx D_1\sum_{ii'}\sum_{jj'}\sum_{g_1h_1}\sum_{g_2h_2}[\mathrm{B.W.}]_1[\mathrm{Inelastic}]_1[\mathrm{Thickness}]_1 \tag{10.23a}$$

where

$$\begin{aligned}[\text{B.W.}]_1 =& C_0^{(i)}(\mathbf{K})C_{g_1}^{(i)*}(\mathbf{K})C_0^{(i')*}(\mathbf{K})C_{g_2}^{(i')}(\mathbf{K}) \\ & \times C_0^{(j)*}(\mathbf{K}_0)\, C_{h_1}^{(j)}(\mathbf{K}_0)\, C_0^{(j')}(\mathbf{K}_0)\, C_{h_2}^{(j')*}(\mathbf{K}_0)\end{aligned} \quad (10.23\text{b})$$

$$[\text{Inelastic}_{\,1}] = S_{\text{TDS}}\,(\mathbf{g}_1 - \mathbf{h}_1 + \mathbf{u}_b, \mathbf{g}_2 - \mathbf{h}_2 + \mathbf{u}_b) \quad (10.23\text{c})$$

and

$$[\text{Thickness}]_1 = \frac{\sin\left\{2\pi\left[\nu^{(j)} - \nu^{(j')}\right]d\right\} - \sin\left\{2\pi\left[\nu^{(i)} - \nu^{(i)'}\right]d\right\}}{2\pi\left[\nu^{(j)} - \nu^{(j')} - \nu^{(i)} + \nu^{(i')}\right]} \quad (10.23\text{d})$$

providing an approximation of the weak dependence of S_{TDS} on q_z is made and a relation of

$$C_{-g}^{(i)}(-\mathbf{K}) = C_g^{(i)*}(\mathbf{K})$$

was used, with $\mathbf{K} = \mathbf{K}_0 + \mathbf{u}_b$. The kinematic scattering result is obtained under the small thickness approximation of (10.23),

$$I_1\,(\mathbf{u}_b) \approx d D_1 S_{\text{TDS}}\,(\mathbf{u}_b, \mathbf{u}_b) \quad (10.24)$$

which means that the electron is directly scattered to $\mathbf{u}_b$ by TDS without Bragg scattering.

10.4 Atomic Inner Shell Single-Inelastic Excitation

Diffraction of single-inelastically scattered electrons due to atomic inner shell or valence band excitation can be derived from the first term of (10.7). Using (10.16), the intensity distribution at $\mathbf{u}$ in the diffraction plane is

$$\begin{aligned}I_s(\mathbf{u}) =& \frac{D_1}{e^2}\sum_{q_z}\int d\mathbf{r}_1 \int d\mathbf{r}_2\, \Psi_0(-\mathbf{K}_0 - \mathbf{u}, \mathbf{r}_1)\Psi_0^*(-\mathbf{K}_0 - \mathbf{u}, \mathbf{r}_2) \\ & \sum_n \left[H_{n0}(\mathbf{q}, \mathbf{r}_1)H_{n0}^*(\mathbf{q}, \mathbf{r}_2)\right] \times \Psi_0(\mathbf{K}_0, \mathbf{r}_1)\Psi_0^*(\mathbf{K}_0, \mathbf{r}_2)\end{aligned} \quad (10.25)$$

where the sum over n is the incoherent summation over contributions made by different crystal excited states with the same momentum transfer but different energy losses. Equation (10.25) can also be converted into the Bloch wave form as listed in (10.23a)–(10.23d) except with a different expression for [Inelastic]. For single-electron excitations,

$$[\text{ Inelastic }]_1 = \int_{E_c}^{E_c+\Delta E} dE\ H_{g_1-h_1}(-\mathbf{u})H^*_{g_2-h_2}(-\mathbf{u}) \tag{10.26}$$

where integration of E includes electrons with different energy losses but the same momentum.

For valence excitations from (6.113),

$$[\text{Inelastic}]_1 \approx \frac{\hbar e^2}{8\pi^3\varepsilon_0}\left(\frac{1}{|\mathbf{g}_1-\mathbf{h}_1+\mathbf{u}|^2+q_0^2}+\frac{1}{|\mathbf{g}_2-\mathbf{h}_2+\mathbf{u}|^2+q_0^2}\right)\int_0^\infty d\omega\ \text{Im}\left[-\frac{1}{\varepsilon(\omega)}\right] \tag{10.27}$$

It is necessary to point out that Green's function theories for first-order TDS (10.23a)–(10.23d) and single inelastic atomic inner shell or valence band scattering (10.26)–(10.27) are applicable to any order of Laue zone reflections. This is different from the assumption made for the double inelastic scattering in the next section.

10.5 Double Inelastic Electron Scattering

Using the second and the third terms of (10.7), the intensity scattered to $\mathbf{u}_b$ in the diffraction plane due to TDS-atomic inner shell and atomic inner shell TDS, double inelastic excitations may be directly written as Wang (1994)

$$\begin{aligned} I_2(\mathbf{u}_b) \approx e^2\gamma^2 \int d\mathbf{r}_1 \int d\mathbf{r}_1' \int d\mathbf{r}_2 \int d\mathbf{r}_2' \Big\{ \hat{G}(\mathbf{u}_b,\mathbf{r}_1)\hat{G}(\mathbf{r}_1,\mathbf{r}_1')\hat{G}^*(\mathbf{u}_b,\mathbf{r}_2)\hat{G}^*(\mathbf{r}_2,\mathbf{r}_2') \\ \times \Big[\langle \Delta V(\mathbf{r}_1)\Delta V(\mathbf{r}_2)\rangle H_{n0}(\mathbf{q}_1,\mathbf{r}_1')H^*_{n0}(\mathbf{q}_1,\mathbf{r}_2') \\ + H_{n0}(\mathbf{q}_2,\mathbf{r}_1)H^*_{n0}(\mathbf{q}_2,\mathbf{r}_2)\langle \Delta V(\mathbf{r}_1')\Delta V(\mathbf{r}_2')\rangle \Big] \Psi_0(\mathbf{K}_0,\mathbf{r}_1')\Psi_0^*(\mathbf{K}_0,\mathbf{r}_2') \Big\} \end{aligned} \tag{10.28}$$

where $\mathbf{q}_1$ and $\mathbf{q}_2$ are introduced to correspond to the change of incidence electron wave vectors due to the first and the second inelastic excitations, respectively, and an approximation of $\hat{G}_n \approx \hat{G}$ is made for small energy loss. Using (10.16) for $\hat{G}$ $(\mathbf{u}_b, z, \mathbf{r}_1)$, the two-dimensional intensity distribution of double inelastic electron scattering is

$$I_2(\mathbf{u}_b) = D_1 \sum_{q_z} \sum_{q_{1z}} \sum_{q_{1b}} \int d\mathbf{r}_1\, d\mathbf{r}_1'\, d\mathbf{r}_2\, d\mathbf{r}_2'\, \Psi_0(-\mathbf{K}_0 - \mathbf{u}_b, \mathbf{r}_1)\Psi_0^*(-\mathbf{K}_0 - \mathbf{u}_b, \mathbf{r}_2)$$
$$\times \hat{G}(\mathbf{r}_1, \mathbf{r}_1')\hat{G}^*(\mathbf{r}_2, \mathbf{r}_2')\Big[\langle \Delta V(\mathbf{r}_1)\Delta V(\mathbf{r}_2)\rangle H_{n0}(\mathbf{q}_1, \mathbf{r}_1')H_{n0}^*(\mathbf{q}_1, \mathbf{r}_2')$$
$$+ H_{n0}(\mathbf{q}_2, \mathbf{r}_1)H_{n0}^*(\mathbf{q}_2, \mathbf{r}_2)\langle \Delta V(\mathbf{r}_1')\Delta V(\mathbf{r}_2')\rangle\Big]\Psi_0(\mathbf{K}_0, \mathbf{r}_1')\Psi_0^*(\mathbf{K}_0, \mathbf{r}_2') \tag{10.29}$$

where Σ_{q_z} sum over contributions from final states with different z-component momentum transfers, and $\Sigma_{q_{1z}}$ and Σ_{q_b} represent the sums over intermediate (or bridge) states. The $\mathbf{q}_b$ is the projection of $\mathbf{q}$ in the $\mathbf{b}$ plane. In principle, (10.29) can be applied to any of crystal shapes as long as the G function can be found.

For transmission electron diffraction of a thin crystal slab, Green's function solution of (10.6) can be found from the discussion given in Sect. 2.9.2. For ZOLZ reflections, we assume that the crystal potential V_0 is z independent. By writing

$$\hat{G}\,(\mathbf{r}, \mathbf{r}_1) = \exp\left[2\pi i \mathbf{K}_n \cdot (\mathbf{r} - \mathbf{r}_1)\right] \psi\,(\mathbf{r}, \mathbf{r}_1) \tag{10.30}$$

under the small-angle scattering approximation, (10.6) becomes

$$\left[-\frac{\hbar^2}{2m_0}\nabla_b^2 - e\gamma V_0(\mathbf{b}) - 4\pi i K_{nz}\frac{\partial}{\partial z}\right]\psi\,(\mathbf{r}, \mathbf{r}_1) \approx \delta\,(\mathbf{b} - \mathbf{b}_1)\,\delta\,(z - z_1) \tag{10.31}$$

Before we start to look for the solution to (10.31), it is helpful to examine the eigen states of (10.31) if $\mathbf{b} \neq \mathbf{b}_1$, which are

$$\left[-\frac{\hbar^2}{2m_0}\nabla_b^2 - e\gamma V_0(\mathbf{b}) - 4\pi i K_{nz}\frac{\partial}{\partial z}\right]\psi\,(\mathbf{r}, \mathbf{r}_1) = 0 \tag{10.32}$$

Since the projection approximation is assumed, the eigen solution of (10.32) can be directly written as

$$\psi_j = \exp\left[-i s_j' z/4\pi K_{nz}\right]\tau_j'(\mathbf{b}) \tag{10.33a}$$

where $\tau_j(\mathbf{b})$ is a two-dimensional Bloch wave solution of

$$\left[-\frac{\hbar^2}{2m_0}\nabla_b^2 - e\gamma V(\mathbf{b})\right]\tau_j'(\mathbf{b}) = s_j'\tau_j'(\mathbf{b}) \tag{10.33b}$$

In Sect. 2.9.2, we proved that $\tau_j(\mathbf{b})$ is a complete orthonormal function set, and the following relation holds

$$\sum_i \tau_i'(\mathbf{b})\tau_i'^*(\mathbf{b}) = \delta\left(\mathbf{b} - \mathbf{b}'\right) \tag{10.34}$$

Therefore Green's function can be written as a superposition of Bloch states ψ_j. By using (10.34), we can prove directly

$$\left[-\frac{\hbar^2}{2m_0}\nabla_b^2 - e\gamma V(\mathbf{b}) - 4\pi i K_{nz}\frac{\partial}{\partial z}\right]\times$$
$$\times\left\{\sum_i \tau_i'(\mathbf{b})\tau_i'^*(\mathbf{b}_1)\exp\left[2\pi i\nu^{(i)}(z-z_1)\right]\right\} = \delta(\mathbf{b}-\mathbf{b}_1) \qquad (10.35)$$

where $\nu^{(i)} = -s_i/8\pi^2 K_{nz}$. Comparing (10.31) and (10.35) leads to

$$\psi(\mathbf{r}-\mathbf{r}_1) = \frac{i}{4\pi K_{nz}}\left\{\sum_i \tau_i'(\mathbf{b})\tau_i'^*(\mathbf{b}_1)\exp\left[2\pi i\nu^{(i)}(z-z_1)\right]\right\}\Theta(z-z_1) \qquad (10.36)$$

where $\Theta(z-z_1)$ is a step function satisfying $d\Theta/dz = \delta(z-z_1)$. Substituting (10.36) into (10.30), Green's function is

$$\hat{G}(\mathbf{r},\mathbf{r}_1) = \frac{i}{4\pi K_{nz}}\exp\left[2\pi i\mathbf{K}_0\cdot(\mathbf{r}-\mathbf{r}_1)\right]$$
$$\times\left\{\sum_i \tau_i'(\mathbf{b})\tau_i'^*(\mathbf{b}_1)\exp\left[2\pi i\nu^{(i)}(z-z_1)\right]\right\}\Theta(z-z_1) \qquad (10.37)$$

Therefore the calculation using (10.29) is feasible as long as ZOLZ reflections are concerned. The $< \Delta V(\mathbf{r}_1)\Delta V(\mathbf{r}_2) >$ term is given by (10.19).

For diffraction pattern calculations, we intend to express (10.29) in the Bloch wave form. For ZOLZ reflections, quantities in (10.29) are expanded in the Bloch wave representation as follows:

$$\Psi_0(-\mathbf{K},\mathbf{r}_1) = \sum_i C_0^{(i)*}(-\mathbf{K})\sum_{g_1} C_{g_1}^{(i)}(-\mathbf{K})$$
$$\times\exp\left[-2\pi i\mathbf{K}\cdot\mathbf{r}_1 - 2\pi i\mathbf{g}_1\cdot\mathbf{b}_1\right]\exp\left[-2\pi i\nu^{(i)}(z_1-d)\right] \qquad (10.38a)$$

which satisfies the boundary condition $\Psi_0(-\mathbf{K},\mathbf{b}_1,d) = \exp(-2\pi i\mathbf{K}\cdot\mathbf{r}_1)$;

$$\hat{G}(\mathbf{r},\mathbf{r}_1') = \frac{i}{2K_0}\exp\left[2\pi i\mathbf{K}_1\cdot(\mathbf{r}_1-\mathbf{r}_1')\right]\left\{\sum_{i'}\sum_{g_1'}\sum_{h_1'} C_{g_1'}^{(i')}(\mathbf{K}_1)C_{h_1'}^{(i')*}(\mathbf{K}_1)\right.$$
$$\left.\times\exp\left(2\pi i\mathbf{g}_1'\cdot\mathbf{b}_1 - 2\pi i\mathbf{h}_1'\cdot\mathbf{b}_1'\right)\exp\left[2\pi i\nu^{(i')}(z_1-z_1')\right]\right\}\Theta(z_1-z_1') \qquad (10.38b)$$

with $\mathbf{K}_1 = \mathbf{K}_0 + \mathbf{q}_{1b}$. From (6.8),

$$H_{n0}\left(\mathbf{r}_1'\right) = \exp\left(2\pi i \mathbf{q}_1 \cdot \mathbf{b}_1'\right) \sum_{g_3} H_{g_3}^{n0}\left(-\mathbf{q}_1\right) \exp\left(2\pi i \mathbf{g}_3 \cdot \mathbf{b}_1'\right) \tag{10.38c}$$

$$\Psi_0\left(-\mathbf{K}, \mathbf{r}_1'\right) = \sum_f C_0^{(f)^*}\left(\mathbf{K}_0\right) \sum_{g_2} C_{g_2}^{(f)}\left(\mathbf{K}_0\right) \times \exp\left(2\pi i \mathbf{K}_0 \cdot \mathbf{r}_1' + 2\pi i \mathbf{g}_2 \cdot \mathbf{b}_1'\right) \exp\left[-2\pi i \nu^{(f)} z_1'\right] \tag{10.38d}$$

Equation (10.38b) satisfies the boundary condition $\Psi_0\left(\mathbf{K}_0, \mathbf{b}_1, 0\right) = \left[\exp\left(2\pi i \mathbf{K}_0 \cdot \mathbf{r}_1\right)\right]$. Substituting (10.37) and (10.38a)–(10.38d) into (10.29), summing over q_z and q_{1z} first by using (8.12), (10.29) is finally converted (Wang and Li 1994) into

$$I_2(\mathbf{u}_b) \approx \Omega D_1 \sum_{ii'} \sum_{jj'} \sum_{ff''} \sum_{g_1h_1} \sum_{g_1'h_1'} \sum_{g_2h_2} \sum_{g_2'h_2'} \int_{\mathrm{BZ}} d\mathbf{q}_{1b}[\mathrm{B.W.}]_2[\mathrm{Inelastic}]_2[\mathrm{Thickness}]_2 \tag{10.39a}$$

where

$$[\mathrm{B.W.}]_2 = C_0^{(i)*}(-\mathbf{K})C_{g_1}^{(i)}(-\mathbf{K})C_0^{(j)}(-\mathbf{K})C_{h_1}^{(j)*}(-\mathbf{K})C_{g_1'}^{(i')}(\mathbf{K}_1)C_{h_1'}^{(i')*}(\mathbf{K}_1) \times C_{g_2'}^{(j')*}(\mathbf{K}_1)C_{h_2'}^{(j')}(\mathbf{K}_1)C_0^{(f)*}(\mathbf{K}_0)C_{g_2}^{(f)}(\mathbf{K}_0)C_0^{(f'')}(\mathbf{K}_0)C_{h_2}^{(f'')*}(\mathbf{K}_0) \tag{10.39b}$$

with $\mathbf{K} = \mathbf{K_0} + \mathbf{u}_b$ and $\mathbf{K}_1 = \mathbf{K_0} + \mathbf{q}_{1b}$

$$[\mathrm{Inelastic}]_2 = \left\{\left[S_{\mathrm{TDS}}(\mathbf{g}_1 - \mathbf{g}_1' + \mathbf{u} - \mathbf{q}_{1b}, \mathbf{h}_1 - \mathbf{g}_2' + \mathbf{u} - \mathbf{q}_{1b})\right] \times H_{h_1'-g_2}^{n0}(-\mathbf{q}_{1b})\left[H_{h_2'-h_2}^{n0}(-\mathbf{q}_{1b})\right]^* + H_{g_1-g_2'}^{n0}(-\mathbf{u}+\mathbf{q}_{1b})\left[H_{h_1-g_2'}^{n0}(-\mathbf{u}+\mathbf{q}_{1b})\right]^* \times \left[S_{\mathrm{TDS}}(\mathbf{h}_1' - \mathbf{g}_2 + \mathbf{q}_{1b}, \mathbf{h}_2' - \mathbf{h}_2 + \mathbf{q}_{1b})\right]\right\} \tag{10.39c}$$

and

$$[\,\mathrm{Thickness}\,]_2 = \left(\frac{\cos\left\{2\pi\left[\nu^{(j')} - \nu^{(i')}\right]d\right\} - \cos\left\{2\pi\left[\nu^{(i)} - \nu^{(j)}\right]d\right\}}{2\pi\left[\nu^{(j)} + \nu^{(j')} - \nu^{(i')} - \nu^{(i)}\right]} - \frac{\cos\left\{2\pi\left[\nu^{(f)} - \nu^{(f'')}\right]d\right\} - \cos\left\{2\pi\left[\nu^{(i)} - \nu^{(j)}\right]d\right\}}{2\pi\left[\nu^{(j)} + \nu^{(f)} - \nu^{(i)} - \nu^{(f'')}\right]}\right) \times \left\{2\pi\left[\nu^{(i')} - \nu^{(j')} + \nu^{(f)} - \nu^{(f'')}\right]\right\}^{-1} \tag{10.39d}$$

Equation (10.39) is the formal dynamic theory of double inelastic electron diffraction.

Under the small thickness approximation, (10.39) reduces to

$$\begin{aligned}I_2(\mathbf{u}_b) \approx &\frac{d^2}{2}\Omega D_1 \sum_{g_1}\sum_{g_2}\int_{BZ} d\mathbf{q}_{1b} \left\{\left[S_{TDS}(-\mathbf{g}_1+\mathbf{u}_b-\mathbf{q}_{1b}-\mathbf{g}_2+\mathbf{u}_b-\mathbf{q}_{1b})\right]\right.\\ &\times H_{g_1}^{n0}(-\mathbf{q}_{1b})\left[H_{g_2}^{n0}(-\mathbf{q}_{1b})\right]^*\\ &\left.+H_{-g_1}^{n0}(-\mathbf{u}_b+\mathbf{q}_{1b})\left[H_{-g_2}^{n0}(-\mathbf{u}_b+\mathbf{q}_{1b})\right]^*\left[S_{TDS}(\mathbf{g}_1+\mathbf{q}_{1b},\mathbf{g}_2+\mathbf{q}_{1b})\right]\right\}\end{aligned} \tag{10.40}$$

The physical meaning of the first term, for example, in (10.40) follows: The electron is scattered to $\mathbf{Q}_1 = \mathbf{g}_1 - (-\mathbf{q}_{1b})$ by single-electron excitation; it is then scattered for $\mathbf{Q}_2 = (-\mathbf{g}_1 + \mathbf{u}_b - \mathbf{q}_{1b})$ due to TDS, and the resultant change of wave vector is $\mathbf{Q} = \mathbf{Q}_1 + \mathbf{Q}_2 = \mathbf{u}$, as expected. By substituting $\mathbf{g}_1 \to -\mathbf{g}_1$ and $\mathbf{g}_2 \to -\mathbf{g}_2$, the second term in (10.40) is exactly the same as the first term, which means that the same scattering intensity is obtained for TDS-atomic inner shell processes as for atomic inner shell-TDS processes under the small thickness approximation. In other words, there is no difference in choosing either of the two inelastic events as the first process in double inelastic electron diffraction calculations under the small thickness approximation. Thus (10.40) is rewritten as

$$\begin{aligned}I_2(\mathbf{u}_b) = &d^2\Omega D_1 \sum_{g_1}\sum_{g_2}\int_{BZ} d\mathbf{q}_{1b}\\ &\times\left[S_{TDS}(-\mathbf{g}_1+\mathbf{u}_b-\mathbf{q}_{1b},-\mathbf{g}_2+\mathbf{u}_b-\mathbf{q}_{1b})\right] H_{g_1}^{n0}(-\mathbf{q}_{1b})\, H_{g_2}^{n0*}(-\mathbf{q}_{1b})\big]\end{aligned} \tag{10.41}$$

This is the kinematic scattering expression of double inelastic electron scattering.

We now proceed with a further derivation of (10.41) to examine its physical meaning. For simplification, we assume that TDS is described by the Einstein model, so that S_{TDS} is given by (10.22). Using the definition of H_g^{nm} given by (6.8b), we have

$$\begin{aligned}H_g^{n0}(\mathbf{q}) &= \int d\mathbf{r}\exp[-2\pi i(\mathbf{g}-\mathbf{q})\cdot\mathbf{r}]H_{n0}(\mathbf{r})\\ &= \int d\mathbf{r}\exp[-2\pi i(\mathbf{g}-\mathbf{q})\cdot\mathbf{r}]\sum_{\kappa'} H_{\kappa'}^{n0}(\mathbf{r}-\mathbf{r}_{\kappa'})\\ &= \sum_{\kappa'}\exp\left[-2\pi i(\mathbf{g}-\mathbf{q})\cdot\mathbf{r}_{\kappa'}\right] H_{\kappa'}^{n0}(\mathbf{g}-\mathbf{q})\end{aligned} \tag{10.42}$$

If the incident beam direction is far from the crystal zone axes, we need to consider only the central transmitted beam (000), so that $\mathbf{g}_1 = \mathbf{g}_2 = (000)$. Substituting (10.22) and (10.42) into (10.41), we have

$$I_2(\mathbf{u}_b) \approx d^2 \Omega D_1 \sum_{\kappa} \sum_{\kappa'} \int_{BZ} d\mathbf{q}_{1b} \left| f_{\kappa}^{e}(\mathbf{u}_b - \mathbf{q}_{1b}) \right|^2 \\ \times \left\{ 1 - \exp\left[-2W_{\kappa}(\mathbf{u}_b - \mathbf{q}_{1b}) \right] \right\} \left| H_{\kappa'}^{n0}(\mathbf{q}_{1b}) \right|^2 \tag{10.43}$$

where the single-electron excitation from different atomic sites is treated as incoherent. Equation (10.43) can be rewritten as

$$I_2(\mathbf{u}_b) \approx d^2 \Omega D_1 \sum_{\kappa} \int_{BZ} d\mathbf{q}_{1b} \left| f_{\kappa}^{\ell}(\mathbf{u}_b - \mathbf{q}_{1b}) \right|^2 \\ \times \left\{ 1 - \exp\left[-2W_{\kappa}(\mathbf{u}_b - \mathbf{q}_{1b}) \right] \right\} \left| H_{\kappa}^{n0}(\mathbf{q}_{1b}) \right|^2 \\ + d^2 \Omega D_1 \sum_{\kappa \neq \kappa'} \sum \int_{BZ} d\mathbf{q}_{1b} \left| f_{\kappa}^{\ell}(\mathbf{u}_b - \mathbf{q}_{1b}) \right|^2 \\ \times \left\{ 1 - \exp\left[-2W_{\kappa}(\mathbf{u}_b - \mathbf{q}_{1b}) \right] \right\} \left| H_{\kappa'}^{n0}(\mathbf{q}_{1b}) \right|^2 \tag{10.44}$$

In (10.44), the first term is contributed by the coherent inelastic double scattering, in which TDS and single-electron ionization are excited simultaneously in the scattering of the same atom; the second term is contributed by the incoherent inelastic double scattering, in which TDS and single-electron ionization are excited individually in the scattering of different atoms.

10.6 Summary

Green's function theory of inelastic electron diffraction was introduced in this chapter. The theory is best suited for thermal diffuse scattering because the time average over scattering intensities made by crystal lattices of various thermal vibration configurations can be performed before any numerical calculation. Phase correlations among atomic vibrations were included without introducing simplified models.

The most important advantage of Green's function approach is that excitations of all possible multiphonon processes are comprehensively included in a single formula; thus calculated results are expected to be valid for both low- and high-angle TDS. But the disadvantage of this approach is that Green's function solution for double inelastic electron scattering can be easily solved only for ZOLZ reflections. Proper perturbation operation is needed to include higher order Laue zone reflections.

Chapter 11
Dynamic Inelastic Electron Scattering III: Multislice Theory (1)

The multislice theory of elastic-scattering theory was outlined in Chap. 3. The unique advantage of this theory is its convenience of including defects in computations. The equivalent method has been applied to calculate the scattering of inelastic electrons in crystals. In this chapter, we first illustrate the multislice solution of Yoshioka's equations. Then we discuss the applications of the theory for calculating the images of high-angle phonon-scattered electrons in STEM and TEM. Finally, we present a real-space multislice theory which is more convenient for calculating the images and diffraction patterns of TDS electrons. The exact theory based on the real-space multislice approach for simulating the composition sensitive image using diffusely scattered electrons is also outlined.

11.1 Multislice Solution of Yoshioka's Equations

Our first goal is to find the multislice solution of Yoshioka's equations (Wang 1989a, 1990a). For convenience of the analysis, we write (6.4a) and (6.4b) into a matrix equation:

$$\nabla^2 \begin{pmatrix} \Psi_0 \\ \Psi_1 \\ \vdots \\ \Psi_m \end{pmatrix} + 4\pi^2 \begin{pmatrix} K_0^2 \Psi_0 \\ K_1^2 \Psi_1 \\ \vdots \\ K_m^2 \Psi_m \end{pmatrix} = \frac{2m_e}{\hbar^2} \begin{pmatrix} H_{00} & H_{01} & \cdots & H_{0m} \\ H_{10} & H_{11} & \cdots & H_{1m} \\ \vdots & \vdots & \ddots & \vdots \\ H_{m0} & H_{m1} & \cdots & H_{mm} \end{pmatrix} \cdot \begin{pmatrix} \Psi_0 \\ \Psi_1 \\ \vdots \\ \Psi_m \end{pmatrix} \qquad (11.1)$$

H is an $m \times m$ transition matrix. By defining $\Psi_n = \exp(2\pi i \mathbf{K}_n \cdot \mathbf{r})\, \Phi_n$, using Green's function method (see Appendix D), and assuming $\varepsilon_n \ll E_0$, (11.1) is converted into an integral form of (11.2)

Z. L. Wang, *Elastic and Inelastic Scattering in Electron Diffraction and Imaging*, Springer Series in Solid-State Sciences 205, https://doi.org/10.1007/978-3-031-90819-4_11

$$\begin{pmatrix} \Phi_0(\mathbf{r}) \\ \Phi_1(\mathbf{r}) \\ \vdots \\ \Phi_m(\mathbf{r}) \end{pmatrix} = \mathbf{I_0} - \frac{2m_e}{4\pi\hbar^2}\int d\mathbf{r}' \begin{pmatrix} F(\mathbf{r}-\mathbf{r}',k_0) & 0 & \cdots & 0 \\ 0 & F(\mathbf{r}-\mathbf{r}',k_1) & \cdots & 0 \\ \vdots & \vdots & \ddots & \vdots \\ 0 & 0 & \cdots & F(\mathbf{r}-\mathbf{r}',k_m) \end{pmatrix}$$

$$\times \begin{pmatrix} H_{00}(\mathbf{r}') & H_{01}(\mathbf{r}') & \cdots & H_{0m}(\mathbf{r}') \\ H_{10}(\mathbf{r}') & H_{11}(\mathbf{r}') & \cdots & H_{1m}(\mathbf{r}') \\ \vdots & \vdots & \ddots & \vdots \\ H_{m0}(\mathbf{r}') & H_{m1}(\mathbf{r}') & \cdots & H_{mm}(\mathbf{r}') \end{pmatrix} \cdot \begin{pmatrix} \Phi_0(\mathbf{r}') \\ \Phi_1(\mathbf{r}') \\ \vdots \\ \Phi_m(\mathbf{r}') \end{pmatrix} \tag{11.2}$$

where $\mathbf{I}_0$ is a matrix with elements $(\mathbf{I}_0)_n = \delta_{n0}$. The F is related to the propagator, and it is defined as

$$F\left(\mathbf{r}-\mathbf{r}',\mathbf{K}_n\right) = \frac{\exp\left\{2\pi i\left[K_n\left|\mathbf{r}-\mathbf{r}'\right| - \mathbf{K}_n\cdot\left(\mathbf{r}-\mathbf{r}'\right)\right]\right\}}{\left|\mathbf{r}-\mathbf{r}'\right|} \tag{11.3}$$

When high-energy electron scattering satisfies the following conditions:

$$\theta^2 \ll 1$$

$\left|\mathbf{r}-\mathbf{r}'\right| \approx z - z'$ for nondeflection approximation and neglecting the backscattering

$$\mathrm{K}_n\left|\mathbf{r}-\mathbf{r}'\right| - \mathbf{K}_n\cdot\left(\mathbf{r}-\mathbf{r}'\right) \approx K_n\frac{\left|\mathbf{b}-\mathbf{b}'\right|^2}{2\left(z-z'\right)}$$

the F function is related to the propagation function P_n by

$$F\left(\mathbf{r}-\mathbf{r}',\mathbf{K}_n\right) = i\lambda_n P_n\left(\mathbf{b}-\mathbf{b}',\Delta z\right) \tag{11.4}$$

with

$$P_n(\mathbf{b},\Delta z) = \frac{1}{i\lambda_n\Delta z}\exp\left(\frac{i\pi b^2}{\lambda_n\Delta z}\right) \tag{11.5}$$

where λ_n is the wavelength of the electron with energy E_n. Equation (11.2) becomes

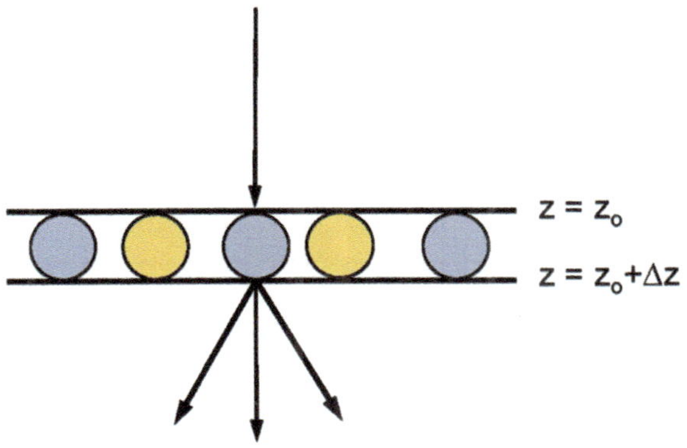

Fig. 11.1 A schematic diagram illustrating the electron wave before and after transmitting a thin crystal slice

$$\begin{pmatrix} \Phi_0(\mathbf{b}, z) \\ \Phi_1(\mathbf{b}, z) \\ \vdots \\ \Phi_m(\mathbf{b}, z) \end{pmatrix} = \mathbf{I}_0 + \left(-\frac{i}{\hbar v}\right) \int d\mathbf{b}' \int_{z'=-\infty}^{z'=z} dz'$$
$$\times \begin{pmatrix} P_0(\mathbf{b}-\mathbf{b}', z-z') & 0 & \cdots & 0 \\ 0 & P_1(\mathbf{b}-\mathbf{b}', z-z') & \cdots & 0 \\ \vdots & \vdots & \ddots & \vdots \\ 0 & 0 & \cdots & P_m(\mathbf{b}-\mathbf{b}', z-z') \end{pmatrix} \tag{11.6}$$
$$\times \begin{pmatrix} H_{00}(\mathbf{b}', z') & H_{01}(\mathbf{b}', z') & \cdots & H_{0m}(\mathbf{b}', z') \\ H_{10}(\mathbf{b}', z') & H_{11}(\mathbf{b}', z') & \cdots & H_{1m}(\mathbf{b}', z') \\ \vdots & \vdots & \ddots & \vdots \\ H_{m0}(\mathbf{b}', z') & H_{m1}(\mathbf{b}', z') & \cdots & H_{mm}(\mathbf{b}', z') \end{pmatrix} \cdot \begin{pmatrix} \Phi_0(\mathbf{b}', z') \\ \Phi_1(\mathbf{b}', z') \\ \vdots \\ \Phi_m(\mathbf{b}', z') \end{pmatrix}$$

To determine the multislice solution of this equation, we need to derive the relationship between the waves $\Phi_n(\mathbf{b}, z_0)$ going into a crystal slice at $z = z_0$ and the waves $\Phi_n(\mathbf{b}, z)$ after penetrating the slice at $z = z$ if $\Delta z = z - z_0$ is very small (Fig. 11.1). We now consider the nth row of (11.6)

$$\begin{aligned} \Phi_n(\mathbf{b}, z) &= \delta_{n0} + \left(-\frac{i}{\hbar v}\right) \int d\mathbf{b}' \int_{z'=-\infty}^{z'=z} dz'\, P_n(\mathbf{b}-\mathbf{b}', z-z') \sum_m H_{nm}(\mathbf{b}', z')\Phi_m(\mathbf{b}', z') \\ &= \delta_{n0} + \left(-\frac{i}{\hbar v}\right) \int d\mathbf{b}'' \int_{z'=-\infty}^{z'=z_0} dz''\, P_n(\mathbf{b}-\mathbf{b}'', z-z'') \sum_m H_{nm}(\mathbf{b}'', z'')\Phi_m(\mathbf{b}'', z'') \\ &\quad + \left(-\frac{i}{\hbar v}\right) \int d\mathbf{b}' \int_{z'=z_0}^{z'=z} dz'\, P_n(\mathbf{b}-\mathbf{b}', z-z') \sum_m H_{nm}(\mathbf{b}', z')\Phi_m(\mathbf{b}', z') \end{aligned} \tag{11.7}$$

For the normal incident electron beam case, we have the following relations, as proved in Sect. 3.2.1

$$\int d\mathbf{b} P_n(\mathbf{b}, z) = 1 \tag{11.8a}$$

and

$$P_n\left(\mathbf{b} - \mathbf{b}', z - z'\right) = \int_{\Sigma} d\mathbf{b}'' P_n\left(\mathbf{b} - \mathbf{b}'', z - z''\right) P_n\left(\mathbf{b}'' - \mathbf{b}', z'' - z'\right) \tag{11.8b}$$

where Σ is a plane positioned at z'' between z and z'. Then (11.7) can be rewritten as

$$\begin{aligned}\Phi_n(\mathbf{b}, z) = &\int d\mathbf{b}_0 \left[\delta_{n0} + \left(-\frac{i}{\hbar v}\right) \int d\mathbf{b}'' \int_{z''=-\infty}^{z''=z_0} dz'' P_n\left(\mathbf{b}_0 - \mathbf{b}'', z_0 - z''\right)\right.\\ &\left.\times \sum_m H_{nm}\left(\mathbf{b}'', z''\right) \Phi_m\left(\mathbf{b}'', z''\right)\right] P_n\left(\mathbf{b} - \mathbf{b}_0, z - z_0\right)\\ &+ \left(-\frac{i}{\hbar v}\right) \int d\mathbf{b}' \int_{z'=z_0}^{z'=z} dz' P_n\left(\mathbf{b} - \mathbf{b}', z - z'\right) \sum_m H_{nm}\left(\mathbf{b}', z'\right) \Phi_m\left(\mathbf{b}', z'\right)\end{aligned} \tag{11.9}$$

Comparing the first term in (11.9) with (11.7) for $z = z_0$, (11.9) can be rewritten as

$$\begin{aligned}\Phi_n(\mathbf{b}, z) = &\int d\mathbf{b}_0 P_n\left(\mathbf{b} - \mathbf{b}_0, z - z_0\right) \phi_n\left(\mathbf{b}_0, z_0\right)\\ &+ \left(-\frac{i}{\hbar v}\right) \int d\mathbf{b}' \int_{z'=z_0}^{z'=z} dz' P_n\left(\mathbf{b} - \mathbf{b}', z - z'\right) \sum_m H_{nm}\left(\mathbf{b}', z'\right) \Phi_m\left(\mathbf{b}', z'\right)\end{aligned} \tag{11.10}$$

To solve (11.10), we expand Φ_n in the power of $\left(-\frac{i}{\hbar v}\right)$

$$\Phi_n(\mathbf{b}, z) = \sum_{L=0}^{\infty} \left(-\frac{i}{\hbar v}\right)^L Y_L^{(n)}(\mathbf{b}, z) \tag{11.11}$$

Substituting (11.11) into (11.10), and equating the coefficients of $\left(-\frac{i}{\hbar v}\right)^L$, we have

$$Y_0^{(n)}(\mathbf{b}, z) = \int d\mathbf{b}_0 P_n\left(\mathbf{b} - \mathbf{b}_0, z - z_0\right) \Phi_n\left(\mathbf{b}_0, z_0\right) \tag{11.12a}$$

and

$$Y_L^{(n)}(\mathbf{b}, z) = \int d\mathbf{b}' \int_{z'=z_0}^{z'=z} dz' P_n\left(\mathbf{b} - \mathbf{b}', z - z'\right) \sum_m H_{nm}\left(\mathbf{b}', z'\right) Y_{L-1}^{(m)}\left(\mathbf{b}', z'\right) \tag{11.12b}$$

If H_{nm} varies slowly in the region of $\Delta z = z - z_0$ and also $K_0 \approx K_n$, then the solution of (11.12a) and (11.12b) is

$$\begin{pmatrix} Y_L^{(0)}(\mathbf{b}, z) \\ Y_L^{(1)}(\mathbf{b}, z) \\ \vdots \\ Y_L^{(m)}(\mathbf{b}, z) \end{pmatrix} = \frac{1}{L!} \int d\mathbf{b}_0 \begin{pmatrix} P_0\left(\mathbf{b} - \mathbf{b}_0, \Delta z\right) & 0 & \dots & 0 \\ 0 & P_1\left(\mathbf{b} - \mathbf{b}_0, \Delta z\right) & \dots & 0 \\ \vdots & \vdots & \ddots & \vdots \\ 0 & 0 & \dots & P_m\left(\mathbf{b} - \mathbf{b}_0, \Delta z\right) \end{pmatrix}$$
$$\times \begin{pmatrix} H_{00}\left(\mathbf{b}_0, \Delta z\right) & H_{01}\left(\mathbf{b}_0, \Delta z\right) & \dots H_{0m}\left(\mathbf{b}_0, \Delta z\right) \\ H_{10}\left(\mathbf{b}_0, \Delta z\right) & H_{11}\left(\mathbf{b}_0, \Delta z\right) & \dots H_{1m}\left(\mathbf{b}_0, \Delta z\right) \\ \vdots & \vdots & \vdots \\ H_{m0}\left(\mathbf{b}_0, \Delta z\right) & H_{m1}\left(\mathbf{b}_0, \Delta z\right) & \dots H_{mm}\left(\mathbf{b}_0, \Delta z\right) \end{pmatrix}^L \cdot \begin{pmatrix} \Phi_0\left(\mathbf{b}_0, z_0\right) \\ \Phi_1\left(\mathbf{b}_0, z_0\right) \\ \vdots \\ \Phi_m\left(\mathbf{b}_0, z_0\right) \end{pmatrix} \tag{11.13a}$$

This equation is conveniently written as a product of matrices in a shortened form as

$$\mathbf{Y}_L(\mathbf{b}, z) = \frac{1}{L!} \int d\mathbf{b}_0 \mathbf{P}\left(\mathbf{b} - \mathbf{b}_0, \Delta z\right) \mathbf{h}\left(\mathbf{b}_0, \Delta z\right) \Phi\left(\mathbf{b}_0, z_0\right) \tag{11.13b}$$

where h_{nm} is defined as the projected matrix elements of H_{nm}

$$h_{nm} = \int_{z_0}^{z} dz' \, H_{nm}(\mathbf{b}, z') \tag{11.14a}$$

for $(z - z_0) \to 0$,

$$h_{nm} \approx H_{nm}\left(\mathbf{b}, z'\right) \Delta z \tag{11.14b}$$

We now use the method of mathematical induction to prove that (11.13a) and (11.13b) is the solution of (11.12a) and (11.12b). First we consider the solution for $L = 1$. From (11.12a) and (11.12b),

$$\begin{aligned} Y_1^{(n)}(\mathbf{b}, z) = & \int d\mathbf{b}_0 \int d\mathbf{b}' \int_{z'=z_0}^{z'=z} dz' \, P_n(\mathbf{b} - \mathbf{b}', z - z') \\ & \times \sum_m H_{nm}(\mathbf{b}', z') \Phi_m(\mathbf{b}_0, z_0) P_m(\mathbf{b}' - \mathbf{b}_0, z' - z_0) \end{aligned} \tag{11.15}$$

By using the method of stationary phase to integrate the propagation function P, i.e., for a slow variation function V in the region $(z - z_0)$, the following approximation holds (Ishizuka and Uyeda 1977):

$$\int d\mathbf{b}' V\left(\mathbf{b}', z'\right) P_n\left(\mathbf{b} - \mathbf{b}', z - z'\right) P_n\left(\mathbf{b}' - \mathbf{b}_0, z' - z_0\right) \approx V\left(\mathbf{b}_0, z'\right) P_n\left(\mathbf{b} - \mathbf{b}_0, z - z_0\right) \quad (11.16)$$

Integrating over $\mathbf{b}'$ in (11.15) and assuming the slow variation of H_{nm} in the region $(z - z_0)$ as well as $K_1 \approx K_m$, (11.15) becomes

$$Y_1^{(n)}(\mathbf{b}, z) = \int d\mathbf{b}_0\, P_1(\mathbf{b} - \mathbf{b}_0, z - z_0) \sum_m h_{nm}(\mathbf{b}_0, \Delta z)\Phi_m(\mathbf{b}_0, z_0) \quad (11.17)$$

It is obvious that (11.17) is the second row of (11.13a) and (11.13b) for $L = 1$. If $Y_L^{(n)}$ satisfies (11.13a) and (11.13b), then $Y_{L+1}^{(n)}$ is obtained from (11.13a) and (11.13b) as

$$\begin{pmatrix} Y_{L+1}^{(0)}(\mathbf{b}, z) \\ Y_{L+1}^{(1)}(\mathbf{b}, z) \\ \vdots \\ Y_{L+1}^{(m)}(\mathbf{b}, z) \end{pmatrix} = \int d\mathbf{b}' \int_{z'=z_0}^{z'=z} dz' \times \begin{pmatrix} P_0(\mathbf{b} - \mathbf{b}', z - z') & 0 & \cdots & 0 \\ 0 & P_1(\mathbf{b} - \mathbf{b}', z - z') & \cdots & 0 \\ \vdots & \vdots & \ddots & \vdots \\ 0 & 0 & \cdots & P_m(\mathbf{b} - \mathbf{b}', z - z') \end{pmatrix}$$
$$\times \begin{pmatrix} H_{00}(\mathbf{b}', z') & H_{01}(\mathbf{b}', z') & \cdots & H_{0m}(\mathbf{b}', z') \\ H_{10}(\mathbf{b}', z') & H_{11}(\mathbf{b}', z') & \cdots & H_{1m}(\mathbf{b}', z') \\ \vdots & \vdots & \ddots & \vdots \\ H_{m0}(\mathbf{b}', z') & H_{m1}(\mathbf{b}', z') & \cdots & H_{mm}(\mathbf{b}', z') \end{pmatrix}$$
$$\times \frac{1}{L!} \int d\mathbf{b}_0 \begin{pmatrix} P_0(\mathbf{b}' - \mathbf{b}_0, z' - z_0) & 0 & \cdots & 0 \\ 0 & P_1(\mathbf{b}' - \mathbf{b}_0, z' - z_0) & \cdots & 0 \\ \vdots & \vdots & \ddots & \vdots \\ 0 & 0 & \cdots & P_m(\mathbf{b}' - \mathbf{b}_0, z' - z_0) \end{pmatrix}$$
$$\times \begin{pmatrix} h_{00}(\mathbf{b}_0, \Delta z') & h_{01}(\mathbf{b}_0, \Delta z') & \cdots & h_{0m}(\mathbf{b}_0, \Delta z') \\ h_{10}(\mathbf{b}_0, \Delta z') & h_{11}(\mathbf{b}_0, \Delta z') & \cdots & h_{1m}(\mathbf{b}_0, \Delta z') \\ \vdots & \vdots & \ddots & \vdots \\ h_{m0}(\mathbf{b}_0, \Delta z') & h_{m1}(\mathbf{b}_0, \Delta z') & \cdots & h_{mm}(\mathbf{b}_0, \Delta z') \end{pmatrix}^L$$
$$\times \begin{pmatrix} \Phi_0(\mathbf{b}_0, z_0) \\ \Phi_1(\mathbf{b}_0, z_0) \\ \vdots \\ \Phi_m(\mathbf{b}_0, z_0) \end{pmatrix} \quad (11.18a)$$

where $\Delta z' = z' - z_0$. Equation (11.18a) is rewritten in a shortened form as

$$\mathbf{Y}_{L+1}(\mathbf{b}, z) = \int d\mathbf{b}' \int_{z'=z_0}^{z'=z} dz' \mathbf{P}\left(\mathbf{b} - \mathbf{b}', z - z'\right) \mathbf{H}\left(\mathbf{b}', z'\right)$$
$$\times \frac{1}{L!} \int d\mathbf{b}_0 \mathbf{P}\left(\mathbf{b}' - \mathbf{b}_0, z' - z_0\right) \left[\mathbf{h}\left(\mathbf{b}_0, z' - z_0\right)\right]^L \mathbf{\Phi}\left(\mathbf{b}_0, z_0\right) \tag{11.18b}$$

Multiplying the first three matrices, integrating over $\mathbf{b}'$ and using (11.16), then (11.18b) becomes

$$\mathbf{Y}_{L+1}(\mathbf{b}, z) = \frac{1}{L!} \int d\mathbf{b}_0 \mathbf{P}\left(\mathbf{b} - \mathbf{b}_0, z - z_0\right) \int_{z'=z_0}^{z'=z} dz' \mathbf{H}\left(\mathbf{b}_0, z'\right) \left[\mathbf{h}\left(\mathbf{b}_0, z' - z_0\right)\right]^L \mathbf{\Phi}\left(\mathbf{b}_0, z_0\right) \tag{11.19}$$

Noting that $h_{nm} \approx H_{nm} \cdot \left(z' - z_0\right)$ for $\Delta z' \to 0$ and then integrating by parts in (11.19) for z', we obtain

$$\mathbf{Y}_{L+1}(\mathbf{b}, z) = \frac{1}{(L+1)!} \int d\mathbf{b}_0 \mathbf{P}\left(\mathbf{b} - \mathbf{b}_0, \Delta z\right) \left[\mathbf{h}\left(\mathbf{b}_0, \Delta z\right)\right]^{L+1} \mathbf{\Phi}\left(\mathbf{b}_0, z_0\right) \tag{11.20}$$

Equations (11.17) and (11.20) complete the proof of mathematical induction. Combining (11.11) and (11.13a) and (11.13b), the final solution is

$$\Phi(\mathbf{b}, z) = \int d\mathbf{b}_0 \mathbf{P}\left(\mathbf{b} - \mathbf{b}_0, \Delta z\right) \exp\left[-\frac{i}{\hbar v}\mathbf{h}\left(\mathbf{b}_0, \Delta z\right)\right] \Phi\left(\mathbf{b}_0, z_0\right)$$
$$= \mathbf{P}(\mathbf{b}, \Delta z) \otimes \left\{\exp\left[-\frac{i}{\hbar v}\mathbf{h}(\mathbf{b}, \Delta z)\right] \Phi\left(\mathbf{b}, z_0\right)\right\} \tag{11.21}$$

Equation (11.21) gives a general relationship between all elastic and inelastic waves before and after penetrating a crystal slice (Wang 1989a, 1990a). The exponential matrix indicates multiple scattering of electrons among all these states; the Lth-order expansion term of the exponential matrix describes the Lth-order multiple scattering. This equation is actually the multiple elastic and multiple inelastic electron scattering in the crystal, which is essentially a theory of Kikuchi patterns.

11.2 Conservation of Total Electrons

The total intensity of the electrons, elastically and inelastically scattered by a crystal as governed by (11.21), before and after penetrating a crystal slice, is conserved. To prove this conservation law, we first examine the following relations. It can be directly proved from the definition of $\mathbf{H}$:

$$\mathbf{h}^{\dagger}(\mathbf{b}, \Delta z) = \mathbf{h}(\mathbf{b}, \Delta z) \tag{11.22}$$

where † denotes a Hermitian conjugate (i.e., the complex conjugate plus matrix transpose). Also the propagation function obeys

$$\int d\mathbf{b}\mathbf{P}^{\dagger}(\mathbf{b} - \mathbf{b}_0, \Delta z)\, \mathbf{P}(\mathbf{b} - \mathbf{b}_1, \Delta z) = \delta(\mathbf{b}_1 - \mathbf{b}_0) \tag{11.23}$$

Then the total electron intensity after penetrating a crystal slice can be calculated according to (11.21)

$$\begin{aligned}\sum_{n=0} \int |\Phi_n(\mathbf{b}, z)|^2\, d\mathbf{b} &= \int d\mathbf{b}\Phi^{\dagger}(\mathbf{b}, z)\Phi(\mathbf{b}, z) \\ &= \int d\mathbf{b} \int d\mathbf{b}_1 \Phi^{\dagger}(\mathbf{b}_1, z_0) \exp\left[\frac{i}{\hbar v}\mathbf{h}^{\dagger}(\mathbf{b}_1, \Delta z)\right] \mathbf{P}^{\dagger}(\mathbf{b} - \mathbf{b}_1, \Delta z) \\ &\times \int d\mathbf{b}_0 \mathbf{P}(\mathbf{b} - \mathbf{b}_0, \Delta z) \exp\left[-\frac{i}{\hbar v}\mathbf{h}(\mathbf{b}_0, \Delta z)\right] \Phi(\mathbf{b}_0, z_0)\end{aligned} \tag{11.24}$$

Using (11.22) and (11.23), integrating first over $\mathbf{b}$, then integrating over $\mathbf{b}_1$, (11.24) becomes

$$\int d\mathbf{b}\Phi^{\dagger}(\mathbf{b}, z)\Phi(\mathbf{b}, z) = \int d\mathbf{b}_0 \boldsymbol{\Phi}^{\dagger}(\mathbf{b}, z_0)\, \Phi(\mathbf{b}_0, z_0) \tag{11.25a}$$

or

$$\sum_{n=0} \int d\mathbf{b}\, |\Phi_n(\mathbf{b}, z)|^2 = \sum_{n=0} \int d\mathbf{b}\, |\Phi_n(\mathbf{b}, z_0)|^2 \tag{11.25b}$$

Therefore the total intensity of the electrons before and after being scattered by a crystal slice is conserved (Wang 1990a).

11.3 First-Order Results

To obtain some simple results, we consider the first-order approximation of (11.20). Assuming

$$|H_{nm}| \ll |H_{nn}| \quad \frac{\sigma\, |H_{nm}|}{e} \ll 1 \quad n \neq m$$

with $\sigma/e = 1/\hbar v$, then the exponential term of the matrix can be written as a summation of a diagonal matrix and a nondiagonal matrix.

$$\exp\left(-i\frac{\sigma}{e}\mathbf{H}\right) = \exp\left[-i\frac{\sigma}{e}\left(\begin{pmatrix} h_{00} & 0 & \dots & 0 \\ 0 & h_{11} & \dots & 0 \\ \vdots & \vdots & \dots & \vdots \\ 0 & 0 & \dots & h_{mm} \end{pmatrix} + \begin{pmatrix} 0 & h_{01} & \dots & h_{0m} \\ h_{10} & 0 & \dots & h_{1m} \\ \vdots & \vdots & \dots & \vdots \\ h_{m0} & h_{m1} & \dots & 0 \end{pmatrix}\right)\right] \tag{11.26}$$

Using the approximation of $h_{00} \approx h_{11} \approx \cdots \approx h_{mm} = -e\,V(\mathbf{b})$, the first matrix is a unit matrix, and we readily have

$$\exp\left(-i\frac{\sigma}{e}\mathbf{h}\right) \approx \begin{bmatrix} \exp(i\sigma V) & -i\frac{\sigma}{e}h_{01} & \dots & -i\frac{\sigma}{e}h_{0m} \\ \vdots & \vdots & \dots & \vdots \\ -i\frac{\sigma}{e}h_{m0} & -i\frac{\sigma}{e}h_{m1} & \dots & \exp(i\sigma V) \end{bmatrix} \tag{11.27}$$

where the single-scattering approximation was made. Substituting (11.27) into (11.21) yields

$$\Phi_0(\mathbf{b}, z) = \left\{\exp(i\sigma V)\left[\Phi_0(\mathbf{b}, z_0) - i\frac{\sigma}{e}\sum_{n\neq 0} h_{0n}\Phi_n(\mathbf{b}, z_0)\right]\right\} \otimes P_0(\mathbf{b}, \Delta z) \tag{11.28a}$$

$$\Phi_m(\mathbf{b}, z) = \left\{\exp(i\sigma V)\left[\Phi_n(\mathbf{b}, z_0) - i\frac{\sigma}{e}\sum_{n\neq m} h_{mn}\Phi_n(\mathbf{b}, z_0)\right]\right\} \otimes P_m(\mathbf{b}, \Delta z) \tag{11.28b}$$

The physical meaning of (11.28a) and (11.28b) can be described as follows. For the elastic scattered wave Φ_0, the first term is the phase-grating result of the crystal slice, which is the elastic transmission of the incident elastic wave. Terms containing H_{0n} are transitions of electrons from excited states to the ground state. This is effectively the virtual inelastic-scattering process in electron diffraction, i.e., the electron loses energy first (inelastic scattering), then regains the same amount of energy to become elastic again. This process can be characterized by a complex correction potential related to the dielectric excitation property of the solid (Wang 1990b; see Chap. 6). For the excited states Φ_n, the first term is the elastic scattering of the incident inelastic wave in the slice; the h_{mn} terms are transitions from other states, including the ground state, to the nth excited state. This term can be considered the generation of the inelastic wave when the electron penetrates the slice. Also it is obvious that (11.28a) and (11.28b) reduces to the multislice formula of elastic scattering in (3.8) if all inelastic transitions vanish, i.e., $h_{mn} = 0$ for $n \neq m$.

If we know the elastic incident wave before the electron strikes the crystal entrance surface, then all generated elastic and inelastic waves after the first slice can be determined by (11.28a) and (11.28b). These waves can then be taken as incident waves for the second slice, and waves after penetrating the second slice can also be calculated from (11.28a) and (11.28b). So all waves after penetrating the crystal can be obtained. This is how the multislice theory works.

11.4 Special Cases of Only One Excited State

Let us consider the case of only one excited state. This case occurs in most plasmon and single-electron excitations. The crystal can be considered either in its excited state or its ground state. By taking $m = 1$ and using some algebra, (11.28a) and (11.28b) reduces to the following forms:

$$\Phi_0(\mathbf{b}, z) = \{\exp(i\sigma V)[\cos\left(\frac{\sigma|h_{10}|}{e}\right)\Phi_0(\mathbf{b}, z) - i\sin\left(\frac{\sigma|h_{10}|}{e}\right)\frac{h_{01}}{|h_{01}|}\Phi_1(\mathbf{b}, z_0)]\} \otimes P_0 \tag{11.29a}$$

and

$$\Phi_1(\mathbf{b}, z) = \{\exp(i\sigma V)[\cos(\sigma \mid h_{10} \mid /e)\,\Phi_1(\mathbf{b}, z) - i\sin(\mid h_{10} \mid /e)\frac{h_{10}}{|h_{10}|}\Phi_0(\mathbf{b}, z0)]\} \otimes P_1 \tag{11.29b}$$

The second term in (11.29a) is the result of the electron virtual inelastic-scattering effect. We now apply (11.29a) and (11.29b) for plasmon losses and single-electron excitations.

11.4.1 Valence-Loss Scattering

In (11.29a) and (11.29b), since

$$\cos(\sigma|h_{10}|/e) \approx 1 - \frac{(\sigma|h_{10}|/e)^2}{2} \approx \exp\left(\frac{(\sigma|h_{10}|/e)^2}{2}\right), \quad \text{for} \quad \sigma|h_{10}|/e \ll 1$$

then we can define an absorption function μ as given by (11.30a) and an electron inelastic mean free path length Λ as given by (11.30b); that is,

$$\mu(\mathbf{b}, z_0) = \frac{(\sigma\,|h_{10}|)^2}{\Delta z} \tag{11.30a}$$

and

$$\Lambda(\mathbf{b}, z_0) = \frac{1}{\mu(\mathbf{b}, z_0)} \tag{11.30b}$$

Then (11.29b) can be written as

$$\Phi_1(\mathbf{b}, z) = \left\{ \exp(i\sigma V) \left[\exp\left(-\frac{\mu \Delta z}{2}\right) \Phi_1(\mathbf{b}, z_0) - i\frac{\sigma}{e} h_{10} \Phi_0(\mathbf{b}, z_0) \right] \right\} \otimes P_1 \tag{11.31}$$

In (11.31), the term $\exp(-\mu\Delta z/2)$ describes the absorption effect from inelastic scattering. For plasmon excitations, h_{01} can be taken as a product of a complex constant with a real function. Then after properly choosing the constant phase factor of the wave function Φ_0, (11.29a) and (11.31) can be written in the form:

$$\Phi_0(\mathbf{b}, z) = \left\{ Q(\mathbf{b}) \left[\exp\left(-\frac{\mu \Delta z}{2}\right) \Phi_0(\mathbf{b}, z_0) \right] \right\} \otimes P_0 \tag{11.32a}$$

and

$$\Phi_1(\mathbf{b}, z) = \left\{ Q(\mathbf{b}) \left[\exp\left(-\frac{\mu \Delta z}{2}\right) \Phi_1(\mathbf{b}, z_0) + \left(\frac{\Delta z}{\Lambda}\right)^{1/2} \Phi_0(\mathbf{b}, z_0) \right] \right\} \otimes P_1 \tag{11.32b}$$

where $Q(\mathbf{b})$ is the phase-grating function defined in (3.3). Equation (11.32a) and (11.32b) is the exact form of the proposed theory, based on the physical optics approach for calculating the energy-filtered inelastic images and diffraction patterns (Cowley and Pogany 1968; Doyle 1971; Wang 1989b). The term $\Delta z/\Lambda$ is the probability of generating the plasmon loss while the electron travels through the slice. This means that the electron energy loss is a mean process depending on its average traveling distance.

11.4.2 Thermal Diffuse Scattering

Thermal diffuse scattering involves many excited states with different momentum transfers, but the transition probability of each is very small. In this case, we can neglect inelastic transitions among all these excited states and also consider the fact that TDS produces large amounts of momentum transfer but almost no energy loss. Then multislice solutions of the electron wave functions are

$$\Phi_0(\mathbf{b}, z) = \left\{ Q(\mathbf{b}) \left[\Phi_0(\mathbf{b}, z_0) - i\frac{\sigma}{e} \sum_{n\neq 0} h_{0n} \Phi_n(\mathbf{b}, z_0) \right] \right\} \otimes P_0 \tag{11.33a}$$

and

$$\Phi_n(\mathbf{b}, z) = \left\{Q(\mathbf{b})\left[\Phi_n(\mathbf{b}, z_0) - i\frac{\sigma}{e}h_{n0}\Phi_0(\mathbf{b}, z_0)\right]\right\} \otimes P_0 \tag{11.33b}$$

In electron diffraction patterns, intensities of diffusely phonon-scattered electrons with different momentum transfers are added incoherently in the diffraction plane. The nth excited state in TDS indicates a process of definite momentum transfer, which is a point in the first Brillouin zone. Since there is also no overlay among these points, electrons of different momentum transfers in TDS can be added coherently in real space because there is no interference between them after reconverting to reciprocal space. By defining a total TDS wave Φ_{TDS} as a coherent sum of all TDS waves of different momentum transfers (Wang 1989a),

$$\Phi_{TDS} = \sum_{m=1} \Phi_m \tag{11.34}$$

where m is summed over all phonon modes (i.e., ω_j and $\mathbf{q}$), from (11.33b), we have

$$\Phi_{TDS}(\mathbf{b}, z) = \{Q(\mathbf{b})\,[\Phi_{TDS}(\mathbf{b}, z_0) + i\Gamma(\mathbf{b}, z)\Phi_0(\mathbf{b}, z_0)]\} \otimes P_0 \tag{11.35a}$$

where the TDS generation function is defined by

$$\Gamma(\mathbf{b}, z) = -\frac{\sigma}{e}\sum_{m=1} h_{m0}(\mathbf{b}) \tag{11.35b}$$

In (11.35a), the first term is the elastic scattering of incident TDS electrons generated from previous slices; the second term is the newly generated TDS electron from the diffuse scattering of the slice. We now determine the transition matrix element H_{m0} for the case of TDS. From (6.68), the interaction Hamiltonian for creating a single phonon is

$$\begin{aligned} &H_{n0}(\mathbf{r}, \mathbf{q}, \omega_j) \\ &= e\sum_{\alpha}\sum_{n} A_\alpha(\omega_j(\mathbf{q}))\left[\mathbf{e}(\alpha|_j^q)\cdot\nabla\left(\int d\mathbf{u}\, f_\alpha^e(\mathbf{u})\exp\{2\pi i\mathbf{u}\cdot[\mathbf{r}-\mathbf{R}(n)-\mathbf{r}(\alpha)]\}\right)\right] \\ &\times exp\{2\pi i\mathbf{q}\cdot[\mathbf{R}(n)+\mathbf{r}(\alpha)]\} \\ &= e\sum_{\alpha}\sum_{n} A_\alpha(\omega_j(\mathbf{q}))\left[\mathbf{e}(\alpha|_j^q)\cdot\nabla V_\alpha(\mathbf{r}-\mathbf{R}(n)-\mathbf{r}(\alpha))\right] \\ &\times \exp\{2\pi i\mathbf{q}\cdot[\mathbf{R}(n)+\mathbf{r}(\alpha)]\} \end{aligned} \tag{11.36}$$

The sum over m actually adds up contributions from different phonon modes, so that

$$\begin{aligned}\Gamma(\mathbf{b}, z) &= -\frac{\sigma}{e}\Delta z \sum_{m=1} H_{m0} \\ &= -\frac{\sigma}{e}\Delta z \sum_j \sum_q H_{m0}\left(\mathbf{r}, \mathbf{q}, \omega_j\right) \\ &= -\sigma\Delta z \sum_j \sum_q \sum_\alpha \sum_n A_\alpha\left(\omega_j(\mathbf{q})\right)\left[\mathbf{e}\left(\alpha\big|_j^{\mathbf{q}}\right) \cdot \nabla V_\alpha(\mathbf{r} - \mathbf{r}(n) - \mathbf{r}(\alpha))\right] \\ &\quad \exp\{2\pi i\mathbf{q} \cdot [\mathbf{R}(n) + \mathbf{r}(\alpha)]\}\end{aligned} \tag{11.37}$$

We now discuss applications of (11.35a) and (11.37) for calculating images of TDS electrons.

Equation (11.37) was derived from the single-phonon scattering approximation. In practice, it is possible to calculate quasi-elastic diffraction patterns of crystal lattices with different thermal vibration configurations using the multislice method. The final intensity is the average of the scattering intensity from all possible lattice configurations. This calculation has been performed by Loane et al. (1991), and a good agreement with experimental results has been obtained. Figure 11.2 shows the fit of the observed low-angle (Fig. 11.2a) and high-angle (Fig. 11.2b) coherent CBED patterns with the calculated patterns of a silicon crystal with (Fig. 11.2d) and without (Fig. 11.2c) introducing the time-dependent lattice vibration. The pattern shown in Fig. 11.2c was calculated considering only the equilibrium lattice configuration. The pattern shown in Fig. 11.2d was calculated from the frozen lattice model for a total of 64 lattice vibration configurations. It is apparent that high-angle diffuse scattering is dramatically increased in Fig. 11.2d, which shows the best fit to the experimentally observed pattern in Fig. 11.2b. Intensity distributions within Bragg diffraction disks are best fit by the frozen lattice calculation.

The agreement between observed and computed CBED patterns in Fig. 11.2 shows two points. First high-angle diffuse scattering is primarily generated by TDS. Second the frozen lattice model introduced in Chap. 7 accurately describes scattering behavior of crystal lattices due to thermal vibrations.

11.5 Imaging with TDS Electrons in STEM

11.5.1 Image Formation

Recent instrumental progress made it possible to obtain atomic resolution lattice images using high-angle scattered electrons in STEM when a small probe, less than the size of the unit cell, is used (Pennycook and Boatner 1988; Pennycook and Jesson 1990). This type of images is called the HAADF-STEM image (Liu and Cowley 1991; Xu et al. 1990) or Z-contrast (Pennycook and Jesson 1990). The image contrast is of particular interest because of its strong dependence on atomic number

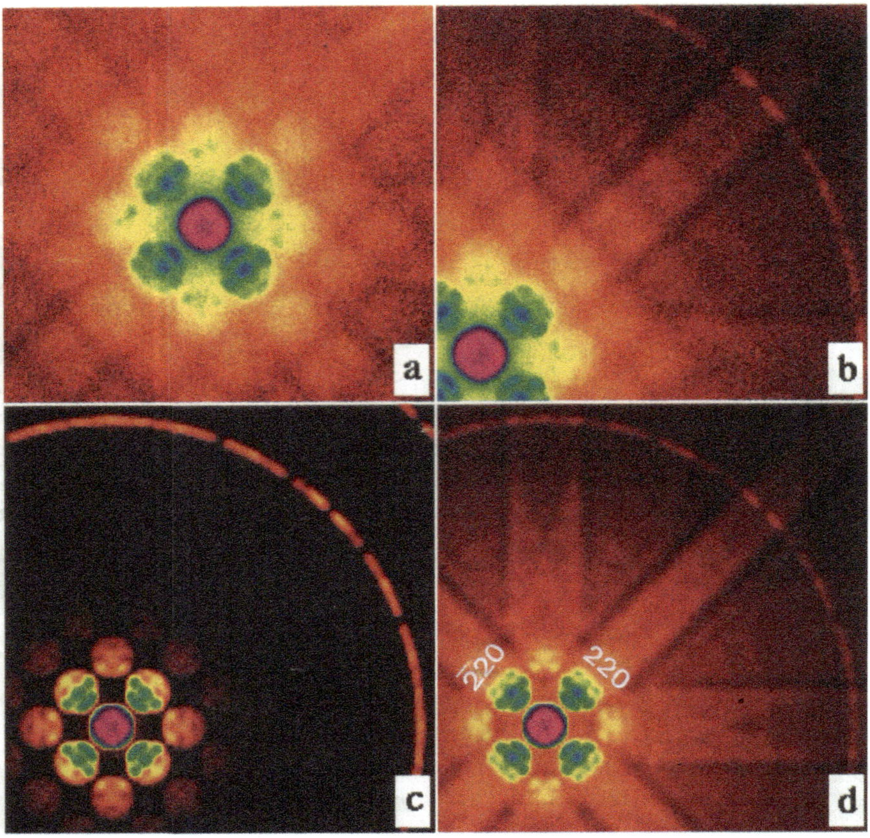

Fig. 11.2 Comparison of experimental and calculated silicon [100] coherent CBED patterns. **a** low- and **b** high-angle experimental patterns recorded at 100 kV using a VG HB501 STEM. Pattern **c** was calculated with the standard multislice algorithm, and pattern **d** was calculated with the frozen lattice multislice method. Specimen thickness was 30 nm. The frozen lattice calculation is the ensemble average of 64 lattice vibration configurations. The mean vibration amplitude of Si was 0.085Å. Courtesy of Dr. Loane et al. (1991)

(for perfect crystals) or equivalently projected mass thickness, so that each image provides a two-dimensional chemical map of the specimen at atomic resolution. Also the image contrast does not show reversal with the change of either specimen thickness or defocus, making it possible to identify atom columns directly from the image. Therefore, a HAADF-STEM image provides structural information not only on atomic arrangement but also on local chemistry. This technique is becoming a powerful tool for characterizing atomic structures at interfaces.

The image is formed by collecting high-angle scattered electrons using a ring-shaped detector in the diffraction plane when the small electron probe is scanned across the crystal lattice (Fig. 11.3). The electron detector is defined by inner (u_1) and outer (u_2) angular cutoffs, and it is called an annular dark-field (ADF) detector. The

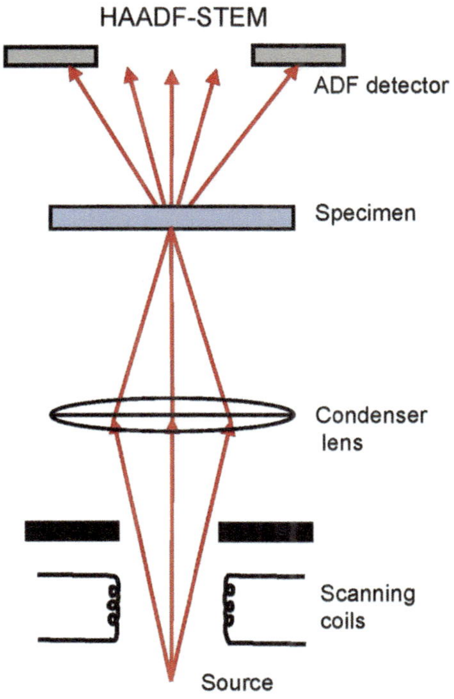

Fig. 11.3 A ray diagram showing the formation of atomic-number-sensitive images using high-angle TDS electrons in STEM

inner cutoff is chosen large enough to avoid the diffraction and phase contrast of Bragg reflected beams with consideration for the signal-to-noise ratio. The electron probe is scanned over the specimen with deflection coils. In any subsequent deflection plane, a nanoprobe CBED pattern of the specimen region illuminated by the beam is formed for each scanning position. The signal detected by the ADF detector is displayed as the HAADF-STEM image on a cathode ray tube with a scan synchronized with that of the incident beam on the specimen.

Figure 11.4a shows a HAADF-STEM image of a GaAs/In_{1-x} $Al_xAs/In_{1-y}Ga_yAs$ multiquantum-well sample viewed along [110], where $x = y = 0.75$. The image has three characteristics. First atomic resolution is achieved, making it possible to image the atomic structure of the sample. Second the image contrast is sensitive to the local projected mass thickness; thus the image contrast is sensitive to the variation in chemical composition (i.e., Z-contrast). Finally bright spots are projected strings of dumbbells, which can easily be interpreted, so that the dark-atom phenomenon observed in conventional HRTEM is absent. These latter two features represent unique advantages of HAADF-STEM. Figure 11.4a reveals that the $In_{0.25}Al_{0.75}As/In_{0.25}Ga_{0.75}As$ interface is not symmetric on both sides; this can be seen more clearly in Fig. 11.4b. The InGaAs/InAlAs interface is chemically sharper than the InAlAs/InGaAs interface. Systematic studies of the interfacial structure could provide important information regarding the growth of the film.

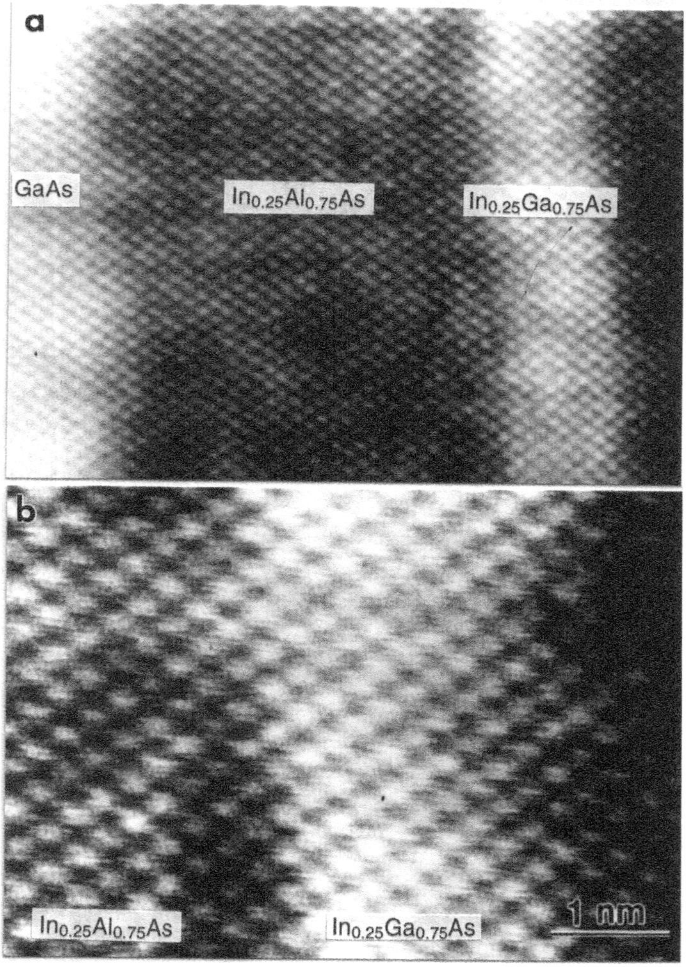

Fig. 11.4 The HAADF-STEM images of GaAs/$In_{0.25}Al_{0.75}As$/$In_{0.25}Ga_{0.75}As$ multiquantum-well structures showing the direct correlation between image contrast and local chemical composition. The image was recorded at 100 kV in a VG HB5 STEM (C_c = 0.88 mm). The inner and outer cutoff angles of the ADF detector are 70 and 135 mrad, respectively. Courtesy of Drs. Liu and Cowley (1991)

For perfect crystals without defects or grain boundaries, there are three possible sources for generating high-angle scattered electrons. The first one is large-angle Rutherford scattering, which is an elastic-scattering process particularly important in scattering of a single atom or a small atom cluster (Crewe 1978–79). The image contrast produced purely by Rutherford scattering is simply proportional to Z_0^2. The HOLZ reflections (Wang and Cowley 1989; Spence et al. 1989), which are due to purely elastic-scattering processes and have all the characteristics of phase contrast imaging, are the second source. The contrast produced by this process usually does not have a simple relation with atomic number. The contribution of HOLZ reflections becomes significant for very thin specimens because of the elongated reflection rods in reciprocal space due to the crystal shape factor. Thus the Bragg condition could be satisfied simultaneously by all first-order Laue zone rods. Finally TDS is the third source for generating high-angle scattered electrons (Wang and Cowley 1989, 1990; Hall and Hirsch 1965). For high-angle scattering, the contrast produced by TDS is related to Z_0, but it is not necessarily proportional to Z_0^2 because of dynamic scattering and possible beam-broadening effects (Wang and Cowley 1990; Amali and Rez 1992; Hillyard et al. 1993). Theoretical calculations and experimental measurements have found that the contrast of HAADF-STEM images is dominated by TDS (Wang and Cowley 1990; Loane et al. 1991). Therefore the contrast of HAADF-STEM images of a perfect crystal is directly related to differences in the atomic number or projected specimen mass thickness.

For crystals containing defects, dislocations, or grain boundaries, however, the HAADF-STEM image contrast showed some anomalies in a way that the defected area shows brighter contrast. This is because diffuse scattering produced from local regions due to their deviation from a three-dimensional periodic structure. This type of scattering may destroy the Z_0 dependence of the HAADF-STEM image contrast. In such a case, the contribution of Huang scattering is greater than that of TDS; thus the contrast mechanism is called Huang-scattering contrast (Wang 1994). Further discussion is given in Sect. 11.8.

Diffuse scattering at high angles can also be generated by electron Compton scattering (Eaglesham and Berger 1994; Bleloch et al. 1994; see Sect. 6.1.4). This scattering is a less localized scattering process in comparison to TDS, thus the $e - e$ scattered electrons mainly contribute to the background in the image. High-resolution, structural-related information is produced by localized TDS and Huang scattering. Therefore, our analysis is focused on the images formed by TDS and Huang-scattered electrons.

We now use dynamic theory to describe the contribution of each component in HAADF-STEM imaging.

11.5.2 Contribution of Bragg-Reflected Electrons

In STEM image simulation, the incident wave probe, centered at a point $\mathbf{b}_p = (x_p, y_p)$ on the specimen surface, is defined by the condenser lens aperture; that is:

$$\Phi_0\left(\mathbf{b}, \mathbf{b}_p, 0\right) = \int d\mathbf{u} A_{\text{con}}(\mathbf{u}) \exp\left[2\pi i \mathbf{u} \cdot \left(\mathbf{b} - \mathbf{b}_p\right) + i T_{\text{obj}}(\mathbf{u})\right] \quad (11.38)$$

where A_{con} is the shape function of the condenser lens. Elastic scattering of the probe by each slice is governed by (3.8). Scattering of the probe is determined by specimen properties.

At the exit face of the crystal, the probe function is $\Phi_0\left(\mathbf{b}-\mathbf{b}_p, d\right)$ after interacting with atoms in the crystal. Components scattered to the angular range of the detector are collected for imaging. Therefore in the diffraction plane, the contribution from purely elastically scattered electrons to the ADF image can be expressed as

$$I_{\text{Elas}}\left(\mathbf{b}_p\right)=\int d\mathbf{u}\left|\mathbf{FT}\left[\Phi_0\left(\mathbf{b}-\mathbf{b}_p, d\right)\right]\right|^2 D(\mathbf{u}) \tag{11.39}$$

This component includes contributions of high-angle Bragg reflections and Rutherford scattering.

11.5.3 Contribution of TDS Electrons

The HAADF-STEM images are formed primarily by diffusely scattered electrons to high angles. Equation (11.35a) and (11.35b) is the full dynamic theory for TDS if inelastic waves generated from each slice are added incoherently in the image or diffraction plane. This means that each individual TDS wave created from a slice has to be considered as a separated stream that is elastically rescattered through the remaining crystal thickness. In practice, this process requires a large number of calculations. Thus it is necessary to seek simplified theories that provide reasonably accurate results without too much calculation.

To start out, we assume that atoms are vibrating in a random phase. A more sophisticated theory considering the phase coupling of atomic vibrations is given in Sect. 11.5.5.

11.5.3.1 Approximation 1: No Elastic Rescattering After TDS

The calculation can be simplified if the electron elastic scattering after inelastic interaction is neglected. This is a good approximation because electrons diffusely scattered to high angles can be considered kinematically scattered, especially for thin specimens. In this case, the electron angular redistribution due to elastic scattering after TDS is limited to the ADF detector range; thus the intensity detected by the ADF detector is unaffected. This is a reasonable approximation when the detected signal is given by integrating intensities over a wide-angle annular detector. Then intensity detected by the ADF detector for a probe centered at $\mathbf{b}_p$ can be written approximately as an incoherent superposition of TDS waves from each slice

$$I\left(\mathbf{b}_p\right)=\sum_{n_c}\int d\mathbf{u}\left|\Delta\Phi\left(\mathbf{u}, \mathbf{b}_p, z_{n_c}\right)\right|^2 D(\mathbf{u}) \tag{11.40}$$

where $\Delta\Phi$ is the Fourier transform of TDS waves generated from scattering of the n_cth crystal slice, and it is expressed as

$$\Delta\Phi\left(\mathbf{u}, \mathbf{b}_p, z_{n_c}\right) = \mathbf{FT}\left\{iQ\left(\mathbf{b}, z_{n_c}\right)\left[\Gamma\left(\mathbf{b}, z_{n_c}\right)\Phi_0\left(\mathbf{b}, z_{n_c}\right)\right]\right\} P_0(\mathbf{u}) \tag{11.41}$$

and $D(\mathbf{u})$ is the detection function of the ADF detector

$$D(\mathbf{u}) = \begin{cases} 1 \text{ for } u_1 < u < u_2 \\ 0 \text{ otherwise} \end{cases} \tag{11.42}$$

Substituting (11.41) into (11.40) and using the Fourier transforms of Γ and Φ_0 as well as $|P_0(\mathbf{u})| = 1$, intensity detected by the ADF detector when the probe is scanned to $\mathbf{b}_p$ is

$$\begin{aligned} I\left(\mathbf{b}_p\right) =& \sum_{n_c} \int d\mathbf{b}\mathrm{Q}^*\left(\mathbf{b}, z_{n_c}\right)\Gamma^*\left(\mathbf{b}, z_{n_c}\right)\Phi_0^*\left(\mathbf{b} - \mathbf{b}_p, z_{n_c}\right) \\ & \times \int d\mathbf{b}'\mathbf{Q}\left(\mathbf{b}', z_{n_c}\right)\Gamma\left(\mathbf{b}', z_{n_c}\right)\Phi_0\left(\mathbf{b}' - \mathbf{b}_p, z_{n_c}\right) D\left(\mathbf{b} - \mathbf{b}'\right) \end{aligned} \tag{11.43}$$

In practice, the ADF detector is large enough so that elastic rescattering of electrons is still within the detector angular range. By assuming that all TDS electrons are collected by the detector or that the signal collected by the detector is proportional to total TDS scattering, the detector function becomes $D\left(\mathbf{b} - \mathbf{b}'\right) = \delta\left(\mathbf{b} - \mathbf{b}'\right)$; then from (11.43) (Wang and Cowley 1990)

$$I\left(\mathbf{b}_p\right) = \sum_{n_c} \int d\mathbf{b} \left|\Gamma\left(\mathbf{b}, z_{n_c}\right)\right|^2 \left|\Phi_0\left(\mathbf{b} - \mathbf{b}_p, z_{n_c}\right)\right|^2 = \sum_{n_c} \left|\Gamma\left(\mathbf{b}_p, z_{n_c}\right)\right|^2 \otimes \left|\Phi_0\left(\mathbf{b}_p, z_{n_c}\right)\right|^2 \tag{11.44}$$

This result is exact if the ADF detector detects all and only inelastic scattered electrons even for thick crystals. The $|\Gamma|^2$ is the effective scattering cross section of the inelastic process, and $|\Phi_0|^2$ is the distribution of electron probe current at depth z.

For very thin objects, (11.44) can be connected with incoherent imaging theory in optics. The $|\Phi_0|^2$ can then be equated to the modulus square of the impulse response of the lens (i.e., the intensity distribution of the image of a point object), and Γ is the transmission function of the object. We now consider the expression of Γ^2:

$$\begin{aligned} \Gamma^2(\mathbf{b}, z) =& \sigma^2\Delta z^2 \left| \sum_j \sum_q \sum_\alpha \sum_n A_\alpha\left(\omega_j(\mathbf{q})\right)\left[\mathbf{e}\left(\alpha|_j^{\mathbf{q}}\right) \cdot \nabla V_\alpha(\mathbf{r} - \mathbf{R}(n) - \mathbf{r}(\alpha))\right] \right. \\ & \left. \times \exp\{2\pi i\mathbf{q} \cdot [\mathbf{R}(n) + \mathbf{r}(\alpha)]\}\right|^2 \end{aligned} \tag{11.45}$$

Since the atomic potential distribution is a sharply peaked function in comparison to interatomic distances, the overlap effect of the atomic potential is effectively very small; thus (11.45) can be approximated as

$$\Gamma^2(\mathbf{b}, z) \approx \sigma^2 \Delta z^2 \sum_\alpha \sum_n | \sum_j \sum_q A_\alpha (\omega_j(\mathbf{q})) \left[\mathbf{e}\left(\alpha|_j^{\mathbf{q}}\right) \cdot \nabla V_\alpha (\mathbf{r} - \mathbf{R}_n - \mathbf{r}(\alpha)) \right]$$
$$\times \exp\{2\pi i \mathbf{q} \cdot [\mathbf{R}(n) + \mathbf{r}(\alpha)]\}|^2 \tag{11.46}$$

As shown in Sect. 11.7, the phase correlation of atomic vibrations does not affect image calculation. Then we can use the Einstein model in which the phonon frequency ω_j is independent of its wave vector $\mathbf{q}$ and the polarization vector e may also be independent of $\mathbf{q}$; thus it can be taken as the $x-$, $y-$, and z-axes. Equation (11.46) is simplified as

$$\begin{aligned}\Gamma^2(\mathbf{b}, z) &\approx \sigma^2 \Delta z^2 \sum_\alpha \sum_n [A_\alpha (\omega_0)]^2 |\nabla V_\alpha (\mathbf{r} - \mathbf{R}_n - \mathbf{r}(\alpha))|^2 \\ &\quad \times \sum_q \sum_{q'} \exp\left\{2\pi i \left(\mathbf{q} - \mathbf{q}'\right) \cdot [\mathbf{R}_n + \mathbf{r}(\alpha)]\right\} \\ &= \sigma^2 \Delta z^2 \sum_\alpha \sum_n a_\alpha^2 |\nabla V_\alpha (\mathbf{r} - \mathbf{R}_n - \mathbf{r}(\alpha))|^2 \end{aligned} \tag{11.47}$$

where $a_\alpha^2 = N_0 [A_\alpha (\omega_0)]^2$ is the averaged mean square atom vibration amplitude in the Einstein model. Equation (11.47) is the result of the single-phonon scattering model. It is important to note that the TDS generation function is the first-order expansion of the crystal potential via small displacement of atom positions:

$$\begin{aligned}\Gamma^2(\mathbf{b}, z) &= \sigma^2 \Delta z^2 \sum_\alpha \sum_n [< |V_\alpha(\mathbf{r} - \mathbf{R}_n - \mathbf{r}(\alpha) - \mathbf{u}_\alpha) \\ &\quad - < V_\alpha(\mathbf{r} - \mathbf{R}_n - \mathbf{r}(\alpha) - \mathbf{u}_\alpha) >|^2 >] \\ &= \sigma^2 \Delta z^2 \sum_\alpha \sum_n [< |V_\alpha(\mathbf{r} - \mathbf{R}_n - \mathbf{r}(\alpha) - \mathbf{u}_\alpha)|^2 > \\ &\quad - |< V_\alpha(\mathbf{r} - \mathbf{R}_n - \mathbf{r}(\alpha) - \mathbf{u}_\alpha) >|^2] \end{aligned} \tag{11.48}$$

This important extrapolation is used in Sect. 11.5.3 to discuss the contribution of multiphonon scattering in HAADF-STEM image (Konnert and D'Antonio 1991).

11.5.3.2 Approximation 2: A Modified Generation Function Method

To make a better approximation by taking into account the size of the detector, we try to introduce a modified generation function, which is defined by (11.49). In reciprocal space (Wang and Cowley 1990):

$$\Gamma_{\text{new}}(\mathbf{u}, z) = \begin{cases} \Gamma(\mathbf{u}, z) \text{ for } u_1 < u < u_2 \\ 0 \text{ otherwise} \end{cases} \tag{11.49}$$

This newly defined inelastic generation function drops Fourier components out of the angular range of the ADF detector in reciprocal space and picks up remaining components. Then the detected intensity can be approximately written as

$$I\left(\mathbf{b}_p\right) \approx \sum_{n_c} \int d\mathbf{u} \left|\left[Q\left(\mathbf{u}, z_{n_c}\right) \otimes \Gamma_{\text{new}}\left(\mathbf{u}, z_{n_c}\right) \otimes \Phi_0\left(\mathbf{u}, z_{n_c}, \mathbf{b}_p\right)\right] P(\mathbf{u})\right|^2 \tag{11.50}$$

In comparison with (11.43), the detection function D is dropped in (11.50); it is contained in the new generation function (11.49). Following an analysis similar to that in the last section, we have

$$\begin{aligned} I\left(\mathbf{b}_p\right) &= \sum_{n_c} \int d\mathbf{b} \left|\Gamma_{\text{new}}\left(\mathbf{b}, z_{n_c}\right)\right|^2 \left|\Phi_0\left(\mathbf{b} - \mathbf{b}_p, z_{n_c}\right)\right|^2 \\ &= \sum_{n_c} \left|\Gamma_{\text{new}}\left(\mathbf{b}_p, z_{n_c}\right)\right|^2 \otimes \left|\Phi_0\left(\mathbf{b}_p, z_{n_c}\right)\right|^2 \end{aligned} \tag{11.51}$$

Equation (11.51) may significantly simplify the numerical calculation. Equation (11.51) still holds if inelastically scattered electrons are elastically scattered to produce Kikuchi patterns within the detector angular range. Calculations based on (11.44) or (11.51) fully incorporate the dynamic diffraction effect of the electron probe. If we neglect the scattering (or broadening) effect of the probe, so that $\Phi_0\left(\mathbf{b} - \mathbf{b}_p, z\right) = \Phi_0\left(\mathbf{b} - \mathbf{b}_p, 0\right)$, (11.51) becomes a simplified theory proposed by Pennycook and Jesson (1990).

It is important to note that the HAADF-STEM image is a convolution of the probe intensity distribution with the square of the inelastic generation function, which is proportional to the scattering cross section of the inelastic process. The contrast of the HAADF-STEM image depends on specimen thickness (Hillyard et al. 1993) but not so much as does the coherent image.

11.5.4 Effects of Multiphonon and Multiple Phonon Scattering

We showed in Sect. 7.5 that high-angle intensity contributed by multiphonon scattering is comparable with that from single-phonon scattering. In the generation function given by (11.47), however, the contribution of multiphonon and multiple phonon scattering was not considered. We now discuss the contribution of high-order terms. First we examine the phase-grating function of a slice with consideration of the perturbation effect of atomic thermal vibrations:

$$\begin{aligned} Q &= \exp(i\sigma V) = \exp\left[i\sigma\left(V_0 + \Delta V\right)\right] \\ &= \exp\left(i\sigma V_0\right) + \exp\left(i\sigma V_0\right)\left[\exp(i\sigma \Delta V) - 1\right] \end{aligned} \tag{11.52}$$

where the first term is the stationary phase grating of the crystal lattice. The second term, which depends on the instantaneous configuration of the vibrating lattice, is responsible for generating the TDS, and it is a product of the first term with a function equivalent to the TDS generation function defined in (11.37). Thus under the phase object approximation (POA), the TDS generation function is

$$\langle|\Gamma|^2\rangle = \langle|\exp(i\sigma\Delta V) - 1|^2\rangle = 2\left\{1 - \exp\left[-\frac{\sigma^2}{2}\langle(\Delta V)^2\rangle\right]\right\} \tag{11.53}$$

where the time average can easily be performed in reciprocal space. For random-phase vibration, the result is

$$\begin{aligned} < [\Delta V(\mathbf{b}, z, t)]^2 > \approx \sum_{\kappa} \int d\mathbf{u} \int d\tau\, f_{\kappa}^{e}(\mathbf{u}) f_{\kappa}^{e*}(\tau) \{\exp\left[-W_{\kappa}(\mathbf{u} - \tau)\right] \\ - \exp\left[-W_{\kappa}(\mathbf{u}) - W_{\kappa}(\tau)\right]\} \exp\left[2\pi i(\mathbf{u} - \tau) \cdot (\mathbf{r} - \mathbf{r}_{\kappa})\right] \end{aligned} \tag{11.54}$$

The multiphonon and multiple phonon scatterings are included in (11.53). The first-order expansion of (11.54) is just (11.47) (Konnert and D'Antonio 1991). The effect of multiple phonon scattering, i.e., the plural scattering of electrons by phonons, in HAADF-STEM images can also be considered using other theories (Wang 1991; see Sect. 14.4.4). It may be possible that multiphonon diffuse scattering is even more localized than single-phonon diffuse scattering.

11.5.5 Effects of Coherent TDS

Atomic vibrations in crystals are always defined by certain modes, which are called phonons. Diffuse scattering from atoms that are vibrating with a certain phase correlation may retain partial coherence, thus affecting the final image calculation. Based on the frozen lattice model, the elastic electron wave before and after being scattered by a crystal slice of thickness Δz at depth z is correlated by

$$\Phi_0(\mathbf{b}, z + \Delta z) = \left\{\exp\left[i\sigma' V(\mathbf{b})\right] \Phi_0(\mathbf{b}, z)\right\} \otimes P(\mathbf{b}) \tag{11.55}$$

where $\sigma' = (K_0/K_{0z})\,(\pi/\lambda U_0)$. Since the projected potential of the crystal slice contains the time-dependent perturbation of atomic thermal vibration, it can be written as $V = V_0 + \Delta V$. Thus (11.55) is rewritten as

$$\Phi_0(\mathbf{b}, z + \Delta z) = (Q(\mathbf{b}, z)\Phi_0(\mathbf{b}, z) + Q(\mathbf{b}, z)\left\{\exp\left(i\sigma'\left[V(\mathbf{b}) - V_0(\mathbf{b})\right]\right) - 1\right\}\Phi_0(\mathbf{b}, z)) \otimes P(\mathbf{b}) \tag{11.56}$$

where the first term is Bragg scattering, and the second term is the TDS wave generated from the slice

$$\Delta\Phi(\mathbf{b}, z) = iQ(\mathbf{b}, z)\Gamma(\mathbf{b}, z)\Phi_0(\mathbf{b}, z) \tag{11.57a}$$

where

$$\Gamma(\mathbf{b}, z) = -i\left(\exp\left\{i\sigma'\left[V(\mathbf{b}) - V_0(\mathbf{b})\right]\right\} - 1\right) \tag{11.57b}$$

If the subsequent diffraction of the diffusely scattered electrons is ignored, the intensity detected by the ADF detector is written as

$$\begin{aligned} I_{\text{TDS}}(\mathbf{b}_p) &= \int d\mathbf{u} < \left|\sum_{n_c} \Delta\Phi\left(\mathbf{u}, z_{n_c}\right)\right|^2 > D(\mathbf{u}) \\ &= \int d\mathbf{u} < \left|\sum_{n_c} Q\left(\mathbf{u}, z_{n_c}\right) \otimes \Gamma\left(\mathbf{u}, z_{n_c}\right) \otimes \Phi_0\left(\mathbf{u}, z_{n_c}\right)\right|^2 > D(\mathbf{u}) \\ &= \sum_{n_c}\sum_{m_c}\int d\mathbf{u} < \left[Q\left(\mathbf{u}, z_{n_c}\right) \otimes \Gamma\left(\mathbf{u}, z_{n_c}\right) \otimes \Phi_0\left(\mathbf{u}, z_{n_c}\right)\right] \\ &\quad \times \left[Q^*\left(\mathbf{u}, z_{m_c}\right) \otimes \Gamma^*\left(\mathbf{u}, z_{m_c}\right) \otimes \Phi_0^*\left(\mathbf{u}, z_{m_c}\right)\right] > D(\mathbf{u}) \end{aligned} \tag{11.58}$$

Using the Fourier transforms of Γ and Φ_0, the intensity detected by the ADF detector is

$$\begin{aligned} I_{\text{TDS}}(\mathbf{b}_p) = \sum_{n_c}\sum_{m_c}\int d\mathbf{b}\int d\mathbf{b}' < \Gamma\left(\mathbf{b}', z_{n_c}\right)\Gamma^*\left(\mathbf{b}, z_{m_c}\right) > Q\left(\mathbf{b}', z_{n_c}\right) Q^*\left(\mathbf{b}, z_{m_c}\right) \\ \times \Phi_0\left(\mathbf{b}' - \mathbf{b}_p, z_{n_c}\right) \Phi_0^*\left(\mathbf{b} - \mathbf{b}_p, z_{m_c}\right) D\left(\mathbf{b} - \mathbf{b}'\right) \end{aligned} \tag{11.59}$$

Equation (11.59) includes contributions from multiphonon and multiple phonon scattering. In general, the time average can be evaluated before any numerical calculation

$$\begin{aligned} < \Gamma(\mathbf{b}', z')\Gamma^*(\mathbf{b}, z) > &= \left\langle\left\{\exp\left[i\sigma'\Delta V(\mathbf{b}', z')\right] - 1\right\}\left\{\exp\left[-i\sigma'\Delta V(\mathbf{b}, z)\right] - 1\right\}\right\rangle \\ &= \left\langle\exp\left[i\sigma'(\Delta V(\mathbf{b}', z') - \Delta V(\mathbf{b}, z))\right]\right\rangle \\ &\quad - \left\langle\exp\left[i\sigma'\Delta V(\mathbf{b}', z')\right]\right\rangle - \left\langle\exp\left[-i\sigma'\Delta V(\mathbf{b}, z)\right]\right\rangle + 1 \\ &\approx \exp\{-\frac{\sigma'^2}{2}\left[\left\langle|\Delta V(\mathbf{b}', z')|^2\right\rangle + \left\langle|\Delta V(\mathbf{b}, z)|^2\right\rangle - 2\left\langle\Delta V(\mathbf{b}', z')\Delta V(\mathbf{b}, z)\right\rangle\right]\} \\ &\quad - \exp[-\frac{\sigma'^2}{2}\left\langle|\Delta V(\mathbf{b}', z')|^2\right\rangle] - \exp[-\frac{\sigma'^2}{2}\langle|\Delta V(\mathbf{b}, z)|^2\rangle] + 1 \end{aligned} \tag{11.60}$$

where the time average can easily be performed by expressing the potential as the Fourier transform of the atomic-scattering factor, and the results are

$$\begin{aligned}\left\langle[\Delta V(\mathbf{b},z)]^2\right\rangle &\approx \sum_{\kappa}\left[\langle| V_{\mathrm{k}}(\mathbf{b}-\mathbf{b}_{\kappa},z)|^2\rangle - V_{0\mathbf{k}}^2(\mathbf{b}-\mathbf{b}_{\kappa},z)\right]\\ &=\sum_{\kappa}\int d\mathbf{u}\int d\tau f_{\kappa}^{e}(\mathbf{u})f_{\kappa}^{e*}(\tau)\\ &\quad\times\exp[-W_{\kappa}(\mathbf{u})-W_{\kappa}(\tau)]\exp[2\pi i(\mathbf{b}-\mathbf{b}_{\kappa})\cdot(\mathbf{u}-\tau)]\\ &\quad\times\left\{\exp\left[4\pi^2<(\mathbf{u}\cdot\mathbf{u}_{\kappa})(\tau\cdot\mathbf{u}_{\kappa})>\right]-1\right\}\end{aligned} \tag{11.61}$$

and

$$\begin{aligned}&<\Delta V(\mathbf{b}',z')\,\Delta V(\mathbf{b},z)>\\ &=\sum_{\kappa}\sum_{\kappa'}\left[<V_k(\mathbf{b}'-\mathbf{b}_{\kappa},z')\,V_{\kappa'}(\mathbf{b}-\mathbf{b}_{\kappa'},z)> - V_{0k}(\mathbf{b}'-\mathbf{b}_{\kappa},z')\,V_{0k'}(\mathbf{b}-\mathbf{b}_{\kappa'},z)\right]\\ &=\sum_{\kappa}\sum_{\kappa'}\int d\mathbf{u}\int d\tau f_{\kappa}^{e}(\mathbf{u})f_{\kappa'}^{e}(\tau)\exp\left[-W_{\kappa}(\mathbf{u})-W_{\kappa'}(\tau)\right]\\ &\times\exp\left[2\pi i(\mathbf{b}'-\mathbf{b}_{\kappa})\cdot\mathbf{u}-2\pi i(\mathbf{b}-\mathbf{b}_{\kappa'})\cdot\tau\right]\\ &\times\left\{\exp\left[4\pi^2<(\mathbf{u}\cdot\mathbf{u}_{\kappa})(\tau\cdot\mathbf{u}_{\kappa'})>\right]-1\right\}\end{aligned} \tag{11.62}$$

where the sums of κ and κ' are limited to atoms located in the crystal slices at depths z and z', respectively.

The $<(\mathbf{u}\cdot\mathbf{u}_{\kappa})(\tau\cdot\mathbf{u}_{\kappa'})>$ term determines the coherence of TDS. Under the harmonic oscillator's approximation, atom displacement due to thermal vibration is written as a superposition of contributions from all phonon modes (Born 1942a; Born 1942b):

$$\mathbf{u}_{\kappa}=\sum_{q}\sum_{i}A_{\kappa}(\mathbf{q},i)\mathbf{e}_{q,i}\cos\left[\omega_i(\mathbf{q})t-2\pi\mathbf{q}\cdot\mathbf{r}_{0\kappa}+\theta_{q,i}\right] \tag{11.63a}$$

where

$$A_{\kappa}(\mathbf{q},i)=\frac{[\hbar(\langle n_s\rangle+1/2)]^{1/2}}{[N_0M_{\kappa}\omega_i(\mathbf{q})]^{1/2}} \tag{11.63b}$$

It can easily be shown that

$$
\begin{aligned}
< (\tau \cdot \mathbf{u}_\kappa)\left(\tau' \cdot \mathbf{u}_{\kappa'}\right) >= & \sum_q \sum_i A_\kappa(\mathbf{q}, i) A_{\kappa'}(\mathbf{q}, i)\left(\tau \cdot \mathbf{e}_{q,i}\right)\left(\tau' \cdot \mathbf{e}_{q,i}\right) \\
& \times \cos\left[2\pi \mathbf{q} \cdot (\mathbf{r}_{0\kappa} - \mathbf{r}_{0\kappa'})\right] \\
= & \frac{\hbar}{N_0 \left(M_\kappa M_{\kappa'}\right)^{1/2}} \sum_q \sum_i \frac{\left(\langle n_s \rangle + \frac{1}{2}\right)}{\omega_i(\mathbf{q})} \\
& \times \left(\tau \cdot \mathbf{e}_{q,i}\right)\left(\tau' \cdot \mathbf{e}_{q,i}\right) \cos\left[2\pi \mathbf{q} \cdot (\mathbf{r}_{0\kappa} - \mathbf{r}_{0\kappa'})\right] \quad (11.64)
\end{aligned}
$$

This function usually decreases with the increase of interatomic distance $(\mathbf{r}_{0\kappa} - \mathbf{r}_{0\kappa'})$. Thus the evaluation of (11.64) gives the coherent length of TDS. Under the Warren approximation, (11.64) can be evaluated [see Sect. 7.7 and (7.51) and (7.52)].

11.5.5.1 Case 1: Transverse Coherence

The first case we discuss is coherence between TDS waves generated from atoms located in different columns, as seen from the beam direction. For the convenience of our discussion, we examine the approximation of (11.59) for the first-order TDS, so that

$$
\exp\left(i\sigma' V\right) = \exp\left(i\sigma' V_0\right) \exp\left(i\sigma' \Delta V\right) \approx \exp\left(i\sigma' V_0\right)\left[1 + i\sigma' \Delta V\right]
$$

thus

$$
\left\langle \Gamma\left(\mathbf{b}', z'\right) \Gamma^*(\mathbf{b}, z)\right\rangle \approx \sigma'^2 < \Delta V\left(\mathbf{b}', z'\right) \Delta V(\mathbf{b}, z)\rangle \quad (11.65)
$$

If we neglect the phase-grating term and assume $D(\mathbf{b}) = \delta(\mathbf{b})$, (11.59) is approximated as

$$
\begin{aligned}
I_{\text{TDS}}\left(\mathbf{b}_p\right) = & \sum_{n_c} \sum_{m_c} \int d\mathbf{b} < \Gamma\left(\mathbf{b}, z_{n_c}\right) \Gamma^*\left(\mathbf{b}, z_{m_c}\right) > \\
& \times Q\left(\mathbf{b}, z_{n_c}\right) Q^*\left(\mathbf{b}, z_{m_c}\right) \Phi_0\left(\mathbf{b} - \mathbf{b}_p, z_{n_c}\right) \Phi_0^*\left(\mathbf{b} - \mathbf{b}_p, z_{m_c}\right) \quad (11.66)
\end{aligned}
$$

Since the width of the perturbation potential $\Delta U_\kappa = U_\kappa - < U_\kappa >$ is much smaller than the interatomic distance, for atoms located in the same crystal slice (i.e., at the same depth z_κ):

$$
\langle \Delta U_\kappa(\mathbf{b} - \mathbf{b}_\kappa, z_\kappa) \Delta U_{\kappa'}(\mathbf{b} - \mathbf{b}_{\kappa'}, z_{\kappa'})\rangle \approx \left\langle |\Delta U_\kappa(\mathbf{b} - \mathbf{b}_\kappa, z_\kappa)|^2\right\rangle \delta_{\kappa\kappa'} \quad (11.67)
$$

where $\delta_{\kappa\kappa'} = 1$ if $\kappa = \kappa'$ and $\delta_{\kappa\kappa'} = 0$ otherwise. This means that TDS electrons generated from atoms located in different atomic columns of the same slice are approximately incoherent (Wang 1992a; Jesson and Pennycook 1993; Treacy and Gibson

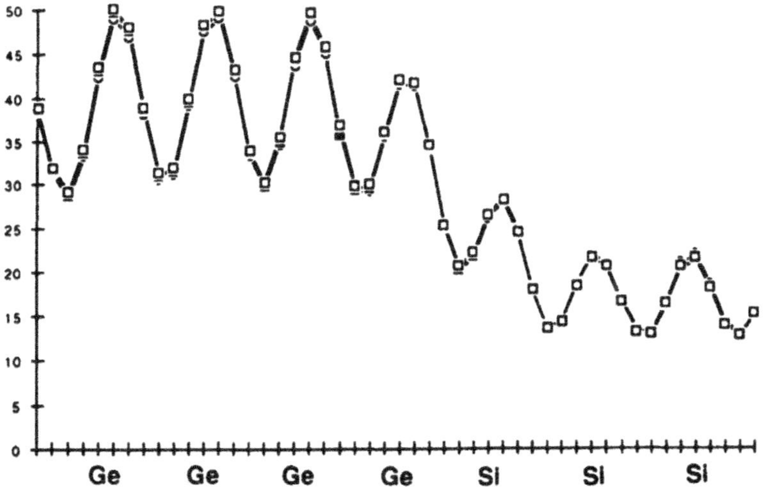

Fig. 11.5 Transverse incoherence in HAADF-STEM; this figure compares the simulated line scan profiles of HAADF-STEM images across the Ge/Si (110) interface for three different atomic vibration phase couplings between atoms distributed in the same slice (see text). The image contrast profiles for the three cases are almost identical. The thermal vibration amplitudes of the Ge and Si atoms are taken as 0.1 Å. The STEM electron probe was generated by an objective aperture of radius 0.5Å^{-1} (corresponding to 18.5 mrad in a semiconvergence angle), $C_s = 0.8$ mm and $\Delta f = -800$Å for 100kV electrons. The angular detection range of the ADF detector is 37–111 mrad; specimen thickness is 10.86 nm

1993). The transverse coherence was first proved by the multislice calculations of Wang and Cowley (1990). The transverse incoherence is the result of assuming that $D(\mathbf{b}) = \delta(\mathbf{b})$, which is determined by the detection geometry of the ADF detector. The condition under which $D(\mathbf{b}) = \delta(\mathbf{b})$ holds is discussed in Sect. 11.5.6.

Figure 11.5 compares multislice, simulated line scan HAADF-STEM images of the Ge/Si (110) interface for the following cases: (1) Atoms within the same slice are vibrating in-phase but with random phases for different slices; (2) atoms are vibrating with random phases within the same slice and randomly for different slices; (3) neighboring atoms within the same slice are vibrating with a phase difference of $2\pi/5$ in the x-direction but in-phase in the y-direction, with atomic vibration treated randomly for different slices; and (4) neighboring atoms within the same slice are vibrating with a phase difference of $2\pi/5$ in the y-direction but in-phase in the x-direction, and atomic vibration is treated randomly for different slices. The ADF imaging intensity profiles not only show exactly the same intensity, but the relative contrast of the Ge and Si atoms is also almost identical. This shows that the ADF imaging contrast is independent on the vibration status of its neighbor atoms within the same slice. In other words, whether the thermal vibration of the atoms are assumed to obey the Einstein model, Debye model, or any other models does not affect the ADF imaging contrast if the probe size is small. This is just the transverse incoherence, as illustrated by (11.67).

11.5.5.2 Case 2: Longitudinal Coherence

For atoms constrained within the same column along the beam direction, however, the coherence of the TDS may be preserved (Fig. 11.6), but the coherent length is determined by phonon dispersion characteristics of the crystal. Therefore based on the approximation of $D(\mathbf{b}) = \delta(\mathbf{b})$, only longitudinal coherence matters to HAADF-STEM. Thus the condition

$$4\pi^2 < (\tau \cdot \mathbf{u}_k)\left(\tau' \cdot \mathbf{u}_{k'}\right) > \approx 0$$

evaluated from (11.64) gives the coherent length of thermal diffuse scattering (see Sect. 7.7), leading to the cigar-shaped coherence volume, as suggested by Treacy and Gibson (1993). Numerical calculations by Wang and Cowley (1990) demonstrated transverse and longitudinal coherences in HAADF-STEM images.

Based on (11.64), it is important to point out that the condition

$$4\pi^2 < (\tau \cdot \mathbf{u}_k)\left(\tau' \cdot \mathbf{u}_{k'}\right) > \approx 0$$

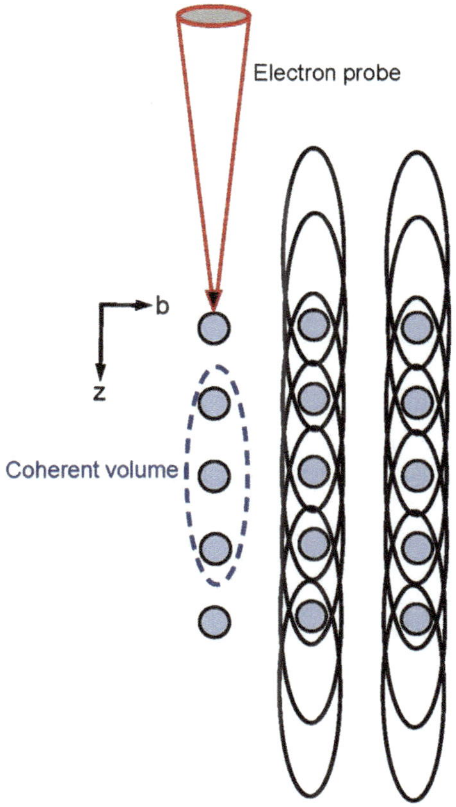

Fig. 11.6 Longitudinal (parallel to the beam) coherence in HAADF-STEM imaging; the coherent volume is approximately indicated by elliptical packets. The incident beam direction is z

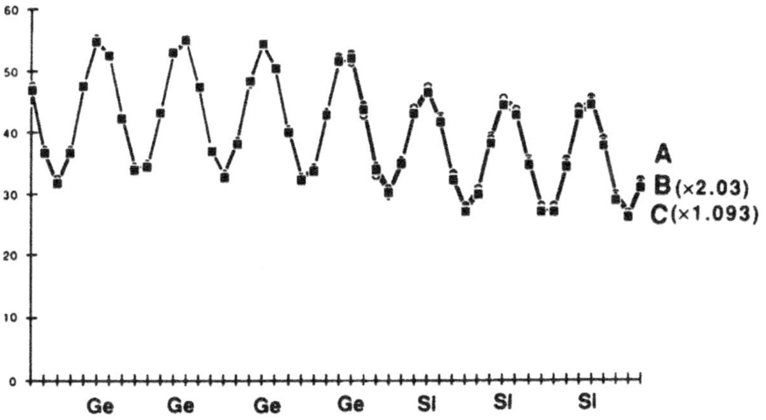

Fig. 11.7 Longitudinal coherence in HAADF-STEM; this figure compares the simulated line scan HAADF-STEM image across the Ge/Si(110) interface for different vibration phase coupling of atoms distributed in different slices (see text). Simulation parameters are the same as in Fig. 11.5, but the mean vibration amplitude of Si is 0.2Å. The image contrast is almost identical for the three-phase coupling cases (see text), but relative intensities are dramatically different. Scaling factors to normalize the three curves are indicated

is easy to satisfy with atoms with high atomic numbers (or large M_{κ}). Therefore coherent length varies with the change of atomic mass in the column. It is thus possible that the z-coherence may disappear for heavy elements.

It was shown in the last section that the ADF imaging contrast is independent of the thermal correlation of vibrations of neighboring atoms within the same slice. Then the imaging simulation can be simply done by assuming that all atoms within the same slice are vibrating in-phase. It is necessary to find out the dependence of the image contrast on atomic vibrations in different crystal slices.

Figure 11.7 compares calculated line scan HAADF-STEM images across the Ge/Si(110) boundary for the following three cases: (A) Random phases are added to each slice; (B) designated phase sequence of $\pi/2, 0, \pi/2, 0, \ldots$; and (C) designated phase sequence of $\frac{\pi}{2}, -\frac{\pi}{2}, \frac{\pi}{2}, -\frac{\pi}{2} \ldots$ are added to successive slices. The thermal vibration amplitude of the Si atoms is doubled in comparison with that of Fig. 11.5. Some differences in the relative magnitudes of ADF images are seen, but contrast profiles for these cases follow almost exactly the same curve after being normalized at the Ge sites (curves B and C are magnified for 2.03 and 1.093 times, respectively). This shows that the vibration status of atoms in different slices may introduce some difference in intensity magnitudes, but the image contrast is almost unaffected under conditions assumed here, since the same coherence length for both Si and Ge was assumed. In practice, however, different contrast could be produced due to differences in coherence lengths across atomic columns. This effect may be significant in imaging interfaces.

Calculation of (11.64) has been performed by Jesson and Pennycook (1993) based on the Debye model and the Warren (1990) approximation (see Sect. 7.7), in which the first BZ is approximated by a sphere of radius q_B. Their results indicate that the number of atoms n_B enclosed in the elliptical coherent volume along the z-direction is determined by the condition:

$$\left(\int_0^{n_B\pi} dx \frac{\sin x}{x} \right) (n_B\pi)^{-1} \ll 1$$

Numerical calculation has shown $n_B < 4$, which means that the diffuse scattering produced by the scattering of four consecutive atoms along the beam direction is coherent.

11.5.6 Detection Geometry and Coherence in HAADF-STEM Imaging

As shown in (11.43), the approximation of $D(\mathbf{b}) = \delta(\mathbf{b})$ is necessary for deriving the incoherent imaging theory. We now discuss conditions under which the $D(\mathbf{b})$ function can be approximated as $\delta(\mathbf{b})$. We first consider the Fourier transform of the detector function:

$$\begin{aligned} D(\mathbf{b}) &= \int d\mathbf{u} \exp(2\pi i \mathbf{u} \cdot \mathbf{b}) D(\mathbf{u}) \\ &= \int_0^{u_2} du \int_0^{2\pi} d\phi u \cos(2\pi u b \cos\phi) - \int_0^{u_1} du \int_0^{2\pi} d\phi u \cos(2\pi u b \cos\phi) \\ &= \frac{1}{b} \left[u_2 J_1 (2\pi u_2 b) - u_1 J_1 (2\pi u_1 b) \right] \end{aligned} \tag{11.68}$$

where $J_1(x)$ is the first-order Bessel function. The outer cutoff of the ADF detector is usually large to detect large-angle scattered electrons. Thus the width of the D function is mainly determined by the inner cutoff of the ADF detector. Since the first zero point of $J_1(x)$ occurs at $x = 3.8317$ (Arfken 1970), the half-width of the D function is approximately

$$b_0 \approx \frac{3.8317}{2\pi u_1} \approx 0.61 \frac{\lambda}{\theta_1} \tag{11.69}$$

Equation (11.69) is just the Rayleigh criterion of the least resolvable distance between two source points of light. For 100 – keV electrons, $2b_0 \approx 0.9$Å for $\theta_1 = 50$ mrad and $2b_0 \approx 0.45$Å for $\theta_1 = 100$ mrad. Since the width of the atomic potential is less than 0.5Å, the width of the D function cannot be represented by a delta function unless the inner detector angle $\theta_1 > 150$ mrad for 100 – keV electrons. This means that incoherent imaging theory is not exact as far as the size of the detector function is concerned.

The exact mathematical condition to make $D(\mathbf{b}) = \delta(\mathbf{b})$ is $u_1 = 0$ and $u_2 = \infty$. This means that the detector detects all and only TDS electrons in the diffraction plane (Wang and Cowley 1990). This is impossible to achieve in practice, because the influence of low-angle Bragg reflections is unavoidable. Therefore the incoherent imaging theory is the extreme case of practical imaging conditions.

The theory presented in Sect. 11.5 ignored electron elastic rescattering after TDS. This is a reasonable approximation provided the ADF detector is large. An exact theory that includes the dynamic scattering effect after TDS is given in Sect. 11.10.2.

11.6 Imaging with TDS Electrons in TEM

We showed the equivalence of TEM and STEM in Chap. 9. Forming the image by phonon-scattered electrons is a dark-field imaging technique that can be more effectively performed in STEM with the use of an ADF detector. However in TEM, it is important not to choose a large objective aperture due to the effect of lens spherical aberration. Therefore this dark-field imaging technique in TEM usually lacks intensity, but it is still possible to obtain some useful information. This section outlines the basic principle of incoherent imaging in TEM.

11.6.1 Image Formation

In high-angle dark-field (HADF) TEM imaging (Fig. 11.8), a plane wave is assumed to strike a crystal entrance face at an angle ϑ with respect to the optic axis. The selection of ϑ must be compromised to avoid both the phase contrast effect at low angles and the weak signal intensity at high angles. The objective aperture is centered on the optic axis, and it allows electrons scattered to a certain angular range to enter. The incident beam can be conically scanned around the optic axis to simulate the configuration of the ADF detector in STEM. The image obtained in this scanning configuration is the average of images obtained when the crystal is illuminated from different incident angles.

The image formed using diffusely scattered electrons in HADF-TEM has shown mass thickness contrast (Bentley et al. 1990). An HADF-TEM image of Ag_2Al γ' plates in Al-17.9 wt.% Ag is shown in Fig. 11.9. The plates oriented parallel to the beam direction [011] show the strongest intensity. Inclined plates with respect to [011] show relatively less contrast, but these can still be easily separated from the matrix. Overlapping regions between plates are apparent. The diffraction contrast effect is minimized due to the averaging effect of the conical scan.

The HAADF-STEM and HADF-TEM are different in the following aspects due to differences in their scattering geometries. First strong incident beam channeling

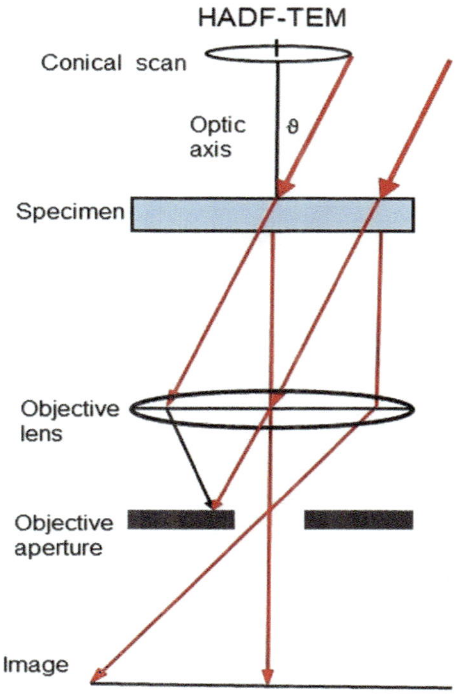

Fig. 11.8 A ray diagram showing the formation of an atomic-number-sensitive image in TEM

can be generated in HAADF-STEM, but not in HADF-TEM, because the primary incident beam travels along the crystal zone axis in the former instead of at a large angle with respect to the zone axis in the latter (Wang 1992a). This difference may greatly reduce the contribution of dechanneling contrast in HADF-TEM imaging. On the other hand, for a zone-axis incident beam, channeling of high-angle diffusely scattered electrons may occur in HADF-TEM but not in HAADF-STEM. Second the contribution of plasmon loss electrons in HADF-TEM images is different than in HAADF-STEM images. For TEM, theoretical calculations have found that the main effect of valence excitation (with energy losses of 5–30 eV) is to introduce a focus shift in the lens transfer function due to chromatic aberration (Wang and Bentley 1991a). The focus shift can be as large as a few tens of nanometers, resulting in contrast reversal in high-resolution phase contrast imaging. The contrast of valence loss electrons is the same as that of elastically scattered electrons in STEM, but not in TEM. This is because the contrast transfer effect of the condenser and objective lenses in the STEM and TEM cases occurs before and after, respectively, electrons interact with the crystal, so that inelastic valence excitation within the specimen has almost no effect on the contrast of high-resolution STEM images but on TEM lattice images. Valence excitations are delocalized inelastic-scattering processes whose effects are thus expected to be noticeable in the image only when the resolution approaches atomic level. Thus the inelastic-scattering effect can be neglected in lower resolution TEM images of dislocations. Finally image contrast can be artificially improved

Fig. 11.9 An HADF-TEM image of Ag_2Al γ' in Al – 17.9wt. % Ag showing mass thickness contrast. The image was recorded at 300 kV, conical scan angle $\theta = 50\,\text{mrad}$. Courtesy of Dr. Bentley et al. (1990)

using an electronic gain-controlling system in STEM, but not in TEM, making it more difficult to enhance the contrast of weak features in HADF-TEM images.

Bearing the preceding points in mind, it is not surprising that there are some differences between images of HADF-TEM and HAADF-STEM. However caution must be exercised if trying to use contrast mechanisms developed for HAADF-STEM to interpret HADF-TEM images.

11.6.2 Incoherent Imaging Theory

We can follow a scheme analogous to that outlined in Sect. 11.5 to derive the incoherent imaging theory of HADF-TEM. The TDS wave generated from a crystal slice is

$$\Delta\Phi\left(\mathbf{b}, z_{n_c}\right) = \left[iQ\left(\mathbf{b}, z_{n_c}\right)\Gamma\left(\mathbf{b}, z_{n_c}\right)\Phi_0\left(\mathbf{b}, z_{n_c}\right)\right] \tag{11.70}$$

If the diffraction effect (or Kikuchi line effect) of TDS electrons is neglected for thin specimens and the Einstein model is valid, the intensity distribution in the image plane formed by an on-axis objective aperture is

$$
\begin{aligned}
I(\mathbf{b}) &= \sum_{n_c} < \left|\left[i Q\left(\mathbf{b}, z_{n_c}\right) \Gamma\left(\mathbf{b}, z_{n_c}\right) \Phi_0\left(\mathbf{b}, z_{n_c}\right)\right] \otimes T_{\text{obj}}(\mathbf{b})\right|^2 > \\
&= \sum_{n_c} \int d\mathbf{b}' \int d\mathbf{b}'' Q\left(\mathbf{b}', z_{n_c}\right) Q^*\left(\mathbf{b}'', z_{n_c}\right) < \Gamma\left(\mathbf{b}', z_{n_c}\right) \Gamma^*\left(\mathbf{b}'', z_{n_c}\right) > \Phi_0\left(\mathbf{b}', z_{n_c}\right) \\
&\quad \times \Phi_0^*\left(\mathbf{b}'', z_{n_c}\right) T_{\text{obj}}\left(\mathbf{b}-\mathbf{b}'\right) T_{\text{obj}}^*\left(\mathbf{b}-\mathbf{b}''\right)
\end{aligned} \tag{11.71}
$$

where $< \Gamma\left(\mathbf{b}', z\right) \Gamma\left(\mathbf{b}'', z\right) >$ is given by (11.60). In (11.71), the contribution of every order of TDS is included, and the single-phonon scattering model is not assumed. To simplify this equation, we assume a case in which the objective lens is near ideal (i.e., small C_s), so that the lens transfer function in real space is approximated by $T_{\text{obj}}\left(\mathbf{b}-\mathbf{b}'\right) \approx \delta\left(\mathbf{b}-\mathbf{b}'\right)$. We approximately have (Wang 1992a)

$$
\begin{aligned}
I(\mathbf{b}) &= \sum_{n_c} \int d\mathbf{b}' < |\Gamma(\mathbf{b}', z_{n_c})|^2 > |\Phi_0(\mathbf{b}', z_{n_c})|^2 \, |T_{\text{obj}}(\mathbf{b}-\mathbf{b}')|^2 \\
&= \sum_{n_c} \left[< |\Gamma(\mathbf{b}, z_{n_c})|^2 > |\Phi_0(\mathbf{b}, z_{n_c})|^2\right] \otimes |T_{\text{obj}}(\mathbf{b})|^2
\end{aligned} \tag{11.72}
$$

where the $\left\langle|\Gamma|^2\right\rangle$ term is given by (11.53). Therefore the image theory is equivalent to incoherent imaging theory if the objective lens is close to perfect (i.e., C_s is small). This type of image does not have the contrast reversal effect when changing focus or specimen thickness. The generation of TDS electrons from each slice is characterized by a Γ^2 function. The image resolution may be higher than that formed by purely elastically scattered electrons (Cowley 1988a; Spence et al. 1989; Xu et al. 1990), because the contrast transfer function is $\left|T_{\text{obj}}(\mathbf{b})\right|^2$ for TDS electrons instead of $T_{\text{obj}}(\mathbf{b})$ for elastically scattered electrons (i.e., phase contrast imaging; see Sect. 11.9).

Figure 11.10 shows simulated HADF-TEM images of a Si/Ge (100) interface viewed along [011]. For simplification, the Si and Ge lattices are assumed to be perfectly match with the same lattice constant. It is clear that atomic-number-sensitive information is contained in the image. Furthermore the contrast of Ge to Si is slightly enhanced with an increase in specimen thickness (Fig. 11.10b). This effect may be related to the stronger channeling effect established along the Ge columns. A channeling effect can enhance the HAADF-STEM image contrast, but it is not necessary for forming the image. Based on the reciprocity theorem, an atomic resolution HADF-TEM image should be obtained as the HAADF-STEM image, but this goal has not been achieved experimentally so far.

The theory described in this section has ignored the coherence of TDS electrons and the elastic rescattering effect of electrons after TDS. An exact theory for calculating HADF-TEM images is given in Sect. 11.10.4.

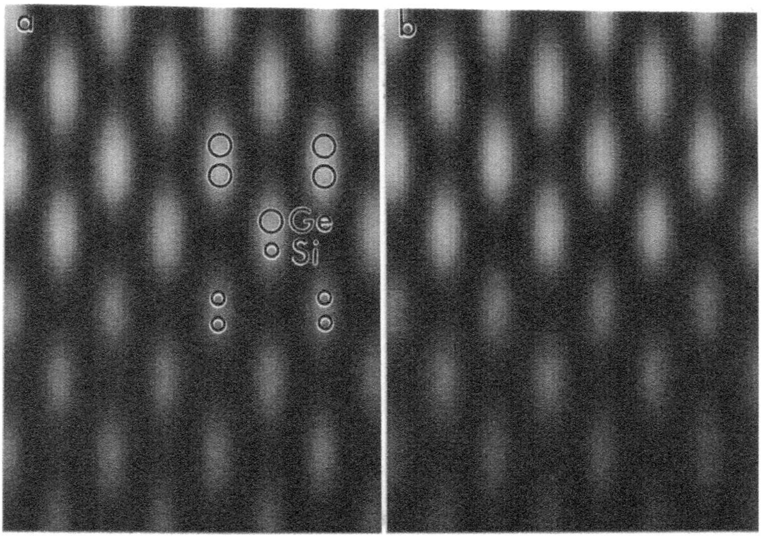

Fig. 11.10 Simulated HADF-TEM images of the Si/Ge (100) interface viewed along [011] for thicknesses of (a) 10.86 nm and (b) 54.3 nm. Conditions for image simulation are selected for a Philip CM30 TEM (300 kV): $C_s = 2$ mm, $\Delta f = -72.45$ nm, conical half-angle $\theta = 50$ mrad; the objective aperture radius is 8.5 mrad. Mean vibration amplitudes of Si and Ge are taken as 0.0077 and 0.0063 nm, respectively

11.7 Effect of Phase Correlation Between Atom Vibrations in TDS Electron Imaging

In HRTEM of thin specimens, if the specimen is viewed along a low-index zone axis, electrons may be channeled along atom rows. It is thus considered that each column of atoms is imaged separately and the wave function in the crystal is a sum of wave functions of individual atomic columns (Van Dyck et al. 1989). Thus the image intensity of each column is almost unaffected by the structures of its neighboring atoms if the image resolution is sufficiently high. Since TDS is a localized scattering process, the image contrast of one row of atoms is almost unaffected by the vibration status of neighboring rows of atoms (Cowley 1988b). Therefore phase correlations between vibrations of atoms in different rows do not affect the image contrast, but these do affect the calculation of diffraction patterns. Interatomic phase correlations may contribute only a weak, featureless background in the image, but longitudinal coherence along the beam may affect the image contrast. Therefore atom vibrations within the same slice can be handled with the Einstein model (Wang and Cowley 1989).

A simple ray diagram in Fig. 11.11 illustrates the physical meaning of transverse incoherence (Wang 1992a). We assume that atoms A and B are widely separated and the objective lens is perfect (i.e., $C_s = 0$); TDS waves generated from atomic

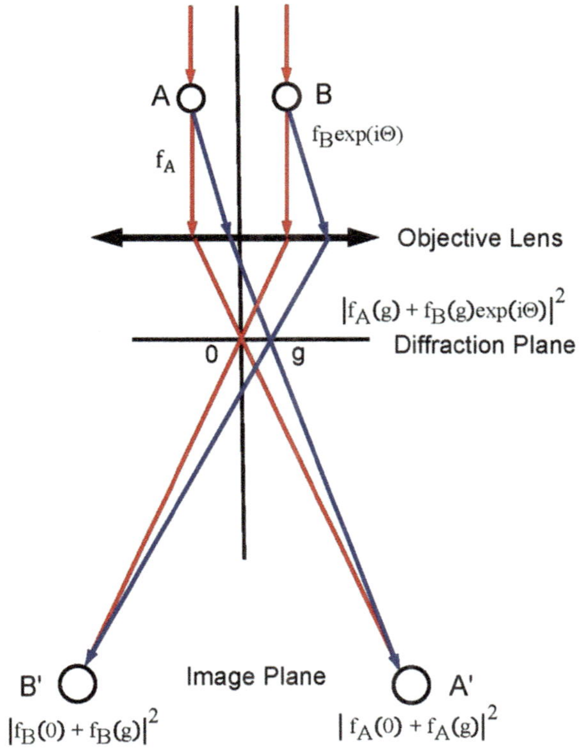

Fig. 11.11 A ray diagram showing the transverse incoherency of thermal diffuse scattering in TEM imaging

site A (in the semiclassical model) are represented by $f_A(0)$ and $f_A(g)$; from Site B, they are denoted by $f_B(0)\exp(i\Theta_{\mathrm{AB}})$ and $f_B(g)\exp(i\Theta_{\mathrm{AB}})$, where Θ_{AB} is the vibrating phase correlation of the A and B atoms. In the diffraction plane, the intensity distribution is determined by the sum of the waves scattered in the same direction from different atomic sites, i.e., $|f_A(g) + f_B(g)\exp(i\Theta_{\mathrm{AB}})|^2$. Thus phase coupling between the localized TDS is very important for calculating TDS diffraction patterns. In the image plane, however, the intensity distribution of the A atom profile depends only on $|f_A(0) + f_A(g)|^2$, and of the B atom profile only on $|f_B(0) + f_B(g)|^2$ if tail components contributed by neighboring atoms are neglected. Therefore the image contrast is not affected by phase coupling in atomic vibrations located in separated columns. (This is just the transverse incoherence.) It is necessary to point out that the preceding discussion is applied only to thin crystals, since the strong dynamic diffraction effect of relatively thicker crystals may complicate the argument.

11.8 Effect of Huang Scattering in Composition-Sensitive Imaging

For crystals containing defects, dislocations, or grain boundaries, diffuse scattering can be generated from the defected regions because of local deviation from the three-dimensional periodic structure. This component usually depends on distortion in crystal lattices produced by impurities, defects, or dislocations, and it has a weak dependence on the atomic number. The contribution of this process can be stronger than that made by TDS, especially near interfaces or defected regions, so that dislocation lines and grain boundaries show bright contrast (Cowley and Huang 1992; Perovic et al. 1993). In general, HAADF-STEM images are taken when the incident beam is parallel or nearly parallel to a crystal zone axis, so that strong channeling effect along atom rows is expected. The following two mechanisms have been proposed for interpreting the observed contrast. The first one is the so-called dechanneling contrast (Cowley and Huang 1992). The lattice distortion due to strain fields near dislocation cores can significantly interrupt electron-channeling propagation along atom rows, resulting in more dechanneled electrons. These electrons are then diffusely scattered by the nonperiodically arranged atoms near dislocation cores, resulting in the enhancement of high-angle scattering signals. The other mechanism is diffuse scattering produced from the static displacement of atoms around dislocation cores (Cowley 1981). This type of scattering is analogous to TDS except that the crystal configuration is time independent and atomic displacement vectors could be quite large, especially near dislocation cores.

Anomalous contrast has also been observed in HAADF-STEM images of a layered silicon specimen, where layers doped with boron showed a brighter contrast than layers without dopants (Perovic et al. 1993). The contrast anomaly may be understood as follows. In general, point defects and site substitutions are produced in a crystalline specimen after ion implantation. These point defects produce diffuse scattering (Huang 1947), which is known as Huang scattering, thus enhancing signals at high angles. But the image contrast produced by this process is weakly dependent on the atomic number.

Therefore caution must be exercised when we try to determine the local compositional variation based purely on image contrast, especially when dislocations or grain boundaries are present. We now use HADF-TEM images to illustrate the contribution of diffuse scattering due to local imperfections.

The contrast of HADF-TEM images depends strongly on the diffracting conditions initially set up for the corresponding on-axis bright-field (BF) TEM image. Figure 11.12 compares BF TEM images (Fig. 11.12a–c) and the corresponding HADF-TEM images (Fig. 11.12d–f) of a thin Ag foil, recorded for three different diffracting conditions. Under conditions of no strongly excited reflections (Fig. 11.12a, d), the visibility of dislocations is poor in both BF TEM and HADF-TEM images. Under the two-beam diffracting condition (Fig. 11.12b, e) both BF TEM and HADF-TEM images show optimum contrast. For the [$\bar{2}$33]-zone-axis case (Fig. 11.12c, f), the image contrast is degraded. It is well known that the optimum

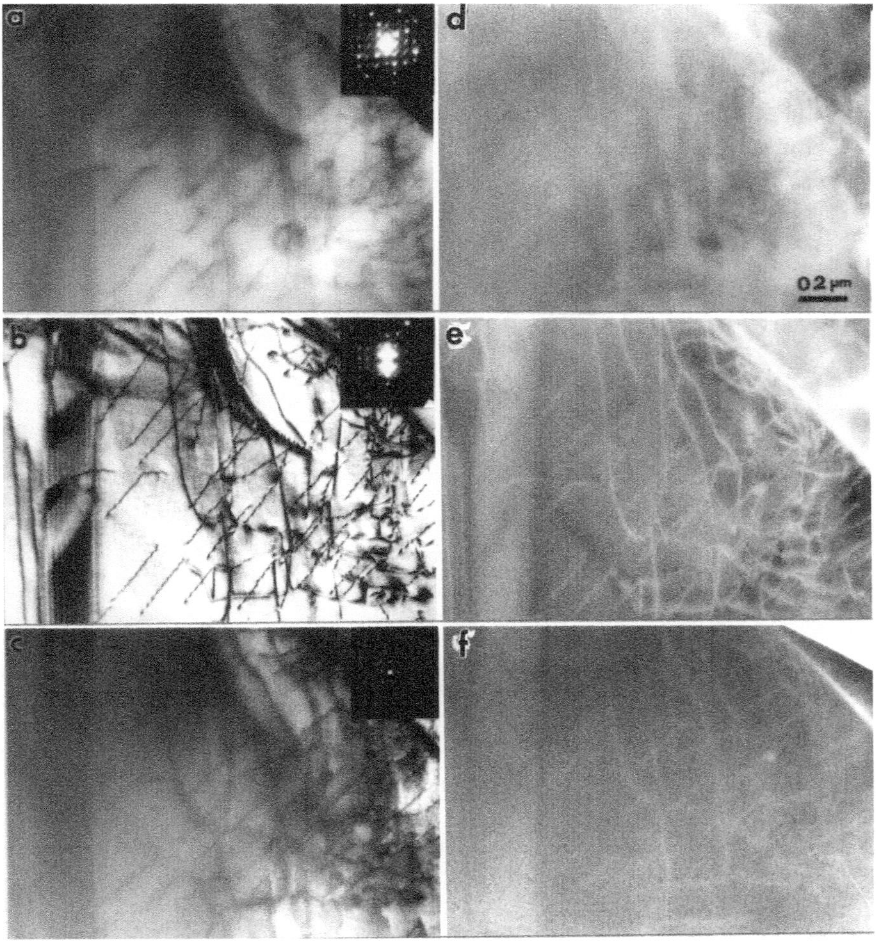

Fig. 11.12 **a–c** The BF TEM images and **d–f** corresponding HADF-TEM images of the same area recorded under conditions of no strong diffraction, two-beam, and $[\bar{2}33]$-zone-axis orientations, respectively, showing the dependence of dislocation contrast in HADF-TEM on electron-channeling conditions. The semiconical angle is $52 < \theta < 58$ mrad. The insets are diffraction patterns (Wang 1994)

diffracting condition for imaging dislocations in BF TEM is generally the two-beam reflection condition, and Fig. 11.12 suggests that HADF-TEM imaging follows the same rule. Since the contrast of dislocations in BF diffraction contrast images generally depends on the value of $\mathbf{g} \cdot \mathbf{b}$, the preceding observation seems to indicate that the brightness of dislocations in HADF-TEM also depends on $\mathbf{g} \cdot \mathbf{b}$. Further experimental observations found that the contrast of dislocation images disappeared if $\mathbf{g} \cdot \mathbf{b} = 0$ is satisfied by the corresponding BFTEM image.

The dislocation contrast of HADF-TEM images is generated by a two-step mechanism—the creation of diffuse scattering due to lattice distortion around dislocations and subsequent channeling propagation of diffusely scattered electrons governed by dynamic diffraction (Wang 1994). The creation of local diffuse scattering is determined by the first step, and the final image contrast is essentially determined by the second step. The last step makes HADF-TEM imaging analogous in many respects to conventional diffraction contrast imaging. In a similar fashion, the bright contrast of grain boundaries in HADF-TEM images is at least in part due to atomic rearrangement and relaxation at the interface producing high-angle, diffusely scattered electrons. This is the result of the Huang-scattering contrast. A rigorous theory for simulating images of Huang-scattered electrons is given in Sect. 11.10.5. We now present a qualitative interpretation of the observed result.

The mechanism for generating a diffraction contrast effect in HADF-TEM is illustrated in Fig. 11.13. In general at a dislocation, the intensity of high-angle scattered electrons increases due to diffuse scattering associated with large static displacements. Electrons that contribute to HADF-TEM images are those distributed within an angular range β around the optic axis. The subsequent scattering of these electrons (parallel or nearly parallel to the optic axis) is determined by diffracting conditions initially set up for the BF TEM imaging. For BF conditions with no strongly excited diffracted beams (Fig. 11.13a), no channeling effect is expected in HADF-TEM, so that the $0'$ beam is broadened, similar to broadening an on-axis electron probe when transmitted through a specimen of thickness $(d - z)$ (z is the depth of dislocation), resulting in weaker and obliterated dislocation contrast; this is shown in Fig. 11.12d.

For BF two-beam conditions (Fig. 11.12b), on-axis diffuse scattering is subject to strong two-beam dynamic diffraction, even though the incident beam is far from the optic axis. Thus channeling propagation is possible for on-axis diffuse scattering. We now use the two Bloch wave theory to simply illustrate this point. In the two-beam (i.e., $0'$ and $\mathbf{g}'$ beams in Fig. 11.13b) approximation with the Bragg reflection condition, two Bloch waves are excited within the crystal [see (2.22) and Fig. 2.4]. The electron current of each Bloch wave flows, on average, parallel to atomic planes. Since the electron current is modulated across atom planes, for Bloch wave B_2, maxima occur between the atomic planes. For Bloch wave B_1, the situation is reversed, and the current is concentrated on the atom planes, i.e., a stronger channeling effect. The channeling propagation of Bloch wave B_1 along atom planes effectively limits broadening of on-axis diffuse scattering, thus preserving the dislocation contrast in HADF-TEM images (see Fig. 11.12e). Bloch wave B_2, which is not channeled, does not preserve the dislocation contrast.

Likewise a stronger channeling effect is expected in HADF-TEM when the corresponding BF TEM image is recorded at a zone axis (Fig. 11.12c). The image contrast is not necessarily the best due to strong dynamic scattering of electrons from the $0'$ beam, but the image contrast is expected to be better than Fig. 11.12a because of a possible residual channeling effect. These expected results are supported by Fig. 11.12f.

Imaging interfacial atomic structures of two phases is an important application of HRTEM and HAADF-STEM. However high-angle diffuse scattering can be gener-

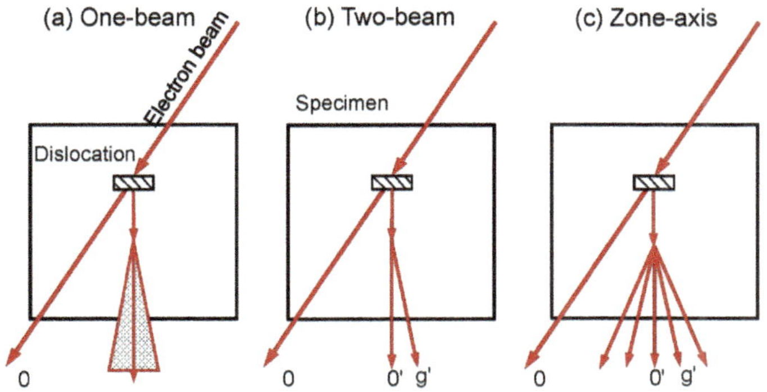

Fig. 11.13 Schematic diagrams showing the two-step contrast mechanism of HADF-TEM, under **a** no strongly diffracted beam, **b** two-beam, and **c** zone-axis diffracting conditions. The image contrast is produced by high-angle diffuse scattering due to static displacements of atoms near the dislocation, and it is followed by channeling (or diffracting) propagation (see text). The primary incident beam is 0. The high-angle diffusely scattered electron on-axis (0′) can be Bragg diffracted (**g**′). The cone in (**a**) represents the broadening effect of the 0′ beam

ated from the interfacial region due to the distortion or relaxation of atoms near the interface, thus enhancing the local contrast of the image. In general, the interface shows bright contrast in HAADF-STEM and HADF-TEM images. But this contrast enhancement does not necessarily mean the segregation of heavier elements at the interface. The Z-contrast characteristic may vanish in such cases. Therefore caution must be exercised in interpreting the HAADF-STEM images of interfaces.

11.9 Resolution of an Incoherent Image

Under the weak phase object approximation (WPOA), the intensity distribution in a conventional phase contrast HRTEM image is (Cowley 1988a)

$$I_p = \left|[1 + i\sigma V(\mathbf{b})] \otimes T_{\text{obj}}(\mathbf{b})\right|^2 = \left|1 + i\sigma V(\mathbf{b}) \otimes T_{\text{obj}}(\mathbf{b})\right|^2 \tag{11.73}$$

If we separate the real and imaginary components of T_{obj} as $T_{\text{obj}}(\mathbf{b}) = T_c(\mathbf{b}) + iT_s(\mathbf{b})$ and ignore second-order terms, (11.74) becomes

$$I_p \approx 1 - 2\sigma V(\mathbf{b}) \otimes T_s(\mathbf{b}) \tag{11.74}$$

In HADF-TEM imaging, however, the contrast transfer function is $\left|T_{\text{obj}}\right|^2$. It is well known that the best phase contrast image can be obtained from a Scherzer focus (Scherzer 1949)

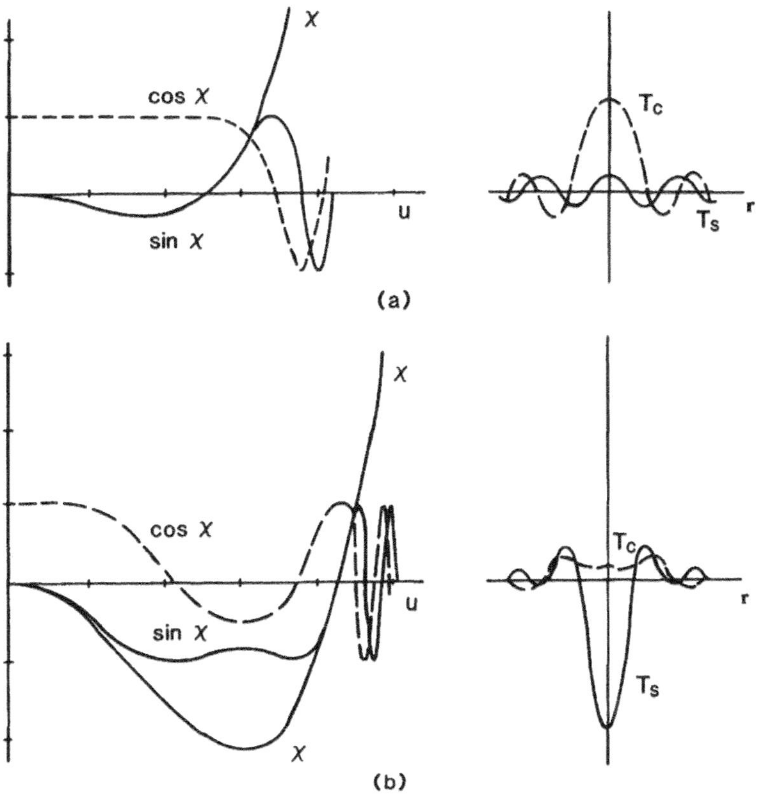

Fig. 11.14 The forms of Re $(T_{\text{obj}}) = \cos\chi$ and Im $(T_{\text{obj}}) = \sin\chi$ of the objective lens transfer function and the corresponding Fourier transform $T_c = \mathbf{FT}^{-1}(\cos\chi)$ and $T_s = \mathbf{FT}^{-1}(\sin\chi)$ in real space, with $\chi = \left(\frac{1}{2}\pi i C_s \lambda^3 u^4 + \pi i \Delta f \lambda u^2\right)$, $C_s = 1$ mm, and $E_0 = 200$ keV. **a** For a defocus value of $\Delta f = -17$ nm; **b** for a defocus value of $\Delta f = -58$ nm (Scherzer defocus). The smallest real-space probe for HAADFSTEM $(T_c^2 + T_s^2)$ is obtained under Scherzer defocus, where $|T_c| \ll |T_s|$

$$\Delta f = -\left(\frac{4}{3}\right)^{1/2} \cdot (C_s\lambda)^{1/2} \tag{11.75}$$

Under this condition, the $T_{\text{obj}}(\mathbf{b})$ function is dominated by its imaginary part T_s (Cowley, 1988a), as shown in Fig. 11.14. Thus at Scherzer focus, the $\left|T_{\text{obj}}\right|^2$ function is dominated by its imaginary part T_s^2. If the $|T_s|$ function in this case can be modeled by a Gaussian function, then the half-intensity width for $|T_s|$ is about $(2)^{1/2}$ times that of $|T_s|^2$. Therefore the resolution limit of HADF-TEM imaging should be about 71% of the resolution limit for phase contrast imaging under equivalent electron microscope conditions. It is suggested that optimum HADF-TEM images can be obtained under Scherzer defocus and the optimum objective aperture angle is

$$\theta_0 = 1.52\,(\lambda/C_s)^{1/4} \tag{11.76}$$

These results can also be applied to HAADF-STEM imaging.

11.10 Real-Space Multislice Theory of TDS

For TDS, the multislice theory introduced in the first part of this chapter involves numerous calculations to consider fully the dynamic scattering of electrons from different lattice vibration configurations. This is the major obstacle to using conventional multislice theory for diffraction pattern calculations. Recently an alternative multislice theory was introduced by Fanidis et al. (1989) for performing the lattice configuration average before numerical calculation. In this section, the major theoretical scheme of this approach and its applications are outlined.

11.10.1 Basic Equations

The theory is purely based on the quasi-elastic scattering model discussed in Chap. 7. The TDS is considered the time-averaged elastic scattering of electrons from a crystal of different instantaneous lattice configurations, so that the electron wave function for one frozen lattice configuration is the solution of the Schrödinger equation:

$$\left[\nabla^2 + 4\pi^2\left(U_0 + \Delta U + K_0^2\right)\right]\Psi_0(\mathbf{r}) = 0 \tag{11.77}$$

Under the high-energy approximation, by writing $\Psi_0(\mathbf{r}) = \exp\left(2\pi i \mathbf{K}_0 \cdot \mathbf{r}\right)\Phi_0(\mathbf{r})$. Equation (11.77) becomes

$$\frac{\partial \Phi_0(\mathbf{r})}{\partial z} \approx \varsigma\left[\Xi + 4\pi^2\left(U_0 + \Delta U\right)\right]\Phi_0(\mathbf{r}) \tag{11.78}$$

where

$$\varsigma = \frac{i}{4\pi K_{0z}} \tag{11.79}$$

and

$$\Xi = \nabla_b^2 + 4\pi i \mathbf{K}_{0b} \cdot \nabla_b \tag{11.80}$$

is an operator. To find the solution of (11.79), a classical method is to separate the wave function into two components (Cowley 1988a); one part is the thermal averaged wave function that is independent of the instantaneous vibration status of the crystal lattice, and the other part is a small deviation that is strongly affected by the crystal vibration

$$\Phi_0(\mathbf{r}) = \bar{\Phi}_0(\mathbf{r}) + \delta\Phi_0(\mathbf{r}) \tag{11.81}$$

where $\bar{\Phi}_0(\mathbf{r}) = \langle\Phi_0(\mathbf{r})\rangle$ and $\langle\delta\Phi_0(\mathbf{r})\rangle = 0$. The boundary conditions are $\bar{\Phi}_0(\mathbf{b}, 0) = 1$ and $\delta\Phi_0(\mathbf{b}, 0) = 0$. Substituting (11.81) and (11.78), we obtain

$$\frac{\partial}{\partial z}\left(\bar{\Phi}_0 + \delta\Phi_0\right) = \varsigma\left[\Xi + 4\pi^2\left(U_0 + \Delta U\right)\right]\left(\bar{\Phi}_0 + \delta\Phi_0\right) \tag{11.82}$$

Taking a time average of (11.82) yields

$$\frac{\partial}{\partial z}\bar{\Phi}_0 = \varsigma\left(\Xi + 4\pi^2 U_0\right)\bar{\Phi}_0 + \varsigma < \Delta U\delta\Phi_0 > \tag{11.83}$$

Subtracting (11.83) from (11.82) gives

$$\frac{\partial}{\partial z}\delta\Phi_0 \approx \varsigma\left\{\left[\Xi + 4\pi^2 U_0\right]\delta\Phi_0 + 4\pi^2\Delta U\bar{\Phi}_0 + 4\pi^2\left(\Delta U\delta\Phi_0 - < \Delta U\delta\Phi_0 >\right)\right\} \tag{11.84}$$

Equations (11.83) and (11.84) are coupled equations. We first consider the solution of the deviation component. Since the term $(\Delta U\delta\Phi_0 - < \Delta U\delta\Phi_0 >)$ corresponds to the effects of multiple diffuse scattering, it can be ignored under the first-order approximation, thus

$$\frac{\partial}{\partial z}\delta\Phi_0 \approx \varsigma\left\{\left[\Xi + 4\pi^2 U_0\right]\delta\Phi_0 + 4\pi^2\Delta U\bar{\Phi}_0\right\} \tag{11.85}$$

In (11.85), the ΔU term is the source for generating TDS and $\left[\Xi + 4\pi^2 U_0\right]$ is responsible for the elastic rescattering of TDS electrons. In dealing with elastic rescattering after TDS, we can assume that the crystal potential is independent of z; thus the solution of (11.84) can be directly written as

$$\delta\Phi_0(\mathbf{b}, z) \approx 4\pi^2\varsigma\int_0^z dz' O_p\left(\mathbf{b}, z - z'\right)\left[\Delta U\left(\mathbf{b}, z', t\right)\bar{\Phi}_0\left(\mathbf{b}, z'\right)\right] \tag{11.86}$$

where the operator O_p is defined as

$$O_p(\mathbf{b}, z) = \exp\left\{\zeta\left[\Xi + 4\pi^2 U_0\right]z\right\} \tag{11.87}$$

and only variable z is shown for simplicity. The physical meaning of (11.86) is stated as follows. The Bragg reflected wave $\bar{\Phi}_0$ is diffusely scattered at z' by ΔU.

The subsequent Bragg rescattering of diffusely scattered electrons from z' to z is contained in the O_p operator. The calculation of the operator in (11.86) is shown in Sect. 11.10.3. We now apply (11.86) to calculate the $< \Delta U \delta \Phi_0 >$ term, which appears in the equation of the averaged wave function (11.83):

$$\langle \Delta U \delta \Phi_0 \rangle = 4\pi^2 \zeta \Delta U(z,t) \int_0^z dz' \exp\left\{\zeta \Xi + 4\pi^2 U_0\right\} (z-z') \left[\Delta U(z',t) \overline{\Phi}_0(z')\right]\rangle \tag{11.88}$$

If z and z' do not belong to the same atom and the displacement of different atoms is assumed to be uncorrelated as in the Einstein model, then $\langle \Delta U \delta \Phi_0 \rangle = \langle \Delta U \rangle \langle \delta \Phi_0 \rangle = 0$. If z and z' belong to the same atom, it can be assumed that $\bar{\Phi}_0(z) \approx \bar{\Phi}_0\left(z'\right)$. In this case, the exponential operator has no important effect on its operants, since $z - z' \approx 0$. Also the integration of z' is actually over a single atomic thickness on the order of $(\Omega_0)^{1/3}$. Thus (11.88) becomes

$$< \Delta U \delta \Phi_0 > \approx 4\pi^2 \left(\Omega_0\right)^{1/3} \varsigma < [\Delta U(z,t)]^2 > \bar{\Phi}_0(z) \tag{11.89}$$

Substituting this relation into (11.83), we obtain

$$\frac{\partial}{\partial z} \bar{\Phi}_0 = \varsigma \left[\Xi + 4\pi^2 \left(U_0 + iU'\right)\right] \bar{\Phi}_0 \tag{11.90}$$

where the imaginary component of the crystal potential is defined as

$$U' \approx (\Omega_0)^{1/3} |\zeta| \left\langle |\Delta U(z,t)|^2 \right\rangle = (\Omega_0)^{1/3} |\zeta| \sum_\kappa \left\langle [\Delta U_\kappa(z - z_\kappa)]^2 \right\rangle \tag{11.91}$$

If U' is approximated to be independent of z, the solution of (11.90) is

$$\bar{\Phi}_0(z) = \exp\left\{\varsigma \left[\Xi + 4\pi^2 \left(U_0 + U'\right)\right] z\right\} \bar{\Phi}_0(0) \tag{11.92}$$

Equation (11.86) is applied to calculate the image and diffraction pattern of phonon-scattered electrons.

11.10.2 Atomic-Number-Sensitive Imaging in STEM—The Exact Theory

The HAADF-STEM image is formed by collecting high-angle scattered electrons using a ring-shaped detector in the diffraction plane when the small electron probe is scanned across the crystal lattice, as introduced in Sect. 11.5.1. The image is produced by high-angle Bragg-scattered, Huang-scattered, and thermal diffusely scattered electrons. The contribution of Huang scattering from defective regions can be introduced in the multislice calculation by using the modified phase-grating function for each slice. We now use real-space multislice theory to calculate the contribution made by TDS electrons (Wang 1995).

For a scan position $\mathbf{b}_p$ of the incident electron probe, the intensity distribution in the diffraction plane is the modulus square of the Fourier transform of (11.81)

$$I(\tau) = \left|\bar{\Phi}_0(\tau, d)\right|^2 + \left\langle|\delta\Phi_0(\tau, d)|^2\right\rangle \tag{11.93}$$

where the first term is the Bragg reflections, and the second term is TDS. We now use (11.86) to obtain the angular distribution of TDS electrons

$$\begin{aligned} I_{\mathrm{TDS}}(\tau) &=< |\delta\Phi_0(\tau, d)|^2 > \\ &= |4\pi^2\zeta|^2 < |\sum_\kappa \mathbf{FT}\{\int_0^d dz'\, O_p(\mathbf{b}, d - z') \\ &\quad \times [\Delta U_\kappa(\mathbf{b} - \mathbf{b}_\kappa, z' - z_k, t)\bar{\Phi}_0(\mathbf{b}, z')]\}|^2 > \end{aligned} \tag{11.94}$$

Equation (11.94) can be simplified using the approximation shown below. In general, the distribution of the electron wave may not be significantly affected by scattering a single atom layer. Also the deviation potential ΔU_κ is a localized function with a spatial variation of subatomic dimension. Thus along the beam direction (z-axis), ΔU_κ is the only rapidly varying function within the thickness of a thin slice. Therefore we can neglect dynamic diffraction for the single atom layer, so that the integration of z' is considered to act only on ΔU_κ. Equation (11.94) is approximated as

$$\begin{aligned} I_{\mathrm{TDS}}(\tau) &\approx |4\pi^2\varsigma|^2 < |\sum_\kappa \mathbf{FT}\{O_p\,(\mathbf{b}, d - z_\kappa)\,[\Delta U_\kappa\,(\mathbf{b} - \mathbf{b}_\kappa, t)\,\bar{\Phi}_0\,(\mathbf{b}, z_\kappa)]\}|^2 > \\ &= |4\pi^2\varsigma|^2 \sum_\kappa \sum_{\kappa'} \int d\mathbf{b} \int d\mathbf{b}' \exp[-2\pi i\tau \cdot (\mathbf{b} - \mathbf{b}')] \\ &\quad \times < \left\{O_p\,(\mathbf{b}, d - z_\kappa)\left[\Delta U_\kappa\,(\mathbf{b} - \mathbf{b}_\kappa, t)\,\bar{\Phi}_0\,(\mathbf{b}, z_\kappa)\right]\right\} \\ &\quad \times \left\{O_p^*\left(\mathbf{b}', d - z_{\kappa'}\right)\left[\Delta U_{\kappa'}\left(\mathbf{b}' - \mathbf{b}_{\kappa'}, t\right)\bar{\Phi}_0^*\left(\mathbf{b}', z_{\kappa'}\right)\right]\right\} > \end{aligned} \tag{11.95}$$

where

$$\Delta U_\kappa\,(\mathbf{b} - \mathbf{b}_\kappa, t) = \int dz' \Delta U_\kappa\left(\mathbf{b} - \mathbf{b}_\kappa, z' - z_\kappa, t\right)$$

is the projected atomic potential. Since $\mathbf{b}$ and $\mathbf{b}'$ are two distinct variables, so that $O_p\,(\mathbf{b}, d - z_\kappa)$ and $O_p^*\left(\mathbf{b}', d - z_{\kappa'}\right)$ commute

$$
\begin{aligned}
I_{\mathrm{TDS}}(\tau) \approx & |4\pi^2\varsigma|^2 \left| \sum_{\kappa}\sum_{\kappa'} \int d\mathbf{b} \int d\mathbf{b}' \exp\left[-2\pi i\tau\cdot(\mathbf{b}-\mathbf{b}')\right] \right. \\
& \times \left\{ O_p(\mathbf{b}, d-z_\kappa)\, O_p^*(\mathbf{b}', d-z_{\kappa'}) \right. \\
& \times \left[< \Delta U_\kappa(\mathbf{b}-\mathbf{b}_\kappa, t)\, \Delta U_{\kappa'}(\mathbf{b}'-\mathbf{b}_{\kappa'}, t) > \bar{\Phi}_0(\mathbf{b}, z_\kappa)\, \bar{\Phi}_0^*(\mathbf{b}', z_{\kappa'}) \right] \Big\}
\end{aligned}
\tag{11.96}
$$

Equation (11.96) gives the angular distribution of electrons in the diffraction pattern. The physical meaning of (11.96) can be described as follows. The real-space wave function of the elastically scattered wave at depth z is $\bar{\Phi}_0$; the source for generating diffuse scattering at the κth atom site is $(\Delta U_\kappa \bar{\Phi}_0)$. The subsequent dynamic diffraction of the diffusely scattered electrons for a thickness $d - z_\kappa$ is included in the O_p operator (see Sect. 2.1.3). The time-averaged term $< \Delta U_\kappa \Delta U_{\kappa'} >$ is the result of phase coupling between atomic vibrations, thus determining the coherent scattering behavior of the diffusely scattered electrons generated from atomic sites κ and κ'.

In STEM, if the incident electron probe is centered at $\mathbf{b}_p$ and the elastic wave at depth z is denoted by $\bar{\Phi}_0(\mathbf{b}-\mathbf{b}_p, z)$, the contribution of TDS electrons to the HAADF-STEM image is

$$
\begin{aligned}
I_{\mathrm{TDS}}(\mathbf{b}_p) &= \int d\tau\, I_{\mathrm{TDS}}(\tau) D(\tau) \\
&= |\, 4\pi^2\varsigma\, |^2 \sum_{\kappa}\sum_{\kappa'} \int d\mathbf{b} \int d\mathbf{b}' D(\mathbf{b}-\mathbf{b}') \left\{ O_p(\mathbf{b}, d-z_\kappa)\, O_p^*(\mathbf{b}', d-z_{\kappa'}) \right. \\
&\quad \times \left[< \Delta U_\kappa(\mathbf{b}-\mathbf{b}_\kappa, t)\, \Delta U_{\kappa'}\left(\mathbf{b}'-\mathbf{b}_{\kappa', t}\right) > \right. \\
&\quad \times \bar{\Phi}_0(\mathbf{b}-\mathbf{b}_p, z_\kappa)\, \bar{\Phi}_0^*(\mathbf{b}'-\mathbf{b}_p, z_{\kappa'})] \}
\end{aligned}
\tag{11.97}
$$

This expression fully incorporates the dynamic scattering of electrons before and after TDS. No simplification was made in considering the phonon dispersion relation. Approximations made in deriving (11.97) are high-energy electron diffraction without backscattering and first-order thermal diffuse scattering. Equation (11.97) is the unified imaging theory of HAADF-STEM. We now consider a few simplified cases of (11.97).

11.10.2.1 Case 1

No dynamic diffraction after TDS, $O_p(\mathbf{b}, d - z_\kappa) = 1$. Thus

$$
\begin{aligned}
I_{\mathrm{TDS}}(\mathbf{b}_p) = & |4\pi^2\zeta|^2 \sum_{\kappa}\sum_{\kappa'} \int d\mathbf{b} \int d\mathbf{b}'\, D(\mathbf{b}-\mathbf{b}')[< \Delta U_\kappa(\mathbf{b}-\mathbf{b}_\kappa, t) \Delta U_{\kappa'}(\mathbf{b}'-\mathbf{b}_{\kappa'}, t) > \\
& \times \overline{\Phi}_0(\mathbf{b}-\mathbf{b}_p, z_\kappa) \overline{\Phi}_0^*(\mathbf{b}'-\mathbf{b}_p, z_{\kappa'})]
\end{aligned}
\tag{11.98}
$$

It is apparent that the coherence of the HAADF-STEM image is partly determined by phase coupling between atomic vibrations and partly by the detection geometry of the ADF detector, as expected from kinematic scattering theory (Jesson and Pennycook 1993; Treacy and Gibson 1993).

11.10.2.2 Case 2

No dynamic diffraction after TDS and all and only TDS electrons are detected, i.e., $D(\mathbf{b}) = \delta(\mathbf{b})$. Thus

$$I_{\text{TDS}}\left(\mathbf{b}_p\right) = \mid 4\pi^2\varsigma \mid^2 \sum_{\kappa}\sum_{\kappa'} \int d\mathbf{b} [< \Delta U_{\kappa}(\mathbf{b}-\mathbf{b}_{\kappa}, t)\, \Delta U_{\kappa'}(\mathbf{b}-\mathbf{b}_{\kappa'}, t)\rangle \times \bar{\Phi}_0\left(\mathbf{b}-\mathbf{b}_p, z_{\kappa}\right) \bar{\Phi}_0^*\left(\mathbf{b}-\mathbf{b}_p, z_{\kappa'}\right)] \tag{11.99}$$

Since the width of ΔU_{κ} is much smaller than the interatomic distance, then for atoms located in the same crystal slice (i.e., at the same depth z_{κ})

$$< \Delta U_{\kappa}(\mathbf{b}-\mathbf{b}_{\kappa}, t)\, \Delta U_{\kappa'}(\mathbf{b}-\mathbf{b}_{\kappa'}, t) > \approx \langle |\Delta U_{\kappa}(\mathbf{b}-\mathbf{b}_{\kappa}, t)|^2 > \delta_{\kappa\kappa'} \tag{11.100}$$

This means that TDS electrons scattered by atoms in different atomic columns (along the beam) of the same slice are approximately incoherent (Wang 1992a; Jesson and Pennycook 1993; Treacy and Gibson 1993).

For atoms constrained within the same column along the beam direction, however, the coherence of TDS may be preserved, but coherent length is determined by phonon dispersion characteristics of the crystal, as discussed in Sect. 11.5.5. Thus (11.99) becomes

$$I_{\text{TDS}}\left(\mathbf{b}_p\right) = \mid 4\pi^2\varsigma \mid^2 \sum_{\kappa} \int d\mathbf{b} \left[< |\Delta U_{\kappa}(\mathbf{b}-\mathbf{b}_{\kappa}, t)|^2 > \left|\bar{\Phi}_0\left(\mathbf{b}-\mathbf{b}_p, z_{\kappa}\right)\right|^2\right] + \mid 4\pi^2\varsigma \mid^2 \sum_{\kappa}\sum_{\kappa'\neq\kappa} \int d\mathbf{b} [< \Delta U_{\kappa}(\mathbf{b}-\mathbf{b}_{\kappa}, t)\, \Delta U_{\kappa'}(\mathbf{b}-\mathbf{b}_{\kappa'}, t) > \bar{\Phi}_0\left(\mathbf{b}-\mathbf{b}_p, z_{\kappa}\right) \times \bar{\Phi}_0^*\left(\mathbf{b}-\mathbf{b}_p, z_{\kappa'}\right)] \tag{11.101}$$

where the sum of κ' is limited to atoms in the same column as the κ atom. The first term in (11.101) is the result of incoherent scattering, and the second term is the coherent scattering from atoms in the same column but located at a different depth z. It is important to note that the first term is obtained without making an assumption regarding the thermal vibration model of the crystal.

11.10.2.3 Case 3

No dynamic diffraction after TDS and no phase coupling between atomic vibrations (i.e., the Einstein model). Thus

$$I_{\mathrm{TDS}}(\mathbf{b}_p) = |4\pi^2\varsigma|^2 \sum_{\kappa} \int d\mathbf{b} \int d\mathbf{b}' D(\mathbf{b}-\mathbf{b}') \left[\langle \Delta U_{\kappa}(\mathbf{b}-\mathbf{b}_{\kappa}, t)\, \Delta U_{\kappa}(\mathbf{b}'-\mathbf{b}_{\kappa}, t) \rangle \right. \\ \left. \times \bar{\Phi}_0(\mathbf{b}-\mathbf{b}_p, z_{\kappa})\, \bar{\Phi}_0^*(\mathbf{b}'-\mathbf{b}_p, z_{\kappa}) \right] \tag{11.102}$$

Equation (11.102) is still unlike the incoherent imaging theory, even though the Einstein model has been assumed. Thus the finite width of the D function in real space partly determines the coherence of an HAADF-STEM image.

11.10.2.4 Case 4

No dynamic diffraction after TDS, no phase coupling between atom vibrations, and $D(\mathbf{b}-\mathbf{b}') = \delta(\mathbf{b}-\mathbf{b}')$. Thus

$$\begin{aligned} I_{\mathrm{TDS}}(\mathbf{b}_p) &= |4\pi^2\varsigma|^2 \sum_{\kappa} \int d\mathbf{b} \left[\langle |\Delta U_{\kappa}(\mathbf{b}-\mathbf{b}_{\kappa}, t)|^2 \rangle \left| \bar{\Phi}_0(\mathbf{b}-\mathbf{b}_p, z_k) \right|^2 \right] \\ &= \sum_{n_c} \int d\mathbf{b} |\Gamma(\mathbf{b}, z_{n_c})|^2 |\bar{\Phi}_0(\mathbf{b}-\mathbf{b}_p, z_k)|^2 \\ &= \sum_{n_c} |\Gamma(\mathbf{b}_p, z_{n_c})|^2 \otimes |\Phi_0(\mathbf{b}_p, z_{n_c})|^2 \end{aligned} \tag{11.103}$$

where the generation function given in (11.48) was used. It is apparent that Eq. (11.103) is identical to (11.44). The atomic-number-sensitive information is contained in term $\langle |\Delta U_{\kappa}|^2 \rangle$, which is a measure of the scattering power of each atomic column. The incoherent scattering of each column actually maximizes the Z sensitivity and makes it possible to identify atomic columns directly in the image.

11.10.2.5 Case 5

No dynamic diffraction after TDS, no phase coupling between atom vibrations $D(\mathbf{b}) = \delta(\mathbf{b})$, and no beam broadening [*i.e.*, $\bar{\Phi}_0(\mathbf{b}-\mathbf{b}_p, z_k) = \bar{\Phi}_0(\mathbf{b}-\mathbf{b}_p, 0)$]. Thus

$$I_{\mathrm{TDS}}\left(\mathbf{b}_p\right) = \left|4\pi^2\varsigma\right|^2 \sum_{\kappa} \int d\mathbf{b}\left[< |\Delta U_{\kappa}\left(\mathbf{b}-\mathbf{b}_{\kappa}, t\right)|^2 > |\bar{\Phi}_0\left(\mathbf{b}-\mathbf{b}_p, 0\right)|^2\right]$$
$$= \left[\sum_{n_c} |\Gamma(\mathbf{b}_p, z_{n_c})|^2\right] \otimes |\Phi_0\left(\mathbf{b}_p, 0\right)|^2 \tag{11.104}$$

This equation is the result of incoherent imaging theory, in which $\left(\sum_{n_c} |\Gamma|^2\right)$ is the projected TDS generation function for the entire crystal and $|\Phi_0(\mathbf{b}, 0)|^2$ is the shape function of the incident probe. It is thus clear that the incoherent imaging model of HAADF-STEM is a very simplified case of the practical situation. It also appears to us that the condition $D(\mathbf{b}) = \delta(\mathbf{b})$ is vital to the incoherent imaging model.

11.10.3 Multislice Calculation of Dynamic Scattering Operator

The introduction of the operator O_p has greatly simplified our derivation of the final expression. Image simulations require numerical calculations of a function in the form:

$$O_p\left(\mathbf{b}, d - z_{\kappa}\right) X\left(\mathbf{b}, z_{\kappa}\right) = \exp\left[\varsigma\left[\Xi + 4\pi^2 U_0\right]\left(d - z_{\kappa}\right)\right] X\left(\mathbf{b}, z_{\kappa}\right) \tag{11.105}$$

where $X(\mathbf{b}, z)$ is assumed to be an arbitrary function. We now use the Cowley-Moodie multislice method to perform this calculation. If the crystal is cut into a total of M slices of equal thickness Δz (i.e., $d = M\Delta z$) and we assume that the κth atom is located in the n_cth slice, then

$$O_p(\mathbf{b}, d - z_{n_c}) X(\mathbf{b}, z_{n_c}) = \exp\left\{\varsigma\left(\nabla_{\mathbf{b}}^2 + 4\pi i \mathbf{K}_{\mathbf{ob}} \cdot \nabla_{\mathbf{b}} + 4\pi^2 U_0\right) \times [(M - n_c)\Delta z]\right\} X(\mathbf{b}, z_{n_c}) \tag{11.106}$$

If the slice thickness is sufficiently small so that for each slice, the $\varsigma\Delta z\left(\nabla_b^2 + 4\pi i \mathbf{K}_{\mathrm{ob}} \cdot \nabla_b\right)$ operator and the scattering function $4\pi^2\varsigma\Delta z U_0$ approximately commute, then the second-order term is proportional to Δz^2; thus for the n_cth slice:

$$\exp\left[\varsigma\Delta z\left(\nabla_b^2 + 4\pi i \mathbf{K}_{\mathrm{ob}} \cdot \nabla_b + 4\pi^2 U_0\right)\right] X\left(\mathbf{b}, z_{n_c}\right)$$
$$\approx \exp\left[\varsigma\Delta z\left(\nabla_b^2 + 4\pi i \mathbf{K}_{\mathrm{ob}} \cdot \nabla_b\right)\right] \exp\left(4\pi^2\varsigma\Delta z U_0\right) X\left(\mathbf{b}, z_{n_c}\right)$$
$$= \exp\left[\varsigma\Delta z\left(\nabla_b^2 + 4\pi i \mathbf{K}_{\mathrm{ob}} \cdot \nabla_b\right)\right]\left[Q\left(\mathbf{b}, z_{n_c}\right) X\left(\mathbf{b}, z_{n_c}\right)\right] \tag{11.107}$$

where $Q\left(\mathbf{b}, z_{n_c}\right) = \exp\left(4\pi^2\varsigma\Delta z U_0\right)$ is the phase-grating function of the slice. We now consider the following operation on an arbitrary function $f(b)$, for which the Fourier transform is $f(\mathbf{u})$. We find that

$$
\begin{aligned}
&\exp\left[\zeta\Delta z\left(\nabla_b^2 + 4\pi i\mathbf{K}_{\text{ob}}\cdot\nabla_b\right)\right] f(\mathbf{b}) \\
&= \exp\left[\zeta\Delta z\left(\nabla_b^2 + 4\pi i\mathbf{K}_{\text{ob}}\cdot\nabla_b\right)\right]\int d\mathbf{u}\, f(\mathbf{u})\exp(2\pi i\mathbf{u}\cdot\mathbf{b}) \\
&= \int d\mathbf{u}\, f(\mathbf{u})\left\{\exp\left[\zeta\Delta z\left(\nabla_b^2 + 4\pi i\mathbf{K}_{\text{ob}}\cdot\nabla_b\right)\right]\exp(2\pi i\mathbf{u}\cdot\mathbf{b})\right\} \\
&= \int d\mathbf{u}\, f(\mathbf{u})\left\{\sum_{m=0}\frac{\left[\zeta\Delta z\left(\nabla_b^2 + 4\pi i\mathbf{K}_{\text{ob}}\cdot\nabla_b\right)\right]^m}{m!}\right\}\exp(2\pi i\mathbf{u}\cdot\mathbf{b}) \\
&= \int d\mathbf{u}\, f(\mathbf{u})\left\{\sum_{m=0}\frac{\left[-4\pi^2\zeta\Delta z\left(u^2 + 2\mathbf{K}_{\text{ob}}\cdot\mathbf{u}\right)\right]^m}{m!}\right\}\exp(2\pi i\mathbf{u}\cdot\mathbf{b}) \\
&= \int d\mathbf{u}\, f(\mathbf{u})\exp\left[-4\pi^2\zeta\Delta z\left(u^2 + 2\mathbf{K}_{\text{ob}}\cdot\mathbf{u}\right)\right]\exp(2\pi i\mathbf{u}\cdot\mathbf{b}) \\
&= \int d\mathbf{u}\, f(\mathbf{u})\exp\left[-\pi i\left(u^2 + 2\mathbf{K}_{\text{ob}}\cdot\mathbf{u}\right)\Delta z\lambda\right]\exp(2\pi i\mathbf{u}\cdot\mathbf{b}) \\
&= f(\mathbf{b})\otimes\left(\mathbf{FT}\left\{\exp\left[-\pi i\left(u^2 + 2\mathbf{K}_{\text{ob}}\cdot\mathbf{u}\right)\Delta z\lambda\right]\right\}\right) \\
&= f(\mathbf{b})\otimes P(\mathbf{b},\Delta z)
\end{aligned}
\tag{11.108}
$$

where P is a propagation function

$$
P(\mathbf{b},\Delta z) = \frac{1}{i\lambda\Delta z}\exp\left(\frac{\pi i\left|\mathbf{b}{-}\mathbf{K}_{0b}\lambda\Delta z\right|^2}{\lambda\Delta z}\right) \tag{11.109}
$$

Thus

$$
\exp\left[\zeta\Delta z\left(\nabla_b^2 + 4\pi i\mathbf{K}_{\text{ob}}\cdot\nabla_b\right)\right] X\left(\mathbf{b}, z_{n_c}\right) \approx \left[Q\left(\mathbf{b}, z_{n_c}\right) X\left(\mathbf{b}, z_{n_c}\right)\right]\otimes P(\mathbf{b},\Delta z) \tag{11.110}
$$

It is important to note that the right-hand side of (11.110) is just the Cowley-Moodie multislice theory. Finally we have approximately

$$
\begin{aligned}
&O_p\left(\mathbf{b}, d - z_{n_c}\right) X\left(\mathbf{b}, z_{n_c}\right) \\
&= \exp\{\varsigma\Delta z\sum_{s=n_c}^{M}\left[\left(\nabla_b^2 + 4\pi i\mathbf{K}_{\text{ob}}\cdot\nabla_b\right) + 4\pi^2 U_0\right]\} X\left(\mathbf{b}, z_s\right) \\
&\approx \prod_{s=n_c}^{M}\exp\left\{\varsigma\Delta z\left[\left(\nabla_b^2 + 4\pi i\mathbf{K}_{\text{ob}}\cdot\nabla_b\right) + 4\pi^2 U_0\right]\right\} X\left(\mathbf{b}, z_s\right) \\
&\approx \{Q_0(\mathbf{b}, z_M)\{Q_0\left(\mathbf{b}, z_{M-1}\right)\cdots\{\left[Q_0\left(\mathbf{b}, z_{n_c}\right) X\left(\mathbf{b}, z_{n_c}\right)\right]\otimes P_0(\mathbf{b},\Delta z)\} \\
&\cdots\}\otimes P_0(\mathbf{b},\Delta z)\}\otimes P_0(\mathbf{b},\Delta z)
\end{aligned}
\tag{11.111}
$$

which is just the multislice calculation of the electron wave $X(\mathbf{b}, z)$ scattered from the n_cth slice to the exit face of the crystal. Thus the $O_p\left(\mathbf{b}, d - z_\kappa\right)$ operator represents dynamic diffraction of the electron wave from $z = z_\kappa$ to $z = d$. This calculation does not involve a time-dependent quantity. The diffraction effect of the electron probe and thickness-dependent probe broadening are completely covered in the multislice calculation. Since the phase-grating function Q_0 can be chosen to be different for each slice, it is thus possible to introduce static defects or dislocation in the multislice calculation. Therefore Huang scattering is automatically included. For atoms located in the n_cth slice, the total number of slice calculations is $M - n_c$. For the entire crystal, the total number of slice calculations to be performed is

$$M_t = \sum_{n_c=1}^{M} (M - n_c) = \frac{M(M-1)}{2}$$

For a crystal less than 10 − nm thick, we can take $M = 50$, $M_t = 1225$. The calculation becomes very large if crystal thickness exceeds 100 nm.

It is important to note that the theory has fully incorporated phonon dispersion relations of the crystal without assuming the Einstein model. This is an important advantage of the theory. In the dynamic theory just proposed, the amount of calculation is minimized for the time average on the TDS process, but the amount of elastic multislice calculation is increased by a factor of $M/2$. Therefore the amounts of calculation for TDS cannot be dramatically reduced without making further approximations.

11.10.4 Atomic-Number-Sensitive Imaging in TEM—The Exact Theory

Details regarding the formation of HADF-TEM images were illustrated in Sect. 11.6.1, but the theory in Sect. 11.6.2 ignored the dynamic diffraction effect of electrons after TDS. This may produce a large error in practice due to the small size of the objective aperture. Here we present the real-space multislice calculation of the HADF-TEM image, including the full dynamic diffraction effect.

Based on Abbe's imaging theory, the contribution of TDS electrons to the HADF-TEM image is

$$I_{\text{TDS}}(\mathbf{b}) = < \left| \int d\mathbf{b}' \delta\Phi_0\left(\mathbf{b}', d\right) T_{\text{obj}}\left(\mathbf{b}' - \mathbf{b}\right) \right|^2 > \tag{11.112}$$

Substituting the wave function of the diffusely scattered electrons given by (11.86) into (11.112), we have

$$I_{\mathrm{TDS}}(\mathbf{b}) = \left|4\pi^2\varsigma\right|^2 < \left|\sum_{\kappa}\{\int d\mathbf{b}' \int_0^d dz' T_{\mathrm{obj}}\left(\mathbf{b}'-\mathbf{b}\right) O_p\left(\mathbf{b}', d-z'\right)\right.$$
$$\left. \times \left[\Delta U_{\kappa}\left(\mathbf{b}'-\mathbf{b}_{\kappa}, z'-z_{\kappa}, t\right) \bar{\Phi}_0\left(\mathbf{b}', z'\right)\right]\}\right|^2 > \tag{11.113}$$

If the dynamic diffraction effect of a single atomic layer is neglected, following the same argument as for (11.94) and (11.95), the integration of z' can be directly applied to ΔU_{κ}. Thus (11.113) is approximated as

$$\begin{aligned} I_{\mathrm{TDS}}(\mathbf{b}) &= \left|4\pi^2\varsigma\right|^2 < \left|\sum_{\kappa}\{\int d\mathbf{b}' T_{\mathrm{obj}}\left(\mathbf{b}'-\mathbf{b}\right) O_p\left(\mathbf{b}', d-z_{\kappa}\right)\right. \\ &\left. \times \left[\Delta U_{\kappa}\left(\mathbf{b}'-\mathbf{b}_{\kappa}, t\right) \bar{\Phi}_0\left(\mathbf{b}', z_{\kappa}\right)\right]\}\right|^2 > \\ &= \left|4\pi^2\varsigma\right|^2 \sum_{\kappa}\sum_{\kappa'} \int d\mathbf{b}' \int d\mathbf{b}''\{T_{\mathrm{obj}}\left(\mathbf{b}'-\mathbf{b}\right) T_{\mathrm{obj}}^{*}\left(\mathbf{b}''-\mathbf{b}\right) \\ &\times < \{O_p\left(\mathbf{b}', d-z_{\kappa}\right) [\Delta U_{\kappa}\left(\mathbf{b}'-\mathbf{b}_{\kappa}, t\right) \bar{\Phi}_0\left(\mathbf{b}', z_{\kappa}\right)] O_p^{*}\left(\mathbf{b}'', d-z_{\kappa'}\right) \\ &\times [\Delta U_{\kappa'}\left(\mathbf{b}''-\mathbf{b}_{\kappa'}, t\right) \bar{\Phi}_0^{*}\left(\mathbf{b}'', z_{\kappa'}\right)]\} >\} \\ &= \left|4\pi^2\varsigma\right|^2 \sum_{\kappa}\sum_{\kappa'} \int d\mathbf{b}' \int d\mathbf{b}'' \left\{T_{\mathrm{obj}}\left(\mathbf{b}'-\mathbf{b}\right) T_{\mathrm{obj}}^{*}\left(\mathbf{b}''-\mathbf{b}\right)\right. \\ &\times O_p\left(\mathbf{b}', d-z_{\kappa}\right) O_p^{*}\left(\mathbf{b}'', d-z_{\kappa'}\right) \\ &\times [< \Delta U_{\kappa}\left(\mathbf{b}'-\mathbf{b}_{\kappa}, t\right) \Delta U_{\kappa'}\left(\mathbf{b}''-\mathbf{b}_{\kappa'}, t\right) > \bar{\Phi}_0\left(\mathbf{b}', z_{\kappa}\right) \bar{\Phi}_0^{*}\left(\mathbf{b}'', z_{\kappa'}\right)]\} \end{aligned} \tag{11.114}$$

Calculation of the HADF-TEM image is identical to that of HAADF-STEM (11.97) if the transfer function $T_{\mathrm{obj}}\left(\mathbf{b}'-\mathbf{b}\right) T_{\mathrm{obj}}^{*}\left(\mathbf{b}''-\mathbf{b}\right)$ is replaced by the detector function $D\left(\mathbf{b}'-\mathbf{b}''\right)$. This is the result of the reciprocity theorem. Therefore Cases 1–5 in Sect. 11.10.2 also apply to the TEM case. But the approximation that ignores the Bragg scattering after TDS (i.e., take $O_p = 1$) does not hold in TEM due to the small size of the objective aperture; thus the electron dynamic scattering of the objective aperture becomes important.

Equation (11.114) can also be applied to calculate the high-resolution image formed by TDS electrons. In this case, the single-phonon approximation in (11.114) is sufficient.

11.10.5 Dislocation Contrast Due to Huang Scattering

Diffuse scattering can be generated by both TDS and Huang scattering. The TDS can be excited from wherever there are atoms, but Huang scattering can be generated only from regions containing imperfections, such as defects or dislocations. For a perfect crystal, if the image resolution is not sufficient to resolve lattice planes, the contribution from TDS is a background in the HADF-TEM image. In this case, contrast can be introduced due to Huang scattering from regions containing defects or dislocations. This result has been observed experimentally, and a image contrast mechanism has been proposed (see Sect. 11.8).

Equation (11.114) can be modified to calculate the image formed by Huang scattering. For simplification, we assume that the spherical aberration coefficient of the objective lens $C_s = 0$, i.e., $T_{\text{obj}}\left(\mathbf{b}' - \mathbf{b}\right) = \delta\left(\mathbf{b}' - \mathbf{b}\right)$. This is an excellent approximation for diffraction contrast imaging. Huang scattering differs from TDS, since scattering from one atomic site has no time-dependent phase correlation with scattering from other atomic sites. Based on (11.114), the image formed by Huang-scattered electrons can be calculated by

$$I_H(\mathbf{b}) \approx |4\pi^2\varsigma|^2 \left| \int_0^d dz\, O_p(\mathbf{b}, d-z)\left[\Delta U(\mathbf{b}, z)\Phi_0(\mathbf{b}, z)\right] \right|^2 \tag{11.115a}$$

with

$$\Delta U(\mathbf{r}) = \sum_{\kappa} \{U_\kappa\left(\mathbf{r} - \mathbf{r}_\kappa - \mathbf{R}(\mathbf{r})\right) - U_\kappa\left(\mathbf{r} - \mathbf{r}_\kappa\right)\} \tag{11.115b}$$

where Φ_0 is the wave function of the incident electron; ΔU is the distorted crystal potential due to local lattice displacement $\mathbf{R}(\mathbf{r})$ introduced by defects; the product $[\ 4\pi^2\varsigma\Delta\Phi_0\ ]$ is the diffuse scattering generated at depth z; and the subsequent dynamic scattering of Huang-scattered electrons is governed by the O_p operator (see Sect. 11.10.3). The integration of z sums over all diffuse scattering generated in the entire crystal thickness. We now apply the Bloch wave theory to perform the O_p calculation.

In HADF-TEM, since the angle between the incident electron beam and the optic axis of the microscope is large, no strong dynamic scattering is expected if the beam direction is far from the crystal zone axes. Thus the incident beam is still a plane wave when it strikes the defect, i.e.:

$$\Psi_0(\mathbf{b}, z) \approx \exp\left(2\pi i \mathbf{K}_b \cdot \mathbf{b} + 2\pi i K_z z\right)$$

where $\mathbf{K}_b$ is the projection of the electron wave vector in the $x - y$ plane. After being diffusely scattered by the defect region, electrons that contribute to the HADF-TEM image propagate parallel or nearly parallel to the optic axis, so that the column approximation can be made. Figure 11.15 shows a model for image calculation in **HADF-TEM**. The column starts from depth z_1 of the dislocation till the exit face of the crystal $z = d$. For each column, diffuse scattering at the entrance of the column is $\left[4\pi^2\varsigma\Delta U\,\Phi_0(\mathbf{b}, z)\right]$, and the wave in the column can be expressed in Bloch waves:

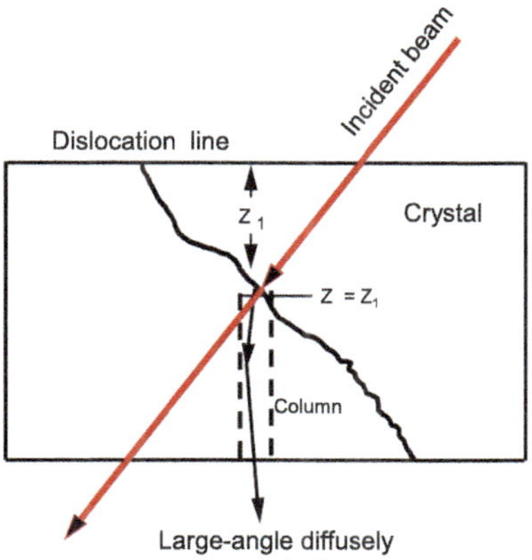

Fig. 11.15 A schematic diagram showing the electron-scattering model in HADF-TEM. Diffuse scattering occurs at $z = z_1$. Subsequent diffraction of the diffusely scattered electrons from z_1 to d is determined by the diffracting condition set up for on-axis imaging. The dashed line indicates the column to be used in image calculation

$$\Psi(\mathbf{K}, \mathbf{r}) = \sum_i \alpha_i(\mathbf{b}) \sum_g C_g^{(i)} \exp\left[2\pi i K_z \cdot z + 2\pi i \mathbf{g} \cdot \mathbf{b} + 2\pi i v_i (z - z_1)\right] \tag{11.116}$$

where z_1 is the depth of the defect region. Matching the boundary condition

$$\Psi(\mathbf{K}, \mathbf{r}) = 4\pi^2 \varsigma \Delta U(\mathbf{b}) \Psi_0(\mathbf{b}, z)$$

at the top of the column $z = z_1$, where

$$\Delta U(\mathbf{b}) = \int dz \Delta U(\mathbf{b}, z)$$

is the projected perturbation potential, the $\alpha_i(\mathbf{b})$ coefficient can be determined

$$\alpha_i(\mathbf{b}) = 4\pi^2 \varsigma C_0^{(i)*} \Delta U(\mathbf{b}) \tag{11.117}$$

If the on-axis objective aperture selects only the $g = 0$ reflection, the intensity of the image is

$$\begin{aligned} I_H &= |\sum_i \alpha_i(\mathbf{b}) C_0^{(i)} \exp[2\pi i \nu_i (d - z_1)]|^2 \\ &= |4\pi^2 \varsigma|^2 [\Delta U(\mathbf{b})]^2 |\sum_i C_0^{(i)*} C_0^{(i)} \exp[2\pi i \nu_i (d - z_1)]|^2 \end{aligned} \tag{11.118}$$

To see the meaning of (11.118), we consider the two-beam approximation. Using the $C_0^{(i)}$ coefficients given in (2.21),

$$I_H = | 4\pi^2 \varsigma |^2 [\Delta U(\mathbf{b})]^2 \left\{1 - \sin^2 \gamma_0 \sin^2 [\pi \Delta \nu (d - z_1)]\right\} \tag{11.119a}$$

where

$$\Delta\nu = \nu_1 - \nu_2 = \frac{[(K\,S_g)^2 + |U_g|^2]^{1/2}}{K_z} \qquad \gamma_0 = \text{arcctg}\left(\frac{K\,S_g}{|U_g|}\right) \tag{11.119b}$$

U_g is the Fourier coefficient of U and S_g is the excitation error. Equation (11.119a) shows that the contrast in the HADF-TEM image can be produced by four sources: (1) The spatial variation of ΔU is due to different degrees of local lattice distortion. This effect is useful for identifying dislocation cores. (2) Change in the diffracting condition due to variation in crystal orientation as a result of crystal bending, for example, leads to a change in the excitation error S_g (or γ_0). This effect has been observed experimentally. (3) The depth z_1 of the dislocation can introduce a thickness-dependent effect, making it possible to identify the head and tail of a dislocation line (Wang 1994). (4) Variation in the incident beam direction introduces a slight change in K_z. This may also give some contrast according to (11.119a).

For the Bragg reflection, so that the excitation error $S_g = 0$, (11.119a) reduces to

$$I_H = \{|\ 4\pi^2\varsigma\ |^2\ [\Delta U(\mathbf{b})]^2\}\cos^2[\pi\Delta\nu\,(d - z_1)] \tag{11.120}$$

Equation (11.120) has the same form as that for conventional on-axis bright-field diffraction contrast imaging [see (2.22)]:

$$I_0 = \cos^2(\pi d\,\Delta\nu)$$

Therefore the HADF-TEM shares many characteristics with conventional diffraction contrast imaging, in agreement with Fig. 11.12. But the thickness dependence of the two types of images may be different.

The theories presented so far are formal dynamic approaches that emphasize diffraction effects in imaging using diffusely scattered electrons. Here we present some experimental observations that demonstrate the importance of effects. The dynamic diffraction effect can be used in HADF-TEM by controlling the diffracting condition, so that only diffusely scattered electrons are selected to form the image. Figure 11.16a shows a double-exposed electron diffraction pattern of a single crystalline gold foil oriented near the [100]-zone axis. Figure 11.16b is a bright-field image of the foil, and Fig. 11.16c is the corresponding dark-field image recorded with the diffusely scattered electrons, as indicated by an arrowhead in Fig. 11.16a. In the diffraction pattern, <110> streaks produced by TDS are clearly seen. The bright-field image shows some bending and strain contrast. Most features observed in the bright-field image are shown in the dark-field image of the diffusely scattered electrons. The contrast in Fig. 11.16b and c is reversed, and the figures appear to be complementary to each other, in agreement to the theoretically expected result just presented. This observation clearly shows the importance of the diffraction effect in the imaging of diffusely scattered electrons.

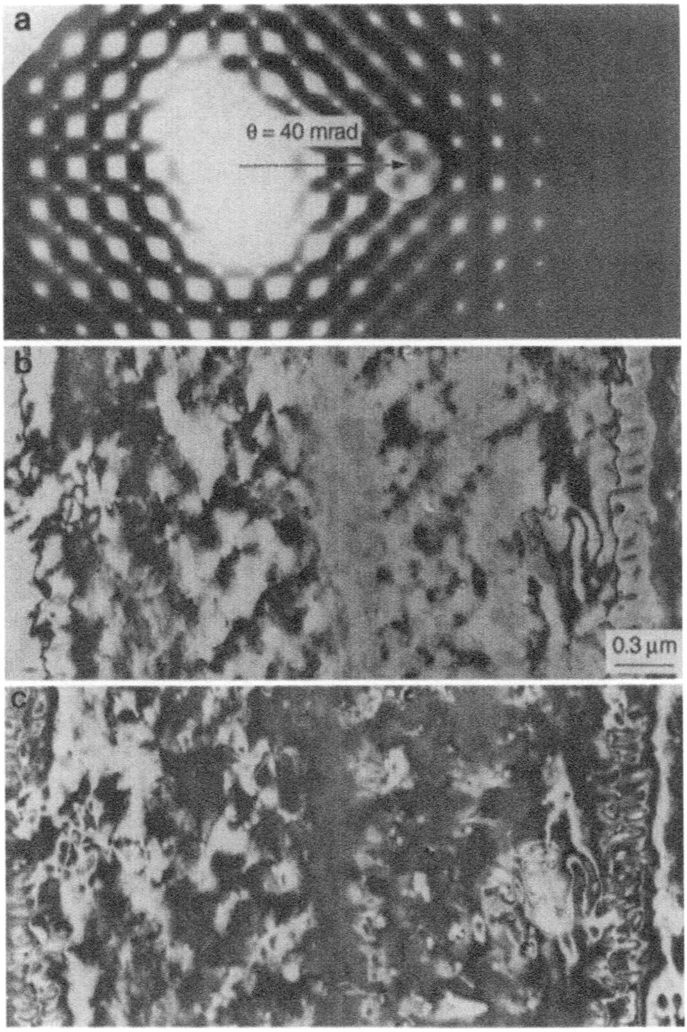

Fig. 11.16 The diffraction effect in dark-field image using diffusely scattered electrons. **a** Double-exposed electron diffraction pattern of a gold film. **b** and **c** Bright-field and dark-field images of the film, respectively; electron beam energy 300 keV (Wang 1995)

In this section, formal dynamic theories of atomic-number-sensitive images formed by thermal diffusely scattered electrons are proposed from first principles. Theories are derived for simulating images obtained either in STEM using a high-angle annular dark-field detector or in TEM using an on-axis objective aperture under the hollow cone beam illumination. No approximation was made in treating the thermal vibration of crystal atoms. The time average of signal intensities contributed by thermal diffuse scattering of various instantaneous crystal lattice configurations was performed before numerical calculation. The following effects are comprehensively covered in the theory: dynamic diffraction of the beam before and after TDS, thickness-dependent beam broadening or channeling, Huang scattering from defected regions, coherence of thermal diffusely scattered electrons generated from atomic layers packed within the coherent length, multiphonon excitations, and detection geometry. The theory is formulated in the multislice scheme, which is best suited for practical numerical calculations, especially when defects and interfaces are present.

11.11 Summary

As a summary of this chapter, multislice theories were proposed from quantum mechanics to approach the multiple elastic and multiple inelastic scattering of high-energy electrons in a crystal that contains defects. Applications of theories for calculating images of phonon-scattered electrons were described. The multislice theory introduced in the first part of this chapter can, in principle, be applied to calculate the diffraction pattern of electrons after inelastic excitation of any kind. Practical examples demonstrated that multislice simulations do, in fact, agree rather well with experimental images.

The real-space multislice theory introduced at the last part is more powerful for calculating diffraction patterns and images of **TDS** electrons. This theory allows the time average over the scattering intensity to be performed analytically before making numerical calculation. Exact theories are introduced for simulating atomic-number-sensitive images (or Z-contrast) formed by TDS electrons in STEM and TEM. The most important advantage of these theories is the inflexibility for introducing desired defects in numerical calculations. This is unique, especially for image simulations.

Chapter 12
Dynamic Inelastic Electron Scattering IV: Multislice Theory (2)

In Chap. 11, we introduced the dynamic multislice theory that fully incorporates elastic scattering of electrons before and after inelastic excitations. The theory can, in principle, be applied to any inelastic excitation and the possible multiple scattering effects in crystals containing defects. Large amounts of calculations are usually required to include dynamic scattering of electrons after inelastic scattering, because the excitation of each state and the subsequent diffraction of the electron associated with this excitation must be calculated individually due to incoherence. The major obstacle to this approach is the involvement of many crystal excited states.

In this chapter, we introduce a modified multislice approach, which is derived from a simplified model of elastic scattering of electrons after an inelastic event, so that the number of calculations has been greatly reduced (Wang and Bentley 1991a, b; Wang 1992a, b). Final calculated results show reasonable agreement with experimental observations. We first derive a generalized equation of the theory and apply it to the diffraction of TDS electrons.

12.1 General Theory

The following approach was initiated by modifying the Bloch wave approach in Chap. 8. Instead of writing the inelastic wave function as a sum of products of z-dependent coefficients with Bloch waves (8.2), we assume that the solution to (6.4a) and (6.4b) can be written in the following form (Wang and Bentley 1991b):

$$\Psi_n \equiv \phi_n(\mathbf{r})\Psi_n^0(\mathbf{r}) \tag{12.1}$$

where $\Psi_n^0(\mathbf{r})$ is the solution of (12.2) for an incident plane wave of wave vector $\mathbf{K}_n$, such that

Z. L. Wang, *Elastic and Inelastic Scattering in Electron Diffraction and Imaging*, Springer Series in Solid-State Sciences 205, https://doi.org/10.1007/978-3-031-90819-4_12

$$\left[\nabla^2 + 4\pi^2\left(K_n^2 + U\right)\right]\Psi_n^0 = 0 \tag{12.2}$$

It is important to point out that $\Psi_n^0(\mathbf{r})$ is the full solution of (12.2) and satisfies the boundary condition of

$$\Psi_n^0(\mathbf{b}, 0) = \exp\left(2\pi i \mathbf{K}_n \cdot \mathbf{r}\right)$$

Simply speaking, $\Psi_n^0(\mathbf{r})$ characterizes elastic scattering of electrons after an inelastic event, and $\phi_n(\mathbf{r})$ is considered a localization function of the inelastic excitation, which satisfies the boundary conditions $\phi_n(\mathbf{b}, 0) = \delta_{n0}$. The principle for introducing the solution of (12.1) is illustrated in Fig. 12.1, in which the final inelastic wave is expressed as the localization function multiplied by the elastic-scattering wave resulting from the incident wave of $\left[\exp\left(2\pi i \mathbf{K}_n \cdot \mathbf{r}\right)\right]$. Since the elastic rescattering of an inelastic wave generated at $\mathbf{r}$ inside the crystal with wave vector $\mathbf{K}_n$ is similar to the elastic scattering of an incident wave with wave vector $\mathbf{K}_n$, then the $\Psi_n^0(\mathbf{r})$ function, in principle, covers part of the elastic rescattering effect of electrons after inelastic excitation. This model works if we can find the solution to $\phi_n(\mathbf{r})$.

Substituting (12.1) into (6.6), since the elastic rescattering of the inelastic electron has partly been included in $\Psi_n^0(\mathbf{r})$, then the $\nabla^2\phi_n(\mathbf{r})$ term may be ignored (this is the first major approximation of this modified multislice method). Thus for transmission electron diffraction satisfying $K_{x,y} \ll K_z$, we have

$$\frac{\partial \phi_n(\mathbf{r})}{\partial z} \approx \beta_0 \sum_{m \neq n} X_{nm}(\mathbf{r})\phi_m(\mathbf{r}) \tag{12.3}$$

where

$$X_{nm} \equiv \frac{H_{nm}(\mathbf{r})\Psi_m^0(\mathbf{r})}{\frac{\partial \Psi_n^0(\mathbf{r})}{\partial z}}, \quad \beta_0 = \frac{m_e}{\hbar^2} \tag{12.4}$$

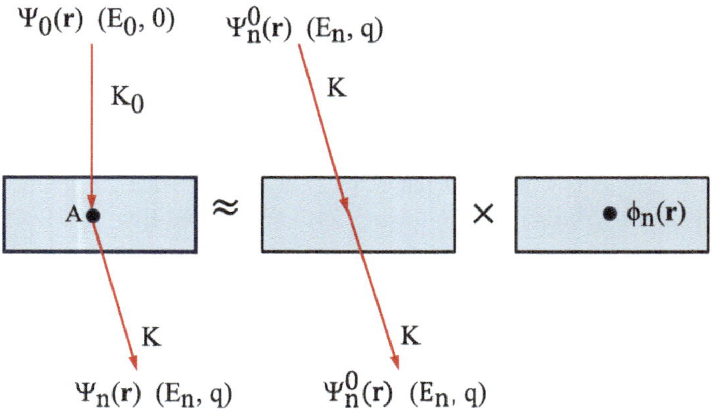

Fig. 12.1 The principle for introducing the solution in (12.1) (see text)

For the convenience of the following calculation, we write (12.3) in matrix form:

$$\frac{\partial\phi(\mathbf{r})}{\partial z} = \beta_0 \mathbf{X}(\mathbf{r})\phi(\mathbf{r}) \tag{12.5}$$

where diagonal elements of $\mathbf{X}$ are all zero. To treat a crystal with defects, the crystal is cut into many slices perpendicular to or nearly perpendicular to the incident beam direction. The atomic arrangement in each slice can be different, and the crystal is sliced so thin that the variation of $\mathbf{X}$ in the z-direction within a crystal slice is negligible. This is analogous to the slice method proposed by Cowley and Moodie (1957, see Chap. 3), but the mathematical operation here is different. We can now use this method to find the solution of (12.5) for transmission electron scattering of a thin crystal slice. The waves $\phi_n(\mathbf{b}, z_0)$ going into a crystal slice and those $\phi_n(\mathbf{b}, z)$ after being scattered by the slice are related by

$$\phi(\mathbf{b}, z) = \exp\left[\beta_0 \int_{z_0}^{z} dz \mathbf{X}(\mathbf{r})\right] \phi(\mathbf{b}, z_0) \tag{12.6}$$

If the crystal is cut into M slices from $z = 0$ to $z = d$, the application of (12.6) to each successive slice gives

$$\phi(\mathbf{b}, d) = \exp\left[\beta_0 \mathbf{X}_M(\mathbf{b})\right] \cdots \exp\left[\beta_0 \mathbf{X}_{n_c}(\mathbf{b})\right] \cdots \exp\left[\beta_0 \mathbf{X}_1(\mathbf{b})\right] \phi(\mathbf{b}, 0) \tag{12.7}$$

where

$$\mathbf{X}_{n_c}(\mathbf{b}) = \int_{z_{n_c-1}}^{z_{n_c}} dz \mathbf{X}(\mathbf{r})$$

is the $x - y$ plane projected matrix and $\phi_n(\mathbf{b}, 0) = \delta_{n0}$.

12.2 Single Inelastic Scattering

Now consider the case of only one excited state. The crystal can be either in its ground state of energy 0 and momentum 0 or in its excited state of energy ε_1 and momentum $\mathbf{q}$. By neglecting the absorption effect to the elastic wave (i.e., taking $H_{01} = 0$), then

$$\mathbf{X}_{n_c} \approx \begin{pmatrix} 0 & 0 \\ X_m & 0 \end{pmatrix} \tag{12.8a}$$

so that

$$\exp\left(\beta_0 \mathbf{X}_{n_c}\right) = \begin{pmatrix} 0 & 0 \\ \exp(\beta_0 X_{n_c}) & 0 \end{pmatrix} \tag{12.8b}$$

Substituting (12.8b) into (12.7), and assuming the inelastic scattering is much smaller than elastic scattering:

$$\left|\beta_0 \int_0^d dz X_{10}\right| \ll 1$$

We have

$$\phi_0 \approx 1 \tag{12.9a}$$

$$\phi_1 = \beta_0 \sum_{n_c=1}^{M} X_{n_c}(\mathbf{b}) \approx \beta_0 \int_0^d dz X_{10}(\mathbf{r})\phi_0 \tag{12.9b}$$

This is the criterion that limits the crystal thickness to ensure the validity of the single inelastic-scattering approximation. Substituting (12.9b) into (12.1), we obtain

$$\Psi_1(\mathbf{b}, \mathbf{q}, d) = \beta_0 \int_0^d dz \left[H_{10}(\mathbf{r}, \mathbf{q})Y(\mathbf{b}, \mathbf{q}, z)\right] \tag{12.10a}$$

where

$$Y(\mathbf{b}, \mathbf{q}, z) \equiv \Psi_0^0(\mathbf{r}) \left[\partial\Psi_0^0(\mathbf{r}, \mathbf{q})/\partial z\right]^{-1} \Psi_0^0(\mathbf{b}, \mathbf{q}, d) \tag{12.10b}$$

and $\Psi_0^0(\tau, \mathbf{q}, d)$ is the solution of (12.2) for an electron with energy loss ε_1 and momentum transfer $\mathbf{q}$. For the nth state, we analogously have

$$\Psi_n(\mathbf{b}, \mathbf{q}, d) = \beta_0 \int_0^d dz \left[H_{n0}(\mathbf{r}, \mathbf{q})Y(\mathbf{b}, \mathbf{q}, z)\right] \tag{12.10c}$$

This is a general solution for single inelastic scattering. The physical meaning of (12.10) can be simply stated as follows. Inelastic waves can be generated at any point inside the specimen with a probability function proportional to $|H_{10}|^2$. Inelastic events occurring at different atom locations are marked with a historic tag

$$\Psi_0^0(\mathbf{r}) \left\{\partial\Psi_1^0(\mathbf{r})/\partial z\right\}^{-1}$$

Elastic scattering of those electrons after a single inelastic excitation is the same as the elastic scattering of incident electrons with equivalent energy and momentum.

Equation (12.10) was derived for thin specimens satisfying

$$\int_0^d dz \beta_0 H_{10} \Psi_0^0 \left[\partial\Psi_0^0/\partial z\right]^{-1} | \ll 1$$

This condition can be further simplified as

$$d \ll \frac{\lambda}{\pi} \frac{E}{H}$$

where H is approximately taken as the average interaction energy of the incident electron with the crystal. For phonon scattering, H can be replaced by

$$H \approx eA_\kappa \left| \frac{dV_\kappa}{dr} \right|$$

If the atomic potential can be replaced by a Gaussian function

$$V_\kappa = V_0 \exp\left(-\frac{r^2}{r_0^2}\right)$$

where V_0 and r_0 are approximately the depth and half-FWHM of the potential well, respectively, we have

$$d \ll \frac{\lambda}{2.7} \frac{r_0}{A_\kappa} \frac{E}{eV_0}$$

This condition is believed to be satisfied if the specimen is in the thickness range of a few tens of nanometers. For valence excitation, crystal thickness can be much larger.

12.3 Equivalence with Multislice Theory

In this section, we prove the equivalence of (12.10) with multislice theory (11.33b) (Wang and Bentley 1991b; Wang 1992a). For easy notation, we separate the wave function from its plane wave component by defining

$$\Psi_n(\mathbf{r}) = \Phi_n(\mathbf{b}, z) \exp(2\pi i \mathbf{K}_n \cdot \mathbf{r}) \tag{12.11a}$$

$$\Psi_n^0(\mathbf{r}) = \Phi_n^0(\mathbf{b}, z) \exp(2\pi i \mathbf{K}_n \cdot \mathbf{r}) \tag{12.11b}$$

The $\Psi_n^0(\mathbf{r})$ is the solution of Schrödinger (12.2); it can be solved using the multislice method (see Chap. 3), so that

$$\Phi_n^0(\mathbf{b}, z + \Delta z) \approx \left[\Phi_n^0(\mathbf{b}, z) Q(\mathbf{b})\right] \otimes P_n(\mathbf{b}) \tag{12.12}$$

We now derive the relation that governs the inelastic wave $\Phi_n(\mathbf{b}, z)$ before and after being scattered by a thin crystal slice of thickness Δz. From (12.10c), we can directly write

$$\begin{aligned}
&\Phi_n(\mathbf{b}, z+\Delta z) - \Phi_n(\mathbf{b}, z) \\
&= \beta_0 \int_0^z dz' \left\{ H_{n0}(\mathbf{b}, z') \Psi_0^0(\mathbf{b}, z') \left[\partial \Psi_n^0(\mathbf{b}, z')/\partial z\right]^{-1} \right\} \left[\Phi_n^0(\mathbf{b}, z+\Delta z) - \Phi_n^0(\mathbf{b}, z)\right] \\
&+ \beta_0 \int_z^{z+\Delta z} dz' \left\{ H_{n0}(\mathbf{b}, z') \Psi_0^0(\mathbf{b}, z') \left[\partial \Psi_n^0(\mathbf{b}, z')/\partial z\right]^{-1} \right\} \Phi_n^0(\mathbf{b}, z+\Delta z) \\
&\approx \Phi_n(\mathbf{b}, z) \left[\frac{\Phi_n^0(\mathbf{b}, z+\Delta z)}{\Phi_n^0(\mathbf{b}, z)} - 1 \right] \\
&+ \beta_0 \left\{ h_{n0}(\mathbf{b}, z) \Psi_0^0(\mathbf{b}, z) \left[\partial \Psi_n^0(\mathbf{b}, z)/\partial z\right]^{-1} \right\} \Phi_n^0(\mathbf{b}, z+\Delta z)
\end{aligned} \tag{12.13}$$

This equation can be conveniently written as

$$\begin{aligned}
\Phi_n(\mathbf{b}, z+\Delta z) &= \Phi_n(\mathbf{b}, z) \left[\frac{\Phi_n^0(\mathbf{b}, z+\Delta z)}{\Phi_n^0(\mathbf{b}, z)} \right] \\
&+ \beta_0 \left\{ h_{n0}(\mathbf{b}, z) \Psi_0^0(\mathbf{b}, z) \left[\frac{\partial \Psi_n^0(\mathbf{b}, z)}{\partial z} \right]^{-1} \right\} \Phi_n^0(\mathbf{b}, z+\Delta z)
\end{aligned} \tag{12.14}$$

Using (12.11b), we consider the following calculation:

$$\begin{aligned}
\frac{\partial \Psi_n^0(\mathbf{b}, z)}{\partial z} &= \left[2\pi i K_{nz} \Phi_n^0(\mathbf{b}, z) + \frac{\partial \Phi_n^0(\mathbf{b}, z)}{\partial z} \right] \exp(2\pi i \mathbf{K}_n \cdot \mathbf{r}) \\
&\approx 2\pi i K_{nz} \Phi_n^0(\mathbf{b}, z) \exp(2\pi i \mathbf{K}_n \cdot \mathbf{r})
\end{aligned} \tag{12.15}$$

where the forward scattering approximation was made. Substituting (12.15) into (12.14), we have

$$\Phi_n(\mathbf{b}, z+\Delta z) \approx \left[\Phi_n(\mathbf{b}, z) - i\sigma h_{n0}(\mathbf{b}, z) \Phi_0^0(\mathbf{b}, z) \right] \left[\frac{\Phi_n^0(\mathbf{b}, z+\Delta z)}{\Phi_n^0(\mathbf{b}, z)} \right] \tag{12.16}$$

The physical meaning of (12.16) can be expressed as follows. The first term $\Phi_n(\mathbf{b}, z)$ is the inelastic wave generated before the wave arrives at the slice entrance face located at $z = z$. In the single inelastic-scattering model, the part of the wave responsible for the formation of Kikuchi patterns is elastically scattered only when it penetrates the crystal slice. The second term $\sigma h_{n0} \Phi_0$ is the newly generated inelastic wave when the elastic wave (Φ_0^0) penetrates the slice. Elastic scattering of these two parts within the slice is included in the term:

$$\frac{\Phi_n^0(\mathbf{b}, z+\Delta z)}{\Phi_n^0(\mathbf{b}, z)}$$

For a very thin slice ($\Delta z \to 0$), the propagation function P_n is approximated by a δ function; using (12.12), (12.16) can thus be approximately written as

$$\begin{aligned}\Phi_n(\mathbf{b}, z+\Delta z) &= \left[\Phi_n(\mathbf{b}, z) - i\sigma h_{n0}(\mathbf{b}, z)\Phi_0^0(\mathbf{b}, z)\right]\frac{\left[\Phi_n^0(\mathbf{b}, z)Q(\mathbf{b}, z)\right]\otimes P_n(\mathbf{b})}{\Phi_n^0(\mathbf{b}, z)} \\ &\approx \left\{Q(\mathbf{b}, z)\left[\Phi_n(\mathbf{b}, z) - i\sigma h_{n0}(\mathbf{b}, z)\Phi_0^0(\mathbf{b}, z)\right]\right\}\otimes P_n(\mathbf{b}) \quad (12.17)\end{aligned}$$

Equation (12.17) has the exact form of (11.33b); thus the equivalence of the modified multislice theory and the conventional multislice theory is proven. This equivalence also confirms the validity of approximations made when deriving (12.2).

12.4 Absorption Function

We fully described the absorption effect of inelastic scattering in Chap. 6. We now consider an alternative expression developed in real space to describe the absorption effect of inelastic excitations. From (12.2) for $n = 0$, the amplitude modulation function of the elastic wave is determined by

$$\frac{\partial \phi_0(\mathbf{r})}{\partial z} = \beta_0 \sum_{m\neq 0} X_{0m}(\mathbf{r})\phi_m(\mathbf{r}) \quad (12.18)$$

Since the elastic rescattering of electrons after inelastic scattering (i.e., the process for forming the Kikuchi pattern) does not affect the calculation of the total absorption coefficient, inelastic waves in (12.18) can be approximately represented by the localized generation function. Thus (12.18) becomes

$$\frac{\partial \phi_0(\mathbf{b}, z)}{\partial z} \approx -\frac{\beta_0^2}{4\pi^2 K_0^2}\sum_{m\neq 0} H_{0m}(\mathbf{b}, z)\int_0^z dz' H_{m0}\left(\mathbf{b}, z'\right) \quad (12.19)$$

Using $H_{m0} = H_{0m}^*$, integrating (12.19) over z, and using the boundary condition $\phi_0(\mathbf{b}, 0) = 1$, we obtain

$$\begin{aligned}\phi_0(\mathbf{b}, z) &= 1 - \frac{{\beta_0}^2}{8\pi^2 {K_{0z}}^2}\sum_{m\neq 0}\left|\int_0^z dz' H_{m0}\left(\mathbf{b}, z'\right)\right|^2 \\ &\approx \exp\left[-\frac{\beta_0^2}{8\pi^2 K_{0z}^2}\sum_{m\neq 0}\left|\int_0^z dz' H_{m0}\left(\mathbf{b}, z'\right)\right|^2\right] \quad (12.20)\end{aligned}$$

where the crystal is assumed to be so thin that the absorption factor is small. Equation (12.20) is a function that decreases in magnitude with the increasing crystal thickness; thus, it is equivalent to the absorption function of inelastic scattering, as described in Chap. 6 (Wang and Bentley 1991c).

12.5 Localized Inelastic Scattering

We apply the newly introduced theory to localized scattering (Wang and Bentley 1991b). From the result of (6.8), we can generally assume that localized inelastic scattering occurring at an atomic site $\mathbf{r}_k$ has an arbitrary phase θ_j. Then H can be written as a superposition of scattering from each atomic site

$$H_{10}(\mathbf{r},\mathbf{q}) = \sum_{\kappa} h^{\kappa}{}_{10}\,(\mathbf{r}-\mathbf{r}_{\kappa},\mathbf{q})\exp\left(-2\pi i\mathbf{q}\cdot\mathbf{r}_{\kappa}\right)\exp\left(i\theta_{\kappa}\right) \tag{12.21}$$

where an arbitrary phase is introduced to evaluate incoherent scattering properties of different atomic sites. Now consider the diffraction pattern formed by the inelastically scattered electrons. Taking a two-dimensional Fourier transform of (12.10a) yields

$$\begin{aligned}|\Psi_1(\tau,\mathbf{q},d)|^2 =&\beta_0^2 \mid \sum_{\kappa}\exp\left(-2\pi i\mathbf{q}\cdot\mathbf{r}_{\kappa}\right)\int_0^d dz\left\{\left[h^{\kappa}_{10}\left(\tau,\mathbf{q},z-z_{\kappa}\right)\right.\right.\\ &\times\exp\left(-2\pi i\tau\cdot\mathbf{r}_{\kappa}\right)\exp\left(i\theta_{\kappa}\right)\big]\\ &\otimes Y(\tau,\mathbf{b},\mathbf{q},z)\}\mid^2\end{aligned} \tag{12.22}$$

The incoherent effect is equivalent to taking a time average over the arbitrary phase factor, and thus (12.22) becomes

$$\begin{aligned}|\Psi_1(\tau,\mathbf{q},d)|^2 =&\beta_0^2\sum_{\kappa}\sum_{\kappa'}\langle\exp\left[i\left(\theta_{\kappa'}-\theta_{\kappa}\right)\right]\rangle\exp\left[-2\pi i\mathbf{q}\cdot\left(\mathbf{r}_{\kappa'}-\mathbf{r}_{\kappa}\right)\right]\\ &\times\left(\int_0^d dz\left\{\left[h^{\kappa'}_{10}\left(\tau,\mathbf{q},z-z_{\kappa'}\right)\exp\left(-2\pi i\tau\cdot\mathbf{r}_{\kappa'}\right)\right]\otimes Y(\tau,\mathbf{q},z)\right\}\right)\\ &\times\left(\int_0^d dz'\left\{\left[h^{\kappa*}_{10}\left(\tau,\mathbf{q},z'-z_{\kappa}\right)\exp\left(2\pi i\tau\cdot\mathbf{r}_{\kappa}\right)\right]\otimes Y^*(\tau,\mathbf{q},z')\right\}\right)\end{aligned} \tag{12.23}$$

It is important to note that the phase correlation of the inelastic wave generated from each atomic site can be evaluated before the numerical calculation. This is the most important characteristic of this theory.

For single-electron excitation, a random-phase approximation was made by assuming the incoherent excitation of each atom site, i.e., $\langle \exp\left[i\left(\theta_\kappa - \theta_{\kappa'}\right)\right]\rangle = \delta_{\kappa\kappa'}$. Thus (12.23) is simplified as

$$|\Psi_1(\tau, \mathbf{q}, d)|^2 = \beta_0^2 \sum_\kappa \left| \int_0^d dz \left[h^\kappa{}_{10}(\tau, \mathbf{q}, z - z_k) \exp\left(-2\pi i \tau \cdot \mathbf{r}_\kappa\right)\right] \otimes Y(\tau, \mathbf{q}, z) \right|^2 \quad (12.24)$$

or in the multislice scheme

$$|\Psi_1(\tau, \mathbf{q}, d)|^2 \approx \beta_0^2 \Delta z^2 \sum_{n_c} \sum_{j'} \left| \left[h^{j'}_1{}^0\left(\tau, \mathbf{q}, z_{n_c}\right) \exp\left(-2\pi i \tau \cdot \mathbf{r}'_j\right)\right] \otimes Y\left(\tau, \mathbf{q}, z_{n_c}\right) \right|^2 \quad (12.25)$$

where j' is summed over all atoms within the n_cth slice. In (12.25), the scattered intensity in direction τ is a sum of all contributions made by each atomic site. Elastic scattering of the generated inelastic wave responsible for forming the Kikuchi patterns is embedded in the Y function.

Equation (12.25) can be further simplified if we assume that the entire two-dimensional space in the plane perpendicular to the incident beam direction is filled with a periodic repetition of a super unit cell with reciprocal lattice vectors **g**. After expanding the convolution operation (Wang and Bentley 1991b), we have

$$\begin{aligned} |\Psi_1(\tau, \mathbf{q}, d)|^2 \approx \beta_0^2 \Delta z^2 \textstyle\sum_{n_c} \sum_{\mathbf{g}} \left\{ \sum_\alpha \exp\left(-2\pi i \mathbf{g} \cdot \mathbf{r}_\alpha\right) \left[h_{10}^\alpha(\tau, \mathbf{q}, z_{n_c})\right] \left[h_{10}^\alpha(\tau - \mathbf{g}, \mathbf{q}, z_{n_c})\right]^* \right\} \\ \otimes \left[Y(\tau, \mathbf{q}, z_{n_c}) Y^*(\tau - \mathbf{g}, \mathbf{q}, z_{n_c})\right] \end{aligned} \quad (12.26)$$

If only the $g = 0$ term is important, (12.26) reduces to

$$|\Psi_1(\tau, \mathbf{q}, d)|^2 \approx \beta_0^2 \Delta z^2 \sum_{n_c} \left[\sum_\alpha \left| h_{10}^\alpha\left(\tau, \mathbf{q}, z_{n_c}\right)\right|^2 \right] \otimes \left| Y\left(\tau, \mathbf{q}, z_{n_c}\right)\right|^2 \quad (12.27)$$

Therefore in the localized inelastic incoherent scattering, the intensity distribution in the diffraction pattern is the sum of all contributions from each atomic site. It is worth pointing out that three-dimensional periodicity was not assumed in the preceding derivations; thus (12.25) can be applied to crystals with defects.

12.6 Diffraction of TDS Electrons Semiclassical Approach

12.6.1 Basic Equations

We now use the harmonic oscillator approximation to calculate the angular distribution of TDS electrons. Under the first-order approximation (or single-phonon scattering approximation), we can write the interaction Hamiltonian of each atom as

$$h^{\kappa}_{10}(\mathbf{r}, \mathbf{q}) = -eA_{\kappa}(\omega)\cos(\omega t + \theta_{\kappa})\,\mathbf{e}_{q,p}\cdot\nabla V_{\kappa}(\mathbf{r}) \tag{12.28a}$$

or in reciprocal space

$$h^{\kappa}{}_{10}(\tau, z, \mathbf{q}) = -2\pi\mathrm{ie}\ A_{\kappa}(\omega)\cos(\omega t + \theta j)\left(\mathbf{e}_{q,p}\cdot\tau\right) f^{e}_{\kappa}(\tau, z) \tag{12.28b}$$

Now consider the diffraction pattern formed by TDS electrons. Substituting (12.28b) into (12.23), taking the time average of the phase coupling, incoherently summing (or integrating) over contributions made by all possible phonon modes, and using a new notation $\Psi_{\mathrm{T}} = \Psi_1$, we have

$$\begin{aligned}|\Psi_T(\tau, d)|^2 =& N_0\Omega\frac{\alpha_0^2}{2}\sum_{\kappa'}\sum_{\kappa}\sum_{p}\sum_{i=1}^{3n_0}\int_{\Sigma} d\mathbf{q}A_{\kappa'}(\omega_i)\,A_{\kappa}(\omega_i) < \cos\left(\theta_{\kappa'}-\theta_{\kappa}\right) > \\ &\times \exp\left[-2\pi i\mathbf{q}\cdot(\mathbf{r}_{\kappa'}-\mathbf{r}_{\kappa})\right] \\ &\times\left(\int_0^d dz\left\{\left[\left(\mathbf{e}_{q,p}\cdot\tau\right) f^{e}_{\kappa'}\left(\tau, z-z_{\kappa'}\right)\exp\left(-2\pi i\tau\cdot\mathbf{r}_{\kappa'}\right)\right]\otimes Y(\tau, \mathbf{q}, z)\right\}\right) \\ &\times\left(\int_0^d dz'\left\{\left[\left(\mathbf{e}_{q,p}\cdot\tau\right) f^{e*}_{\kappa}\left(\tau, z'-z_{\kappa}\right)\exp\left(2\pi i\tau\cdot\mathbf{r}_{\kappa}\right)\right]\otimes Y^*\left(\tau, \mathbf{q}, z'\right)\right\}\right)\end{aligned} \tag{12.29}$$

For convenience, $\alpha_0 \equiv 2\pi e\beta_0$ is defined. In (12.29), the phase correlation of atomic vibrations of near neighbors can be introduced in the calculation according to the assumed phonon excitation model.

In practice, the determination of $\Psi^0_1(\tau, \mathbf{q}, d)$ for different momentum transfers (or different incident beam wave vectors) involves a large number of calculations. For simplification, the effect of $\mathbf{q}$ in $Y(\tau, \mathbf{q}, z)$ is approximately characterized by the effect of a mean value $\mathbf{q}_0$ (this is the second major approximation of this theory), because Y is mainly responsible for the formation of Kikuchi band background. A proper selection of $\mathbf{q}_0$ can give reasonably good agreement with the experimental observations (Wang 1992b). Therefore we can simply use the notation

$$Y(\tau, z) \equiv Y(\tau, \mathbf{q}_0, z) \approx Y(\tau, \mathbf{q}, z)$$

in the following analysis. Now separating (12.29) into two parts for $\kappa = \kappa'$ and $\kappa \neq \kappa'$, denoting $\tau_p =< \mathbf{e}_{q,p} \cdot \tau >_p$ (where $<>_p$ means the average over different polarization directions), we have

$$|\Psi_T(\tau, d)|^2 \approx \frac{\alpha_0^2}{2} \sum_{\kappa} (a_{\kappa,\kappa})^2 \left| \int_0^d dz \left[\tau_p f_\kappa^e (\tau, z - z_\kappa) \exp(-2\pi i \tau \cdot \mathbf{r}_\kappa)\right] \otimes Y(\tau, z) \right|^2$$
$$+ \frac{\alpha_0^2}{2} \sum_{\kappa' \neq \kappa} \sum (a_{\kappa',k})^2 \int_0^d dz \left\{ \left[\tau_p f_{\kappa'}^e (\tau, z - z_{\kappa'}) \exp(-2\pi i \tau \cdot \mathbf{r}_{\kappa'})\right] \otimes Y(\tau, z) \right\}$$
$$\times \int_0^d dz' \left\{ \left[\tau_p f_\kappa^{e*} (\tau, z' - z_\kappa) \exp(2\pi i \tau \cdot \mathbf{r}_\kappa)\right] \otimes Y^* (\tau, z') \right\} \tag{12.30a}$$

where

$$(a_{\kappa',\kappa})^2 \equiv N_0 \Omega \sum_p \sum_{i=1}^{3n_0} \int_{BZ} d\mathbf{q} A_{\kappa'}(\omega_i) A_\kappa(\omega_i) < \cos(\theta_{\kappa'} - \theta_\kappa) > \exp\left[-2\pi i \mathbf{q} \cdot (\mathbf{r}_{\kappa'} - \mathbf{r}_k)\right] \tag{12.30b}$$

and $a_{\kappa',\kappa}$ is defined to specify the vibration correlation of the two atoms located at $\mathbf{r}_{\kappa'}$ and $\mathbf{r}_\kappa$, which decay rapidly with increasing interatomic distance. Hence two atoms located far apart may have phase coupling in their vibrations, but the correlation coefficient $a_{\kappa',\kappa}$ drops almost to zero. In other words, vibrations of all atoms in the crystal may have a certain phase relationship, but only interactions from the nearest neighbors need to be considered in diffraction calculations. In the multislice format, (12.30a) is expressed as

$$|\Psi_T(\tau, d)|^2 \approx \frac{\alpha_0^2}{2} \sum_{n_c} \sum_{i'} (a_{i',i'})^2 \left| \left[\tau_p f_{i'}^e(\tau) \exp(-2\pi i \tau \cdot \mathbf{r}_{i'})\right] \otimes Y(\tau, z_{n_c}) \right|^2$$
$$+ \frac{\alpha_0^2}{2} \sum_{n_c} \sum_{i' \neq j'} \sum (a_{i',j'})^2 \left\{ \left[\tau_p f_{j'}^e(\tau) \exp(-2\pi i \tau \cdot \mathbf{r}_{j'})\right] \otimes Y(\tau, z_n) \right\}$$
$$\times \left\{ \left[\tau_p f_{i'}^{e*}(\tau) \exp(2\pi i \tau \cdot \mathbf{r}_{i'})\right] \otimes Y^*(\tau, z_{n_c}) \right\} \tag{12.31}$$

where i' and j' are summed over all atoms within the same slice.

Equation (12.30b) can be further simplified if we assume that the entire two-dimensional space in the plane perpendicular to the incident beam direction is filled with a periodic repetition of a super unit cell. We can prove (Wang and Bentley 1991b) that

$$
\begin{aligned}
|\Psi_T(\tau,d)|^2 \approx & \frac{\alpha_0^2}{2}\sum_{n_c}\sum_{\alpha}\sum_{g}\exp(-2\pi i\mathbf{g}\cdot\mathbf{r}_\alpha)\{\tau\cdot(\tau-\mathbf{g})f_\alpha^e(\tau)[(a_{\alpha,\alpha})^2 f_\alpha^{e*}(\tau-\mathbf{g}) \\
& + (a_{\alpha,\alpha_1})^2\sum_{\alpha_1} f_{\alpha 1}^{e*}(\tau-\mathbf{g})\exp\left[2\pi i(\tau-\mathbf{g})\cdot\Delta\mathbf{r}_{\alpha 1}\right] \\
& + (a_{\alpha,\alpha_2})^2\sum_{\alpha_2} f_{\alpha 2}^{e*}(\tau-\mathbf{g})\exp[2\pi i(\tau-\mathbf{g})\cdot\Delta\mathbf{r}_{\alpha 2}] + \cdots]\} \\
& \otimes\left\{Y(\tau,z_{n_c})\,Y^*(\tau+\mathbf{g},z_{n_c})\right\}
\end{aligned}
\tag{12.32}
$$

where α_1 and α_2 are summed over the first and second nearest neighbors of the αth atom with relative position $\Delta\mathbf{r}_{\alpha 1}=\mathbf{r}_{\alpha 1}-\mathbf{r}_\alpha$ and $\Delta\mathbf{r}_{\alpha 2}=\mathbf{r}_{\alpha 2}-\mathbf{r}_\alpha$, respectively; $a_{\alpha,\alpha 1}$ and $a_{\alpha,\alpha 2}$ are correlation parameters of the αth atoms with its first and second nearest neighbors, respectively, as defined in (12.30b). The contribution from the third and most distant neighbors can be introduced in a similar way.

In (12.32), the terms related only to phonon scattering are included in the first { } bracket, which represents the intensity distribution of the TDS within the first BZ. Terms related through dynamic electron scattering appear in the second { } bracket; these terms are determined only by elastic electron scattering. The convolution operation introduces dynamic effects. Therefore TDS streaks are totally determined by the coupling of atomic vibrations in the crystal; line shapes of TDS are not produced by dynamic effects. Equation (12.32) can be further simplified if only the $g=0$ term is important

$$
\begin{aligned}
|\Psi_T(\tau,d)|^2 \approx & \frac{\alpha_0^2}{2}\sum_{n_c}\sum_{\alpha}\Big\{\tau^2 f_\alpha^e(\tau)(a_{\alpha,\alpha})^2 f_\alpha^{e*}(\tau) \\
& +(a_{\alpha,\alpha_1})^2\sum_{\alpha_1} f_{\alpha_1}^{e*}(\tau)\exp\left(2\pi i\boldsymbol{\tau}\cdot\Delta\boldsymbol{r}_{\alpha_1}\right) \\
& \sum_{\alpha_2}+(a_{\alpha,\alpha_2})^2 f_{\alpha_2}^{e*}(\tau)\exp\left(2\pi i\boldsymbol{\tau}\cdot\Delta\boldsymbol{r}_{\alpha_2}\right)+\cdots\Big\}\otimes\left|Y(\tau,z_{n_c})\right|^2
\end{aligned}
\tag{12.33}
$$

Therefore, the TDS diffraction pattern is the result of the convolution of the elastic diffraction pattern with the TDS scattering function contained in the first [ ] bracket in (12.33). This type of expression simply separates the kinematic scattering part from the dynamic scattering part, making it rather easy to describe TDS.

12.6.2 *Streaks in TDS Electron Diffraction Patterns*

Based on structure factor analysis, Honjo et al. (1964) suggested that streaking in diffraction patterns is associated with small patterned displacements of atoms from their regular positions. Features appearing in the diffraction pattern are signs of ordered structure in real space. We now apply (12.32) to predict directions of TDS streaks qualitatively.

Considering only the intensity distribution within the ZOLZ and bearing in mind the law of conservation of momentum, TDS streaks are mainly generated by acoustic phonon modes with wave vectors parallel to the diffraction plane, because momentum transferred from incident electrons is almost restricted to this plane and

$$q_z \approx K_0 \frac{\hbar\omega}{2E} \approx 0 \quad \hbar\omega = 0.1\text{eV} \quad E = 100\text{keV}$$

(see Sect. 6.8). These modes can be considered approximately generated by atomic vibrations within the plane perpendicular to the incident beam direction, which is actually a two-dimensional lattice vibration model. Interaction between atoms located in separated layers parallel to the diffraction plane can be considered weak. Phonon modes whose wave vectors qs are not confined to the diffraction plane contribute only a diffuse background. Therefore the sum of α_1, α_2, and α_3 can be restricted to nearest neighbors in the same atomic plane with the αth atom perpendicular to the incidence beam direction. The validity of this two-dimensional plane model is verified experimentally in Sect. 12.7.3. Now consider the following simple cases.

12.6.3 Case 1

For random-phase correlation in the Einstein model, (12.32) becomes

$$|\Psi_T(\tau, d)|^2 \approx \frac{\alpha_0^2}{2} \sum_{n_c} \sum_{\alpha} \sum_{g} \exp(-2\pi i \mathbf{g} \cdot \mathbf{r}_\alpha) \left(a_{\alpha,\alpha}\right)^2 \times \left[\tau \cdot (\tau - \mathbf{g}) f_\alpha^e(\tau) f_\alpha^{e*}(\tau - \mathbf{g})\right] \otimes \left[Y\left(\tau, z_{n_c}\right) Y^*\left(\tau + \mathbf{g}, z_{n_c}\right)\right] \tag{12.34}$$

If only the $g = 0$ term is important,

$$|\Psi_T(\tau, d)|^2 \approx \frac{\alpha_0^2}{2} \sum_{n_c} \sum_{\alpha} \left(a_{\alpha,\alpha}\right)^2 \left[\tau^2 \left|f_\alpha^e(\tau)\right|^2 \otimes \left|Y\left(\tau, z_{n_c}\right)\right|^2\right] \tag{12.35}$$

where the $|Y|^2$ term is approximately the intensity distribution of Bragg reflections in the diffraction pattern. In this random-phase approximation, TDS gives only a diffusely scattered background without sharp streaks in the diffraction pattern.

12.6.4 Case 2

For a monatomic diamond cubic perfect crystal oriented along [001], if only interactions between atom (000) and its first nearest neighbors located at

$$\pm\frac{\mathrm{a}}{2}[110] \text{ and } \pm\frac{\mathrm{a}}{2}[1\bar{1}0]$$

in the same (001) plane are considered, (12.33) can be simplified as

$$\begin{aligned}|\Psi_T(\tau,d)|^2 \approx &\frac{\alpha_0^2}{4}\sum_{n_c}\sum_{\alpha}\sum_{g}\exp\left(-2\pi i\mathbf{g}\cdot\mathbf{r}_\alpha\right)\left\{[\tau\cdot(\tau-\mathbf{g})]f_\alpha^e(\tau)f_\alpha^{e*}(\tau-\mathbf{g})\right.\\ &\times\left[2\left(a_{\alpha,\alpha}\right)^2+\left(a_{\alpha,\alpha_1}\right)^2\left\{\cos\left[\pi\left(\tau_x+\tau_y\right)a\right]\right.\right.\\ &\left.\left.\left.+\cos\left[\pi\left(\tau_x-\tau_y\right)a\right]\right\}\right]\right\}\otimes\left[Y\left(\tau,z_{n_c}\right)Y^*\left(\tau+\mathbf{g},z_{n_c}\right)\right]\end{aligned}\tag{12.36}$$

Here and following, the first nearest neighbor term refers to those atoms in the same (001) two-dimensional plane and located closest to the αth atom, although these atoms may not be the absolute closest neighbor of the αth atom in three-dimensional space. Equation (12.36) shows that the contribution of TDS can be separated into two parts. The term containing $\left(a_{\alpha,\alpha}\right)^2$, the independent part of the phase correlation, contributes an orientation-independent intensity distribution to the diffraction pattern (i.e., a diffuse scattered background, as expected from the Einstein model). The second part depends on phase coupling of the first nearest neighbor (in-plane) atomic vibrations and gives the orientation-dependent intensity distribution, such as streaking to the diffraction pattern. The $\left(a_{\alpha,\alpha_1}\right)^2$ term gives an intensity streaking along $\langle 110\rangle$, connecting reciprocal lattice points of nonforbidden reflections, such as (040) with (220) and (040) with ($\bar{2}$20), as observed in Fig. 12.2. We do not see streaks connecting diagonal spots, which means there are only very weak interactions between atoms and their second nearest neighbors.

12.6.5 Case 3

For a monatomic face-centered cubic perfect crystal oriented along [001], if there are only interactions with the first nearest neighbor lying in the same (001) plane, (12.32) becomes

$$\begin{aligned}|\Psi_T(\tau,d)|^2 \approx &\frac{\alpha_0^2}{4}\sum_{n_c}\sum_{\alpha}\sum_{g}\exp\left(-2\pi i\mathbf{g}\cdot\mathbf{r}_\alpha\right)\left\{[\tau\cdot(\tau-\mathbf{g})]f_\alpha^e(\tau)f_\alpha^{e*}(\tau-\mathbf{g})\right.\\ &\times\left[2\left(a_{\alpha,\alpha}\right)^2+\left(a_{\alpha,\alpha_1}\right)^2\left\{\cos\left[\pi\left(\tau_x+\tau_y\right)a\right]\right.\right.\\ &\left.\left.\left.+\cos\left[\pi\left(\tau_x-\tau_y\right)a\right]\right\}\right]\right\}\otimes\left[Y\left(\tau,z_{n_c}\right)Y^*\left(\tau+\mathbf{g},z_{n_c}\right)\right]\end{aligned}\tag{12.37}$$

Note that (12.37) is the same as (12.35) for diamond cubic structures.

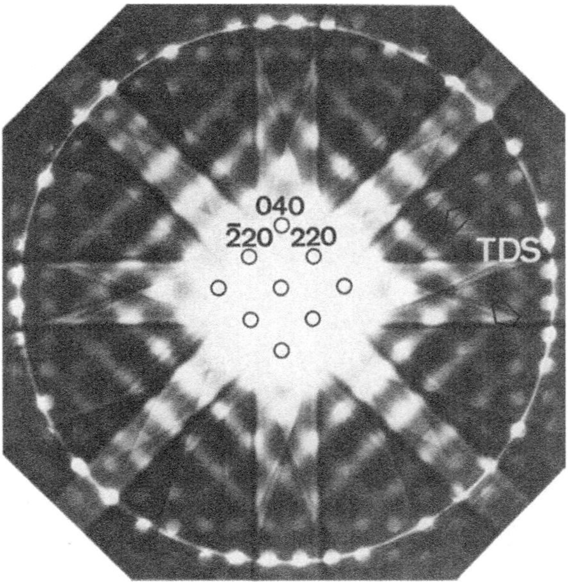

Fig. 12.2 A diffraction pattern from a thin Si[001] crystal showing the formation of streaks along < 110 > due to atomic vibration phase coupling in TDS (Wang and Bentley 1991b)

12.6.6 Case 4

For a monatomic body-centered cubic perfect crystal oriented along [001], if there are only interactions with the first nearest neighbors lying in the same (001) plane, such as those located $\pm a[100]$ and $\pm a[010]$, then (12.32) becomes

$$|\Psi_T(\tau, d)|^2 \approx \alpha_0^2 \sum_{n_c} \sum_{\alpha} \sum_{g} \exp(-2\pi i \mathbf{g} \cdot \mathbf{r}_\alpha) \left([\tau \cdot (\tau - \mathbf{g})] f_\alpha^e(\tau) f_\alpha^{e*}(\tau - \mathbf{g}) \right.$$
$$\left. \left\{ 2 \left(a_{\alpha,\alpha} \right)^2 + \left(a_{\alpha,\alpha_1} \right)^2 \left[\cos(2\pi\tau_x a) + \cos\left(2\pi\tau_y a\right) \right] \right\} \right)$$
$$\otimes \left[Y\left(\tau, z_{n_c}\right) Y^*\left(\tau + \mathbf{g}, z_{n_c}\right) \right] \tag{12.38}$$

We see from (12.32) that for a monatomic system, the possible TDS streaks can be simply predicted by examining a function of

$$\sum_{\alpha} \sum_{\alpha_1} \cos\left[2\pi \boldsymbol{\tau} \cdot (\mathbf{r}_\alpha - \mathbf{r}_{\alpha_1})\right]$$

where the summation of α_1 is restricted to the first nearest neighbor of the αth atom, such that they are located in the atomic plane, which contains the αth atom and is perpendicular to the incident beam direction. The TDS streaks are located on lines satisfying $\tau \cdot \left(\mathbf{r}_\alpha - \mathbf{r}_{\alpha_1}\right) = 0$. This rule can also be applied to multiatomic cases if differences made by atomic structure factors are considered. This rule should also allow the determination of strong atomic interaction relationships from the orientation of TDS in diffraction patterns in a manner similar to a dislocation Burgers vector determination by the $\mathbf{g} \cdot \mathbf{b_B} = 0$ rule. If the electron beam direction $\mathbf{B} = (h_1 k_1 \ell_1)$ and $\mathbf{r}_\alpha - \mathbf{r}_{\alpha 1} = (H_1 \mathbf{a}_1, K_1 \mathbf{a}_2, L_1 \mathbf{a}_3)$, since $\tau \cdot \mathbf{B} = 0$ in the diffraction plane, the TDS streak produced by this nearest neighbor interaction would be along

$$\tau = (\mathbf{r}_\alpha - \mathbf{r}_{\alpha 1}) \times \mathbf{B}$$

for orthonormal crystal structures, i.e.:

$$\tau = \begin{vmatrix} \mathbf{a}^* & \mathbf{b}^* & \mathbf{c}^* \\ H_1 & K_1 & L_1 \\ h_1 & k_1 & \ell_1 \end{vmatrix} \tag{12.39}$$

This provides a very simple method of predicting TDS streaks in electron diffraction patterns (Wang and Bentley 1991b; Wang 1992a).

The experimentally observed TDS streaks are sharper than line shapes described theoretically by the cosine function [e.g., (12.38)]. This discrepancy cannot be accounted for by long-range atomic interactions because long-range interactions produce not only basic streaks expected from the first nearest neighbor interactions, but these also introduce equally intense TDS lines located between and parallel to basic streaks. These extra lines have not been observed experimentally. In fact, the sharpness of the streaks is determined by the phonon dispersion relation, as shown in the next section.

12.7 Diffraction of Phonon-Scattered Electrons—Quantum Mechanical Approach

A phonon is a state of the crystal characterized by frequency, wave vector, and polarization vector. Excitation of a phonon means the transition of the crystal vibration state from $|n\rangle$ to $|n + 1\rangle$. This section illustrates the application of the modified multislice theory in treating phonon excitations (Wang 1992a). The final result is compared with the result obtained from the frozen lattice model presented in Sect. 12.6.1. For simplification, the following discussions are based on the model of single-phonon creation.

12.7.1 Fundamental Treatment

The creation of a phonon by electron scattering is determined by the matrix element defined in (6.68). Substituting (6.68) into (12.10c), the TDS wave at the crystal exit face $z = d$ is

$$\Psi_{\mathrm{T}}(\mathbf{b}, \mathbf{q}, d) = \beta_0 \Delta z \sum_{n_c} H\left(\mathbf{b}, z_{n_c}\right) Y\left(\mathbf{b}, \mathbf{q}, z_{n_c}\right) \tag{12.40}$$

Taking a two-dimensional Fourier transform of (12.40), we have

$$\begin{aligned}\Psi_T(\tau, \mathbf{q}, d) &= \beta_0 \Delta z \sum_{n_c} H\left(\tau, z_{n_c}\right) \otimes Y\left(\tau, \mathbf{q}, z_{n_c}\right) \\ &= 2\pi i \alpha_0 \Delta z \sum_{n_c} \sum_{\alpha} \sum_{g} A_\alpha\left(\omega_j\right) \left[\mathbf{e}\left(\alpha\Big|{}^{\mathbf{q}}_{j}\right) \cdot (\mathbf{g}+\mathbf{q})\right] f_\alpha^e(\mathbf{g}+\mathbf{q}) \\ &\times \exp[-2\pi i \mathbf{g} \cdot \mathbf{r}(\alpha)] Y\left(\tau - \mathbf{q} - \mathbf{g}, \mathbf{q}, z_{n_c}\right)\end{aligned} \tag{12.41}$$

The final TDS patterns are formed by the intensity summation over all TDS electrons in different phonon states (or modes), i.e., an integration over phonon modes of different wave vectors $\mathbf{q}$ and frequencies ω_j, which is

$$\begin{aligned}I_T(\tau) &= N_0 \Omega \sum_{j=1}^{3n_0} \int_{\mathrm{BZ}} d\mathbf{q}\; |\Psi_T(\tau, \mathbf{q}, d)|^2 \\ &= N_0 \Omega (\alpha_0 \Delta z)^2 \sum_{n_c} \sum_{\alpha} \sum_{g} \sum_{n_c'} \sum_{\alpha'} \sum_{g'} \sum_{j=1}^{3n_0} \int_{\mathrm{BZ}} d\mathbf{q}\, A_\alpha(\omega_j) A_{\alpha'}^*(\omega_j) \\ &\times \left[\mathbf{e}(\alpha\, |{}^{q}_{j}) \cdot (\mathbf{g}+\mathbf{q})\right] \\ &\times \left[\mathbf{e}^*(\alpha'\, |{}^{q}_{j}) \cdot (\mathbf{g}'+\mathbf{q})\right] \\ &\times f_\alpha^e(\mathbf{g}+\mathbf{q}) f_{\alpha'}^{e*}(\mathbf{g}'+\mathbf{q}) \exp\left\{-2\pi i\left[\mathbf{g}\cdot\mathbf{r}(\alpha) - \mathbf{g}'\cdot\mathbf{r}(\alpha')\right]\right\} \\ &\times Y(\tau - \mathbf{q} - \mathbf{g}, \mathbf{q}, z_{n_c}) Y^*(\tau - \mathbf{q} - \mathbf{g}', \mathbf{q}, z_{n_c'})\end{aligned} \tag{12.42}$$

Neglecting correlations among different Bragg beams and treating TDS waves generated from different crystal slices as incoherent, (12.42) becomes

$$\begin{aligned}I_T(\tau) &= N_0 \Omega\, (\alpha_0 \Delta z)^2 \sum_{n_c} \sum_{\alpha} \sum_{\alpha'} \sum_{g} \sum_{j=1}^{3n_0} \int_{\mathrm{BZ}} d\mathbf{q} A_\alpha\left(\omega_j\right) A_{\alpha'}^*\left(\omega_j\right) \left[\mathbf{e}\left(\alpha\Big|{}^{\mathbf{q}}_{j}\right) \cdot (\mathbf{g}+\mathbf{q})\right] \\ &\times \left[\mathbf{e}^*\left(\alpha'\Big|{}^{\mathbf{q}}_{j}\right) \cdot (\mathbf{g}+\mathbf{q})\right] f_\alpha^e(\mathbf{g}+\mathbf{q}) f_{\alpha'}^{e*}(\mathbf{g}'+\mathbf{q}) \\ &\times \exp\left\{-2\pi i \mathbf{g} \cdot \left[\mathbf{r}(\alpha) - \mathbf{r}\left(\alpha'\right)\right]\right\} \left|Y\left(\tau - \mathbf{q} - \mathbf{g}, \mathbf{q}, z_{n_c}\right)\right|^2\end{aligned} \tag{12.43}$$

Equation (12.43) can be further simplified by defining a function

$$B_{\alpha\alpha'}(j,\mathbf{q}_b) \equiv N_0\Omega \begin{cases} \int_{BZ} dq_z A_\alpha\left[\omega_j(\mathbf{q})A^*_{\alpha'}\left[\omega_j(\mathbf{q})\right]\right. & \text{if } \mathbf{q}_b \text{ falls within the first BZ} \\ 0 \quad \text{otherwise} \end{cases} \tag{12.44}$$

Substituting (12.44) into (12.43), neglecting the q_z dependence of the remaining terms, and dividing $\mathbf{q}$ into $(\mathbf{q}_b, q_z)$, the integration of $\mathbf{q}_b$ in (12.43) can be extended from the first BZ to the entire reciprocal space, such that

$$\begin{aligned} I_T(\tau) \approx (\alpha_0\Delta z)^2 \sum_{n_c}\sum_{\alpha}\sum_{\alpha'}\sum_{g}\sum_{j=1}^{3n_0} \int_{-\infty}^{\infty} d\mathbf{q}_b B_{\alpha\alpha'}(j,\mathbf{q}_b)\left[\mathbf{e}\left(\alpha|_j^{\mathbf{q}}\right)\cdot(\mathbf{g}+\mathbf{q}_b)\right] \\ \times \left[\mathbf{e}^*\left(\alpha'|_j^{\mathbf{q}}\right)\cdot(\mathbf{g}+\mathbf{q}_b)\right] \\ \times f^e_\alpha(\mathbf{g}+\mathbf{q}_b)\, f^{e*}_{\alpha'}(\mathbf{g}+\mathbf{q}_b)\exp\left\{-2\pi i\mathbf{g}\cdot\left[\mathbf{r}(\alpha)-\mathbf{r}\left(\alpha'\right)\right]\right\} \\ \times \left|Y\left(\tau-\mathbf{q}_b-\mathbf{g},\mathbf{q}_b,z_{n_c}\right)\right|^2 \end{aligned} \tag{12.45}$$

In practice, if the momentum transfer $\mathbf{q}_b$ can be represented by an average value $\mathbf{q}_0$, we may approximately assume

$$|Y(\tau-\mathbf{q}_b-\mathbf{g},\mathbf{q}_b,z_n)|^2 \approx |Y(\tau-\mathbf{q}_b-\mathbf{g},\mathbf{q}_0,z_n)|^2 \tag{12.46}$$

Numerical calculations have shown this is a reasonable approximation (Wang 1992b). Equation (12.45) can be approximately written in convolution form as

$$\begin{aligned} I_T(\tau) \approx amp;\ & (\alpha_0\Delta z)^2 \sum_{n_c}\sum_{\alpha}\sum_{\alpha'}\sum_{g}\sum_{j=1}^{3n_0}\left(B_{\alpha\alpha'}(j,\tau)\left[\mathbf{e}\left(\alpha|_j^{\tau}\right)\cdot(\tau+\mathbf{g})\right]\left[\mathbf{e}^*\left(\alpha'|_j^{\tau}\right)\cdot(\tau+\mathbf{g})\right]\right. \\ amp;\ & \left.\times f^e_\alpha(\tau+\mathbf{g})f^{e*}_{\alpha'}(\tau+\mathbf{g})\exp\left\{-2\pi i\mathbf{g}\cdot\left[\mathbf{r}(\alpha)-\mathbf{r}\left(\alpha'\right)\right]\right\}\right) \\ amp;\ & \otimes \mid Y(\tau-\mathbf{g},\mathbf{q}_0,z_{n_c}) \mid^2 \\ amp;\ & = (\alpha_0\Delta z)^2 \sum_{n_c}\sum_{\alpha}\sum_{\alpha'}\sum_{g}\sum_{j=1}^{3n_0}\left(\left\{B_{\alpha\alpha'}(j,\tau-\mathbf{g})\left[\mathbf{e}\left(\alpha|_j^{\tau-\mathbf{g}}\right)\cdot\tau\right]\left[e^*\left(\alpha'|_j^{\tau-\mathbf{g}}\right)\cdot\tau\right]\right.\right. \\ amp;\ & \left.\left.\times f^e_\alpha(\tau)f^{e*}_{\alpha'}(\tau)\right\}\otimes\delta(\tau+\mathbf{g})\right) \\ amp;\ & \otimes\left(\left|Y\left(\tau-\mathbf{g},\mathbf{q}_0,z_{n_c}\right)\right|^2\exp\left\{-2\pi i\mathbf{g}\cdot\left[\mathbf{r}(\alpha)-\mathbf{r}\left(\alpha'\right)\right]\right\}\right) \\ amp;\ & = (\alpha_0\Delta z)^2 \sum_{n_c}\sum_{\alpha}\sum_{\alpha'}\sum_{j=1}^{3n_0}\left\{B_{\alpha\alpha'}(j,\tau)\left[\mathbf{e}\left(\alpha|_j^{\tau}\right)\cdot\tau\right]\left[\mathbf{e}^*\left(\alpha'|_j^{\tau}\right)\cdot\tau\right]f^e_\alpha(\tau)f^{e*}_{\alpha'}(\tau)\right\} \\ amp;\ & \otimes\left(|Y(\tau,\mathbf{q}_0,z_n)|^2\sum_{g}\exp\left\{-2\pi i\mathbf{g}\cdot\left[\mathbf{r}(\alpha)-\mathbf{r}\left(\alpha'\right)\right]\right\}\right) \\ amp;\ & = (\alpha_0\Delta z)^2 \sum_{n_c}\sum_{\alpha}\sum_{j=1}^{3n_0}\left\{B_{\alpha\alpha'}(j,\tau)\left|\left[\mathbf{e}\left(\alpha|_j^{\tau}\right)\cdot\tau\right]f^e_\alpha(\tau)\right|^2\right\}\otimes\left[\mid Y(\tau,\mathbf{q}_0,z_n)|^2\right] \end{aligned} \tag{12.47}$$

In (12.47), the last step was carried out by using the relationships

$$\mathbf{e}\left(\alpha\big|_j^{\tau-\mathbf{g}}\right) = \mathbf{e}\left(\alpha\big|_j^{\tau}\right), \quad \omega_j(\tau - \mathbf{g}) = \omega_j(\tau)$$

The terms purely related to lattice dynamics, which are included in the first set of bold parentheses and collectively represent the intensity distribution of the TDS within the first BZ, are called the scattering power function. Terms related to dynamic electron scattering, appearing in the second set of bold parentheses, are determined purely by the elastic-scattering Schrödinger equation, and these can be solved using the multislice method (see Chap. 3).

Under the kinematic scattering approximation

$$|Y(\tau, \mathbf{q}_0, z_n)|^2 = \sum_g I_g \delta(\tau - \mathbf{g})$$

where I_g is the intensity of beam $\mathbf{g}$, (12.47) is simplified as

$$I_T(\tau) = (\alpha_0 \Delta z)^2 \sum_{n_c} \sum_{\alpha} \sum_{j=1}^{3n_0} \left\{ B_{\alpha\alpha}(j, \tau) \left| \left[\mathbf{e}\left(\alpha\big|_j^{\tau}\right) \cdot \tau \right] f_\alpha^e(\tau) \right|^2 \right\} \otimes \left[\sum_g I_g \delta(\tau - \mathbf{g}) \right]$$

The TDS streaks, which have been calculated under the kinematic approximation based on the phonon dispersion relations given by lattice dynamics (Hua et al. 1988a, b, 1989), agree well with experimental observations.

12.7.2 Fundamental Treatment

Figure 12.3 shows a comparison of a simulated [using (12.47)] Mo [001] elastic and TDS electron diffraction pattern with an experimentally observed pattern. The experimental pattern in Fig. 12.3b is an energy-unfiltered diffraction pattern containing contributions from electrons that have experienced various elastic- and inelastic-scattering processes, such as plasmon excitations. The TDS streaks are expected to appear sharper if the energy filter is used (Reimer 1990). The observed Kikuchi pattern is the result of valence electron, phonon, and single-electron excitations. This is why crystal thickness assumed in theoretical calculations is considerably larger than the foil thickness used to record the experimental diffraction pattern to obtain a fit, i.e., the scattering result of only one inelastic-scattering process is used theoretically to fit the observed pattern of three inelastic processes. In addition, the simulated diffraction pattern is based only on the single-phonon excitation model, and the contribution of multiphonon processes is not considered. The simulated pattern is the sum of contributions from purely elastically scattered electrons and phonon-scattered electrons. The intensity of the calculated pattern is displayed on a logarithmic scale to enhance the weak features of TDS streaks at large angles. The TDS streaks clearly

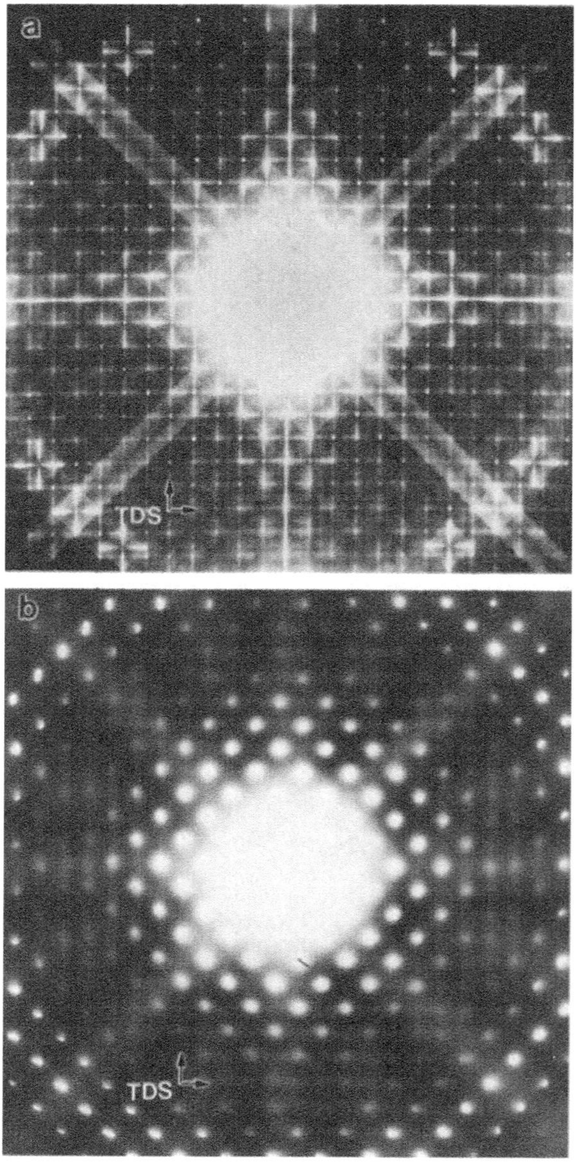

Fig. 12.3 Comparisons of **a** a simulated TDS diffraction pattern of Mo [001] with **b** an observed parallel beam electron diffraction pattern. Electron energy is 100 keV. The simulated diffraction pattern represents total contributions from elastic scattering and phonon scattering. The Mo atomic mean vibration amplitude is 0.008 nm. The mean momentum transfer is taken as $q_0 = 2.25\ \text{nm}^{-1}$. The TDS streaks along <010> and <100> are clearly visible (Wang 1992a)

appear along <010> and <100> corresponding to the observation in Fig. 12.3b. The observed Kikuchi pattern is not so sharp as the calculated one, since the Mo foil specimen contained bends that influence the parallel beam diffraction. In addition, the diffraction pattern was taken without using a select-area aperture because it limits the signal of large-angle scattered TDS electrons. All major features of these results are in good agreement. Similarly good agreement was also achieved for the Si [001] pattern (Fig. 12.4). The TDS streaks along ⟨110⟩ are clearly shown.

It must be emphasized that the preceding simulations are for the parallel beam illumination case, in which TDS streaks show up clearly at high angles. For the convergent beam case, TDS streaks broaden due to the convolution effect of beam convergence. Therefore the applicability of the Einstein model has to be examined experimentally in the nonconvergent beam case.

12.7.3 Directions of TDS Streaks

Dynamic calculation of electron diffraction patterns is quite time-consuming and complex. It is, therefore, necessary to seek simplified models that can give qualitative results without performing any numerical calculations. In this section, we introduce a two-dimensional lattice vibration model for analyzing TDS streaks observed in electron diffraction patterns.

In (12.47), the terms other than $B_{\alpha\alpha}$ in the first { } bracket do not produce sharp intensity variations in the diffraction pattern. Thus the streak directions and line shapes can be qualitatively predicted by examining a function defined as

$$S_{\text{TDS}} \equiv \sum_{j=1}^{3n_0} B_{\alpha\alpha}(j, \mathbf{q}_\perp) \sim \sum_{j=1}^{3n_0} \int_{\text{BZ}} dq_z \frac{1}{\omega_j(\mathbf{q})} \frac{\exp\left(\hbar\omega_j(\mathbf{q})/k_B T\right)}{\exp\left(\hbar\omega_j(\mathbf{q})/k_B T\right) - 1} \quad (12.48)$$

in which phonon dispersion relationship plays the most important role. Optical branches, for which ω varies slowly when $\mathbf{q}$ approaches zero, contribute a diffuse background in the diffraction pattern. Sharp streaks are produced by acoustic branches, for which ω tends to zero when $\mathbf{q}$ approaches zero. For the low-temperature limiting case, (12.48) becomes

$$S_{\text{TDS}}(\tau) \sim \sum_{j=1}^{3n_0} \int_{\text{BZ}} dq_z \frac{1}{\omega_j(\tau, q_z)} \quad (12.49)$$

For the high-temperature limiting case,

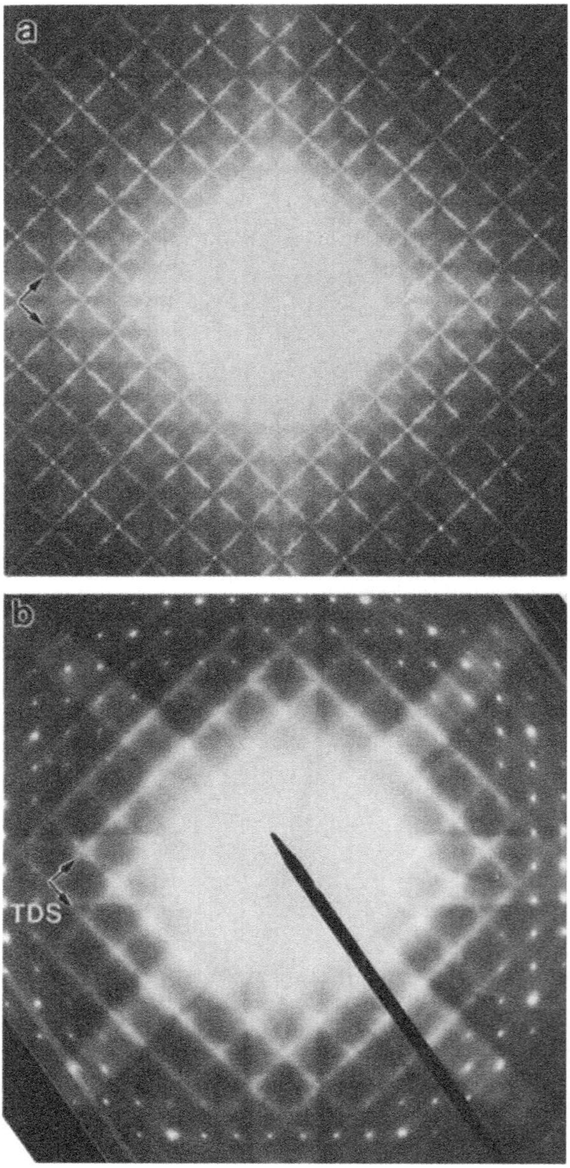

Fig. 12.4 Comparisons of **a** a simulated TDS diffraction pattern of Si[001] with **b** an observed parallel beam electron diffraction pattern; beam energy is 100 keV. The simulated diffraction pattern is the total contribution from elastic scattering and phonon scattering. The Si atom mean vibration amplitude is 0.09Å; mean momentum transfer is taken as $q_0 = 2.6\ \text{nm}^{-1}$; crystal thickness is 162.9 nm. The TDS streaks along <110> are clearly seen (Wang 1992a)

$$S_{\mathrm{TDS}}(\tau) \sim \frac{k_B T}{\hbar} \sum_{j=1}^{3n_0} \int_{\mathrm{BZ}} dq_z \frac{1}{\omega_j^2(\tau, q_z)} \tag{12.50}$$

Equation (12.50) is similar to the TDS intensity in X-ray diffraction derived by Born (1942a). In the phonon spectrum, there are $j = 1, 2 \ldots 3n_0$ branches of modes, among which there are only three acoustic branches. The summation of j can thus be reduced only to a total of three acoustic branches if we are interested only in TDS streaks (Komatsu and Teramoto 1966).

For a monatomic crystal under the central force approximation, the phonon dispersion relation $\omega_j(\mathbf{q})$ is approximately related to atom positions by Born (1942a, b):

$$\omega_j^2(\tau, q_z) \sim \frac{F}{2M} \sum_{\alpha} \sum_{\alpha_1} \sin^2 (\pi \{\tau \cdot [\mathbf{r}(\alpha) - \mathbf{r}(\alpha_1)] + q_z z(\alpha)\}) \tag{12.51}$$

where F is the force constant. For atoms not confined to the same atomic $(hk\ell)$ plane, i.e., $z(\alpha) \neq 0$, $\omega_j^2(\mathbf{q})$ does not approach zero for $\tau \cdot \mathbf{r}(\alpha) = 0$ and $q_z \neq 0$, thus, does not produce sharp intensity walls according to (12.49). Therefore phonon modes with wave vectors $\mathbf{q}$ not confined to the diffraction plane (i.e., $q_z \neq 0$) may contribute only a diffuse background to the diffraction pattern. These modes can be ignored if we are interested only in TDS streaks.

In electron diffraction, TDS is mainly generated by phonon modes with wave vectors $\mathbf{q}$ parallel or almost parallel to the diffraction plane, because momentum transferred from incident electrons is almost restricted to this plane. Conservation of momentum, thus, results in the excitation of phonons with wave vectors lying in the plane. Therefore modes with sharp intensity variations [i.e., $1/\omega_j(\mathbf{q})$] can be considered generated by atomic vibrations within the plane perpendicular to the incident beam direction, which is actually a two-dimensional lattice vibrational model. Thus (12.51) can be approximately written as

$$S_{\mathrm{TDS}}(\tau) \sim \sum_{j=1}^{2} \frac{1}{\omega_j(\tau)} \tag{12.52}$$

where $\omega_j(\tau)$ are defined as the dispersion surface of acoustic branches determined by the two-dimensional atomic vibrations within the $(hk\ell)$ plane when the beam direction is $\mathbf{B} = (hk\ell)$ for orthogonal crystal structures. In the following analyses, only interactions between an atom and its neighbors, located in the same $(hk\ell)$ atomic plane and closest to the atom, are considered. In other words, neighbors closest to the atom but outside the two-dimensional plane are not considered. It is important to note that this model is an approximated case for electron diffraction near a zone axis. We now compare results predicted by this two-dimensional vibration model and results observed experimentally.

12.7.4 Case 1. Monatomic f.c.c. Oriented in [001]

Thermal vibrations of atoms in different crystal slices can be considered independent events. For f.c.c. metals oriented in [001], if only interactions between atom [000] and its first nearest neighbor at $\pm(a/2)[110]$ and $\pm(a/2)\ [1\bar{1}0]$ in the same (001) plane are considered, (6.58) can be written as

$$\begin{pmatrix} D_{xx} & D_{xy} \\ D_{yx} & D_{yy} \end{pmatrix} \begin{pmatrix} e_{xj} \\ e_{yj} \end{pmatrix} = \omega_j^2 \begin{pmatrix} e_{xj} \\ e_{yj} \end{pmatrix} \tag{12.53a}$$

where

$$D_{xx} = D_{yy} = \frac{4\,F}{M}\left[\sin^2\frac{\pi a}{2}\left(q_x + q_y\right) + \sin^2\frac{\pi a}{2}\left(q_x - q_y\right)\right] \tag{12.53b}$$

$$D_{xy} = D_{yx} = \frac{4G}{M}\left[\sin^2\frac{\pi a}{2}\left(q_x + q_y\right) - \sin^2\frac{\pi a}{2}\left(q_x - q_y\right)\right] \tag{12.53c}$$

and F and G are force constants. The nontrivial solution of (12.53) is determined by the following condition:

$$\left(\omega_{1,2}\right)^2 = \frac{4}{M}\left[(F+G)\sin^2\frac{\pi a}{2}\left(q_x \pm q_y\right) + (F-G)\sin^2\frac{\pi a}{2}\left(q_x \mp q_y\right)\right] \tag{12.54}$$

Thus

$$S_{\text{TDS}} \sim \left\{\left[\sin^2\frac{\pi a}{2}\left(\tau_x \pm \tau_y\right) + \frac{F-G}{F+G}\sin^2\frac{\pi a}{2}\left(\tau_x \mp \tau_y\right)\right]^{1/2}\right\}^{-1} \tag{12.55}$$

For simplicity, we assume $F \approx G$. This assumption is exact in the central force model (Born 1942a). For Au, $F = 1.937 \times 10^4$dyn/cm and $G = 2.07 \times 10^4$dyn/cm. Thus

$$S_{\text{TDS}} \sim \left[\left|\ \sin\frac{\pi a}{2}\left(\tau_x \pm \tau_y\right)\ \right|\right]^{-1} \tag{12.56}$$

and it is expected that there are intensity walls located at $\tau_x \pm \tau_y = 0$ (i.e., $\langle 110\rangle$), in agreement with the observation in Fig. 12.5. The sharpness of the walls determines widths of TDS streaks in the diffraction pattern. In practice, it is usually the case that $F \neq G$ and $|F - G| \ll F$. Thus

$$\frac{F-G}{F+G}\sin^2\frac{\pi a}{2}\left(q_x \pm q_y\right)$$

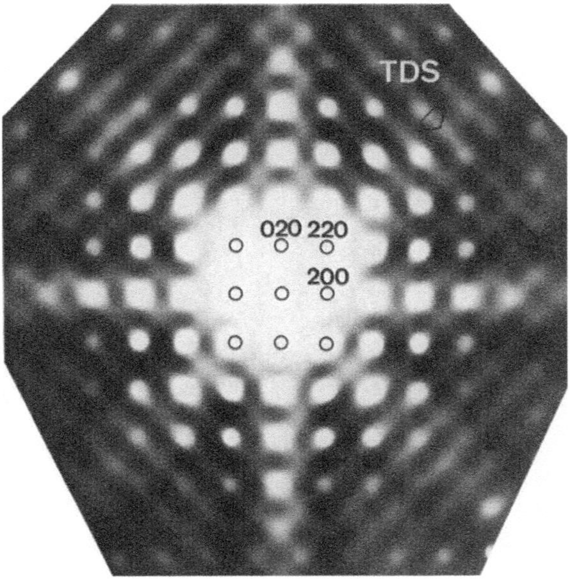

Fig. 12.5 A diffraction pattern from a thin f.c.c. gold crystal oriented in [001] exhibiting the formation of TDS streaks along ⟨110⟩ but no streaking along ⟨100⟩ or ⟨010⟩ (Wang and Bentley 1991b)

terms determine the height and width of TDS streaks. In other words, the width of TDS streaks is determined by the noncentral interaction force between atoms.

12.7.5 Case 2. Monatomic b.c.c. Oriented in [001]

Analogous to Case 1, we can obtain the dispersion relations for b.c.c. metals if only interactions between the first nearest neighbor separated by a in the (001) plane are considered, i.e., interactions of Atom 1 with 2, 2 with 3, 3 with 4, and 4 with 1 (see Fig. 12.6a)

$$\left(\omega_{1,2}\right)^2 = \frac{4}{M}\left[(F+G)\sin^2 \pi a q_{x,y} + (F-G)\sin^2 \pi a q_{y,x}\right] \tag{12.57}$$

For $F = G$,

$$S_{\text{TDS}} \sim \left[\left|\left(\sin \pi a \tau_{x,y}\right)\right|\right]^{-1} \tag{12.58}$$

Sharp TDS streaks are expected along ⟨100⟩ and ⟨010⟩. For Mo, $F = 3.61 \times 10^4$ dyn/cm and $G = 2.79 \times 10^4$dyn/cm; $F \approx G$ is not a good approximation, so TDS streaks may be broadened by the $(F - G)$ term contained in ω^2. This agrees with the experimental observation of Mo [001] shown in Fig. 12.6b. The TDS streaks along ⟨110⟩ would appear in the pattern if the phonon modes generated by the interaction

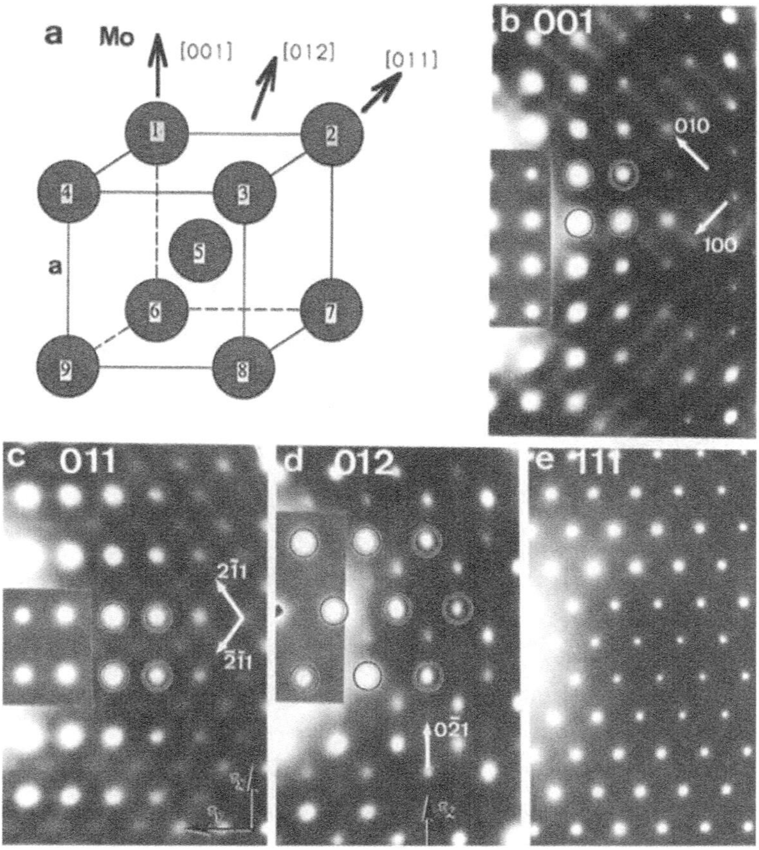

Fig. 12.6 **a** An atomic model of a b.c.c. monatomic lattice; **b–e** electron diffraction patterns taken from an Mo b.c.c. single crystal to show TDS streaks observed when the beam direction is [001], [011], [012], and [111], respectively. The τ_x and τ_y represent vectors used in theoretical investigations. Positions of regular Bragg reflections due to pure elastic scattering are circled (Wang 1992a)

between atom 5 with atoms 1, 2, 3, and 4 were strongly excited. This confirms the validity of the two-dimensional vibration model introduced in the last section.

12.7.6 Case 3. Monatomic b.c.c. Oriented in [011]

For b.c.c. metals oriented in [011], if interactions between Atoms 5 and 1, 4, 7, and 8, separated by (a/2) <111>, are considered (see Fig. 12.6a):

$$(\omega_{1,2})^2 = \frac{4}{M}\left[(F+G)\sin^2\frac{\pi a}{2}\left(q_x \pm \sqrt{2}q_y\right) + (F-G)\sin^2\frac{\pi a}{2}\left(q_x \mp \sqrt{2}q_y\right)\right] \tag{12.59}$$

It must be pointed out that $\tau = (\tau_x, \tau_y)$ is always defined in the plane perpendicular to the incident beam direction, which should be distinguished from indices of the diffraction pattern. Thus

$$S_{\mathrm{TDS}} \sim \left[\left|\sin\frac{\pi a}{2}\left(\tau_x \pm \sqrt{2}\tau_y\right)\right|\right]^{-1} \tag{12.60}$$

and TDS lines satisfy $\tau_x \pm \sqrt{2}\tau_y = 0$ in the diffraction plane. Converting τ into indices of the diffraction pattern, TDS streaks are expected along $\pm[21\bar{1}]$ and $\pm[2\bar{1}1]$. A corresponding experimental observation of Mo(011) is shown in Fig. 12.6c. The joint points of $\pm[21\bar{1}]$ and $\pm[2\bar{1}1]$ streaks show stronger intensity and appear like weak reflected spots. These spots are Bragg reflections in the first-order Laue zone.

12.7.7 *Case 4. Monatomic b.c.c. Oriented in [012]*

If we view the b.c.c. lattice along the [012] direction, the closest atoms are Atoms 1 and 4 (see Fig. 12.6a). Since

$$\omega_1^2 = \frac{4}{M}F\sin^2\pi a q_x \tag{12.61}$$

then

$$S_{\mathrm{TDS}} \sim (|\sin\pi a\tau_x|)^{-1} \tag{12.62}$$

and the TDS line satisfies $\tau_x = 0$ in the diffraction plane. After considering the correspondence of diffraction spots with atomic planes in the crystal unit cell, streaks are expected along $\pm[0\bar{2}1]$, and they are observed experimentally for Mo [012] (Fig. 12.6d). Besides the circled Bragg reflections, some weak streaks are also observed, which are the result of phonon scattering.

12.7.8 *Case 5. Monatomic b.c.c. Oriented in [111]*

For b.c.c. metals oriented in [111], the closest atoms in the (111) plane are separated by a <110>. This distance may be so large that interactions between atoms in the two-dimensional plane become weak. In this case, there may be no TDS streaks but only a diffuse background in the diffraction pattern. The corresponding observation of Mo in Fig. 12.6e shows the important fact that atoms separated by more than $2^{1/2}a$ in Mo do not have strong interactions.

12.7.9 Case 6. Monatomic Diamond-Type Structures Oriented in [001]

For diamond cubic structures, if only interactions between the nearest neighbors in the (001) plane are considered, the result is the same as for f.c.c. oriented in [001]. From (12.56),

$$S_{\text{TDS}} \sim \left[\left|\sin\frac{\pi a}{2}\left(\tau_x \pm \tau_y\right)\right|\right]^{-1} \tag{12.63}$$

Thus TDS streaks are along <110>. Since $F = 2.24 \times 10^4$dyn/cm and $G = 1.94 \times 10^4$dyn/cm for Si , the condition $F \approx G$ is approximately satisfied. Thus TDS streaks are expected to be sharp. This agrees with the experimental observation in Fig. 12.4b.

From the preceding discussion, it seems that a general relationship

$$S_{\text{TDS}}(\tau) \sim \frac{1}{\omega_j} \sim \frac{1}{|\sin\{\pi\tau\cdot[\mathbf{r}(\alpha) - \mathbf{r}(\alpha_1)]\}|}$$

is approximately held for monatomic cubic systems under the central force approximation. Thus S_{TDS} becomes infinite when $\tau \cdot [\mathbf{r}(\alpha - \mathbf{r}(\alpha_1)] = 0$. Therefore directions of TDS streaks predicted according to the preceding procedures are consistent with those predicted according to the $\tau \cdot [\mathbf{r}(\alpha) - \mathbf{r}(\alpha_1)] = 0$ rule, thus leading to the same rule in (12.39).

If the incident electron beam is far from the crystal zone axis, the contribution of optical branches in (12.49) or (12.50) becomes important. The TDS streaks in this case may not be so sharp as those observed in near-zone axis cases, because both acoustic and optical branches have equal importance for the TDS intensity distribution. This may explain the phenomenon that TDS streaks observed in X-ray or neutron diffraction patterns are not so sharp as those observed in electron diffraction patterns. In X-ray or neutron diffraction, there is always a misoriented large, thick crystal with defects, so that local deviation from the main-zone axis may enhance excitation of optical modes, resulting in wider TDS streaks according to (12.49). In electron diffraction, however, this misorientation effect is greatly reduced with a thin crystal; thus TDS streaks are sharper.

To summarize, TDS streaks in a near-zone axis electron diffraction pattern of an orthogonal crystal structure can be predicted as follows: (1) Determine the incident beam direction $\mathbf{B} = (hk\ell)$; (2) define the two-dimensional atomic planes perpendicular to $\mathbf{B}$; (3) consider two-dimensional interactions between an atom and its first nearest neighbor in the $(hk\ell)$ plane (not necessarily the first nearest neighbor in three-dimensional space) under harmonic oscillator and central force (e.g., $F = G$) approximations; (4) determine the dispersion relationships $\omega_j(\mathbf{q})$ for acoustic branches of different atomic planes; (5) determine TDS streaks given by the $\tau_x - \tau_y$ curves satisfying $\omega_j(\tau) \approx 0$; and (6) convert the $\tau_x - \tau_y$ curves to the corresponding indices according to the beam direction and crystal structure.

12.7.10 Equivalence of Frozen Lattice Model and Phonon Excitation Theories For TDS

Quantum theory and semiclassical theory can be equivalently applied to calculate TDS electron diffraction patterns. In the quantum mechanical approach, a phonon is created or annihilated by the excitation of an inelastic state characterized by some quantum numbers. The final intensity distribution of TDS electrons is an incoherent sum of electrons scattered by different phonon states (or modes). In the semiclassical (or quasi-elastic) scattering theory of frozen lattices, an electron diffraction pattern or image is considered the sum of scattering intensities for the many instantaneous pictures of displaced atoms. In other words, TDS is actually a statistically averaged quasi-elastic scattering of electrons from crystals with different thermal vibrational configurations. The quantum and semiclassical approaches appear to be different, but essentially they should be equivalent. We now prove this equivalence.

We begin with the frozen lattice model of semiclassical theory. The elastic scattering of electrons from a frozen lattice configuration is governed by

$$\left\{\nabla^2 + 4\pi^2\left[U_0(\mathbf{r}) + \Delta U(\mathbf{r},\mathbf{q},t) + \mathrm{K}^2\right]\right\}\Psi = 0 \tag{12.64}$$

where $\Delta U(\mathbf{r},\mathbf{q},t)$ is the time (or configuration)-dependent perturbation to the crystal potential from the vibration mode $(\mathbf{q}, j)$. From (6.63), under the single-phonon approximation:

$$\Delta U(\mathbf{r},\mathbf{q},t) = -\frac{2m_e e}{h^2}\sum_n\sum_\alpha \mathbf{u}\begin{pmatrix} n & q \\ \alpha & j \end{pmatrix} \cdot \nabla V_\alpha\left(\mathbf{r} - \mathbf{R}_n - \mathbf{r}(\alpha)\right) \tag{12.65a}$$

where

$$\mathbf{u}\begin{pmatrix} n & q \\ \alpha & j \end{pmatrix} = \sqrt{2}A_\alpha^0\left(\omega_j\right)\mathbf{e}\left(\alpha\big|{}^{\mathbf{q}}_{j}\right)\exp\left\{2\pi i\mathbf{q}\cdot\left[\mathbf{R}_n + \mathbf{r}(\alpha)\right]\right\}\cos\left(\omega_j t\right) \tag{12.65b}$$

is the time-dependent lattice displacement for atom $\binom{n}{\alpha}$ in phonon mode $\left(\mathbf{q}, \omega_j(\mathbf{q})\right)$, and $A_\alpha^0\left(\omega_j\right)$ is the atomic vibration amplitude

$$A_\alpha^0\left(\omega_j\right) = \frac{\hbar^{1/2}}{\left[2\omega_j(\mathbf{q})M_\alpha N_0\right]^{1/2}} \tag{12.65c}$$

which is the limiting result of (6.69b) when T tends to zero. If the solution of (12.64) can be written as

$$\Psi = \phi(\mathbf{r})\Psi_0(\mathbf{r}) \tag{12.66}$$

where $\Psi_0(\mathbf{r})$ is the solution of (12.64) for $\Delta U = 0$ under the small-angle scattering approximation, we have the following results by using the same approximations leading to (12.3):

$$\frac{\partial\phi(\mathbf{r})}{\partial z} \approx \frac{2\pi^2 \Delta U(\mathbf{r}, \mathbf{q}, t)\Psi_0(\mathbf{r})}{\frac{\partial\Psi_0(\mathbf{r})}{\partial z}}\phi(\mathbf{r}) \tag{12.67}$$

thus

$$\phi(\mathbf{r}) = \exp\left(2\pi^2 \int_0^z dz \left\{\Delta U(\mathbf{r}, \mathbf{q}, t)\Psi_0(\mathbf{r}) \cdot \left[\frac{\partial\Psi_0(\mathbf{r})}{\partial z}\right]^{-1}\right\}\right) \tag{12.68}$$

Substituting (12.68) into (12.66) for thin crystals

$$\left|2\pi^2 \int_0^d dz \Delta U \Psi_0 \left[\partial\Psi_0/\partial z\right]^{-1}\right| \ll 1$$

the pure TDS wave in mode $\left[\mathbf{q}, \omega_j(\mathbf{q})\right]$ at the crystal exit face is

$$\Psi_T(\mathbf{b}, \mathbf{q}, d, t) = \Psi - \Psi_0 = [\phi(\mathbf{b}, d) - 1]\Psi_0 \approx 2\pi^2 \int_0^d dz \Delta U(\mathbf{r}, \mathbf{q}, t) Y(\mathbf{b}, 0, z) \tag{12.69}$$

In the multislice scheme,

$$\Psi_T(\mathbf{b}, \mathbf{q}, d, t) = 2\pi^2 \Delta z \sum_{n_c} \Delta U(\mathbf{r}, \mathbf{q}, t) Y\left(\mathbf{b}, 0, z_{n_c}\right) \tag{12.70}$$

Taking a two-dimensional Fourier transform of (12.70) and using (12.65a), we have

$$\begin{aligned}\Psi_T(\tau, \mathbf{q}, d, t) =& i\sqrt{2}\alpha_0 \Delta z \cos\left(\omega_j t\right) \sum_{n_c}\sum_{\alpha}\sum_{m} A_\alpha^0\left(\omega_j\right)\left(\mathbf{e}\left(\alpha|_j^{\mathbf{q}}\right) \cdot \tau f_\alpha^e(\tau)\right.\\ &\times \exp\{-2\pi i(\tau - \mathbf{q}) \cdot [\mathbf{R}_m + \mathbf{r}(\alpha)]\}) \otimes Y\left(\tau, 0, z_{n_c}\right)\\ =& i\sqrt{2}\alpha_0 \Delta z \cos\left(\omega_j t\right) \sum_{n_c}\sum_{\alpha}\sum_{g} A_\alpha^0\left(\omega_j\right)\left[\mathbf{e}\left(\alpha|_j^{\mathbf{q}}\right) \cdot (\mathbf{g} + \mathbf{q})\right]\\ &\times f_\alpha^e(\mathbf{g} + \mathbf{q}) \exp[-2\pi i \mathbf{g} \cdot \mathbf{r}(\alpha)] Y\left(\tau - \mathbf{q} - \mathbf{g}, 0, z_{n_c}\right)\end{aligned} \tag{12.71}$$

The final TDS patterns are formed by the incoherent sum over all TDS electrons in different phonon modes; after taking the time average, we have

$$I_t(\tau) = N_0\Omega \sum_{j=1}^{3n_0} \int_{BZ} d\mathbf{q} < |\Psi_T(\tau, \mathbf{q}, d)|^2 >$$

$$= N_0\Omega (\alpha_0 \Delta z)^2 \sum_{n_c} \sum_{\alpha} \sum_{g} \sum_{n_c'} \sum_{\alpha'} \sum_{g'} \sum_{j=1}^{3n_0} \int_{BZ} d\mathbf{q} A_\alpha^0 (\omega_j) A_{\alpha'}^{0*} (\omega_j) \left[\mathbf{e}\left(\alpha|_j^{\mathbf{q}}\right) \cdot (\mathbf{g}+\mathbf{q})\right]$$
$$\times \left[\mathbf{e}^*\left(\alpha'|_j^{\mathbf{q}}\right) \cdot (\mathbf{g}'+\mathbf{q})\right]$$
$$\times f_\alpha^e(\mathbf{g}+\mathbf{q}) f_{\alpha'}^{e*}(\mathbf{g}'+\mathbf{q}) \exp\left\{-2\pi i \left[\mathbf{g}\cdot\mathbf{r}(\alpha) - \mathbf{g}'\cdot\mathbf{r}(\alpha')\right]\right\}$$
$$\times Y\left(\tau - \mathbf{q} - \mathbf{g}, 0, z_{n_c}\right) Y^*\left(\tau - \mathbf{q} - \mathbf{g}', 0, z_{n_c}'\right) \tag{12.72}$$

Equation (12.72) is the same as (12.42) except that A_α is replaced by A_α^0 and $Y(\tau - \mathbf{q} - \mathbf{g}, \mathbf{q}_0, z_n)$ is replaced by $Y(\tau - \mathbf{q} - \mathbf{g}, 0, z_n)$. The latter means that the effect of phonon momentum transfer, which is equivalent to an inclined incident wave vector, is neglected in semiclassical theory. Therefore semiclassical theory is approximately equivalent to quantum theory if the following two conditions are met (Wang, 1992a): (1) low-temperature limit so that $k_B T \ll \hbar\omega$ and (2) the mean momentum transfer $\mathbf{q}_0$ is small. This work provides a mathematical proof of the frozen lattice model of quasi-elastic scattering for TDS, as introduced in Chap. 7.

12.7.11 Diffraction of Atomic Inner Shell Scattered Electrons

We now apply (12.10a) for calculating energy-filtered diffraction patterns formed by inelastically scattered electrons in a periodic crystal. For easy notation, we replace $\Psi_1^0(\mathbf{r})$ by $\Psi_1^0(\mathbf{r}, \mathbf{q})$ to represent the inelastic incident electron wave with wave vector $\mathbf{K} - \mathbf{q}$. Taking a two-dimensional Fourier transform of (12.10a) yields

$$\Psi_1(\tau, d) \approx \beta_0 \int_0^d dz H_{10}(\tau, z, \mathbf{q}) \otimes Y(\tau, \mathbf{q}, z)$$
$$\approx \beta_0 \Delta z \sum_{n_c} H_{10}\left(\tau, z_{n_c}, \mathbf{q}\right) \otimes Y\left(\tau, \mathbf{q}, z_{n_c}\right) \tag{12.73}$$

For the transition matrix elements, we take a Fourier transform of (6.8)

$$H_{10}(\tau, z, \mathbf{q}) = \sum_g H_g^{(10)}(\mathbf{q}) \delta(\tau - \mathbf{g} + \mathbf{q}) \tag{12.74}$$

Summing the intensities contributed by inelastic-scattering processes of different $\mathbf{q}$ incoherently, and assuming that inelastic waves generated from different slices are incoherent, we get

$$
\begin{aligned}
I_1(\tau) &= N_0\Omega \int_{BZ} d\mathbf{q}\, |\Psi_1(\tau, d)|^2 \\
&\approx N_0\Omega\, (\beta_0\Delta z)^2 \sum_{n_c} \int_{BZ} d\mathbf{q} \left| H_{10}\left(\tau, z_{n_c}, \mathbf{q}\right) \otimes Y\left(\tau, \mathbf{q}, z_{n_c}\right)\right|^2
\end{aligned}
\tag{12.75a}
$$

To further simplify (12.75), we ignore the interference effect of waves scattered to different **g**s and substitute (12.74) into (12.75a)

$$
I_1(\tau) \approx N_0\Omega\, (\beta_0\Delta z)^2 \sum_{n_c} \sum_{g} \int_{BZ} d\mathbf{q} \left| H_g^{(10)}(\mathbf{q}) Y\left(\tau - \mathbf{g} + \mathbf{q}, \mathbf{q}, z_{n_c}\right)\right|^2 \tag{12.75b}
$$

Equation (12.75b) can be further simplified by separating the integration of $\mathbf{q}$ into $\mathbf{q}_b = (q_x, q_y)$ and q_z. Assuming that Y is almost independent of q_z and using the approximated relation in (12.46) for Y:

$$
I_1(\tau) \approx N_0\Omega(\beta_0\Delta z)^2 \sum_{n_c} \sum_{\mathbf{g}} \left[\int_{BZ} dq_z \left| H_{\mathbf{g}}^{10}(-\tau + \mathbf{g}, q_z)\right|^2 \right] \otimes |Y(\tau - \mathbf{g}, \mathbf{q}_0, zn_c)|^2 \tag{12.76}
$$

Equation (12.76) can be applied to different inelastic-scattering processes. For single-electron excitation, the transition matrix is given by (6.120).

A calculated Si K edge ($\Delta E = 1850$eV) energy-filtered [001] pattern is shown in Fig. 12.7. High-index Bragg reflections are broadened by the angular distribution of inelastic electrons. It appears that the channeling effect is strong, resulting in the formation of sharp intensity streaks located midway between and parallel to the Kikuchi lines, as indicated by arrowheads. This may be because the K shell ionization edge can be excited only by electrons channeling close to the atomic nuclei.

For delocalized scattering, the approximation due to ignoring the $\nabla^2\phi_n(\mathbf{r})$ term may hold for relatively thick crystals. Applications of the modified multislice theory for dealing with diffraction and imaging of valence loss electrons are described in Chap. 13.

12.7.12 Summary

Theory introduced in this chapter was proposed by simplifying dynamic elastic rescattering of inelastically scattered electrons, so that the number of numerical calculations is greatly reduced. Some useful analytic solutions were obtained for TDS electrons. A simple rule, introduced to predict directions of TDS streaks in electron diffraction patterns, was tested in reference to several monatomic materials. We derived the conditions under which equivalent results are obtained based on either

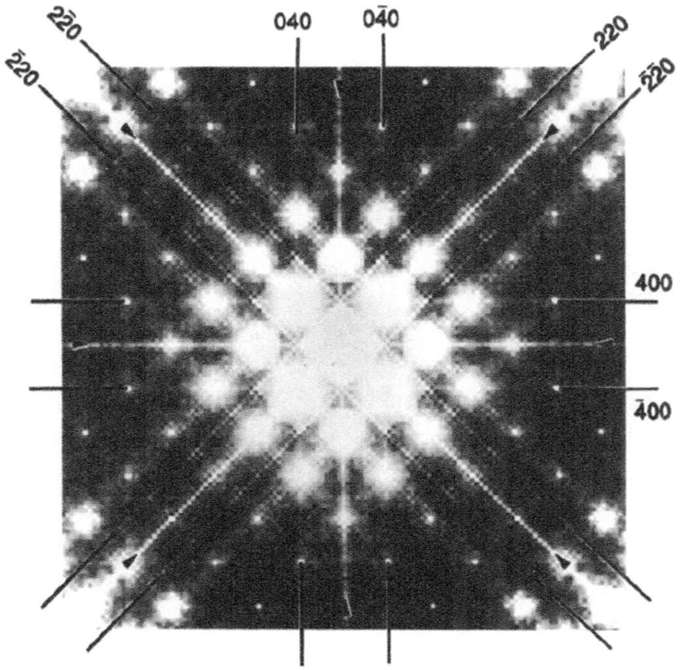

Fig. 12.7 A simulated [001] Si K ionization edge (ΔE = 1839eV) energy-filtered electron diffraction pattern. Beam convergence is zero, crystal thickness 162.9 nm, mean momentum transfer $q_0 = q_E = 7.73 \times 10^{-3} K_0$, and beam energy 100 keV. Sharp streaks appearing midway between the Kikuchi lines are indicated by arrowheads (Wang 1992a)

the frozen lattice model or the quantum mechanical phonon excitation theories. This work provides a theoretical basis for treating TDS based on the frozen lattice model, as introduced in Chap. 7.

Chapter 13
Inelastic Scattering in Sub-Angstrom Electron Imaging and Holography

Plasmon (or valence) excitation is an important source of inelastic scattering in high-resolution TEM because valence loss electrons are distributed very near to Bragg beams. Atomic core-shell excitations are localized scattering processes whose probability is very small. Thermal diffuse scattering is also a localized inelastic scattering but the large momentum transfer in this process results in the majority of TDS electrons being distributed in the angular range out of the objective aperture. This is partially equivalent to introducing an absorption factor in the elastic HRTEM image simulation, as illustrated in Chap. 6, with electrons enclosed by the objective aperture contributing to the image. Therefore, valence excitation and phonon scattering are inelastic-scattering processes that must be considered in HRTEM image simulations.

In inelastic scattering, electrons that have excited the same crystal state of energy loss ε_n and wave vector change $\boldsymbol{q}$ are considered to be coherent (Kainuma and Kogiso 1968). The elastic rescattering of inelastically scattered electrons forms lattice images. Electrons that have excited different crystal states are incoherent. In practice, HRTEM images are formed by the interferences of electrons elastically scattered to different Bragg beams (Wang and Bentley 1991a; Rez 1993). Therefore lattice images can be formed by diffracted inelastic electrons after exciting the same crystal state of $(\varepsilon_n, \boldsymbol{q})$ (Shuman et al. 1986; Hashimoto et al. 1992). It has been shown experimentally that lattice images can be formed by plasmon (Ajika et al., 1984; Wang and Shapiro, 1995) and core-less electrons (Hashimoto et al., 1994). In HRTEM imaging, since a finite-sized objective aperture is used, inelastically scattered electrons (including TDS electrons) distributed within the allowed angular range of the objective aperture contribute to the image, and those scattered outside the aperture can be treated as being absorbed.

There are several main advances in the instrumentation of electron microscopy in recent few decades years. First, the development of monochromator narrows the electron energy spread to the order of meV range, which not only drastically reduces the chromatic aberration effect but also allows much improved beam coherence. The implementation of the energy filter gives us the capability of selecting the energy

Z. L. Wang, *Elastic and Inelastic Scattering in Electron Diffraction and Imaging*, Springer Series in Solid-State Sciences 205, https://doi.org/10.1007/978-3-031-90819-4_13

of the output signals for imaging and diffraction. The energy filter filters off the contribution made by inelastically scattered electrons with energy losses larger than a few eV; thus, the calculated results based on elastic-scattering theory can be directly compared to the experimental data. The Cs corrector (Batson et al. 2002), the electron holography (Fu et al. 1991), and image reconstruction technique (Zandbergen and Van Dyck 2000) are powerful approaches not only for extension of the resolution of the electron imaging, but also for retrieving electron phase image. Quantitative electron microscopy with sub-Angstrom image resolution is now experimentally feasible.

With the vast development in experimental approaches, one must re-examine the theories developed before the 1960s for image and diffraction calculations to see if any effect that was not important back to then but is vital in today's super-resolution microscopy. We know that an energy filter filters away most of the inelastically scattered electrons, but electrons with an energy loss less than a sub-eV remain in the image formation process. This means that phonon scattered, or thermal diffuse scattered (TDS), electrons are included in imaging. The theories most popularly used for today's image simulations, however, are for elastic scattering, and the contribution made phonon scattering is excluded. This is not a main problem if the image resolution is poorer than 0.2 nm. But as the image resolution is reaching 0.1 nm and even better, an inclusion of phonon scattering in image calculation is mandatory.

In this chapter, we first outline the theory for calculating HRTEM images of valence loss electrons. Secondly, in quantitative high-resolution transmission electron microscopy, the theoretically calculated images usually give better contrast than the experimentally observed ones, although all of the factors have been accounted for. This discrepancy is suggested due to phonon scattered or thermal diffusely scattered (TDS) electrons, which were not included in the image calculation. The contribution from TDS electrons is especially important if the image resolution is approaching 0.1 nm and beyond with the introduction of Cs corrected microscopes. Furthermore, we proved that the off-axis holography is an ideal energy filter that even filters away the contribution made by TDS electrons in the electron wave function, but conventional high-resolution microscopy do contain the contribution made by phonon-scattered electrons.

13.1 Contribution of Valence Loss Electrons

13.1.1 Diffraction of Valence Loss Electrons

Before considering images formed by other inelastically scattered electrons, we first calculate the diffraction pattern formed by valence loss electrons (Wang and Bentley 1991a). This section describes image calculation (used in the next section), because the effect of the objective lens does not need to be considered in diffraction pattern calculation. We assume that H_{nm} does not depend on z for valence excitations. We now use the modified multislice theory in Chap. 12 to calculate the diffraction pattern of valence loss electrons. At the crystal exit face $z = d$ and under the single inelastic-scattering approximation, (12.10a) is

$$\Psi_n(\mathbf{b}, d) = \beta_0 H_{n0}(\mathbf{b}) \int_0^d dz \left\{ \Psi_0^0(\mathbf{r}) \left[\partial \Psi_n^0(\mathbf{r}) / \partial z \right]^{-1} \right\} \Psi_n^0(\mathbf{b}, d) \tag{13.1}$$

In valence excitation, since momentum transfer is very small, the most significant quantity is the electron energy loss. For easy notation, we use ω as a quantum number to represent an excited state, i.e., $\Psi_n^0(\mathbf{b}, z) = \Psi_0^0(\mathbf{b}, z, \omega)$; therefore (13.1) becomes

$$\Psi_n(\mathbf{b}, d) = \beta_0 H_{n0}(\mathbf{b}) Y'(\mathbf{b}, \omega) \tag{13.2}$$

where

$$Y'(\mathbf{b}, \omega) = \int_0^d dz \Psi_0^0(\mathbf{b}, z) \left[\partial \Psi_0^0(\mathbf{b}, z, \omega) / \partial z \right]^{-1} \Psi_0^0(\mathbf{b}, d, \omega) \tag{13.3}$$

and $\hbar\omega$ is the electron energy loss. The symbol change simply neglects the $\boldsymbol{q}$ dependence of Ψ_n^0 (i.e., all states are presupposed to be nondegenerate). The consequence of this approximation is discussed in Sect. 13.1.3. For small-angle plasmon excitation, the introduction of small-angle spreading (less than 0.2 mrad) actually does not affect the final calculated results. We now consider the energy-filtered diffraction pattern formed by valence-excited electrons. Taking a Fourier transform of (13.2), contributions of all valence-excited states to the final electron diffraction pattern are added incoherently, thus

$$\begin{aligned} I_v(\mathbf{u}) &= \sum_n |\Psi_n(\mathbf{u}, d)|^2 = \beta_0^2 \sum_n \left| H_{n0}(\mathbf{u}) \otimes Y'(\mathbf{u}, \omega) \right|^2 \\ &= \beta_0^2 \int d\tau_b \int d\tau' \sum_n \left[H_{n0}(\tau) H_{n0}^*(\tau') \right] Y'(\mathbf{u} - \tau, \omega) Y'^* \left(\mathbf{u} - \tau', \omega \right) \end{aligned} \tag{13.4}$$

where n is summed over all possible valence states. We now correlate transition matrix elements with the mixed dynamic form factor defined in (6.109a), (6.109b), and (6.110). First, consider

$$\sum_n \left[H_{n0}(\tau) H_{no}^*(\tau') \right] = \sum_n \left[H_{n0}(\tau) H_{0n}(-\tau') \right] = \frac{e^4}{16\pi^4 \varepsilon_0^2} \frac{1}{\tau^2 \tau'^2} S\left(-\tau', -\tau\right) \tag{13.5}$$

where (6.109a) and (6.109b) was used. By using (6.112) for $S\left(-\tau', -\tau\right)$ and using the approximation $|\hbar\omega| >> k_B T$, (13.4) can be rewritten as

$$I_v(\mathbf{u}) = i\,\text{Co} \int d\tau_b \int d\tau_b' Y'(\mathbf{u}-\tau,\omega) Y'^*\left(\mathbf{u}-\tau',\omega\right) \times \int_0^{\infty} d\omega' \left[\frac{1}{\tau'^2 \varepsilon_{\tau\tau'}(\omega')} - \frac{1}{\tau^2 \varepsilon^*_{\tau'\tau}(\omega')} \right] \tag{13.6}$$

where

$$C_0 = \frac{\beta_0^2 e^2 \hbar}{8\pi^3 \varepsilon_0} \quad \tau^2 = \tau_b^2 + \tau_z^2$$

and the z component of the momentum transfer $t_z = (q_{n0})_z$ is related only to the electron energy loss by

$$(q_{n0})_z \approx \frac{K_0 \Delta E}{2\gamma E_0} = \frac{\omega}{2\pi v}$$

(see Sect. 6.8).

If the electron wave function is not significantly affected by a small energy loss, then $Y(\boldsymbol{u}-\tau,\omega)$ can be approximately included in the integration of ω' (In numerical calculations, this is actually an excellent approximation). Thus (13.6) becomes

$$I_\nu(u) \approx 2C_0 \int d\tau_b \int d\tau_b' \int_0^{\infty} d\omega \frac{1}{\tau'^2} \times \text{Im}\left[-\frac{1}{\varepsilon_{\tau\tau'}(\omega)} Y'(u-\tau,\omega) Y'^*\left(u-\tau',\omega\right) \right] \tag{13.7}$$

For a homogeneous medium, the dispersion of the dielectric function is negligible, so that (13.7) is simplified as

$$I_v(\mathbf{u}) = 4\pi^2 K_0^2 \int_0^{\infty} d\omega \int_{-\infty}^{\infty} d\tau_b \frac{d^3 P(\tau_b,\omega)}{d\omega d\tau_b dz} \left| Y'(\mathbf{u}-\tau,\omega) \right|^2 \tag{13.8}$$

where the differential electron excitation probability per unit distance is [see (6.95)]

$$\frac{d^3 P(\tau_b,\omega)}{d\omega d\tau_b dz} = \frac{e^2}{4\pi^3 \varepsilon_0 \hbar v^2} \frac{1}{\tau_b^2 + q_n^2} \text{Im}\left[-\frac{1}{\varepsilon(\omega)} \right] \tag{13.9}$$

Since the integration of (13.9) over τ_b is just (6.96), the diffraction pattern is the sum of electrons with different energy losses $\hbar\omega$ and momentum transfers, as weighted by the double differential scattering cross section. Equation (13.8) is the basic equation for calculating inelastic diffraction patterns formed by valence-excited electrons.

13.1.2 Energy-Filtered HRTEM Images of Valence Loss Electrons

For image simulations, we first consider an ideal case in which the objective lens is assumed as perfect, so there are no spherical nor chromatic aberrations. Thus image intensity is an incoherent summation of electrons after exciting different valence states

$$I_v(\mathbf{b}) = \beta_0^2 \sum_n \left|H_{n0}(\mathbf{b})Y'(\mathbf{b},\omega)\right|^2 \tag{13.10}$$

If H_{n0} is written in the form of its inverse Fourier transformation, (13.10) becomes

$$\begin{aligned} I_v(\mathbf{b}) = \beta_0^2 \int d\tau_b \int d\tau_b' \exp\left[2\pi i\left(\tau - \tau'\right)\cdot \mathbf{b}\right] \\ \times \sum_n \left[H_{n0}(\tau)H_{n0}^*\left(\tau'\right)\right]\left|Y'(\mathbf{b},\omega)\right|^2 . \end{aligned} \tag{13.11}$$

Using (13.5) and following the analogous procedures used to derive (13.7), (13.11) becomes

$$\begin{aligned} I_v(\mathbf{b}) \approx 2C_0 \int_0^\infty d\omega \int d\tau_b \int d\tau_b' \\ \times \operatorname{Im}\left\{-\frac{\exp\left[2\pi i\left(\tau - \tau'\right)\cdot \mathbf{b}\right]}{\tau'^2 \varepsilon_{\tau\tau'}(\omega)}\right\}\left|Y'(\mathbf{b},\omega)\right|^2 . \end{aligned} \tag{13.12}$$

Equation (13.12) has to be generalized to include aberration effects of the objective lens. In practice, a nonideal objective lens has different contrast transfer functions for electrons with different energy losses; therefore (13.12) is generalized as

$$\begin{aligned} I_v(\mathbf{b}) \approx 2C_0 \int_0^\infty d\omega \int d\tau_b \int d\tau_b' \operatorname{Im}\left\{-\frac{\exp\left[2\pi i\left(\tau - \tau'\right)\cdot \mathbf{b}\right]}{\tau'^2 \varepsilon_{\tau\tau'}(\omega)}\right\} \\ \times \left|Y'(\mathbf{b},\omega)\otimes T_{\text{obj}}(\mathbf{b},\omega)\right|^2 \end{aligned} \tag{13.13}$$

where $T_{obj}(\boldsymbol{b},\omega)$ represents the optic transfer function of the objective lens for electrons with energy $E_0 - \hbar\omega$. The Fourier transform of this function is approximately a product of the lens transfer function for electrons of energy E_0 and an exponential factor that characterizes the chromatic aberration effect

$$T_{\text{obj}}(\tau,\omega) = T_{\text{obj}}(\tau)\exp\left[\pi i \lambda \tau^2 \Delta f_c(\omega)\right] \tag{13.14a}$$

with

$$\Delta f_c = \frac{C_c \hbar\omega}{E_0} \tag{13.14b}$$

where Δf_c is the focus shift introduced by chromatic aberration and C_c is the chromatic aberration coefficient of the objective lens. For $C_c = 1.2$ mm, $\Delta E = 15$ eV, and $E_0 = 200$ keV, (13.14) yields $\Delta f_c = 90$ nm. This is a relatively large change. For a homogeneous medium, (13.13) becomes

$$I_v(\mathbf{b}) \approx 4\pi^2 K_0^2 \int_0^\infty d\omega \frac{d^2 P}{d\omega dz} \left|Y'(\mathbf{b}, \omega) \otimes T_{\text{obj}}(\mathbf{b}, \omega)\right|^2 \tag{13.15}$$

Equation (13.15) is the principle equation for calculating HRTEM images of valence loss electrons. The $Y(\boldsymbol{b}, \omega)$ term contains the perturbation of the electron wave function by energy losses; the calculation of this function is discussed next. It is important to point out that (13.7) and (13.15) were derived from the single inelastic-scattering model. In practice, plural plasmon excitation obeys Poisson law (see Sect. 14.1), hence the relative intensity of the mth plasmon loss to the pure elastic peak is

$$\frac{(d/\Lambda)^m}{m!}$$

where Λ is the mean free path of plasmon excitation, typically about 200–300 nm for 200-keV electrons (Raether 1980). In the crystal thickness range of about 50 nm, the condition $(d/\Lambda)^2 << 1$ is always satisfied. Therefore in the thickness range where an HRTEM image is obtainable, high-order plasmon excitations can be neglected. In other words, the single-plasmon scattering model is a good approximation.

13.1.3 Approaching the Completely Delocalized Scattering Model

To see the physical meaning of (13.7) and (13.15), we consider a case in which the crystal thickness is small. For delocalized valence excitations with energy loss less than about 30 eV and angular spread less than 0.5 mrad, the perturbation on the wave function of high-energy electrons may be neglected if the specimen is thin, i.e., $\Phi_n^0 \approx \Phi_0^0$ (for specimens thinner than 20 nm, this is actually an excellent approximation). The dependence of the Y' function on the electron energy loss is contained in the ratio function

$$\Psi_0^0(\mathbf{b}, z)\left[\partial\Psi_n^0(\mathbf{b}, z)/\partial z\right]^{-1}$$

We now consider the approximated expression of this ratio function. Since Ψ_n^0 is the solution of the Schrödinger equation, we can substitute $\Psi_n^0(\boldsymbol{b}, z) = \Phi_n^0(\boldsymbol{b}, z) \exp(2\pi i \mathbf{K}_n \cdot \mathbf{r})$. Neglecting the $\frac{\partial^2 \Phi_n^0}{\partial z^2}$ term under the small-angle scattering approximation, (12.2) becomes

$$\frac{\partial \Phi_n^0(\mathbf{r})}{\partial z} \approx \zeta \left[4\pi^2 U(\mathbf{r}) \Phi_n^0 + \nabla_b^2 \Phi_n^0 + 2\pi i \mathbf{K}_b \cdot \nabla_b \Phi_n^0\right] \tag{13.16}$$

so that

$$\frac{\Psi_0^0(\mathbf{b}, z)}{\partial \Psi_n^0(\mathbf{b}, z)/\partial z} = \frac{\Phi_0^0(\mathbf{r}) \exp(2\pi i \mathbf{q} \cdot \mathbf{r})}{2\pi i \mathbf{K}_{nz} \Phi_n^0 + \zeta \left[4\pi^2 U(\mathbf{r}) \Phi_n^0 + \nabla_b^2 \Phi_n^0 + 2\pi i \mathbf{K}_b \cdot \nabla_b \Phi_n^0\right]} \tag{13.17}$$

On the right-hand side of (13.17), the first term in the denominator is much larger than the remaining terms, especially for valence excitations, so that

$$\frac{\Psi_0^0(\mathbf{b}, z)}{\partial \Psi_n^0(\mathbf{b}, z)/\partial z} \approx \frac{\Phi_0^0(\mathbf{r}) \exp(2\pi i \mathbf{q} \cdot \mathbf{r})}{2\pi i \ \mathrm{K}_{nz} \Phi_n^0} \tag{13.18}$$

If the specimen thickness is sufficiently small so that $\Phi_n^0 \approx \Phi_0^0$, the ratio function in (13.18) actually reduces to

$$\frac{\Psi_0^0(\mathbf{b}, z)}{\partial \Psi_n^0(\mathbf{b}, z)/\partial z} \approx \frac{\exp(2\pi i \mathbf{q} \cdot \mathbf{r})}{2\pi i \ \mathrm{K}_{0z}} \approx \frac{1}{2\pi i \ \mathrm{K}_{0z}} \tag{13.19}$$

where the electron momentum transfer is assumed to be small. Therefore the Y' function is approximated by

$$Y'(\mathbf{b}, \omega) = \frac{d}{2\pi i K_{0z}} \Psi_0^0(\mathbf{b}, d, \omega) \tag{13.20}$$

Substituting (13.20) into (13.15), we have

$$I_\nu(\mathbf{b}) \approx d^2 \int_0^\infty d\omega \frac{d^2 P}{d\omega dz} I_{\text{els}}\left(\mathbf{g}_b, E_0 - \hbar\omega\right) \tag{13.21a}$$

with

$$I_{\text{els}}\left(\mathbf{g}_b, E_0 - \hbar\omega\right) = \left|\Psi_0^0(\mathbf{b}, \omega) \otimes T_{\text{obj}}(\mathbf{b}, \omega)\right|^2 \tag{13.21b}$$

where $I_{els}(\mathbf{g}_b, E)$ is the image formed by electrons of incident energy E. According to (13.21a), the image formed by valence-excited electrons is an incoherent superposition of those formed by electrons with different incident energies, as weighted by the intensity distribution of electron energy loss.

However the thickness dependence of the image formed by single-inelastically scattered electrons should be d instead of d^2. This unphysical result is the consequence of presupposing that all states are nondegenerate energy states in (13.2). This assumption simply treats the states of the same energy loss but different momentum transfers as coherent, resulting in the d^2 dependence. In practice, the incoherence of these states makes the image intensity proportional to d.

Equation (13.21) indicates that nonlocalized valence excitations can be considered to occur at the entrance face of the crystal for thin specimens. The diffraction of valence loss electrons can be approximately taken as the diffraction of elastic electrons with incident energy $E = E_0 - \hbar\omega$. It is not necessary to consider detailed excitation processes of valence electrons. This is the so-called perfect delocalization model (Krivanek et al. 1990; Stobbs and Saxton 1989; Tang et al. 1989) that holds for thin specimens only.

13.1.4 Perturbation Theory for Calculating Ψ

The Y' function defined in (13.3) depends on two functions, $\Psi_0^0(\mathbf{b}, z, E_0)$ and $\Psi_0^0(\mathbf{b}, z, E)$, with $E = E0 - \hbar\omega$. Since $\hbar\omega << E_0$, it would be useful if we could calculate $\Psi_0^0(\mathbf{b}, z, E)$ from $\Psi_0^0(\mathbf{b}, z, E_0)$ using the perturbation theory. We now consider equations satisfied by $\Psi_0^0(\mathbf{r}, E - \Delta E)$ and $\Psi_0^0(\mathbf{r}, E)$,

$$\left[\nabla^2 + \frac{2m_0}{\hbar^2}(E_0 - \Delta E)\right]\Psi_0^0(\mathbf{r}, E - \Delta E) = -4\pi^2 U \Psi_0^0(\mathbf{r}, E_0 - \Delta E) \quad (13.22)$$

and

$$\left[\nabla^2 + \frac{2m_0}{\hbar^2}E_0\right]\Psi_0^0(\mathbf{r}, E_0) = -4\pi^2 U \Psi_0^0(\mathbf{r}, E_0) \quad (13.23)$$

We write $\Psi_0^0(\mathbf{r}, E - \Delta E)$ in the form

$$\Psi_0^0(\mathbf{r}, E_0 - \Delta E) = \delta\rho(\mathbf{r})\Psi_0^0(\mathbf{r}, E_0) \quad (13.24)$$

where $\delta\rho(r)$ is a slow varying function. By neglecting the $\nabla^2\delta\rho$ term, substituting (13.24) into (13.22), and using (13.23), the equation satisfied by $\delta\rho(r)$ is

$$\frac{\partial\delta\rho(\mathbf{r})}{\partial z} \approx \frac{2\pi^2 K^2 \Delta E}{E_0}\Psi_0^0(\mathbf{r}, E_0)\left[\partial\Psi_0^0(\mathbf{r}, E_0)/\partial z\right]^{-1}\delta\rho(\mathbf{r}) \quad (13.25)$$

The solution of (13.25) is

$$\delta\rho(\mathbf{r}) = \exp\left\{\frac{2\pi^2 K^2 \Delta E}{E_0}\int_0^z dz' \Psi_0^0(\mathbf{b}, z', E_0)\left[\partial\Psi_0^0(\mathbf{b}, z', E_0)/\partial z\right]^{-1}\right\} \quad (13.26)$$

Therefore

$$\Psi_0^0(\mathbf{r}, E_0 - \Delta E) = \exp\left\{\frac{2\pi^2 K^2 \Delta E}{E_0} \int_0^z dz' \Psi_0^0(\mathbf{b}, z', E_0) \times \left[\frac{\partial \Psi_0^0(\mathbf{b}, z', E_0)}{\partial z}\right]^{-1}\right\} \Psi_0^0(\mathbf{r}, E_0). \tag{13.27}$$

Numerical calculations show that almost no visible difference is made in the electron wave function if an energy loss of about 10 eV is introduced in the wave function calculation for 200-keV electrons (Wang and Bentley 1991a). Therefore in HRTEM images, the perfect delocalization model usually holds.

13.1.5 Effect of Surface Plasmon Excitation

In HRTEM, the excitation of surface plasmons may become important in comparison to that of volume plasmons in thin specimens. In this case, the total electron energy loss intensity is determined by

$$P(\omega) = \frac{d^2 P_v(\varepsilon)}{d\omega dz} d + \frac{dP_s(\varepsilon)}{d\omega} \tag{13.28}$$

where $\frac{dP_v}{dz}$ is the excitation probability of volume plasmon per unit distance and P_s is the excitation probability of the surface plasmon in the geometry of electron transmission through a thin crystal slab of thickness d (Ritchie 1957). Refer to Ritchie (1957) for a detailed derivation whose result is

$$\frac{dP_s(\varepsilon)}{d\omega} = \frac{e^2}{2\pi^2 \varepsilon \varepsilon_0 \hbar v^2} \int_0^{q_c} dq \frac{q^2}{\left[q^2 + (\omega/2\pi v)^2\right]^2} \times \mathrm{Im}\left[-\frac{1-\varepsilon}{\varepsilon} \cdot \frac{2(\varepsilon - 1)\cos(\omega d/2\pi v) + (\varepsilon - 1)\exp(-2\pi q d)}{(\varepsilon - 1)^2 \exp(-2\pi q d) - (\varepsilon + 1)^2 \exp(2\pi q d)} + \frac{(1 - \varepsilon^2)\exp(2\pi q d)}{(\varepsilon - 1)^2 \exp(-2\pi q d) - (\varepsilon + 1)^2 \exp(2\pi q d)}\right] \tag{13.29}$$

Based on (13.28) we use the delocalized scattering approximation to calculate images formed by all volume plasmon loss and surface plasmon loss electrons according to

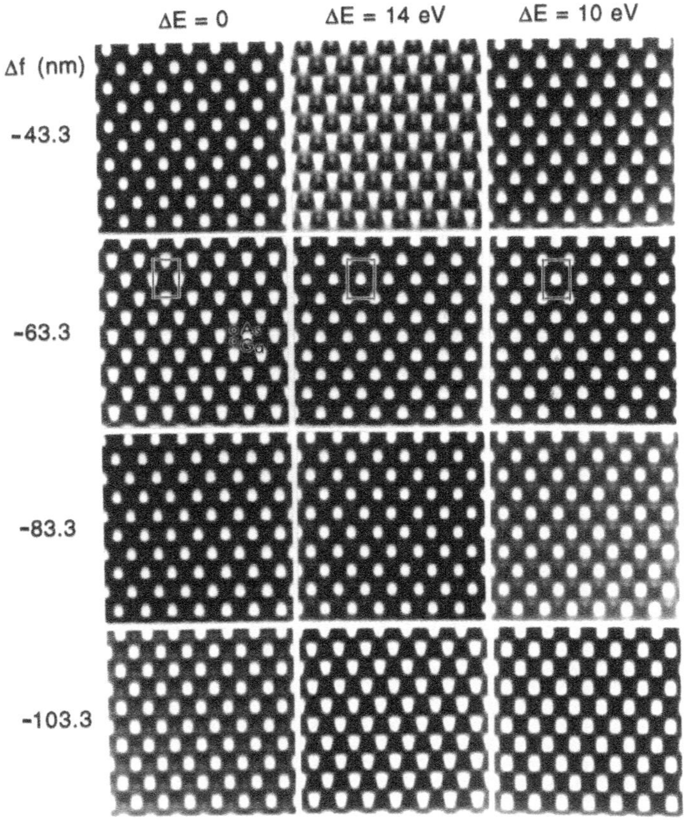

Fig. 13.1 Simulated energy-filtered HRTEM images of a GaAs thin foil for elastic (ΔE = 0), volume plasmon loss (ΔE = 14 eV) and surface plasmon loss (ΔE = 10 eV) electrons at different defocus values (see text for details), beam azimuth is [011], and electron energy is 200 keV (Wang and Bentley 1991a)

$$I(\mathbf{b}) \approx \int_0^\infty d\omega \left[\delta(\omega) + \frac{d^2 P_v(\varepsilon)}{d\omega dz} d + P_s(\varepsilon) \right] \times \left| \Psi_0^0 (\mathbf{b}, d, E_0 - \hbar\omega) \otimes F_{\text{OB}}(\mathbf{b}, \omega) \right|^2 . \tag{13.30}$$

where the first term is the image formed by purely elastically scattered electrons; the second and third terms are contributions from electrons after exciting the volume and surface plasmons, respectively. Terms included in [ ] are the energy loss spectrum of incident electrons.

Figure 13.1 compares simulated images of a thin GaAs [110] slab by electrons with no energy loss ($\Delta E = 0$), volume plasmon loss ($\Delta E = 14$ eV), and surface plasmon loss ($\Delta E = 10$ eV). Simulations were performed under the following conditions:

$C_s = 1.2\,\text{mm}$, $C_c = 1.5\,\text{mm}$, specimen thickness 20 nm, electron energy 200 keV, objective aperture radius 4.0 nm^{-1}, energy filter width 2 eV, at $\Delta f = -43.3\,\text{nm}$. The image formed by elastic electrons is significantly different from the image formed by volume plasmon-loss electrons, while only slightly different from the image formed by surface plasmon loss electrons. At Schertzer defocus, $\Delta f = -63.3\,\text{nm}$, the elastic image has the optimum contrast, with the nonsymmetric GaAs dumbbell identified by a dark contrast. Inelastic images are reversed in contrast, so it is not easy to distinguish the Ga from the As atomic columns. At $\Delta f = -83.3\,\text{nm}$, elastic and inelastic images are still reversed in contrast, with some residual intensity appearing near Ga columns observed in the energy loss images. At $\Delta f = -103.3\,\text{nm}$, the 14–eV energy loss image has the optimum contrast, and the nonsymmetric GaAs dumbbell can be identified. In the 10–eV energy loss image, however, the GaAs dumbbell symmetric contrast is symmetric. This suggests that inelastic scattering could complicate the image interpretation, and it is necessary to use an energy filter for quantitative image analysis.

The dominant effect of inelastic scattering introduces a focus shift in the objective lens transfer function due to chromatic aberration. Calculated images using (13.15) and (13.22) agree very well for specimens even as thick as 40 nm. This means that the perfect delocalization model is a good approximation in practice.

For diffraction contrast imaging, it was illustrated by Howie (1963) that the image formed by plasmon loss electrons preserves the contrast of elastically scattered electrons. This is probably because the smaller objective aperture used in diffraction contrast (or amplitude contrast) imaging limits the effect of chromatic aberration. However in phase contrast HRTEM, since image resolution is determined by the transfer function of the objective lens, a small chromatic aberration significantly affects phase transfer properties of the lens and results in contrast reversal (Fig. 13.1). Therefore the preservation of contrast by inelastically scattered electrons in HRTEM does not hold.

13.1.6 Energy-Filtered Inelastic Images of Interfaces

A more complex situation involves the inelastic imaging of an interface of two different materials. This case occurs in the surface profile and grain boundary images, as shown in Fig. 13.2. Since the two media are characterized by different dielectric functions, ε_2 (left-hand side) and ε_1 (right-hand side), excitation of the interface depends on the distance of the electron from the interface. The full dielectric resonance excitation of the entire foil is complicated. Hence for simplification, we assume that excitations of the top and bottom surfaces and the interface are not strongly coupled, so they can be treated separately. The excitation of entrance and exit surfaces can be approximately described by (13.29). By neglecting the finite size of the slab, the interface excitation can be treated using the method for an electron moving parallel to a large plane, as illustrated in Sect. 6.5.3. The image formed by all transmitted electrons can be written

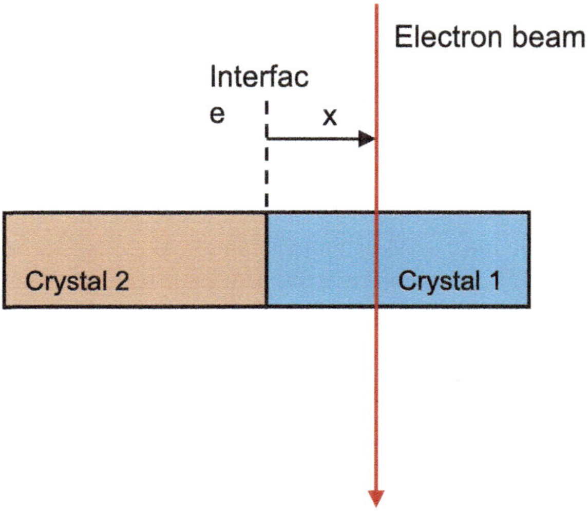

Fig. 13.2 The geometry of an interface between two adjacent thin foils. The x indicates the position of the incident electron with respect to the interface

$$I(\mathbf{b}) \approx \int_0 d\omega \left\{ \delta(\omega) + \left[\frac{d^2 P_v(\varepsilon_1, x)}{d\omega dz} d + \frac{P_s(\varepsilon_1)}{d\omega} \right] \Theta(x) + \left[\frac{d^2 P_v(\varepsilon_2, x)}{d\omega dz} d + \frac{dP_s(\varepsilon_1)}{d\omega} \right] \Theta(-x) + \frac{d^2 P_i(\varepsilon_2, \varepsilon_1, x)}{d\omega dz} d \right\} \times \left| \Psi_0^0(\mathbf{b}, d, E_0 - \hbar\omega) \otimes F_{\mathrm{OB}}(\mathbf{b}, \omega) \right|^2 \tag{13.31}$$

where the excitation probability of the interface [from (6.108)] is

$$\frac{d^2 P_i(\varepsilon_1, \varepsilon_2, x)}{d\omega dz} = \frac{e^2}{2\pi^2 \varepsilon_0 \hbar v^2} \mathrm{Im}\left(-\frac{2}{\varepsilon_1 + \varepsilon_2} \right) K_0(2\omega x/v) \tag{13.32}$$

the volume plasmon excitation probability, which depends on the distance of electrons from the interface, is

$$\frac{d^2 P_v(\varepsilon, x)}{d\omega dz} = \frac{e^2}{2\pi^2 \varepsilon_0 \hbar v^2} \mathrm{Im}\left(-\frac{1}{\varepsilon} \right) [\ln(2\pi q_c v/\omega) - K_0(2\omega x/v)] \tag{13.33}$$

and $\Theta(x)$ is a step function. It is important to note that excitations of the volume (P_v) and interface (P_i) plasmons depend on the relative position of electrons with respect to the interface. This effect introduces some additional contrast into the inelastic images of interfaces, as described in the following paragraphs.

To simplify the discussion, we take the surface profile image of GaAs (100) as an example. The variations of $\frac{dP_v}{dz}$ and $\frac{dPi}{dz}$ are quite slow when an electron moves close to the interface (or surface) from the GaAs side ($x < 0$). The only sharp change occurs when the electron is about 0.3 nm away from the interface. On the vacuum side ($x > 0$), the interaction of the electron beam with the interface is a long-range

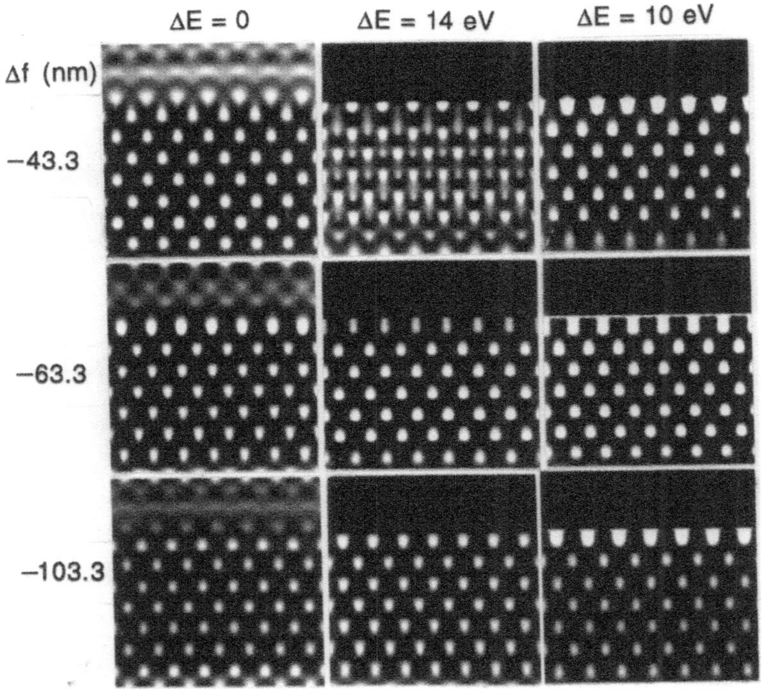

Fig. 13.3 Simulated energy-filtered HRTEM profile images of a GaAs(100) surface for elastic ($\Delta E = 0$), volume plasmon loss ($\Delta E = 14$eV), and surface plasmon loss ($\Delta E = 10$eV) electrons, beam azimuth is [011]. The foil thickness is 50 nm, and electron energy is 200 keV (Wang and Bentley 1991a)

interaction, resulting in some residual surface plasmon excitation even at $x > 3$ nm. This is probably the nature of the delocalized inelastic excitation. On the vacuum side, only surface excitation occurs. This localization effect affects the local image contrast. Figure 13.3 shows the simulated profile images of a GaAs (100) surface viewed along [011] for different energy losses. For simplification, the GaAs (100) is assumed to be a perfect surface without atomic reconstruction. For elastically scattered electrons, besides the regular lattice images at the GaAs side, some residual contrast due to Fresnel fringes is also observed on the vacuum side. For inelastic images, due to variations of excitation probabilities close to the surface (or interface), as shown in Fig. 13.3, the decrease and increase in image contrast for 14–eV (volume plasmon) and 10–eV (surface plasmon) losses, respectively, are significant. This may be important in image interpretation. Also the fringe-type contrast observed in zero-loss electron image disappears in the image of plasmon loss electrons. This effect may be used to locate positions of surface atoms accurately.

In addition, a sharp line contrast is observed at the surface in the 10–eV loss images in Fig. 13.3. This effect is an artifact produced in the following way. In classical dielectric theory, a surface is treated as a plane without thickness, and surface charges are assumed to be distributed in this plane. This is actually not precise at the atomic level. Ignoring the dispersion dependence of $\varepsilon(\omega)$ actually sets the value of the K_0 Bessel function very high at x = 0, resulting in a sharp intensity wall at x = 0. This artifact can be removed by using more rigorous dielectric theory (Zabala and Echenique 1990).

Following the preceding theoretical scheme, the excitation of interfaces in finite semiconductor quantum wells or superlattices can also be included in image simulation by using the corresponding excitation probability function (Nkoma 1987).

13.2 Effects of Phonon Scattering on Elastic Wave

Phonon scattering exists in any electron microscopy experiment performed at ambient temperatures. In a general approach, the crystal potential is written as a sum of time-independent and time-dependent components (Takagi 1958a, b)

$$V(\mathbf{r}, t) = V_0(\mathbf{r}) + \Delta V(\mathbf{r}, t), \tag{13.34}$$

where $V_0(\mathbf{r}) =< V(\mathbf{r}, t) >$ represents the time-averaged crystal potential, and it is determined by the equilibrium positions of the atoms and is responsible for the Bragg-reflected intensities; and the ΔV term is the perturbation of the thermal vibration on the crystal potential and it is the source of diffuse scattering

$$\Delta V(\mathbf{r}, t) = \sum_i [\, V_i\,(\mathbf{r} - \mathbf{r}_i - \mathbf{u}_i(t)) - V_{i0}\,(\mathbf{r} - \mathbf{r}_i)]\,, \tag{13.35}$$

where $\mathbf{r}_i$ is the equilibrium position of the ith atom (with potential V_i), and $\mathbf{u}_i$ is the time-dependent displacement of the atom from its equilibrium position.$\mathbf{u}_i$ is usually a nonperiodic function, and so is ΔV.

To see the relative magnitude of ΔV to V_0, Fig. 13.4 shows the calculated instantaneous potential V, thermal equilibrium potential V_0, and the difference $\Delta V = V - V_0$ for a silicon atom using the experimentally determined mean square root vibration amplitude of Si. If the instantaneous position of the atom is its equilibrium lattice site, ΔV is a symmetric, sharp function. If the atom is displaced to an instantaneous position located at 0.068 Å on the left-hand side from the equilibrium site, ΔV is no longer symmetric (Fig. 13.4b). Two striking characteristics are noticed. Firstly, the magnitude of ΔV is comparable with that of V_0, and this is true for each atom present in the specimen. Secondly, ΔV is considerably narrower than V_0, indicating that TDS is a rather localized scattering process, with the majority of the electrons being generated from the nuclear sites. It is apparent that the TDS is a very localized scattering process. In high-resolution transmission electron microscopy (HRTEM), as the image resolution is approaching 1 Å, the image contrast at the atomic columns is likely to be affected by TDS. It is necessary to take this component into consideration in image calculation if quantitative data analysis is needed.

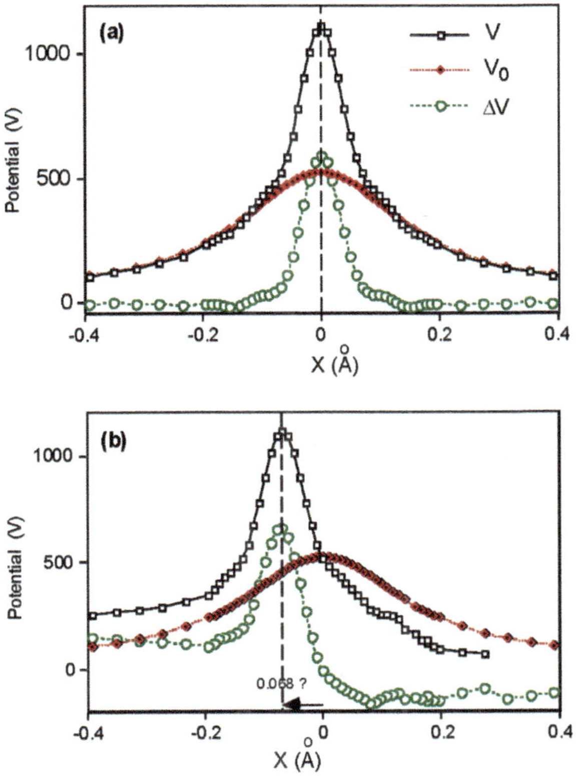

Fig. 13.4 Potential of a silicon atom with (V_0) and without (V) including the Debye-Waller factor. V is the instantaneous atomic potential, and V_0 is the time-averaged atomic potential. The equilibrium position of the atom is b = 0. $\Delta V = V - V_0$ (dash-dotted line) is the deviation of the atomic potential from the time-averaged potential V_0 when the atomic displacement is **a** $u_\kappa = 0$ or **b** $u_\kappa = -0.068$ Å. The arrowhead in (**b**) indicates the displacement of the atom. The mean square root atom vibration amplitude was taken as 0.07 Å in the calculation

13.3 Effect of Phonon Scattering on Phase Contrast Lattice Imaging

As the image resolution is approaching 1 Å, quantitative matching of experimental images with the theoretically calculated images is needed for quantitative structure determination. Most of the available image calculation packages were developed with consideration of the contribution made by elastically scattered electrons, and we are always puzzled as to why the calculated image, with consideration of all the possible instabilities in the experiments, has a better contrast than the observed image. One of the unresolved questions is how the TDS electron affects the image.

Before answering this question, one must examine the importance of TDS in electron diffraction and imaging. Shown in Fig. 13.5 is a group of energy-filtered electron diffraction patterns at high scattering angles, where the Kikuchi lines are clearly present. The patterns were formed by the electrons with different energy losses so that the contributions made by the different processes can be separated. A line scan is made across the diffraction pattern for quantitative comparison purposes. In the pattern formed by the electrons with energy losses smaller than the energy spread of

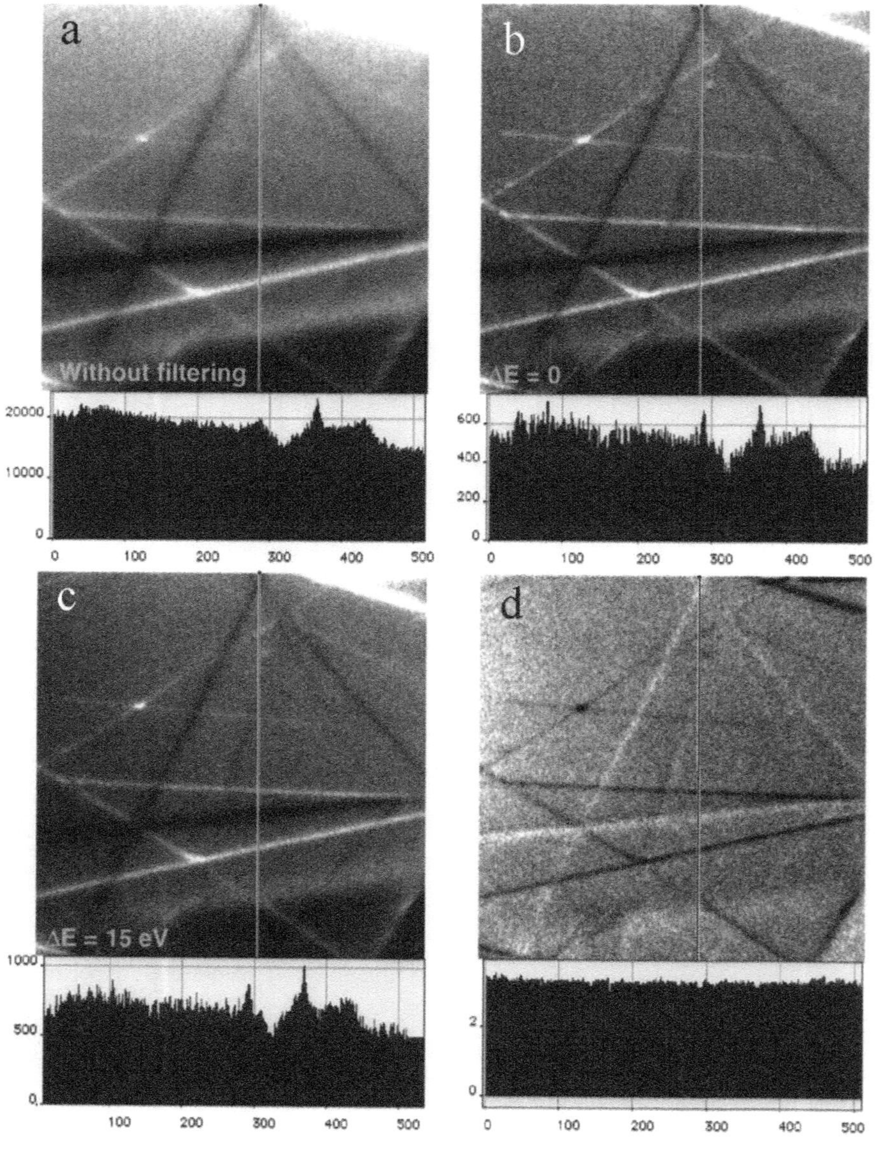

Fig. 13.5 Large-angle Kikuchi patterns of Si recorded **a** without energy filtering, **b** using the electrons without energy loss, and **c** with plasmon loss. **d** A thickness/mean free path image calculated from (**a**) and (**b**). The line scan profiles of the images are also shown to compare the absolute intensities

the filament, typically 1.5 eV, the dominant contribution is TDS (Fig. 13.5b) because the elastic Bragg scattering does not fall into the angular range shown. The intensity taken by the first plasmon peak of GaAs is ~10% of the total intensity (Fig. 13.5c). The specimen thickness can be seen in the t/Λ image, where t is specimen thickness and Λ the electron mean free path length, which is ~80 nm for GaAs. For a specimen of thickness ~100 nm, the average intensity contributed by pure TDS electrons at high angles is ~27% of the total intensity excluding the component with larger energy losses. These electrons come from the nuclear sites, and they can strongly affect the image contrast, especially when the image resolution is high.

To simplify our discussion, the weak phase object approximation is used to illustrate the contribution made by TDS electrons in HRTEM images. The electron wave after exiting the crystal surface can be represented by

$$\Phi(\mathbf{b}) \approx 1 + i\sigma\, V_p(\mathbf{b}) = 1 + i\sigma\, V_{p0}(\mathbf{b}) + i\sigma\Delta\, V_p(\mathbf{b}, t), \tag{13.36}$$

where $\sigma = \pi/\lambda U_0$, $\mathbf{b} = (x, y)$, λ the electron wave length, U_0 the electron acceleration voltage, and V_p the instantaneous crystal projected potential. Taking a Fourier transform of the wave function

$$\Phi(\mathbf{u}) \approx \delta(\mathbf{u}) + i\sigma FT\left[\, V_{p0}(\mathbf{b})\right] + i\sigma FT\left[\Delta\, V_p(\mathbf{b}, t)\right], \tag{13.37}$$

where the first term is the center transmitted beam, the second term is the Bragg-reflected beams, and the third term is the diffusely scattered electrons. From Abbe's imaging theory, the intensity seen in the image plane is

$$I(\mathbf{b}) = <\left|\left[1 + i\sigma\, V_{p0}(\mathbf{b}) + i\sigma\Delta\, V_p\,(\mathbf{b}, t)\right] * t_{obj}(\mathbf{b})\right|^2 >, \tag{13.38}$$

where $t_{obj}(x, y)$ characterizes the information transfer of the objective lens, and it is a Fourier transform of the objective lens transfer function $T_{obj}(\mathbf{u})$, and $<>$ represents the time average of the image intensity. With the use of $< \Delta V >= 0$, the result is

$$\begin{aligned} I(\mathbf{b}) &= 1 - 2\sigma V_{p0}(\mathbf{b}) * \mathrm{Im}\left[t_{\mathrm{obj}}(\mathbf{b})\right] \\ &+ \sigma^2\left|V_{p0}(\mathbf{b}) * t_{\mathrm{obj}}(\mathbf{b})\right|^2 \\ &+ \sigma^2\left\langle\left|\Delta V_p(\mathbf{b}, t) * t_{\mathrm{obj}}(\mathbf{b})\right|^2\right\rangle. \end{aligned} \tag{13.39}$$

where the first term is the incident beam, the second term is the interference between the center beam with the Bragg diffracted beams (the first-order effect), e.g., the bright-field lattice image, the third term is the interference between the Bragg-reflected beams (the second-order effect), and the last term is the contribution made by TDS (the second-order effect). This equation clearly shows that the contribution made by TDS is on the same order of magnitude as the cross-interference term between Bragg beams (Fig. 13.6a) excluding the central transmitted beam (Wang 1998c), since ΔV is comparable to V_0.

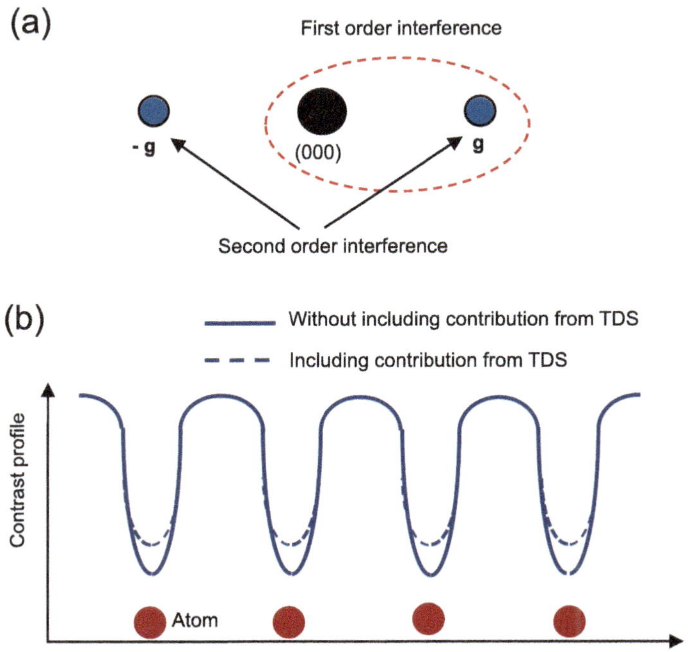

Fig. 13.6 **a** Lattice images can be formed by the interference of the central transmitted beam (000) and the diffracted beams. The cross-interference between the Bragg beams is the second-order effect, which is equivalent to the contribution made by TDS in the imaging. **b** The reduction of the image contrast at the atom sites due to TDS under the Schertzer defocus

On the image contrast, the second term (the bright-field term) can produce contrast reversal as the lens defocus is changed, but the intensity contributed by TDS is always the strongest at the atom sites. Therefore, at the Schertzer defocus condition under which the atomic columns show dark contrast, adding the TDS contribution reduces the darkness of the atom columns, resulting in a decrease in the image contrast (see Fig. 13.6b), possibly giving a better fit to the experimental image. This is the importance of TDS in HRTEM. This simple physical illustration can be spelled out using a full quantum mechanical approach, leading to a new theory for calculating the images formed by phonon-scattered electron in HRTEM with the inclusion of full dynamic effects (Wang 1998b). Most of the software packages currently available for image and diffraction pattern simulations, however, do not include the contribution made by TDS.

13.4 Effect of Phonon Scattered on High-Angle Annular Dark-Field STEM Imaging

Determining the lattice structure and the atom types filling the lattices are two critical steps in the characterization of functional materials. Most of the chemical imaging has been performed in scanning electron microscopy (SEM) and scanning transmission electron microscopy (STEM). The characteristic inelastic-scattering signals emitted from the interaction volume of the electron probe with the specimen are determined by the thickness-projected elemental concentrations, provided there is no beam broadening and no fluctuation in incident beam current. X-rays and Auger electrons are acquired sequentially as a function of the electron scanning position. Chemical images obtained using X-ray signals can be directly correlated with the thickness-projected elemental maps in the specimen. The images are more sensitive to elements with large atomic numbers than those with small Z's. The images formed using Auger signals are sensitive to light elements distributed within 1–2 nm into the surface of the specimen.

Figure 13.7 shows the principle of STEM (Pennycook et al. 2018). Like a scanning electron microscope, an incident probe is scanned across a sample, but it is thin enough so that the beam is transmitted, then several signals can be detected simultaneously and used to form images with complementary characteristics. The high-angle annular dark-field (HAADF) detector collects Rutherford scattering from the atomic nuclei, producing an image with strong sensitivity to atomic number Z, often called a Z-contrast image. Light columns such as O are only weakly visible in the Z-contrast image, but are seen clearly in a simultaneous bright-field (BF) image or annular bright-field (ABF) image. In perovskites and related materials, this allows the octahedral rotations to be determined, which are crucial to understanding their properties. If the bright-field detector is removed, electrons can be passed through a magnetic sector electron energy loss spectrometer (EELS) to provide elemental maps and electronic structure information. The resolution is determined by the size of the electron probe.

The dominant contribution of TDS to the annular dark-field STEM image was first proposed by Wang and Cowley (1989, 1990). The localization of phonon excitation can be seen in the calculated single atom-scattering factor for elastic and thermal diffuse scattering (Fig. 13.8). It is apparent that the low-angle scattering is dominated by elastic scattering, while the high-angle scattering is dominated by TDS. This dominant effect is apparent when the scattering angle is larger than that for the (880) reflection from Si. At the first-order Laue zone (FOLZ) position, TDS is about 6 times stronger than the elastic scattering. In STEM, if an annular dark-field (ADF) detector is positioned to collect the electrons that are scattered to high angles, the image contrast is insensitive to diffraction and phase contrast but is sensitive to the projected atomic number, because at high scattering angles, the atomic-scattering factor is approximated by

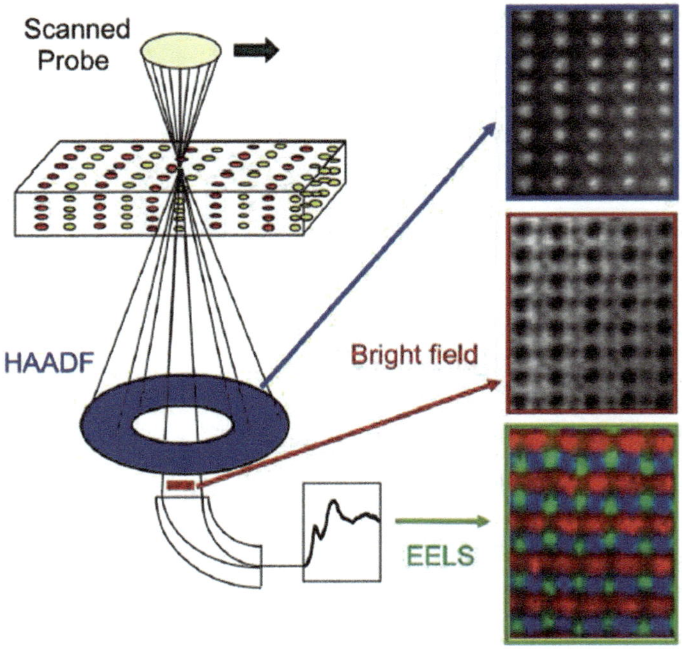

Fig. 13.7 Schematic showing the multiple imaging and analytical signals available in STEM. Top right is a HAADF image of $BiFeO_3$; the brightest atom columns are Bi and the less bright ones Fe, which are slightly shifted to the right due to ferroelectric polarization. The central bright-field image shows the arrangement of oxygen atom columns. The lower color-composite map is an EELS elemental distribution of $LaMnO_3$: red represents Mn, green denotes O (showing displacements arising from octahedral rotations), and blue corresponds to La. Adapted from Borisevich et al. (2010); Varela et al. (2012), copyright American Physical Society

$$f_K^e(s) = \frac{e}{16\pi^2\varepsilon_0}\frac{\left[Z - f_K^x(s)\right]}{s^2} \approx \frac{e}{16\pi^2\varepsilon_0}\frac{Z}{s^2} \quad \text{for large } s, \tag{13.40}$$

where f^x is the X-ray scattering factor. The scaling of $f_k^e(s)$ with Z is the basis of Z-contrast imaging (Crewe 1978–79; Pennycook and Jesson 1990), provided there is no complication from the dynamic electron diffraction effect.

In general, images calculated using elastic-scattering theory give much better contrast than the experimental images. This difference is called the "Stobbs factor." This discrepancy was first suggested by Wang (2003) due to the contribution of TDS electrons that are not included in the calculation. Quantitative calculations using the Frozen lattice model (see Chap. 14) have been carried out to compare the HAADF-STEM images by including TDS with experiments, excellent agreement within about a difference of five percent is achieved between theory and experiments (LeBeau et al. 2009), confirming the contribution of TDS to the final image. Figure 13.9 shows experimental and simulated images for different thicknesses for $PbWO_4$, confirming the excellent agreement for the entire thickness range. Although

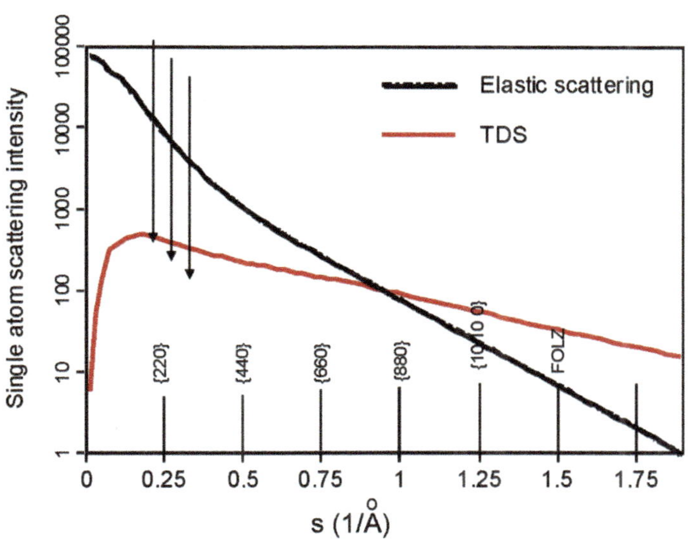

Fig. 13.8 Theoretically calculated atom-scattering factors for elastic scattering $\left[(f^e)^2\{\exp(-2W)]\right\}$ and thermal diffuse scattering $\left[(f^e)^2\{1-\exp(-2W)]\right\}$ of a Si atom, showing the dominant contribution of TDS at high scattering angles, where W is the Debye-Waller factor. The scattering vector s=g/2. The mean square root atomic vibration amplitude was taken as 0.07 Å in the calculation

the oxygen columns are not directly visible in the images in Fig. 13.9, their arrangement is intimately connected to the change in background intensity as a function of thickness. In particular, for thicknesses greater than about 40 nm the background exhibits an alternating pattern, being highest in the locations where the oxygen atoms are located furthest from one another in the projected structure but closest to the heavy W columns. The scattering potential of oxygen is too weak to contribute much directly to the HAADF images or to significantly bind the channeling probe. But the oxygen atoms can have an appreciable effect on the evolving electron intensity distribution. Basically, a probe positioned between an oxygen column pair will spread preferentially in the direction of the oxygen columns.

13.5 Effect of Phonon Scattering on High-Angle Annular Dark-Field Stem Imaging

Electron holography was first introduced by Gabor (1949) to retrieve the electron phase and to exceed the point-to-point resolution of an electron microscope. Holography is an imaging process based on the interference properties of waves, which permit both the amplitude and the phase of the object wave to be recovered rather than the intensity alone as in conventional HRTEM images. Electron nongraphy requires

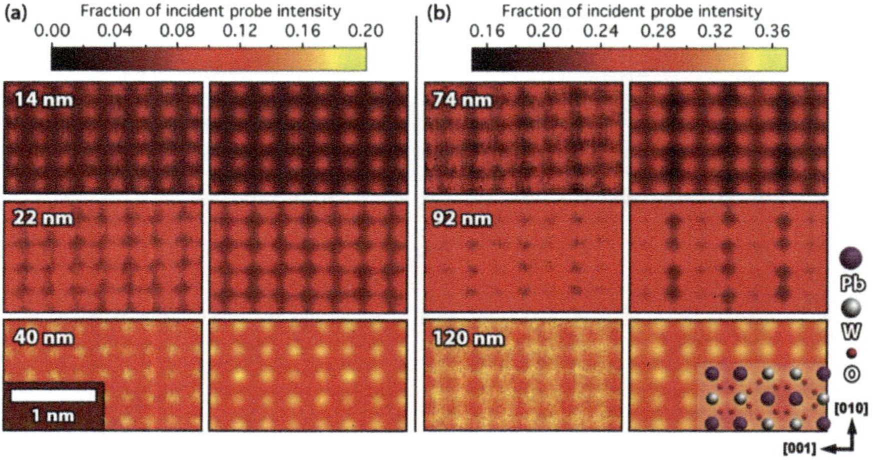

Fig. 13.9 Experimental (left panels) and simulated (right panels) images for different thicknesses (see labels). The simulated images account for a finite source size characterized by a 0.115 nm FWHM Gaussian distribution. Note that different intensity scales are used in (**a**) and (**b**) for clarity. The uncertainties in the experimental thicknesses are 2âŁ"4 nm and the simulated thicknesses are within 2 nm of the value given in the figure. Adopted from LeBeau et al. (2009)

a highly coherent incident beam, now available using high-brightness field emission electron sources in TEM (Tonomura 1992; Lichte 1986). The potential for correcting spherical aberration effects in the complex image wave, either by optical or mathematical reconstruction, offers the possibility of achieving 0.1 nm point resolution in TEMs (Völkl and Lichte 1990).

The breakthrough of the 1-Å image resolution limit has a great impact on the application of electron microscopy in materials science. However, it also raises some concerns in quantitative data analysis, such as including TDS in image simulation. In general, the atomic mean vibration amplitude is between 0.005 and 0.015 nm, giving a perturbation effect to the atomic potential, which may be significant when the image resolution is better than about 0.15 nm. The purpose of this section is to consider the effect of TDS in quantitative image simulations for high-resolution electron holography (Wang 1993b).

13.5.1 Electron Holography with Time-Dependent Perturbation

An important characteristic of holography is its ability to recover both the amplitude and phase components of the object's complex exit wave. For this purpose, off-axis electron holography can be regarded as a two-step imaging technique (Gabor 1949), as shown in Fig. 13.10. The first step is to form a hologram by means of an electrostatic biprism, inserted, for example, between the back focal plane of the objective lens and the intermediate image plane (Möllenstedt and Düker 1956). With respect to the image plane, the specimen is positioned to cover half of the image plane, leaving the other half of the incident wave for the reference wave $\exp(2\pi i\mathbf{K}_0 \cdot \mathbf{r})$. In the following discussion, the complex image, which is the exit wave of the object after modification by the aberration effect of the objective lens, is assumed to be

$$\Psi_{0b} = \exp(2\pi i\mathbf{K} \cdot \mathbf{r})A(x, y, t)\exp[i\varphi(x, y, t)] \tag{13.41}$$

where $A(x, y, t)$ and $\phi(x, y, t)$are real functions describing the amplitude and phase, respectively, of the image wave, and t indicates the time dependence of the exit wave function on the perturbation effect of atomic thermal vibrations in crystals. By applying a positive voltage to the filament of the biprism, reference and object waves on both sides of the biprism are deflected toward each other, to directions $\mathbf{K}_L$ and $\mathbf{K}_R$, respectively, forming an interference pattern in the image plane, i.e., a hologram whose measured intensity distribution is

$$\begin{aligned} I_{\text{hol}}(x, y) &= \left\langle |\exp(2\pi i\mathbf{K}_L \cdot \mathbf{r}) + \exp(2\pi i\mathbf{K}_R \cdot \mathbf{r})\, A(x, y, t)\exp[i\varphi(x, y, t)]|^2 \right\rangle \\ &= 1 + \left\langle A^2(x, y, t) \right\rangle \\ &\quad + 2\left\langle A(x, y, t)\cos[\varphi(x, y, t) + 2\pi\Delta\mathbf{K} \cdot \mathbf{r}] \right\rangle, \end{aligned} \tag{13.42}$$

where $\Delta\mathbf{K} = \mathbf{K}_L - \mathbf{K}_R$. The recorded intensity distribution is the result of electron scattering by the specimen at different times (or with different thermal vibrational configurations of the lattice).

The second step is to reconstruct the hologram to obtain the object wave. Using the computing reconstruction method, we take a Fourier transform of the hologram,

$$\begin{aligned} \text{FT}[I_{\text{hol}}(x, y)] = \delta(\mathbf{u}) &+ \text{FT}\left[\left\langle A^2(x, y, t) \right\rangle\right] \\ &+ \text{FT}\left[\langle A(x, y, t)\exp(i\varphi(x, y, t))\rangle\right] \otimes \delta(\mathbf{u} + \Delta\mathbf{K}) \\ &+ \text{FT}\left[\langle A(x, y, t)\exp(-i\varphi(x, y, t))\rangle\right] \otimes \delta(\mathbf{u} - \Delta\mathbf{K}), \end{aligned} \tag{13.43}$$

where the $\delta(\mathbf{u})$ function is the transmitted (000) beam; the second term is the auto-correlation term located near the center of the diffraction pattern; the last two terms are centered at $\mathbf{u} = -\Delta\mathbf{K}$ and $\mathbf{u} = \Delta\mathbf{K}$, respectively; these are the so-called side-bands. To see the effect of TDS, the time-dependent and time-independent parts are separated as Cowley (1988a)

$$\begin{aligned} A(x, y, t)\exp[i\varphi(x, y, t)] = &\; A_0(x, y)\exp[i\varphi_0(x, y)] \\ &+ \Delta A(x, y, t)\exp[i\varphi'(x, y, t)]. \end{aligned} \tag{13.44a}$$

where

$$A_0(x, y)\exp[i\varphi_0(x, y)] =< A(x, y, t)\exp[i\varphi(x, y, t)] > \tag{13.44b}$$

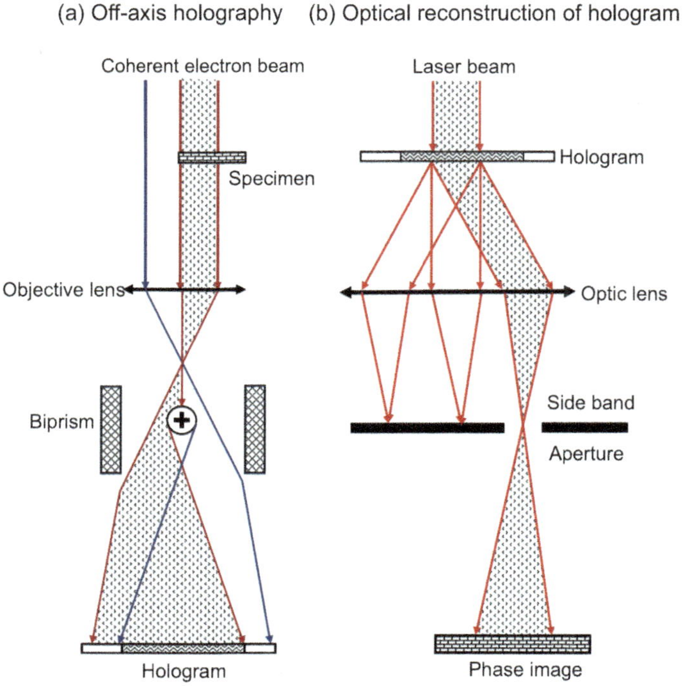

Fig. 13.10 **a** Formation of off-axis electron hologram in TEM using an electrostatic biprism, and **b** reconstruction of electron hologram using laser diffraction

and

$$< \Delta A(x, y, t) \exp\left[i\phi'(x, y, t)\right] >= 0. \tag{13.44c}$$

Using the notations defined in (13.44a) to (13.44c), (13.43) becomes

$$\begin{aligned} \mathrm{FT}\left[I_{\mathrm{hol}}(x, y)\right] = \delta(\mathbf{u}) + \mathrm{FT}\left[A_0^2(x, y)\right] + \mathrm{FT}\left[\langle\Delta A^2(x, y, t)\rangle\right] \\ + \mathrm{FT}\left[A_0(x, y)\exp\left(i\varphi_0(x, y)\right)\right] \otimes \delta(\mathbf{u} + \Delta\mathbf{K}) \\ + \mathrm{FT}\left[A_0(x, y)\exp\left(-i\varphi_0(x, y)\right)\right] \otimes \delta(\mathbf{u} - \Delta\mathbf{K}). \end{aligned} \tag{13.45}$$

If the carrier spatial frequency $\Delta\mathbf{K}$ is sufficiently large, sidebands do not overlap with the center band. By selecting only one sideband, say, the band centered at $\mathbf{u} = -\Delta\mathbf{K}$, the electron wave function at the exit face of the specimen can be recovered as

$$< \Psi(x, y, t) >= \exp(2\pi i\mathbf{K}\cdot\mathbf{r}) < \Phi(x, y, t)\rangle \tag{13.46a}$$

with

$$< \Phi(x, y, t) >= \mathrm{FT}^{-1}\left\{\mathrm{FT}\left[\, A_0(x, y) \exp\left(i\varphi_0(x, y)\right)\right] / T_{obj}(\mathbf{u})\right\}. \quad (13.46b)$$

The wave reconstructed from a hologram in this way is the time-averaged complex object wave, and it can therefore be compared directly with the calculated electron wave function.

It is clear that thermal diffusely scattered electrons, represented by the FT[$< \Delta A^2(x, y, t) >$] term in (13.45), are distributed only in the center band. As shown in the next section, the effect of TDS on the sidebands is to introduce an absorption function in the reconstructed elastic wave due to the time average.

It is important to note that the effect of TDS in electron holography (i.e., on sidebands) is contained in $< \Phi >$ rather than $< |\Phi|^2 >$ as in conventional high-resolution electron micrographs. This leads to the following method, which may be different from the conventional treatment of TDS absorption in electron imaging.

If the carrier spatial frequency $\Delta\mathbf{K}$ is sufficiently large, the sidebands do not overlap with the center band. By selecting only one sideband, say the band centered at $\mathbf{u} = -\Delta\mathbf{K}$, the electron wave function at the exit face of the specimen can be recovered to be $< \Phi >$. The reconstructed wave is a time-averaged complex object wave, and can therefore be compared directly with the calculated electron wave function. This indicates that the TDS electrons will not contribute to the finally reconstructed phase image because it comes from the sideband (Wang 1993a, b). The only effect of TDS is to introduce an absorption function in the reconstructed amplitude image due to the time average. Therefore, it is possible that electron holography is an ideal energy filter that even rules out the contribution made by TDS electrons if the electron source is monochromatic.

Experiments have been carried out to verify this prediction (Van Dyck et al. 2000). By simultaneously recording the off-axis electron hologram and the on-axis focal series of high-resolution TEM images from the same specimen region and under identical experimental conditions, the wave functions carried by the hologram and the high-resolution TEM images can be reconstructed, respectively, following the standard procedures (Zandbergen and Van Dyck 2000). Such an experiment has been carried out by Lehmann and Lichte (2002), who found that the lattice fringe contrast reconstructed from the sideband is substantially higher than the contrast from HRTEM, and they suggested that the discrepancy is due to inelastic scattering. This conclusion proves the theoretically predicted result.

13.5.2 Multislice Calculation of Time-Averaged Electron Wave Function

The multislice solution of Φ before and after being scattered by a crystal slice is determined (see Chap. 3) by

$$\Phi_{n_c+1}(\mathbf{b}, t) = \left[\Phi_{n_c}(\mathbf{b}, t) Q_{n_c+1}(\mathbf{b}, t)\right] \otimes P(\mathbf{b}) \quad (13.47)$$

where the time-dependent phase-grating function is

$$Q_{n_c+1}(\mathbf{b}, t) = \exp\{i\sigma[\ V_0(\mathbf{b}) + \Delta V(\mathbf{b}, t)]\} \tag{13.48}$$

For simplification, we assume that there is no phase correlation among the atoms' vibrations (i.e., random-phase vibration). Thus, the phase-grating function Q_{n_c+1} $(\mathbf{b}, t)$ of the slice does not have a time-dependent correlation with the entrance wave $\Phi_{n_c}(\mathbf{b}, t)$ of the slice if backscattering is neglected; therefore

$$< \Phi_{n_c}(\mathbf{b}, t) Q_{n_c+1}(\mathbf{b}, t) > \approx < \Phi_{n_c}(\mathbf{b}, t) > < Q_{n_c+1}(\mathbf{b}, t) > \tag{13.49}$$

Thus taking a time average of (13.49),

$$< \Phi_{n_c+1}(\mathbf{b}, t) > = \left[< \Phi_{n_c}(\mathbf{b}, t) > < Q_{n_c+1}(\mathbf{b}, t) >\right] \otimes P(\mathbf{b}) \tag{13.50}$$

where the time-dependent perturbation of atomic thermal vibration is equivalent to introducing an absorption factor in q_{n+1} (Cowley 1981), i.e.,

$$< Q_{n_c+1}(\mathbf{b}, t) > = \exp[i\sigma\ V_0(\mathbf{b})] \exp[-\mu(\mathbf{b})] \tag{13.51}$$

with

$$\begin{aligned} \mu(\mathbf{b}) &= \frac{\sigma^2}{2} < [\Delta V(\mathbf{b}, t)]^2 > \\ &= \frac{\sigma^2}{2} \sum_{j'} \int d\mathbf{u} \int d\mathbf{u}' f^e_{j'}(\mathbf{u}) f^{e*}_{j'}(\mathbf{u}') \left\{\exp\left[-W_{j'}\left(\mathbf{u} - \mathbf{u}'\right)\right]\right. \\ &\quad \left. - \exp\left[-W_{j'}(\mathbf{u}) - W_{j'}(\mathbf{u})\right]\right\} \exp\left[2\pi i \left(\mathbf{u} - \mathbf{u}'\right) \cdot \left(\mathbf{b} - \mathbf{b}_{j'}\right)\right] \end{aligned} \tag{13.52}$$

the sum of j' is restricted to atoms located within the slice.

To show the importance of TDS in electron holography, calculations were made for an 8 nm thick GaAs [110] crystal with and without TDS absorption. For relatively lower resolution electron holograms, TDS absorption does not show a significant effect in the simulated amplitude and phase. However differences appear when the image resolution is improved to 0.16 nm, as shown in Fig. 13.11. It appears that TDS absorption primarily affects the amplitude but not the phase distribution. This effect becomes more significant when the image resolution is further improved.

The thermally diffuse scattered electrons do not contribute to the sidebands of off-axis holograms, but they do introduce an absorption factor in the simulated elastic wave function. The absorption effect shows up only in the reconstructed amplitude but not in the phase, and it becomes significant only when the image resolution approaches the atomic level. The reconstructed wave function from a hologram is a perfect zero loss, energy-filtered image of the specimen. By contrast, inelastically scattered electrons including TDS electrons do contribute to high-resolution images formed by diffracted beams in the center band. This simply means that quantitative

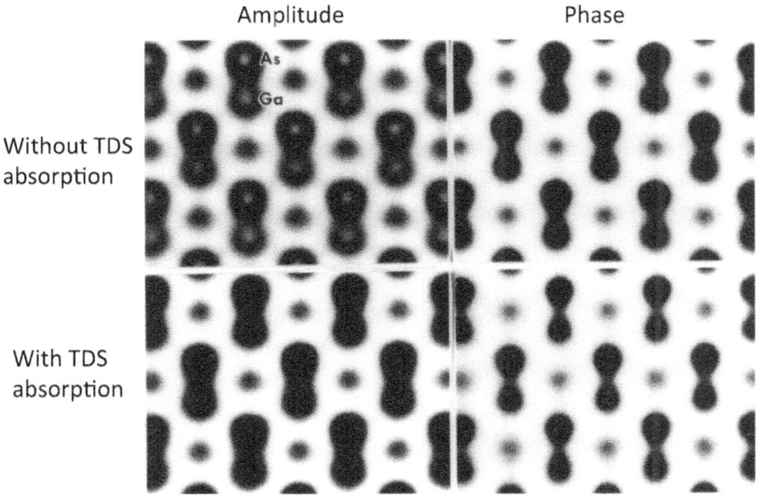

Fig. 13.11 Simulated amplitude and phase distributions of GaAs [110] for the diffracted beams limited by an objective aperture of radius 6.25 nm^{-1}, corresponding to a resolution limit of 0.16 nm, without and with the TDS absorption effect, crystal thickness 80 Å, electron beam energy 200 keV, Cs = 1.2 mm, and mean vibration amplitudes of Ga and As were both taken as 0.0063 nm (Wang 1993b)

analysis of electron holograms may be easier than the analysis of HRTEM images if image conditions, such as defocus and spherical aberration, are precisely known.

13.6 Summary

With the help of an energy filter, most of the inelastic signals can be removed from the image, except phonon-scattered electrons. However, the contribution from TDS electrons is especially important if the image resolution is approaching 0.1 nm and beyond with the introduction of Cs corrected microscopes. It is shown that the contribution of the TDS electrons to the image is of the same order as the cross-interference terms for the Bragg-reflected beams in the dark-field high-resolution TEM imaging. In quantitative high-resolution transmission electron microscopy, the theoretically calculated images usually give better contrast than the experimentally observed ones, although all of the factors have been accounted for. This discrepancy is suggested due to TDS. A more rigorous multislice theory has been developed to account for this effect (Wang 1998c).

Back to 1993 (Wang 1993a, b), we proposed theoretically that the off-axis holography is an ideal energy filter that even filters away the contribution made by TDS electrons in the electron wave function, but conventional high-resolution microscopy do contain the contribution made by phonon-scattered electrons. The only effect of TDS is to introduce an absorption function in the reconstructed amplitude image due to the time average, but no effect on the phase image. This theoretical hypothesis now has been proved experimentally.

Chapter 14
Dynamic Theory of Thermal Diffusely Scattered Electrons

The "frozen lattice" model is a semiclassical approach for calculating electron diffuse scattering in crystals arisen from thermal vibration of crystal atoms. This quasi-elastic scattering approach is, however, concerned since its equivalence with the incoherent phonon excitation model is not yet established. As quantitative electron microscopy is becoming realistic, it is necessary to examine the accuracy of the model. In the first part of this chapter, based on a rigorous quantum mechanical phonon excitation theory, we have proved that an identical result would be obtained using the frozen lattice model and the formal phonon excitation model if (1) the incoherence between different orders of thermal diffuse scattering is considered in the frozen lattice model calculation and (2) the specimen thickness and the mean free path length for phonon excitation both are smaller than the distance traveled by the electron within the lifetime of the phonon ($\tau_0 \nu$) condition (2) is usually absolutely satisfied and condition (1) can be precisely accounted for in the calculation with the introduction of the mixed dynamic form factor $S(\mathbf{Q}, \mathbf{Q}')$. The conclusion holds for each and all of the orders of diffuse scattering, thus, the quantum mechanical basis of the frozen lattice model is established, confirming the validity, reliability, and accuracy of using the frozen lattice model in quantitative dynamical electron diffraction and imaging calculations. It has also been shown that the frozen lattice model is suitable for low-energy electrons.

In the second part, a rigorous theoretical proof is given based on the quantum inelastic excitation theory to show that the inclusion of the optical potential in the dynamic calculation automatically recovers the contributions made by the high-order diffuse scattering although the calculation is done using the equation derived for single diffuse scattering. This conclusion expands the existing first-order diffuse scattering theories to cases in which the incoherent multiple diffuse scatterings are important. It is suggested that multiple thermal diffuse scatterings (or phonon excitations) are likely to be the dominant source for affecting image contrast in quantitative electron microscopy because of the channeling effect and the localized scattering nature of phonon excitation. The optical potential is no longer a sim-

Z. L. Wang, *Elastic and Inelastic Scattering in Electron Diffraction and Imaging*, Springer Series in Solid-State Sciences 205, https://doi.org/10.1007/978-3-031-90819-4_14

ple potential function, rather it is a non-local function strongly depending on the dynamic diffraction in the crystal because of the involvement of Green's function. These characteristics can be properly taken into the computation using a proposed Bloch wave-multislice approach, in which the non-local effect is resolved in the Bloch wave matrix diagonalization and the dynamical effect is taken care using Green's function calculated by the multislice theory using the small-angle (or high-energy) scattering approximation. Under the ground state approximation, by which we mean that the crystal is in its ground state before each and every inelastic excitation, an introduction of the mixed dynamic form factor and the density matrix automatically produces the incoherence between different order and different phonon excitation processes, resolving a problem encountered by many other theories. In conventional dynamical calculation for elastic wave, the inclusion of the Debye-Waller factor and the optical potential accounts only for the effects of the phonon excitation on the elastic wave rather than the contribution made by the diffusely scattered electrons to the image/diffraction pattern.

14.1 The "Frozen Lattice" Model for Thermal Diffuse Scattering in Electron Scattering

In high-energy electron scattering, the "frozen lattice" model is assumed in describing diffuse scattering arisen from thermal vibration of crystal atoms (Hall and Hirsch 1965), which means that, although atom vibration is a time-dependent process, the crystal lattice appears as if in a stationary instantaneous configuration for an incident electron since the interaction time of the electron with the crystal is much shorter than the vibration period of the crystal atom, but the crystal lattice can be in another configuration for the next incoming electron. Thus, for each lattice configuration, the scattering of the electron can be considered as a time-independent, quasi-elastic scattering process, and the finally observed diffraction pattern contributed by millions of electrons is equivalent to a time average on the intensities calculated for the different lattice configurations.

To properly treat TDS using the frozen lattice model, the instantaneous crystal potential $V(\mathbf{r})$ is written into a sum of a time-averaged potential V_0 and a fluctuation term which characterizes the distortion of crystal lattice due to atom displacement from its equilibrium position:

$$V(\mathbf{r}) = V_0(\mathbf{r}) + \Delta V(\mathbf{r}, t) \tag{14.1a}$$

where the instantaneous crystal potential is

$$V(\mathbf{r}) = \sum_{\kappa} V_{\kappa}\left(\mathbf{r} - \mathbf{r}_{\kappa} - \mathbf{u}_k\right), \tag{14.1b}$$

where $\mathbf{r}_\kappa$ is the equilibrium position of the κth atom in the crystal and $\mathbf{u}_\kappa$ is its displacement; the time average of the crystal potential is

$$\begin{aligned} V_0(\mathbf{r}) &= \left\langle \sum_\kappa V_\kappa (\mathbf{r} - \mathbf{r}_\kappa - \mathbf{u}_\kappa) \right\rangle \\ &= \sum_\kappa \int \mathrm{d}\boldsymbol{\tau} f_\kappa^e(\boldsymbol{\tau}) \exp[-W_\kappa(\boldsymbol{\tau})] \exp[2\pi i \boldsymbol{\tau} \cdot (\mathbf{r} - \mathbf{r}_\kappa)] \end{aligned} \tag{14.1c}$$

$W_\kappa = \langle 2\pi^2 |\boldsymbol{\tau} \cdot \mathbf{u}_\kappa|^2 \rangle$ is the Debye-Waller factor, where <> stands for time average, $f_\kappa^e(\boldsymbol{\tau})$ is the electron scattering factor, and $\boldsymbol{\tau}$ is a reciprocal space vector, and the deviation potential depending on the lattice configuration is

$$\Delta V(\mathbf{r}) = V(\mathbf{r}, t) - V_0(\mathbf{r}) \tag{14.1d}$$

The electron wave function scattered by a given crystal lattice is determined by the time-independent Schrödinger equation:

$$\left[\left(\hbar^2/2m_0 \right) \nabla^2 - eV_0 - e\Delta V - E \right] \Psi = 0 \tag{14.2}$$

where E is the electron energy. For electron scattering, (14.2) is converted into an integral equation with the use of Green's function $G(\mathbf{r}, \mathbf{r}_1)$:

$$\Psi(\mathbf{r}) = \Psi_0(\mathbf{r}) + e \int \mathrm{d}\mathbf{r}_1 G(\mathbf{r}, \mathbf{r}_1) \Delta V(\mathbf{r}_1) \Psi(\mathbf{r}_1) \tag{14.3a}$$

where G is the solution of

$$\left[-\left(\hbar^2/2m_0 \right) \nabla^2 - eV_0 - E \right] G(\mathbf{r}, \mathbf{r}_1) = \delta(\mathbf{r} - \mathbf{r}_1) \tag{14.3b}$$

which corresponds to the electron wave coming from a point source at $\mathbf{r}'$ after being scattered by the crystal, and it can be solved for a general case (Wang 1998a), and $\Psi_0(\mathbf{r})$ represents the elastic wave initiated by an incident plane wave of wave vector $\mathbf{K}_0$ after being scattered by the time-independent potential V_0, and it satisfies

$$\left[-\left(\hbar^2/2m_0 \right) \nabla^2 - eV_0 - E \right] \Psi_0 = 0 \tag{14.3c}$$

For the purpose of this paper, we define a density matrix

$$\rho_f(\mathbf{r}, \mathbf{r}') = \langle \Psi(\mathbf{r}) \Psi^*(\mathbf{r}') \rangle \tag{14.4}$$

where represents an average over lattice configurations. The electron diffraction pattern including TDS can be calculated by taking a double Fourier transform of the density matrix:

$$I(\boldsymbol{\tau}) = \int \mathrm{d}\mathbf{r} \int \mathrm{d}\mathbf{r}' \exp\left[-2\pi i \boldsymbol{\tau} \cdot (\mathbf{r} - \mathbf{r}')\right] \rho_f\left(\mathbf{r}, \mathbf{r}'\right) \tag{14.5}$$

The intensity distribution in the electron image can be calculated by

$$I(\mathbf{r}) = \int \mathrm{d}\mathbf{r} \int \mathrm{d}\mathbf{r}' \rho_f\left(\mathbf{r}_1, \mathbf{r}_2\right) t\left(\mathbf{r} - \mathbf{r}_1\right) t^*\left(\mathbf{r} - \mathbf{r}_2\right) \tag{14.6}$$

where $t(\mathbf{r})$ characterizes the information transfer property of the optical system in an electron microscope. Therefore, the entire scattering process can be described by the density matrix equation and the key here is to find its solution. We now derive the equation that governs the evolution of the density matrix.

Starting from (14.3a) and using $\langle \Delta V(\mathbf{r}, t)\rangle = 0$, we have

$$\begin{aligned} \rho_f\left(\mathbf{r}, \mathbf{r}'\right) = & \rho_0\left(\mathbf{r}, \mathbf{r}'\right) + e^2 \int \mathrm{d}\mathbf{r}_1 \int \mathrm{d}\mathbf{r}_2 \left[G\left(\mathbf{r}, \mathbf{r}_1\right) G^*\left(\mathbf{r}', \mathbf{r}_2\right)\right. \\ & \left. \times \left\langle \Delta V\left(\mathbf{r}_1\right) \Delta V\left(\mathbf{r}_2\right) \Psi\left(\mathbf{r}_1\right) \Psi^*\left(\mathbf{r}_2\right)\right\rangle\right] \end{aligned} \tag{14.7}$$

where the density matrix of the Bragg-scattered electrons is $\rho_0\left(\mathbf{r}, \mathbf{r}'\right) = \Psi_0(\mathbf{r}) \Psi_0^*\left(\mathbf{r}'\right)$. Using the iterative solution of (14.3a) and treating the different orders of diffuse scattering incoherently, which means

$$\begin{aligned} & \left\langle \Delta V\left(\mathbf{r}_1\right) \Delta V\left(\mathbf{r}_2\right) \Delta V\left(\mathbf{r}_3\right) \Delta V\left(\mathbf{r}_4\right)\right\rangle \\ & = \left\langle \Delta V\left(\mathbf{r}_1\right) \Delta V\left(\mathbf{r}_2\right)\right\rangle \left\langle \Delta V\left(\mathbf{r}_3\right) \Delta V\left(\mathbf{r}_4\right)\right\rangle, \\ & \left\langle \Delta V \Delta V \Delta V \Delta V \Delta V \Delta V\right\rangle \\ & = \left\langle \Delta V\left(\mathbf{r}_1\right) \Delta V\left(\mathbf{r}_2\right)\right\rangle \left\langle \Delta V\left(\mathbf{r}_3\right) \Delta V\left(\mathbf{r}_4\right)\right\rangle \\ & \quad \times \left\langle \Delta V\left(\mathbf{r}_5\right) \Delta V\left(\mathbf{r}_6\right)\right\rangle \end{aligned} \tag{14.8a}$$

and similar relations hold for higher order terms, it has been proved (Wang 1996b)

$$\begin{aligned} & \left\langle \Delta V\left(\mathbf{r}_1\right) \Delta V\left(\mathbf{r}_2\right) \Psi\left(\mathbf{r}_1\right) \Psi^*\left(\mathbf{r}_2\right)\right\rangle \\ & \quad = \left\langle \Delta V\left(\mathbf{r}_1\right) \Delta V\left(\mathbf{r}_2\right)\right\rangle \left\langle \Psi\left(\mathbf{r}_1\right) \Psi^*\left(\mathbf{r}_2\right)\right\rangle \end{aligned} \tag{14.8b}$$

Thus, (14.7) becomes

$$\begin{aligned} \rho_f\left(\mathbf{r}, \mathbf{r}'\right) = & \rho_0\left(\mathbf{r}, \mathbf{r}'\right) + e^2 \int \mathrm{d}\mathbf{r}_1 \int \mathrm{d}\mathbf{r}_2 \left[G\left(\mathbf{r}, \mathbf{r}_1\right) G^*\left(\mathbf{r}', \mathbf{r}_2\right)\right. \\ & \left. \times \left\langle \Delta V\left(\mathbf{r}_1\right) \Delta V\left(\mathbf{r}_2\right)\right\rangle \rho_f\left(\mathbf{r}_1, \mathbf{r}_2\right)\right] \end{aligned} \tag{14.9}$$

This is the equation that determines the solution of the density matrix and it can be solved iteratively:

$$\begin{aligned}\rho_f\left(\mathbf{r},\mathbf{r}'\right) =& \rho_0\left(\mathbf{r},\mathbf{r}'\right) + e^2 \int \mathrm{d}\mathbf{r}_1 \int \mathrm{d}\mathbf{r}_2 \left[G\left(\mathbf{r},\mathbf{r}_1\right) G^*\left(\mathbf{r}',\mathbf{r}_2\right)\right. \\ & \times \left\langle \Delta V\left(\mathbf{r}_1\right) \Delta V\left(\mathbf{r}_2\right)\right\rangle \rho_0\left(\mathbf{r}_1,\mathbf{r}_2\right)] \\ & + e^4 \int \mathrm{d}\mathbf{r}_1 \int \mathrm{d}\mathbf{r}_2 \int \mathrm{d}\mathbf{r}_3 \int \mathrm{d}\mathbf{r}_4 \\ & \times \left[G\left(\mathbf{r},\mathbf{r}_1\right) G^*\left(\mathbf{r}',\mathbf{r}_2\right) G\left(\mathbf{r}_1,\mathbf{r}_3\right) G^*\left(\mathbf{r}_2,\mathbf{r}_4\right)\right. \\ & \times \left\langle \Delta V\left(\mathbf{r}_1\right) \Delta V\left(\mathbf{r}_2\right)\right\rangle \\ & \times \left\langle \Delta V\left(\mathbf{r}_3\right) \Delta V\left(\mathbf{r}_4\right)\right\rangle \rho_0\left(\mathbf{r}_3,\mathbf{r}_4\right)] + \ldots\end{aligned} \tag{14.10}$$

where the higher orders of TDS are included. For the general purpose of the proof, all the higher order terms are kept. On the other hand, the time (or configurational) average of $\Delta V\left(\mathbf{r}_1\right) \Delta V\left(\mathbf{r}_2\right)$ is directly related to the mixed dynamic form factor $S\left(\mathbf{Q},\mathbf{Q}'\right)$ by Wang (1996a)

$$\begin{aligned}&\left\langle \Delta V(\mathbf{r},t) \Delta V\left(\mathbf{r}_1,t\right)\right\rangle \\ &= \int \mathrm{d}\mathbf{Q} \int \mathrm{d}\mathbf{Q}' \exp\left[2\pi i\left(\mathbf{r}\cdot\mathbf{Q} - \mathbf{r}_1\cdot\mathbf{Q}'\right)\right] S\left(\mathbf{Q},\mathbf{Q}'\right)\end{aligned} \tag{14.11a}$$

and for TDS with consideration of the phase correlation among atom vibrations, the mixed dynamic form factor is given by

$$\begin{aligned}S\left(\mathbf{Q},\mathbf{Q}'\right) =& \sum_{\kappa}\sum_{\kappa'} \exp\left[2\pi i\left(\mathbf{r}_{\kappa'}\cdot\mathbf{Q}' - \mathbf{r}_{\kappa}\cdot\mathbf{Q}\right)\right] \\ & \times f_{\kappa}^{e}(\mathbf{Q}) f_{\kappa'}^{e}\left(\mathbf{Q}'\right) \exp\left[-W_{\kappa}(\mathbf{Q}) - W_{\kappa'}\left(\mathbf{Q}'\right)\right] \\ & \times \left\{\exp\left[2F_{\kappa\kappa'}\left(\mathbf{Q},\mathbf{Q}'\right)\right] - 1\right\}\end{aligned} \tag{14.11b}$$

where

$$F_{\kappa\kappa'}\left(\mathbf{Q},\mathbf{Q}'\right) = 2\pi^2\left\langle\left(\mathbf{u}_{\kappa}\cdot\mathbf{Q}\right)\left(\mathbf{u}_{\kappa'}\cdot\mathbf{Q}'\right)\right\rangle \tag{14.11c}$$

is defined as a correlation function that is determined by the coupling between vibrations of the κ and κ' atoms. The calculation of $W_{\kappa}(\mathbf{Q})$ and $F_{\kappa\kappa'}\left(\mathbf{Q},\mathbf{Q}'\right)$ using the harmonic oscillators model has been given elsewhere (Sears and Shelley 1991; Wang 1995). Equation (14.9) is the main result of the frozen lattice model and the introduction of the mixed dynamic form factor $S\left(\mathbf{Q},\mathbf{Q}'\right)$ [see (14.3a)] automatically considers the incoherence between the different orders of diffuse scattering. We now derive a similar equation using the rigorous quantum-mechanical approach.

14.2 Quantum Mechanical Description of Phonon Excitation in Electron Scattering

14.2.1 Quantum Mechanical Theory

Phonon excitation is an inelastic-scattering process that is most adequately described using (Yoshioka 1957) approach, in which the eigenstate of the incident electron-crystal system is expressed as a product of the eigenstate of the crystal with that of the incident electron:

$$\Phi_s(\mathbf{r}, \mathbf{R}) = \sum_{n=0}^{\infty} a_n(\mathbf{R})\Psi_n(\mathbf{r}) \tag{14.12}$$

where a_n is the nth eigenstate of the crystal of energy ε_n; Ψ_0 is the elastic scattered wave of energy E, and Ψ_n is the inelastically scattered wave of energy $E_n = E - \varepsilon_n$ after exciting the nth crystal state; $\mathbf{R}$ represents the coordinates of crystal electrons $(\mathbf{r}_1 \ldots, \mathbf{r}_M)$. The fundamental equations that govern the generation, scattering, and transition between/among the elastic wave and the inelastic waves are

$$\left[-\left(\hbar^2/2m_0\right)\nabla^2 + H_{00} - E\right]\Psi_0 = -\sum_{m=1} H_{0m}\Psi_m \tag{14.13a}$$

and

$$\left[-\left(\hbar^2/2m_0\right)\nabla^2 + H_{nn} - E_n\right]\Psi_n = -\sum_{m\neq n} H_{nm}\Psi_m \tag{14.13b}$$

where the transition matrix from state a_m to a_n is

$$H_{nm} = \int a_n^* H a_m \, \mathrm{d}\mathbf{R} = \langle a_n| H |a_m\rangle \tag{14.13c}$$

H is the interaction Hamiltonian between the incident electron with the electrons and nuclei in the crystal (Fig. 14.1).

Phonon excitation is a quantum transition process, in which the creation and annihilation of phonons are both possible. The excitation process can be adequately described using the transition matrix, similar to other inelastic electron transitions. Phonon excitation has the following characteristics. Firstly, the phonon energy is on the order of 0.1 eV or less, much smaller than the incident energy of the electron (typically 100 keV), thus, the electron energy can be considered to be unaffected by the energy transfer due to phonon excitation, which means $E_n \simeq E$ but this energy loss has destroyed the coherence of the electron. Hence, the incoherence among all of the crystal states (e.g., different phonon states) must be preserved. Secondly, a phonon is characterized by a group of quantities of $(\mathbf{q}, \omega_j, \mathbf{e})$, where $\mathbf{q}$ is the phonon wave vector, ω_j the phonon frequency belonging to the jth branch, and $\mathbf{e}$ the polarization

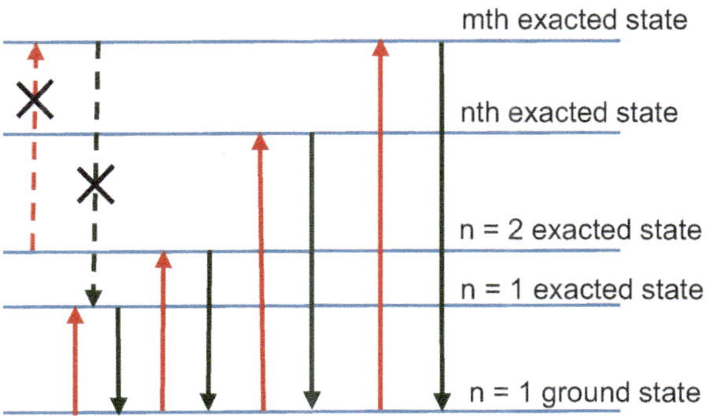

Fig. 14.1 A schematic model showing quantum transition in crystal state under the ground state approximation

vector, which means that an excited state a_n is also characterized by $\left(\mathbf{q}, \omega_j, \mathbf{e}\right)$, which is usually denoted as $|n_s\rangle$ in phonon excitation. Finally, the interaction Hamiltonian of the electron with the crystal is determined by the instantaneous crystal potential V as given by (14.1b), thus, the diagonal matrix element H_{nn} is

$$H_{nn} \simeq H_{00} \tag{14.14a}$$

The approximation of $H_{nn} \simeq H_{00}$ is considered to be almost exact since the excitation of phonons does not affect the time-averaged potential of the crystal. For the statistical distribution of phonons in the ground state at a finite temperature T, a weight factor must be added, which is related to the energy of the harmonic oscillator by

$$\begin{aligned} p\left(n_s\right) &= \frac{\exp\left[-\left(n_s + 1/2\right)\hbar\omega_j/k_B T\right]}{\sum_{n_s=0}\exp\left[-\left(n_s + 1/2\right)\hbar\omega_j/k_B T\right]} \\ &= \left[1 - \exp\left(-\hbar\omega_j/k_B T\right)\right]\exp\left(-n_s\hbar\omega_j/k_B T\right) \end{aligned} \tag{14.14b}$$

A statistical average needs to be added accordingly, thus

$$H_{00} = -e\sum_{n_s}\langle n_s|\, V(\mathbf{r})\, |n_s\rangle\, p\left(n_s\right), \tag{14.14c}$$

where the sum of n_s is over all of the phonon states and the number of phonons. Using the relationship between the electron scattering factor and the atomic potential, one has

$$
\begin{aligned}
H_{00} &= -e \sum_{n_s} \langle n_s | \sum_{\kappa} V_{\kappa} \left(\mathbf{r} - \mathbf{r}_{\kappa} - \mathbf{u}_{\kappa} \right) | n_s \rangle \, p \left(n_s \right) \\
&= e \sum_{\kappa} \int \mathrm{d}\boldsymbol{\tau} \, f_{\kappa}^{e}(\boldsymbol{\tau}) \exp \left[2\pi i \boldsymbol{\tau} \cdot \left(\mathbf{r} - \mathbf{r}_{\kappa} \right) \right] \\
&\quad \times \sum_{n_s} \langle n_s | \exp \left(-2\pi i \boldsymbol{\tau} \cdot \mathbf{u}_{\kappa} \right) | n_s \rangle \, p \left(n_s \right) \\
&= -e \sum_{\kappa} \int \mathrm{d}\boldsymbol{\tau} \, f_{\kappa}^{e}(\boldsymbol{\tau}) \exp \left[2\pi i \boldsymbol{\tau} \cdot \left(\mathbf{r} - \mathbf{r}_{\kappa} \right) \right] \\
&\quad \times \exp \left[-W_{\kappa}(\boldsymbol{\tau}) \right]
\end{aligned}
\tag{14.14d}
$$

where a general relationship used in the above derivation is $\langle \exp(x) \rangle = \exp \left[\langle x^2 \rangle /2 \right]$ if $\langle x \rangle = 0$, which has been proved to be exact (Glauber, 1955), and the Debye-Waller factor is given by

$$
W_{\kappa} = 2\pi^2 \sum_{n_s} \langle n_s | \left(\boldsymbol{\tau} \cdot \mathbf{u}_{\kappa} \right)^2 | n_s \rangle \, p \left(n_s \right) \tag{14.14e}
$$

In comparison to the average crystal potential V_0 given in (14.1c), one has

$$
H_{nn} = -eV_0 \tag{14.14f}
$$

in which the Debye-Waller factor is included. Therefore, (14.13a) and (14.13b) are transformed into

$$
\left[- \left(\hbar^2/2m_0 \right) \nabla^2 - eV_0 - E \right] \Psi_0 = - \sum_{m=1} H_{0m} \Psi_m \tag{14.15a}
$$

and

$$
\left[- \left(\hbar^2/2m_0 \right) \nabla^2 - eV_0 - E \right] \Psi_n = - \sum_{m \neq n} H_{nm} \Psi_m \tag{14.15b}
$$

The density matrix in this case is defined to be

$$
\rho_p \left(\mathbf{r}, \mathbf{r}' \right) = \sum_{n=0} \Psi_n(\mathbf{r}) \Psi_n^* \left(\mathbf{r}' \right) \tag{14.16}
$$

where the sum over n is over all the crystal states including the elastic-scattering state (i.e., n = 0, ground state). We now derive the equation that governs the motion of the density matrix. By using Green's function $G \left(\mathbf{r}, \mathbf{r}' \right)$ introduced in (14.3b), (14.15a), and (14.15b) are transformed into a unified integral equation:

$$\Psi_n(\mathbf{r}) = \Psi_0(\mathbf{r})\delta_{n0} - \sum_{m\neq n} \int \mathrm{d}\mathbf{r}_1 G(\mathbf{r}, \mathbf{r}_1) H_{nm}(\mathbf{r}_1) \Psi(\mathbf{r}_1) \tag{14.17}$$

With consideration the incoherence between the elastic wave $\Psi_0(\mathbf{r})$ and the inelastic wave Ψ_m, a substitution of (14.17) into (14.16) gives

$$\begin{aligned}\rho_p(\mathbf{r}, \mathbf{r}') =& \rho_0(\mathbf{r}, \mathbf{r}') + \int \mathrm{d}\mathbf{r}_1 \int \mathrm{d}\mathbf{r}_2 G(\mathbf{r}, \mathbf{r}_1) G^*(\mathbf{r}', \mathbf{r}_2) \\ &\times \sum_{n\neq 0}\sum_m\sum_{m'} \left[H_{nm}(\mathbf{r}_1) H_{m'n}(\mathbf{r}_2) \Psi_m(\mathbf{r}_1) \Psi^*_{m'}(\mathbf{r}_2)\right]\end{aligned} \tag{14.18}$$

Since the electrons scattered by the m and m' states are incoherent, the sums over m and m' vanish unless $m = m'$, which means the interference terms disappear under the random-phase approximation, resulting in

$$\begin{aligned}\rho_p(\mathbf{r}, \mathbf{r}') =& \rho_0(\mathbf{r}, \mathbf{r}') + \int \mathrm{d}\mathbf{r}_1 \int \mathrm{d}\mathbf{r}_2 G(\mathbf{r}, \mathbf{r}_1) G^*(\mathbf{r}', \mathbf{r}_2) \\ &\times \left\{\sum_m \left[\sum_{n\neq 0} H_{nm}(\mathbf{r}_1) H_{mn}\right] \Psi_m(\mathbf{r}_1) \Psi^*_m(\mathbf{r}_2)\right\}\end{aligned} \tag{14.19}$$

The following approximation is made in order to simplify this equation (Dudarev et al. 1994). In one-particle multiple scattering theory, each time the electron inelastically interacts with the crystal atoms the collision is assumed to take place in such a way as if the crystal is in its ground state, and the effect of previous collisions is very small. This condition is satisfied if $\tau_0 v > \Lambda$, where τ_0 is the lifetime of the crystal excitation state. This approximation is mathematically expressed as

$$\sum_{n\neq 0} \left[H_{nm}(\mathbf{r}_1) H^*_{nm}(\mathbf{r}_2)\right] \simeq \sum_{n\neq 0} \left[H_{n0}(\mathbf{r}_1) H^*_{n0}(\mathbf{r}_2)\right] \tag{14.20}$$

It is simply called the "ground-state" approximation and its meaning will be discussed later. Thus Equation (14.19) is approximated as

$$\begin{aligned}\rho_p(\mathbf{r}, \mathbf{r}') \simeq& \rho_0(\mathbf{r}, \mathbf{r}') + \int \mathrm{d}\mathbf{r}_1 \int \mathrm{d}\mathbf{r}_2 G(\mathbf{r}, \mathbf{r}_1) G^*(\mathbf{r}', \mathbf{r}_2) \\ &\times \left[\sum_{n\neq 0} H_{n0}(\mathbf{r}_1) H_{0n}(\mathbf{r}_2)\right] \left[\sum_m \Psi_m(\mathbf{r}_1) \Psi^*_m(\mathbf{r}_2)\right] \\ =& \rho_0(\mathbf{r}, \mathbf{r}') + \int \mathrm{d}\mathbf{r}_1 \int \mathrm{d}\mathbf{r}_2 G(\mathbf{r}, \mathbf{r}_1) G^*(\mathbf{r}', \mathbf{r}_2) \\ &\times \left[\sum_{n\neq 0} H_{n0}(\mathbf{r}_1) H_{0n}(\mathbf{r}_2)\right] \rho_p(\mathbf{r}_1, \mathbf{r}_2)\end{aligned} \tag{14.21a}$$

which can be solved iteratively:

$$\begin{aligned}\rho_p\left(\mathbf{r},\mathbf{r}'\right) &= \rho_0\left(\mathbf{r},\mathbf{r}'\right) + \int d\mathbf{r}_1 \int d\mathbf{r}_2 \Bigg\{ G\left(\mathbf{r}',\mathbf{r}_1\right) G^*\left(\mathbf{r}',\mathbf{r}_2\right) \\ &\quad\times \left[\sum_{n\neq 0} H_{n0}\left(\mathbf{r}_1\right) H_{0n}\left(\mathbf{r}_2\right)\right] \rho_0\left(\mathbf{r}_1,\mathbf{r}_2\right)\Bigg\} \\ &\quad+ \int d\mathbf{r}_1 \int d\mathbf{r}_2 \int d\mathbf{r}_3 \int d\mathbf{r}_4 \\ &\quad\times \Bigg\{ G\left(\mathbf{r},\mathbf{r}_1\right) G^*\left(\mathbf{r}',\mathbf{r}_2\right) G\left(\mathbf{r}_1,\mathbf{r}_3\right) G^*\left(\mathbf{r}_2,\mathbf{r}_4\right) \\ &\quad\times \left[\sum_{n\neq 0} H_{n0}\left(\mathbf{r}_1\right) H_{0n}\left(\mathbf{r}_2\right)\right] \\ &\quad\times \left[\sum_{n'\neq 0} H_{n'0}\left(\mathbf{r}_3\right) H_{0n'}\left(\mathbf{r}_4\right)\right] \\ &\quad\times \rho_0\left(\mathbf{r}_3,\mathbf{r}_4\right)\Bigg\} + \ldots\end{aligned} \tag{14.21b}$$

14.2.2 The Dynamic Form Factor

Using a completeness equation of

$$\sum_n \left|a_n\left(\mathbf{R}_2\right)\right\rangle \left\langle a_n\left(\mathbf{R}_1\right)\right| = \delta\left(\mathbf{R}_2-\mathbf{R}_1\right),$$

one has

$$\begin{aligned}&\sum_{n\neq 0} H_{n0}\left(\mathbf{r}_1\right) H^*_{n0}\left(\mathbf{r}_2\right) \\ &= \sum_{n\neq 0} \left\langle a_0\left(\mathbf{R}_2\right)\right| H\left(\mathbf{r}_2,\mathbf{R}_2\right) \left|a_n\left(\mathbf{R}_2\right)\right\rangle \\ &\quad\times \left\langle a_n\left(\mathbf{R}_1\right)\right| H\left(\mathbf{r}_1,\mathbf{R}_1\right) \left|a_0\left(\mathbf{R}_1\right)\right\rangle \\ &= \left\langle a_0\left(\mathbf{R}_2\right)\right| H\left(\mathbf{r}_2,\mathbf{R}_2\right) [\sum_n \left|a_n\left(\mathbf{R}_2\right)\right\rangle \left\langle a_n\left(\mathbf{R}_1\right)\right| \\ &\quad- \left|a_0\left(\mathbf{R}_2\right)\right\rangle \left\langle a_0\left(\mathbf{R}_1\right)\right|] H\left(\mathbf{r}_1,\mathbf{R}_1\right) \left|a_0\left(\mathbf{R}_1\right)\right\rangle \\ &= \left\langle a_0\left(\mathbf{R}_2\right)\right| H\left(\mathbf{r}_2,\mathbf{R}_2\right) [\delta\left(\mathbf{R}_2-\mathbf{R}_1\right) \\ &\quad- \left|a_0\left(\mathbf{R}_2\right)\right\rangle \left\langle a_0\left(\mathbf{R}_1\right)\right|] H\left(\mathbf{r}_1,\mathbf{R}_1\right) \left|a_0\left(\mathbf{R}_1\right)\right\rangle\end{aligned}$$

$$
\begin{aligned}
&= \langle a_0 (\mathbf{R}_1)| \, H (\mathbf{r}_2, \mathbf{R}_2) \, H (\mathbf{r}_1, \mathbf{R}_1) \, |a_0 (\mathbf{R}_1)\rangle \\
&\quad - \langle a_0 (\mathbf{R}_2)| \, H (\mathbf{r}_2, \mathbf{R}_2) \, |a_0 (\mathbf{R}_2)\rangle \\
&\quad \times \langle a_0 (\mathbf{R}_1)| \, H (\mathbf{r}_1, \mathbf{R}_1) \, |a_0 (\mathbf{R}_1)\rangle
\end{aligned} \tag{14.22}
$$

Using the Fourier transform of $H = -eV$, which is

$$
H(\mathbf{r}) = -eV(\mathbf{r}) = -e \sum_{\kappa} \int \mathrm{d}\mathbf{Q} \exp\left[2\pi i \left(\mathbf{r} - \mathbf{r}_{\kappa}\right) \cdot \mathbf{Q}\right] f_{\kappa}^{e}(\mathbf{Q}), \tag{14.23}
$$

Equation (14.22) can be converted into the form of

$$
\begin{aligned}
&\sum_{n \neq 0} \left[H_{n0} (\mathbf{r}_1) \, H_{n0}^{*} (\mathbf{r}_2) \right] \\
&= e^2 \int \mathrm{d}\mathbf{Q} \int \mathrm{d}\mathbf{Q}' \exp\left[2\pi i \left(\mathbf{r}_1 \cdot \mathbf{Q} - \mathbf{r}_2 \cdot \mathbf{Q}'\right)\right] \times S' \left(\mathbf{Q}, \mathbf{Q}'\right)
\end{aligned} \tag{14.24}
$$

where the S' function is given by

$$
\begin{aligned}
&S' \left(\mathbf{Q}, \mathbf{Q}'\right) \\
&= \textstyle\sum_{\kappa} \sum_{\kappa'} \exp\left[2\pi i \left(\mathbf{r}_{\kappa'} \cdot \mathbf{Q}' - \mathbf{r}_{\kappa} \cdot \mathbf{Q}\right)\right] f_{\kappa}^{e}(\mathbf{Q}) f_{\kappa'}^{e} \left(\mathbf{Q}'\right) \\
&\times \left\{ \langle a_0| \exp\left[2\pi i \left(\mathbf{Q}' \cdot \mathbf{u}_{\kappa'} - \mathbf{Q} \cdot \mathbf{u}_{\kappa}\right)\right] |a_0\rangle \right. \\
&- \langle a_0| \exp\left(2\pi i \mathbf{Q}' \cdot \mathbf{u}_{\kappa'}\right) |a_0\rangle \\
&\times \left. \langle a_0| \exp\left(-2\pi i \mathbf{Q} \cdot \mathbf{u}_{\kappa}\right) |a_0\rangle \right\} \\
&= \textstyle\sum_{\kappa} \sum_{\kappa'} \exp\left[2\pi i \left(\mathbf{r}_{\kappa'} \cdot \mathbf{Q}' - \mathbf{r}_{\kappa} \cdot \mathbf{Q}\right)\right] f_{\kappa}^{e}(\mathbf{Q}) f_{\kappa'}^{e} \left(\mathbf{Q}'\right) \\
&\times \left\{ \exp\left[-2\pi^2 \left\langle a_0 \mid (\mathbf{Q}' \cdot \mathbf{u}_{\kappa'} - \mathbf{Q} \cdot \mathbf{u}_{\kappa})^2 \mid a_0 \right\rangle\right] \right. \\
&- \exp\left[-2\pi^2 \left\langle a_0 \mid (\mathbf{Q} \cdot \mathbf{u}_{\kappa'})^2 \mid a_0 \right\rangle\right] \\
&\times \left. \exp\left[-2\pi^2 \left\langle a_0 \mid (\mathbf{Q} \cdot \mathbf{u}_{\kappa})^2 \mid a_0 \right\rangle\right] \right\}
\end{aligned} \tag{14.25}
$$

From the notation, it may be misunderstood that there is no phonon in the ground state $|a_0\rangle$. This is true at absolute zero ($T = 0$ K). In practice, any system is not isolated from the environment, thus, energy exchange is involved in electron diffraction (Fanidis et al. 1992; Fanidis et al. 1993). To incorporate the statistical distribution of phonons in the ground state at a finite temperature, the terms in (14.25) need to be modified to represent the expectation value, hence

$$
\langle a_0| \left(\mathbf{Q}' \cdot \mathbf{u}_{k'}\right)^2 |a_0\rangle = \sum_{n_s} \langle n_s| \left(\mathbf{Q}' \cdot \mathbf{u}_{k'}\right)^2 |n_s\rangle \, p \left(n_s\right) \tag{14.26}
$$

Therefore, (14.25) becomes

$$S'\left(\mathbf{Q},\mathbf{Q}'\right) = \sum_{\kappa}\sum_{\kappa'} \exp\left[2\pi i\left(\mathbf{r}_{\kappa'}\cdot\mathbf{Q}' - \mathbf{r}_{\kappa}\cdot\mathbf{Q}\right)\right] f_{\kappa}^{e}(\mathbf{Q}) f_{\kappa'}^{e}\left(\mathbf{Q}'\right) \times \exp\left[-W_{\kappa}(\mathbf{Q}) - W_{\kappa'}\left(\mathbf{Q}'\right)\right] \times \left\{\exp\left[2F_{\kappa\kappa'}\left(\mathbf{Q},\mathbf{Q}'\right)\right] - 1\right\} \quad (14.27)$$

where

$$F_{\kappa k'}\left(\mathbf{Q},\mathbf{Q}'\right) = 2\pi^2 \sum_{n_s} \langle n_s|\left(\mathbf{u}_k\cdot\mathbf{Q}\right)\left(\mathbf{u}_{\kappa'}\cdot\mathbf{Q}'\right)|n_s\rangle\, p\left(n_s\right) \quad (14.28)$$

Equation (14.28) gives the same result as defined in (14.11c) under the harmonic oscillators approximation. A comparison of (14.28) and (14.11b) gives

$$S\left(\mathbf{Q},\mathbf{Q}'\right) = e^2 S'\left(\mathbf{Q},\mathbf{Q}'\right) \quad (14.29)$$

or

$$\sum_{n\neq 0}\left[H_{n0}\left(\mathbf{r}_1\right) H_{n0}^{*}\left(\mathbf{r}_2\right)\right] = e^2\left\langle \Delta V\left(\mathbf{r}_1\right)\Delta V\left(\mathbf{r}_2\right)\right\rangle \quad (14.30)$$

which means the sum over crossed transition matrix elements is the time-averaged cross-perturbation crystal potentials.

14.3 The Approximations and Conditions

14.3.1 The Incoherence Between Different Orders of Diffuse Scattering

The major approximation introduced in the calculation of the density matrix in the frozen lattice model is

$$\left\langle \Delta V\left(\mathbf{r}_1\right)\Delta V\left(\mathbf{r}_2\right)\Delta V\left(\mathbf{r}_3\right)\Delta V\left(\mathbf{r}_4\right)\right\rangle = \left\langle \Delta V\left(\mathbf{r}_1\right)\Delta V\left(\mathbf{r}_2\right)\right\rangle\left\langle \Delta V\left(\mathbf{r}_3\right)\Delta V\left(\mathbf{r}_4\right)\right\rangle$$

In electron diffraction, the first-order scattering intensity is determined by $\langle \Delta V(\mathbf{r}_1)\Delta V(\mathbf{r}_2)\rangle$ and the second order is by $\langle \Delta V(\mathbf{r}_1)\Delta V(\mathbf{r}_2)\rangle \times \langle \Delta V(\mathbf{r}_3)\Delta V(\mathbf{r}_4)\rangle$, which can be interpreted as the second order being a double scattering of two incoherent first-order events. The approximation simply means the scattering produced by the different orders of diffuse scattering (such as ΔV^2, ΔV^4 and ΔV^6 or multiple diffuse scattering) must be treated as incoherent, and the higher order scattering is the result of multiple incoherent first-order scattering events. The second-order scattering, for example, is an incoherent double first-order scattering (e.g., $\langle \Delta V^4\rangle = \langle \Delta V^2\rangle\langle \Delta V^2\rangle$) and $\langle \Delta V^6\rangle = [\langle \Delta V^2\rangle\langle \Delta V^2\rangle\langle \Delta V^2\rangle]$) for the third-order scattering, etc. (Wang, 1996b). This is in fact the practical situation and the approximation holds exactly. This approximation also reduces the coherence effect in the quasi-elastic scattering process. The consequence of this approach means that any theory that has been developed with use of the mixed dynamic form factor $S\left(\mathbf{Q}, \mathbf{Q}'\right)$ instead of using the real space $\Delta V\left(\mathbf{r}_1\right)$ has correctly considered this incoherence. Thus, the corresponding calculation is equivalent to the result of quantum phonon excitation theory.

Calculations using the frozen lattice model have frequently been performed in the multislice theory (Cowley and Moodie 1957) following two different approaches. One approach (Wang 1991; Loane et al. 1991) assumes that the entire crystal in one instantaneous configuration is a frozen lattice and the diffuse scattering generated in the entire volume is coherent but the final intensity is an incoherent sum over the intensities contributed by different lattice configurations. The other approach assumes that the diffuse scattering generated within one thin crystal slice is coherent but incoherence is preserved for the scattering generated in different slices (Dinges et al. 1995; Wang 1995). The former has properly considered the multiple diffuse scattering effect but ignored the incoherence between the different orders of diffuse scattering. This is likely to be an excellent approximation for thin specimens. The latter treats the incoherence properly but the multiple diffuse scattering among slices is neglected. The current study shows that the most significant error in the frozen lattice model calculation is the "quasi-elastic" approximation, which can be compensated for with the introduction of the mixed dynamic form factor.

14.3.2 The "Ground State" Approximation

The approximation made in the phonon excitation model is the so-called "ground state" approximation. The approximation means that each time the electron inelastically interacts with the crystal atoms the collision is assumed to take place as if the crystal is in its ground state and the effect of any previous collision is very small. For a thin crystal of thickness $d = 100$ nm, the duration that the electron interacts with the crystal atoms for a 100 keV electron is on the order of femo-second. The approximation made in the phonon excitation model is the so-called "ground-state" approximation. The approximation means that each time the electron inelastically interacts with the crystal atoms the collision is assumed to take place as if the crystal is in its ground state and the effect of any previous collision is very small. For a thin crystal of thickness $d = 100$ nm, the duration that the electron interacts with the crystal atoms for a 100 keV electron is on the order of $\Delta t \simeq 6 \times 10^{-15}$ s. Theoretical calculation has found that the lifetime of a phonon is on the order of $\tau_0 = 10^{-13}$ s (Björkman et al. 1967; Woll and Kohn 1962). The limited lifetime of phonons is due to anharmonic effects in atom vibrations. The lifetime of a phonon is much longer than the interaction time $Deltat$ of the electron with the crystal, thus, there is no phonon decay during the scattering of an incident electron. On the other hand, for a modern field-emission source, the electron flux is on the order of 10^{12} e s^{-1}, thus, the average time interval between two successive incident electrons is 10^{-12} s, much longer than the lifetime of the phonon. The phonons generated by the previous incident electron have annihilated before the next incoming electron interacts with the crystal. Therefore, the crystal is in its ground state for each incident electron.

The conditions for this approximation to be held are: the lifetime of the phonon is larger than the time required to create a new phonon, e.g., $\tau_0 > \Lambda/v$ (or $\tau_0 v > \Lambda$), and the electron-specimen interaction time is shorter than the lifetime of the phonon, e.g., $d < \tau_0 v$. The first condition depends on the energy of the incident electron and the second condition is satisfied in almost all of our experiments. The conditions $\tau_0 v > \Lambda$ and $d < \tau_0 v$ mean that the specimen thickness and the mean free path length for phonon excitation are both required to be smaller than the distance traveled by the electron within the lifetime of the phonon. For a 100 kV electron, $\tau_0 \nu \simeq 5\ \mu$ m. Therefore, both conditions are absolutely satisfied in all of our current experiments. On the other hand, the average number of phonons in a crystal is rather large at ambient temperature, thus, adding or subtracting a few phonons due to the electron-phonon interaction in the whole system has almost no effect on the potential distribution in the crystal. This is always assumed in electron scattering. Therefore, the ground-state approximation is excellent.

14.4 Equivalence Between the "Frozen Lattice" Model and Quantum Theory

Using the conclusion in (14.30), (14.21a) becomes

$$\rho_p\left(\mathbf{r},\mathbf{r}'\right) \simeq \rho_0\left(\mathbf{r},\mathbf{r}'\right) + e^2 \int \mathrm{d}\mathbf{r}_1 \int \mathrm{d}\mathbf{r}_2 G\left(\mathbf{r},\mathbf{r}_1\right) G^*\left(\mathbf{r}',\mathbf{r}_2\right) \times \left\langle \Delta V\left(\mathbf{r}_1\right) \Delta V\left(\mathbf{r}_2\right)\right\rangle \rho_p\left(\mathbf{r}_1,\mathbf{r}_2\right) \tag{14.31}$$

which can be solved iteratively:

$$\begin{aligned}
\rho_p\left(\mathbf{r},\mathbf{r}'\right) \simeq\ & \rho_0\left(\mathbf{r},\mathbf{r}'\right) + e^2 \int \mathrm{d}\mathbf{r}_1 \int \mathrm{d}\mathbf{r}_2 \Big[G\left(\mathbf{r},\mathbf{r}_1\right) G^*\left(\mathbf{r}',\mathbf{r}_2\right) \\
& \times \left\langle \Delta V\left(\mathbf{r}_1\right) \Delta V\left(\mathbf{r}_2\right)\right\rangle \rho_0\left(\mathbf{r}_1,\mathbf{r}_2\right)\Big] \\
& + e^4 \int \mathrm{d}\mathbf{r}_1 \int \mathrm{d}\mathbf{r}_2 \int \mathrm{d}\mathbf{r}_3 \int \mathrm{d}\mathbf{r}_4 \\
& \times \Big[G\left(\mathbf{r},\mathbf{r}_1\right) G^*\left(\mathbf{r}',\mathbf{r}_2\right) G\left(\mathbf{r}_1,\mathbf{r}_3\right) G^*\left(\mathbf{r}_2,\mathbf{r}_4\right) \\
& \times \left\langle \Delta V\left(\mathbf{r}_1\right) \Delta V\left(\mathbf{r}_2\right)\right\rangle \\
& \times \left\langle \Delta V\left(\mathbf{r}_3\right) \Delta V\left(\mathbf{r}_4\right)\right\rangle \rho_0\left(\mathbf{r}_3,\mathbf{r}_4\right)\Big] + \ldots
\end{aligned} \tag{14.32}$$

This is the identical form of (14.10), therefore, the density matrix calculated based on the frozen lattice model is equal to the density matrix calculated following the phonon excitation model in quantum mechanics:

$$\rho_p(\mathbf{r}, \mathbf{r}') = \rho_f(\mathbf{r}, \mathbf{r}') \tag{14.33}$$

This identity holds for all of the orders and each order of the diffuse scattering. Therefore, the result calculated using the frozen lattice model is identical to the result calculated using phonon excitation model if the approximations made in the following two sections hold (Wang 1998a).

Based on rigorous Green's function and density matrix theories, it has been proved that the calculation based on the "frozen lattice model" for thermal diffuse scattering in electron diffraction gives an identical result as that obtained from the phonon excitation model if the following two conditions are satisfied: (1) the incoherence between different orders of thermal diffuse scattering is considered in the frozen lattice model calculation and (2) the specimen thickness and the mean free path length for phonon excitation both are smaller than the distance traveled by the electron within the lifetime of the phonon. Condition (2) is absolutely satisfied by both low- and high-energy electrons, and satisfying condition (1) is the most critical requirement in the frozen lattice model calculation. More specifically, the diffuse scattering produced by the different orders of diffuse scattering (such as ΔV^2, ΔV^4, and ΔV^6 or multiple diffuse scattering) must be treated incoherent in the frozen lattice model, and the higher order scattering is the result of multiple incoherent first-order scattering events. This incoherence can be precisely accounted for with the introduction of the mixed dynamic form factor $S(\mathbf{Q}, \mathbf{Q}')$, a key quantity for inelastic electron scattering. A new multislice approach has been proposed recently in which the multiple diffuse scattering and the incoherence effects are both accounted for using the $S(\mathbf{Q}, \mathbf{Q}')$ function (Wang 1998a). The conclusion, reached directly using the integral form of the Schrödinger equation without making the high-energy approximation or ignoring the backscattering, holds for each and all of the orders of diffuse scattering if they are treated incoherently, thus, the quantum mechanical basis of the frozen lattice model is established. The lattice model can also be used to calculate the diffuse diffraction of low-energy electron because their effective penetration depth is relatively small. This conclusion confirms the validity, reliability, and accuracy of using the frozen lattice model in numerous dynamical theories of phonon excitation in electron diffraction and imaging of thin specimens.

14.5 The "Absorption" Potential Introduced by Yoshioka

The "absorption" potential, or the so-called optical potential, is an important factor in quantitative dynamical calculations for electron diffraction and imaging. The introduction of optical potential in electron diffraction was over four decades ago, and it can account for the effect of inelastic scattering, such as plasmon excitation, atomic inner-shell excitation, and phonon (or thermal diffuse) scattering, on the elastically scattered wave. This potential is usually approximated as an imaginary function and it is responsible for the reduction of the Bragg-scattered intensity. In recently years, the commercialization of electron energy filter in transmission electron microscopy

(TEM) has greatly impacted quantitative electron microscopy, using which electrons with energy loss larger than a few eV can be removed from the image or diffraction pattern except the phonon-scattered electrons. It is therefore inevitable to involve phonon-scattered electron in data analysis, but the multiple phonon excitation theory in electron diffraction is not yet well established.

In this section, we try to answer three questions: (1) what is the nature of the optical potential? (2) what can it be used for? and (3) how can it be precisely calculated? These questions seem easy at the first sight, but in fact they have profound truth beyond conventional understanding because the approximation made for deriving the potential has never been re-examined. Using the formal inelastic excitation theory of Yoshioka, a rigorous proof is given to show that the high-order diffuse scattering can be recovered in the calculation using the first-order diffuse scattering equation, provided the correct optical potential is included in the calculation of the elastic wave. It is shown that the incoherence between different orders of diffuse scattering is automatically included with the use of the mixed dynamic form factor. The precise form of the optical potential is given and its calculation is proposed based on a Bloch wave-multislice approach.

The absorption potential in electron diffraction was formally introduced by Yoshioka in consideration of the effects of inelastic scattering on elastically scattered wave. Phonon excitation is an inelastic-scattering process that is most adequately described using Yoshioka's approach, in which the quantum nature of phonons and the incoherence effects are adequately treated. In this theory, the eigenstate of the incident electron-crystal system is expressed as a product of the eigenstate of the crystal with that of the incident electron.

The ground-state approximation simply means that we can ignore the transition among different excited states except the ground state. Equation (14.14d) is approximated as

$$\left[\left(-\frac{\hbar^2}{2m_0}\nabla^2 - eV_0 - E\right)\right]\Psi_n \approx -H_{n0}\Psi_0 \tag{14.34a}$$

This is actually the first-order diffuse scattering equation. We first make this approximation, then we will prove that the inclusion of an optical potential in (14.13a)–(14.13b) can recover all of the high-order scattering terms dropped here. Equation (14.34a) can be converted into an integral equation:

$$\Psi_n(\boldsymbol{r}) = -\int \mathrm{d}\boldsymbol{r}_1 G(\boldsymbol{r}, \boldsymbol{r}_1)\, H_{n0}(\boldsymbol{r}_1)\, \Psi_0(\boldsymbol{r}_1), \quad \text{for } n > 0 \tag{14.34b}$$

where G is Green's function satisfying

$$\left(-\frac{\hbar^2}{2m_0}\nabla^2 - eV_0 - E\right) G(\boldsymbol{r}, \boldsymbol{r}_1) = \delta(\boldsymbol{r} - \boldsymbol{r}_1) \tag{14.34c}$$

Substituting (14.34b) into (14.14a) yields

$$\left[\left(-\frac{\hbar^2}{2m_0}\nabla^2 - eV_0 - E\right)\right]\Psi_0(\boldsymbol{r}) = \int \mathrm{d}\boldsymbol{r}_1 G(\boldsymbol{r},\boldsymbol{r}_1)\left[\sum_{m=1} H_{0m}(\boldsymbol{r})H_{m0}(\boldsymbol{r}_1)\right]\Psi_0(\boldsymbol{r}_1) \tag{14.35}$$

where the term on the right-hand side is equivalent to a non-local "potential" V' defined by

$$\left[eV'\Psi_0\right] = \int \mathrm{d}_1 G(\boldsymbol{r},\boldsymbol{r}_1)\left[\sum_{m=1} H_{0m}(\boldsymbol{r})H_{m0}(\boldsymbol{r}_1)\right]\Psi_0(\boldsymbol{r}_1) \tag{14.36a}$$

The optical potential given by (14.36a) has two important characteristics. One the V' function is a non-local function, depending on the "history" of inelastic excitation events. The other is the enrollment of dynamical diffraction in V', so it is not, in general, simply a potential function.

To make V' a potential function independent of dynamic diffraction, Yoshioka made an approximation of replacing Green's function by its form in free space, thus

$$\left[eV'\Psi_0\right] \approx \frac{2m_0}{\hbar^2}\int \mathrm{d}\boldsymbol{r}_1 \frac{\exp\left(2\pi K_0\left|\boldsymbol{r}-\boldsymbol{r}_1\right|\right)}{4\pi\left|\boldsymbol{r}-\boldsymbol{r}_1\right|}\left[\sum_{m=1} H_{0m}(\boldsymbol{r})H_{m0}(\boldsymbol{r}_1)\right]\Psi_0(\boldsymbol{r}_1) \tag{14.36b}$$

This approximation was given in his paper but it has been forgotten in the literature, and it has been believed that (14.36b) is the "true" absorptive potential. The key of this approximation is that the dynamic diffraction effect is dropped in the calculation. On the other hand, Yoshioka's approximation is very reasonable if one is interested only in the elastically scattered electrons because once the electron is diffusely scattered out of the Bragg peaks, it is considered to be "absorbed" and there is no need to consider its subsequent diffraction behavior. The dynamic diffraction after the inelastic scattering can only change the angular distribution of the electrons, but has no effect on the total inelastic-scattering probability. It is sure that this approximation makes the numerical calculation much easier, but the question is what has been dropped by this approximation and what can the dropped effect be used for? This question will be answered in Sect. 14.6.

14.6 The Optical Potential for Recovering Multiple Diffuse Scattering

From the quantum mechanical transition theory, the transition out of the elastic state (or ground state) is characterized by a matrix element H_{n0}, and the transition back to the elastic state is by H_{0n}, and they are correlated by $H_{n0} = H_{0n}^*$. This means that the mechanism that determines the intensity loss of the elastic wave must be directly correlated to the intensity gaining process in the excited state, because the total scattering intensity of elastic and inelastic waves must be conserved. Therefore, the optical potential that represents the effect of inelastic excitation on the elastic wave could be applied, in principle, for recovering the multiple diffuse scattering process in a practical calculation. This is possible only if the total scattering governed by (14.34a) and (14.35) is conserved (see Sect. 14.6.1).

To transform the physical argument above into mathematical expressions, we introduce a density matrix that can be applied conveniently for calculating either images or diffraction patterns:

$$\rho\left(\boldsymbol{r},\boldsymbol{r}'\right)=\sum_{n=0}\Psi_n(\boldsymbol{r})\Psi_n^*\left(\boldsymbol{r}'\right) \tag{14.37}$$

where the sum of n is over all of the crystal states including the elastic-scattering state (ground state). The electron diffraction pattern including TDS can be calculated by taking a double Fourier transform of the density matrix:

$$I(\tau)=\int \mathrm{d}\boldsymbol{r}\int \mathrm{d}\boldsymbol{r}'\exp\left[-2\pi\mathrm{i}\tau\cdot\left(\boldsymbol{r}-\boldsymbol{r}'\right)\right]\rho\left(\boldsymbol{r},\boldsymbol{r}'\right) \tag{14.38}$$

The intensity distribution in the electron image can be calculated by

$$I(\boldsymbol{r})=\int \mathrm{d}\boldsymbol{r}_1\int \mathrm{d}\boldsymbol{r}_2\,\rho\left(\boldsymbol{r}_1,\boldsymbol{r}_2\right)t\left(\boldsymbol{r}-\boldsymbol{r}_1\right)t^*\left(\boldsymbol{r}-\boldsymbol{r}_2\right) \tag{14.39}$$

where $t(\boldsymbol{r})$ characterizes the information transfer property of the optical system in an electron microscope. For the convenience of the discussion, we introduce a real-space dynamic form factor defined by Wang (1996a), Kohl and Rose (1985)

$$\begin{aligned}s\left(\boldsymbol{r},\boldsymbol{r}_1\right)&=\frac{1}{e^2}\sum_{m=1}H_{0m}(\boldsymbol{r})H_{m0}\left(\boldsymbol{r}_1\right)\\&=\int \mathrm{d}\boldsymbol{Q}\int \mathrm{d}\boldsymbol{Q}'\exp\left[2\pi\mathrm{i}\left(\boldsymbol{r}\cdot\boldsymbol{Q}-\boldsymbol{r}_1\cdot\boldsymbol{Q}'\right)\right]S\left(\boldsymbol{Q},\boldsymbol{Q}'\right)\end{aligned} \tag{14.40}$$

The calculation of $s\left(\boldsymbol{r},\boldsymbol{r}_1\right)$ is given in Sect. 14.6.2. Equation (14.35) can be converted into an integral equation:

$$\begin{aligned}\Psi_0(\boldsymbol{r})&=\Psi_0^{(0)}(\boldsymbol{r})\\&\quad+e^2\int \mathrm{d}\boldsymbol{r}_1G\left(\boldsymbol{r},\boldsymbol{r}_1\right)\int \mathrm{d}\boldsymbol{r}_2\,G\left(\boldsymbol{r}_1,\boldsymbol{r}_2\right)s\left(\boldsymbol{r}_1,\boldsymbol{r}_2\right)\Psi_0\left(\boldsymbol{r}_2\right)\end{aligned} \tag{14.41}$$

where $\Psi_0^{(0)}(\boldsymbol{r})$ is the elastic wave without the influence of the V' potential:

$$\left(-\frac{\hbar^2}{2m_0}\nabla^2 - eV_0 - E\right)\Psi_0^{(0)}(r) = 0. \tag{14.42}$$

Equation (14.41) can be solved iteratively

$$\begin{aligned}\Psi_0(\boldsymbol{r}) =& \Psi_0^{(0)}(\boldsymbol{r}) \\ &+ e^2 \int \mathrm{d}\boldsymbol{r}_1 \int \mathrm{d}\boldsymbol{r}_2 G(\boldsymbol{r}, \boldsymbol{r}_2) G(\boldsymbol{r}_1, \boldsymbol{r}_2) s(\boldsymbol{r}_1, \boldsymbol{r}_2) \Psi_0^{(0)}\boldsymbol{r}(\boldsymbol{r}_2) \\ &+ e^4 \int \mathrm{d}\boldsymbol{r}_1 \int \mathrm{d}\boldsymbol{r}_2 \int \mathrm{d}\boldsymbol{r}_3 \int \mathrm{d}\boldsymbol{r}_4 \\ &\times G(\boldsymbol{r}, \boldsymbol{r}_1) G(\boldsymbol{r}_1, \boldsymbol{r}_2) G(\boldsymbol{r}_2, \boldsymbol{r}_3) G(\boldsymbol{r}_3, \boldsymbol{r}_4) \\ &\times s(\boldsymbol{r}_1, \boldsymbol{r}_2) s(\boldsymbol{r}_3, \boldsymbol{r}_4) \Psi_0^{(0)}(\boldsymbol{r}_4) + \ldots\end{aligned} \tag{14.43}$$

Substituting (14.41) and (14.34b) into (14.37) yields

$$\begin{aligned}\rho(\boldsymbol{r}, \boldsymbol{r}') =& \Psi_0(\boldsymbol{r})\Psi_0^*(\boldsymbol{r}') \\ &+ e^2 \int \mathrm{d}\boldsymbol{r}_1 \int \mathrm{d}\boldsymbol{r}_2 G(\boldsymbol{r}, \boldsymbol{r}_1) G^*(\boldsymbol{r}', \boldsymbol{r}_2) \\ &\times s(\boldsymbol{r}_1, \boldsymbol{r}_2) \psi_0(\boldsymbol{r}_1) \Psi_0^*(\boldsymbol{r}_2)\end{aligned} \tag{14.44}$$

From the proof given in Sect. 14.6.3, a substitution of (14.43) into (14.44) recovers all the orders of diffuse scattering terms. Therefore, by inclusion of the optical potential in the dynamical calculation of the elastic wave (14.35), the high-order diffuse scatterings are fully recovered in the calculations made using the equation for the first-order diffuse scattering (14.34a), and, more importantly, the incoherence in phonon excitation is considered properly. This conclusion agrees completely with that obtained previously based on the "frozen" lattice model for TDS and the system containing short-range point defects such as oxygen vacancies (Wang 1996b, c). This is a rather important conclusion which establishes the basis for generalizing the existing first-order scattering theories for calculating high-order diffuse scattering. The density matrix theory is unique to treat the incoherence between different orders of diffuse scattering.

In conventional dynamical diffraction, inclusion of Debye-Waller factor and the optical potential accounts only for the effect of diffuse scattering on the elastic wave. This task is included in the first term at the right-hand side of (14.44), the elastic wave component. The calculations result in Bragg reflections. One must perform the calculation of the second term in order to include the contributions made by the diffusely scattered electrons.

We now come back to the question raised at the end of Sect. 2: what is missing if Green's function is replaced by its form in free space? In (14.44), the $\Psi_0(\boldsymbol{r}_1)$ function is responsible for exciting the diffuse scattering at $\boldsymbol{r}_1$, the $\Psi_0(\boldsymbol{r}_2)$ function is responsible for exciting the diffuse scattering at $\boldsymbol{r}_2$, while the strength of the excitation is determined by a "probability" and spatial correlation (or coherence) function

$s(\boldsymbol{r}_1, \boldsymbol{r}_2)$. The dynamic diffraction of the inelastic wave from its emission point $\boldsymbol{r}_1$ to the observation point $\boldsymbol{r}$ is characterized by Green's function $G(\boldsymbol{r}, \boldsymbol{r}_1)$. If $G(\boldsymbol{r}, \boldsymbol{r}_1)$ is replaced by its form in free space, it means that the subsequent dynamical diffraction of the emitted inelastic wave is ignored, while the multiple diffuse scattering is still preserved. In other words, the calculation covers the dynamic diffraction before the first inelastic event and the multiple inelastic scattering but not Bragg reflections after and between the inelastic events. This is the consequence of Yoshioka's approximation. Unfortunately, this approximation has rarely been re-examined since it was proposed in 1957, and many calculations were performed with the inclusion of the optical potential, but the consequence and real nature of the optical potential were ambiguous. It is hoped that this paper will clear it up.

14.6.1 Conservation of Total Intensity

The numerical calculation is based on (14.34a) and (14.35), which are

$$\left[\left(-\frac{\hbar^2}{2m_0}\nabla^2 - eV_0 - E\right)\right]\Psi_n \approx -H_{n0}\Psi_0, \tag{14.45}$$

or in the integral form

$$\Psi_n(\boldsymbol{r}) = -\int \mathrm{d}\boldsymbol{r}_1 G(\boldsymbol{r}, \boldsymbol{r}_1)\, H_{n0}(\boldsymbol{r}_1)\, \Psi_0(\boldsymbol{r}_1), \tag{14.46}$$

and

$$\left[\left(-\frac{\hbar^2}{2m_0}\nabla^2 - eV_0 - E\right)\right]\Psi_0(\boldsymbol{r}) = e^2\int \mathrm{d}\boldsymbol{r}_1 G(\boldsymbol{r}, \boldsymbol{r}_1)\, s(\boldsymbol{r}, \boldsymbol{r}_1)\, \Psi_0(\boldsymbol{r}_1) \tag{14.47}$$

The conservation of total scattering intensity is proved using the current density defined in quantum mechanics

$$\int_\Sigma \mathrm{d}\boldsymbol{S}\cdot\boldsymbol{j} = \frac{\hbar}{2\mathrm{i}m_0}\int_{\mathrm{vol}} \mathrm{d}\boldsymbol{r} \sum_{n=0} \left[\Psi_n^*\nabla^2\Psi_n - \Psi_n\nabla^2\Psi_n^*\right] \tag{14.48}$$

where the integral is over the crystal volume. From (14.45) and using (14.46)

$$\begin{aligned}
\sum_{n=1}\left[\Psi_n^*\nabla^2\Psi_n - \Psi_n\nabla^2\Psi_n^*\right] &= \frac{2m_0}{\hbar^2}\sum_{n=1}\left[\Psi_n^* H_{n0}\Psi_0 - \Psi_n H_{n0}^*\Psi_0^*\right] \\
&= -\frac{2m_0}{\hbar^2}\sum_{n=1}\Bigg[\int \mathrm{d}\boldsymbol{r}_1 G^*\left(\boldsymbol{r},\boldsymbol{r}_1\right) H_{n0}^*\left(\boldsymbol{r}_1\right)\Psi_0\left(\boldsymbol{r}_1\right) H_{n0}(\boldsymbol{r})\Psi_0(\boldsymbol{r}) \\
&\quad - \int \mathrm{d}\boldsymbol{r}_1 G\left(\boldsymbol{r},\boldsymbol{r}_1\right) H_{n0}\left(\boldsymbol{r}_1\right)\Psi_0\left(\boldsymbol{r}_1\right) H_{n0}^*(\boldsymbol{r})\Psi_0^*(\boldsymbol{r})\Bigg] \\
&= -e^2\frac{2m_0}{\hbar^2}\Bigg[\int \mathrm{d}\boldsymbol{r}_1 G^*\left(\boldsymbol{r},\boldsymbol{r}_1\right) s^*\left(\boldsymbol{r},\boldsymbol{r}_1\right)\Psi_0\left(\boldsymbol{r}_1\right)\Psi_0(\boldsymbol{r}) \\
&\quad - \int \mathrm{d}\boldsymbol{r}_1 G\left(\boldsymbol{r},\boldsymbol{r}_1\right) s\left(\boldsymbol{r},\boldsymbol{r}_1\right)\psi_0\left(\boldsymbol{r}_1\right)\Psi_0^*(\boldsymbol{r})\Bigg] \\
&= -e\frac{2m_0}{\hbar^2}\left\{\left[V'\Psi_0(\boldsymbol{r})\right]^*\psi_0(\boldsymbol{r}) - \left[V'\Psi_0(\boldsymbol{r})\right]\Psi_0^*(\boldsymbol{r})\right\}
\end{aligned} \tag{14.49}$$

On the other hand, the following calculation can be given starting from (14.47)

$$\begin{aligned}
\left[\Psi_0^*\nabla^2\Psi_0 - \Psi_0\nabla^2\Psi_0^*\right] &= -\frac{2m_0}{\hbar^2}\left[\Psi_n^* H_{n0}\Psi_0 - \Psi_n H_{n0}^*\Psi_0^*\right] \\
&= -e^2\frac{2m_0}{\hbar^2}\Bigg[\int \mathrm{d}\boldsymbol{r}_1 G\left(\boldsymbol{r},\boldsymbol{r}_1\right) s\left(\boldsymbol{r},\boldsymbol{r}_1\right)\Psi_0^*(\boldsymbol{r})\Psi_0\left(\boldsymbol{r}_1\right) \\
&\quad - \int \mathrm{d}\boldsymbol{r}_1 G^*\left(\boldsymbol{r},\boldsymbol{r}_1\right) s^*\left(\boldsymbol{r},\boldsymbol{r}_1\right)\Psi_0\left(\boldsymbol{r}_1\right)\Psi_0^*\left(\boldsymbol{r}_1\right)\Bigg] \\
&= -e\frac{2m_0}{\hbar^2}\left\{\left[V'\Psi_0(\boldsymbol{r})\right]\psi_0(\boldsymbol{r}) - \left[V'\Psi_0(\boldsymbol{r})\right]^*\Psi_0(\boldsymbol{r})\right\}
\end{aligned} \tag{14.50}$$

A sum of (14.49) and (14.50) gives

$$\int_{\Sigma} \mathrm{d}\boldsymbol{S}\cdot\boldsymbol{j} = 0$$

which means the incoming intensity is equal to the outgoing intensity, the intensity is conserved.

14.6.2 Full Recovery of All High-Order Scattering

From (14.43), one can perform the following calculation:

$$\begin{aligned}\Psi_0(\boldsymbol{r})\Psi_0^*\left(\boldsymbol{r}'\right) = &\ \Psi_0^{(0)}(\boldsymbol{r})\Psi_0^{(0)*}\left(\boldsymbol{r}'\right)\\ &+ e^2\Psi_0^{(0)*}\left(\boldsymbol{r}'\right)\int \mathrm{d}\boldsymbol{r}_1 \int \mathrm{d}\boldsymbol{r}_2\, G\left(\boldsymbol{r},\boldsymbol{r}_1\right) G\left(\boldsymbol{r}_1,\boldsymbol{r}_2\right) s\left(\boldsymbol{r}_1,\boldsymbol{r}_2\right) \Psi_0^{(0)}\left(\boldsymbol{r}_2\right)\\ &+ e^2\Psi_0^{(0)}(\boldsymbol{r})\int \mathrm{d}\boldsymbol{r}_1 \int \mathrm{d}\boldsymbol{r}_2\, G^*\left(\boldsymbol{r}',\boldsymbol{r}_1\right) G^*\left(\boldsymbol{r}_1,\boldsymbol{r}_2\right) s^*\left(\boldsymbol{r}_1,\boldsymbol{r}_2\right) \Psi_0^{(0)*}\left(\boldsymbol{r}_2\right)\\ &+ e^4\Psi_0^{(0)*}\left(\boldsymbol{r}'\right)\int \mathrm{d}\boldsymbol{r}_1 \int \mathrm{d}\boldsymbol{r}_2 \int \mathrm{d}\boldsymbol{r}_3 \int \mathrm{d}\boldsymbol{r}_4\, G\left(\boldsymbol{r},\boldsymbol{r}_1\right) G\left(\boldsymbol{r}_1,\boldsymbol{r}_2\right)\\ &\times G\left(\boldsymbol{r}_2,\boldsymbol{r}_3\right) G\left(\boldsymbol{r}_3,\boldsymbol{r}_4\right) s\left(\boldsymbol{r}_1,\boldsymbol{r}_2\right) s\left(\boldsymbol{r}_3,\boldsymbol{r}_4\right) \Psi_0^{(0)}\left(\boldsymbol{r}_4\right)\\ &+ e^4\Psi_0^{(0)}(\boldsymbol{r})\int \mathrm{d}\boldsymbol{r}_1 \int \mathrm{d}\boldsymbol{r}_2 \int \mathrm{d}\boldsymbol{r}_3 \int \mathrm{d}\boldsymbol{r}_4\, G^*\left(\boldsymbol{r}',\boldsymbol{r}_1\right) G^*\left(\boldsymbol{r}_1,\boldsymbol{r}_2\right)\\ &\times G^*\left(\boldsymbol{r}_2,\boldsymbol{r}_3\right) G^*\left(\boldsymbol{r}_3,\boldsymbol{r}_4\right) s^*\left(\boldsymbol{r}_1,\boldsymbol{r}_2\right) s^*\left(\boldsymbol{r}_3,\boldsymbol{r}_4\right) \Psi_0^{(0)*}\left(\boldsymbol{r}_4\right) + \ldots\end{aligned} \tag{14.51}$$

In (14.44), the first term represents the elastically scattered wave and the second term is the diffuse scattering term(s). We preserve the form of the Bragg reflections that is given in (14.51) and substitute (14.51) into the second term of (14.44), which gives

$$\begin{aligned}\rho(\boldsymbol{r},\boldsymbol{r}) = &\ \Psi_0(\boldsymbol{r})\Psi_0^*\left(\boldsymbol{r}'\right)\\ &+ e^2\int \mathrm{d}\boldsymbol{r}_1 \int \mathrm{d}\boldsymbol{r}_2\, G\left(\boldsymbol{r},\boldsymbol{r}_1\right) G^*\left(\boldsymbol{r}',\boldsymbol{r}_2\right) s\left(\boldsymbol{r}_1,\boldsymbol{r}_2\right) \Psi_0^{(0)*}\left(\boldsymbol{r}_2\right) \Psi_0^{(0)}\left(\boldsymbol{r}_1\right)\\ &+ e^4\int \mathrm{d}\boldsymbol{r}_1 \int \mathrm{d}\boldsymbol{r}_2 \int \mathrm{d}\boldsymbol{r}_3 \int \mathrm{d}\boldsymbol{r}_4\, G\left(\boldsymbol{r},\boldsymbol{r}_1\right) G^*\left(\boldsymbol{r}',\boldsymbol{r}_2\right)\\ &\times G\left(\boldsymbol{r}_1,\boldsymbol{r}_3\right) G\left(\boldsymbol{r}_3,\boldsymbol{r}_4\right) s\left(\boldsymbol{r}_1,\boldsymbol{r}_2\right) s\left(\boldsymbol{r}_3,\boldsymbol{r}_4\right) \Psi_0^{(0)*}\left(\boldsymbol{r}_2\right) \Psi_0^{(0)}\left(\boldsymbol{r}_4\right)\\ &+ e^4\int \mathrm{d}\boldsymbol{r}_1 \int \mathrm{d}\boldsymbol{r}_2 \int \mathrm{d}\boldsymbol{r}_3 \int \mathrm{d}\boldsymbol{r}_4\, G\left(\boldsymbol{r},\boldsymbol{r}_1\right) G^*\left(\boldsymbol{r}',\boldsymbol{r}_2\right)\\ &\times G^*\left(\boldsymbol{r}_2,\boldsymbol{r}_3\right) G^*\left(\boldsymbol{r}_3,\boldsymbol{r}_4\right) s\left(\boldsymbol{r}_1,\boldsymbol{r}_2\right) s^*\left(\boldsymbol{r}_3,\boldsymbol{r}_4\right) \Psi_0^{(0)}\left(\boldsymbol{r}_1\right) \Psi_0^{(0)*}\left(\boldsymbol{r}_4\right)\\ &+ e^6\int \mathrm{d}\boldsymbol{r}_1 \int \mathrm{d}\boldsymbol{r}_2 \int \mathrm{d}\boldsymbol{r}_3 \int \mathrm{d}\boldsymbol{r}_4 \int \mathrm{d}\boldsymbol{r}_5 \int \mathrm{d}\boldsymbol{r}_6\, G\left(\boldsymbol{r},\boldsymbol{r}_1\right) G^*\left(\boldsymbol{r}',\boldsymbol{r}_2\right)\\ &\times G\left(\boldsymbol{r}_2,\boldsymbol{r}_3\right) G\left(\boldsymbol{r}_3,\boldsymbol{r}_4\right) G\left(\boldsymbol{r}_4,\boldsymbol{r}_5\right) G\left(\boldsymbol{r}_5,\boldsymbol{r}_6\right)\\ &\times s\left(\boldsymbol{r}_1,\boldsymbol{r}_2\right) s\left(\boldsymbol{r}_3,\boldsymbol{r}_4\right) s\left(\boldsymbol{r}_5,\boldsymbol{r}_6\right)\\ &\times \Psi_0^{(0)*}\left(\boldsymbol{r}_2\right) \Psi_0^{(0)}\left(\boldsymbol{r}_6\right)\\ &+ e^6\int \mathrm{d}\boldsymbol{r}_1 \int \mathrm{d}\boldsymbol{r}_2 \int \mathrm{d}\boldsymbol{r}_3 \int \mathrm{d}\boldsymbol{r}_4 \int \mathrm{d}\boldsymbol{r}_5 \int \mathrm{d}\boldsymbol{r}_6\, G\left(\boldsymbol{r},\boldsymbol{r}_1\right) G^*\left(\boldsymbol{r}',\boldsymbol{r}_2\right)\\ &\times G^*\left(\boldsymbol{r}_2,\boldsymbol{r}_3\right) G^*\left(\boldsymbol{r}_3,\boldsymbol{r}_4\right) G^*\left(\boldsymbol{r}_4,\boldsymbol{r}_5\right) G^*\left(\boldsymbol{r}_5,\boldsymbol{r}_6\right)\\ &\times s\left(\boldsymbol{r}_1,\boldsymbol{r}_2\right) s^*\left(\boldsymbol{r}_3,\boldsymbol{r}_4\right) s^*\left(\boldsymbol{r}_5,\boldsymbol{r}_6\right)\\ &\times \Psi_0^{(0)}\left(\boldsymbol{r}_1\right) \Psi_0^{(0)*}\left(\boldsymbol{r}_6\right) + \ldots\end{aligned} \tag{14.52}$$

where the first term is the result of Bragg scattering, the second term is the first-order diffuse scattering, the third and the fourth terms are the second-order diffuse scattering, and the fourth and the sixth terms are the third-order diffuse scattering. The split of the third and the fourth (or the fifth and the sixth) terms is due to the different variables $\boldsymbol{r}$ and $\boldsymbol{r}'$ introduced in the density matrix. All of the higher order terms are included if one continues the iterative calculation.

14.6.3 Theory Under the Small-Angle Approximation

In a practical calculation, one of the keys is to solve (14.35). The first step is to find Green's function. For a general case, Green's function has been given by Dudarev et al. (1994), but the general solution involves a huge amount of computation and it simply becomes impractical. In this section, we introduce an approximated form of Green's function for high-energy electron under the small-angle scattering approximation, which can be reasonably computed in practice. We consider the first-order inelastic-scattering equation as a start:

$$\left[-\frac{\hbar^2}{2m_0}\nabla^2 - eV_0 - E\right]\Psi_n = -H_{n0}\Psi_0 \tag{14.53a}$$

which can be converted into an integral equation with the use of Green's function:

$$\Psi_n(\boldsymbol{r}) = -\int \mathrm{d}\boldsymbol{r}_1 G\left(\boldsymbol{r}, \boldsymbol{r}_1\right) H_{n0}\left(\boldsymbol{r}_1\right) \Psi_0\left(\boldsymbol{r}_1\right) \tag{14.53b}$$

Alternatively, (14.53b) can be directly obtained from (14.34a) using the real-space multislice theory, which was initiated by Van Dyck (1985) and has lately been developed by Chen et al. (1997). Figure 14.2 shows a schematic diagram of a thin foil in TEM, where the incident beam strikes the surface at a small angle. The z-axis is chosen downward normal to the specimen surface. If the wave function is written in a form $\Psi(\boldsymbol{r}) = \exp\left(2\pi \mathrm{i}\boldsymbol{K}_0 \cdot \boldsymbol{r}\right)\Phi(\boldsymbol{r})$, where $\boldsymbol{K}_0$ is the wave vector of the incident beam, the $\partial^2\Phi/\partial z^2$ term can be dropped under the small-angle scattering approximation, hence

$$\frac{\partial \Phi_n(\boldsymbol{r})}{\partial_z} \approx \varsigma\left[\Xi + \frac{2m_0}{\hbar^2}eV_0\right]\Phi_n(\boldsymbol{r}) - \frac{2m_0}{\hbar^2}H_{n0}(\boldsymbol{r})\Phi_0(\boldsymbol{r}) \tag{14.54a}$$

where $\varsigma = \mathrm{i}/4\pi K_{0z}$,

$$\Xi = \frac{\partial^2}{\partial x^2} + \frac{\partial^2}{\partial y^2} + 4\pi\mathrm{i}\left(K_{0x}\frac{\partial}{\partial y} + K_{0y}\frac{\partial}{\partial y}\right) \tag{14.54b}$$

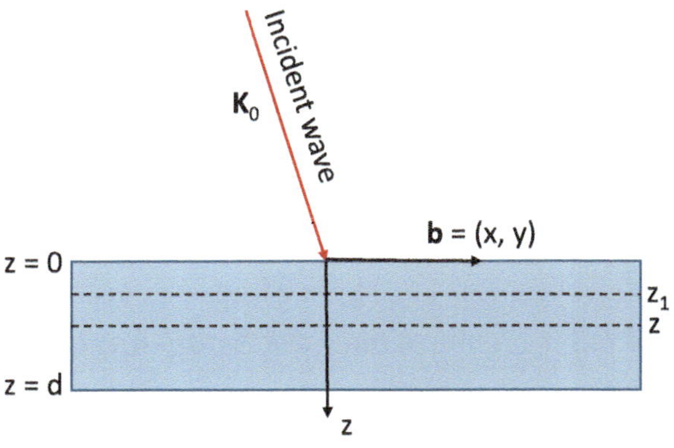

Fig. 14.2 A schematic model showing quantum transition in crystal state under the ground state approximation

is an operator, and $\boldsymbol{b} = (x, y)$ is the coordinate in the image plane. For practical HRTEM dominated by the reflections in the zero-order Laue zone, the crystal potential V_0 is effectively independent of z. Thus the projected potential approximation holds. The solution of (14.54a) can be written in the form:

$$\Phi_n(\boldsymbol{r}) = \xi \int_0^z \mathrm{d}z' O_\mathrm{p}\left(\boldsymbol{b}, z - z'\right)\left[H_{n0}\left(\boldsymbol{b}, z'\right)\Phi_0\left(\boldsymbol{b}, z'\right)\right] \tag{14.55a}$$

where the operator O_p is defined as

$$O_p\left(\boldsymbol{b}, z - z_1\right) = \exp\left\{\varsigma\left[\Xi + \frac{2m_0}{\hbar^2} e V_0\right](z - z_1)\right\} \tag{14.55b}$$

and $\xi = -\left(2m_0/\hbar^2\right)\varsigma$. The operator O_p represents the consecutive phase grating and Fresnel propagation calculations in the multislice theory (Van Dyck 1985).

To match the form of (14.53b), we add the plane wave component back to (14.55a) and perform the following calculation using an identity:

$$\begin{aligned} &\int \mathrm{d}\boldsymbol{r}_1\, \delta\left(\boldsymbol{b} - \boldsymbol{b}_1\right)\delta\left(z_1 - z'\right) H_{n0}\left(\boldsymbol{b}_1, z_1\right)\Phi_0\left(\boldsymbol{b}_1, z_1\right) \\ &\quad = H_{n0}\left(\boldsymbol{b}, z'\right)\Phi_0\left(\boldsymbol{b}, z'\right), \end{aligned} \tag{14.56}$$

$$
\begin{aligned}
\Psi_n(\boldsymbol{r}) &= \exp\left(2\pi \mathrm{i}\boldsymbol{K}_0\cdot\boldsymbol{r}\right)\xi\int_0^z \mathrm{d}z' O_p\left(\boldsymbol{b}, z-z'\right)\\
&\quad\times\left[\int \mathrm{d}\boldsymbol{r}_1\delta\left(\boldsymbol{b}-\boldsymbol{b}_1\right)\delta\left(z_1-z'\right)H_{n0}\left(\boldsymbol{r}_1\right)\exp\left(-2\pi\mathrm{i}\boldsymbol{K}_0\cdot\boldsymbol{r}_1\right)\Psi_0\left(\boldsymbol{r}_1\right)\right]\\
&= \xi\exp\left(2\pi\mathrm{i}\boldsymbol{K}_0\cdot\boldsymbol{r}\right)\int \mathrm{d}\boldsymbol{r}_1 O_p\left(\boldsymbol{b}, z-z_1\right)\delta\left(\boldsymbol{b}-\boldsymbol{b}_1\right)\\
&\quad\times\exp\left(-2\pi\mathrm{i}\boldsymbol{K}_0\cdot\boldsymbol{r}_1\right)H_{n0}\left(\boldsymbol{r}_1\right)\Psi_0\left(\boldsymbol{r}_1\right)
\end{aligned}
\tag{14.57}
$$

A comparison of (14.57) with (14.53b) gives Green's function:

$$
\begin{aligned}
G\left(\boldsymbol{r}, \boldsymbol{r}_1\right) &\approx -\xi\exp\left[2\pi\mathrm{i}\boldsymbol{K}_0\cdot\left(\boldsymbol{r}-\boldsymbol{r}_1\right)\right]O_p\left(\boldsymbol{b}, z-z_1\right)\delta\left(\boldsymbol{b}-\boldsymbol{b}_1\right)\\
&= -\xi\exp\left[2\pi\mathrm{i}\boldsymbol{K}_0\cdot\left(\boldsymbol{r}-\boldsymbol{r}_1\right)\right]\int_0^z \mathrm{d}z' O_p\left(\boldsymbol{b}, z-z'\right)\\
&\quad\times\left\{\delta\left(\boldsymbol{b}-\boldsymbol{b}_1\right)\delta\left(z'-z_1\right)\right\}, \quad \text{for } z > z_1
\end{aligned}
\tag{14.58}
$$

For the convenience of the proof, the plane wave component is separated by defining

$$
G\left(\boldsymbol{r}, \boldsymbol{r}_1\right) \approx \exp\left[2\pi\mathrm{i}\boldsymbol{K}_0\cdot\left(\boldsymbol{r}-\boldsymbol{r}_1\right)\right]\bar{g}\left(\boldsymbol{r}, \boldsymbol{r}_1\right), \tag{14.59}
$$

thus for $z > z_1$ in the forward scattering

$$
\begin{aligned}
\bar{g}\left(\boldsymbol{r}, \boldsymbol{r}_1\right) &= -\xi O_p\left(\boldsymbol{b}, z-z_1\right)\left\{\delta\left(\boldsymbol{b}-\boldsymbol{b}_1\right)\right\}\\
&= -\xi\int_0^z \mathrm{d}z'\, O_p\left(\boldsymbol{b}, z-z'\right)\left\{\delta\left(\boldsymbol{b}-\boldsymbol{b}_1\right)\delta\left(z'-z_1\right)\right\}
\end{aligned}
\tag{14.60}
$$

A direct differentiation of $\bar{g}\left(\boldsymbol{r}, \boldsymbol{r}_1\right)$ yields

$$
\begin{aligned}
\frac{\partial\bar{g}\left(\boldsymbol{r}, \boldsymbol{r}_1\right)}{\partial z} &= -\varsigma\left[\Xi+\frac{2m_0}{\hbar^2}eV_0\right]\xi\int_0^z \mathrm{d}z'\, O_p\left(\boldsymbol{b}, z-z'\right)\\
&\quad\times\left\{\delta\left(\boldsymbol{b}-\boldsymbol{b}_1\right)\delta\left(z'-z_1\right)\right\}-\xi\left\{\delta\left(\boldsymbol{b}-\boldsymbol{b}_1\right)\delta\left(z-z_1\right)\right\}\\
&= \varsigma\left[\Xi+\frac{2m_0}{\hbar^2}eV_0\right]\bar{g}\left(\boldsymbol{r}, \boldsymbol{r}_1\right)-\xi\delta\left(\boldsymbol{r}-\boldsymbol{r}_1\right)
\end{aligned}
\tag{14.61}
$$

This is just the equation satisfied by $\bar{g}\left(\boldsymbol{r}, \boldsymbol{r}_1\right)$ under the small-angle forward scattering approximation. Therefore, Green's function satisfies (14.34c) under the small-angle scattering approximation.

The accurate solution of Green's function is independent of $\boldsymbol{K}_0$. The scattering from a point source should be more symmetric. On the other hand, our observation

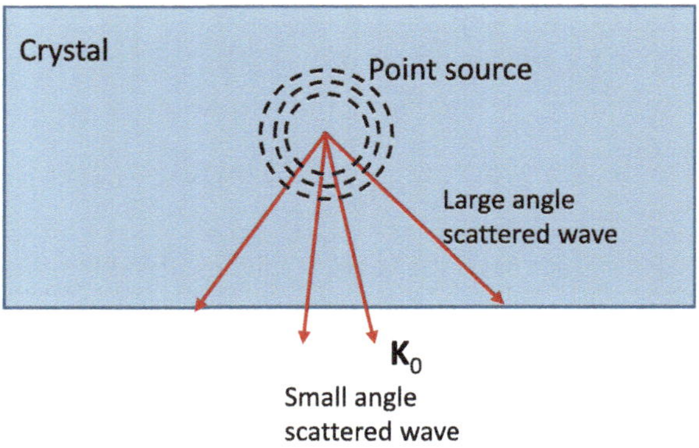

Fig. 14.3 A schematic model showing the spherical wave emitted from a point source in the crystal and the small-angle components near $\boldsymbol{K}_0$ that are likely to contribute to the observed image/diffraction pattern in TEM. The waves distributed at high-angle range make little contribution to the observed image if the backscattering is ignored for high-energy electrons

is only at the small-angle range, the waves scattered to high angles are unlikely to contribute to the image formed by lower angle Bragg reflections (see Fig. 14.3). Thus, the part of Green's function with appreciable contribution to the observed wave is confined within the nearby region of the incident beam $\boldsymbol{K}_0$. This is likely to be the reason for Green's function given in (14.58) to depend on the incident beam direction under the small-angle approximation. A substitution of (14.58) into (14.44) leads to

$$\begin{aligned}
\rho\left(\boldsymbol{r}, \boldsymbol{r}'\right) = {} & \Psi_0(\boldsymbol{r})\Psi_0^*\left(\boldsymbol{r}'\right) \\
& + e^2\xi^2 \int \mathrm{d}\boldsymbol{Q} \int \mathrm{d}\boldsymbol{Q}' S\left(\boldsymbol{Q}, \boldsymbol{Q}'\right) \int \mathrm{d}\boldsymbol{r}_1 \int \mathrm{d}\boldsymbol{r}_2 \exp\left[2\pi \mathrm{i}\boldsymbol{r}_1 \cdot \boldsymbol{Q}\right] \\
& \times \exp\left[-2\pi \mathrm{i}\boldsymbol{r}_2 \cdot \boldsymbol{Q}'\right] \exp\left[2\pi \mathrm{i}\boldsymbol{K}_0 \cdot \left(\boldsymbol{r} - \boldsymbol{r}'\right)\right] \\
& \times \int_0^z \mathrm{d}z' O_p\left(\boldsymbol{b}, z - z'\right)\left\{\delta\left(\boldsymbol{b} - \boldsymbol{b}_1\right)\delta\left(z' - z_1\right)\right\} \\
& \times \int_0^z \mathrm{d}z'' O_p^*\left(\boldsymbol{b}', z - z''\right)\left\{\delta\left(\boldsymbol{b}' - \boldsymbol{b}_2\right)\delta\left(z'' - z_2\right)\right\} \Phi_0\left(\boldsymbol{r}_1\right)\Phi_0^*\left(\boldsymbol{r}_2\right) \\
= {} & \Psi_0(\boldsymbol{r})\Psi_0^*\left(\boldsymbol{r}'\right) + e^2\xi^2 \int \mathrm{d}\boldsymbol{Q} \int \mathrm{d}\boldsymbol{Q}' \Big[S\left(\boldsymbol{Q}, \boldsymbol{Q}'\right) \\
& \times \int_0^z \mathrm{d}z' \Big\{ O_p\left(\boldsymbol{b}, z - z'\right) \Big[\exp\left[2\pi \mathrm{i}\left(\boldsymbol{b} \cdot \boldsymbol{Q}_b + z' Q_z\right)\right] \Phi_0\left(\boldsymbol{b}, z'\right) \Big] \Big\} \\
& \times \int_0^z \mathrm{d}z'' \Big\{ O_p^*\left(\boldsymbol{b}', z - z''\right)
\end{aligned}$$

$$\times\left[\exp\left[-2\pi i\left(\boldsymbol{b}'\cdot\boldsymbol{Q}_b'+z''Q_z'\right)\right]\Phi_0^*\left(\boldsymbol{b}',z''\right)\right]\Big\}\Big]$$
$$=\Psi_0(\boldsymbol{r})\Psi_0^*\left(\boldsymbol{r}'\right)$$
$$\times+e^2\xi^2\int d\boldsymbol{Q}\int d\boldsymbol{Q}'S\left(\boldsymbol{Q},\boldsymbol{Q}'\right)\psi(\boldsymbol{Q},\boldsymbol{b},z)\psi^*\left(\boldsymbol{Q}',\boldsymbol{b}',z\right) \quad (14.62a)$$

where

$$\psi(\boldsymbol{Q},\boldsymbol{b},z)=\int_0^z dz'\left\{O_p\left(\boldsymbol{b},z-z'\right)\left[\exp\left[2\pi i\left(\boldsymbol{b}\cdot\boldsymbol{Q}_b+z'Q_z\right)\right]\Phi_0\left(\boldsymbol{b},z'\right)\right]\right\} \quad (14.62b)$$

The physical meaning of (14.62a) can be given as follows. At point ($\boldsymbol{b}$, z') in the specimen, the initial source of intensity is the elastic wave $\Phi_0\left(\boldsymbol{b},z'\right)$. The effect of momentum transfer $\hbar\boldsymbol{Q}$ in the inelastic scattering is equivalent to multiplying a plane wave component of $\exp\left[2\pi i\left(\boldsymbol{b}\cdot\boldsymbol{Q}_b+z'Q_z\right)\right]$ onto the elastic wave $\Phi_0(\boldsymbol{b},z')$, whose propagation from $(\boldsymbol{b},z')$ to $(\boldsymbol{b},z)$ must suffer from dynamical elastic diffraction characterized by the $O_p\left(\boldsymbol{b},z-z'\right)$ operator. The wave $\psi(\boldsymbol{Q},\boldsymbol{b},z)$ with momentum transfer $\hbar\boldsymbol{Q}$ observed at $(\boldsymbol{b},z)$ is a sum of the waves generated from the crystal entrance face $z=0$ to $z=z$. For the total scattering intensity, on the other hand, the probability for generating the wave with momentum transfer $\hbar\boldsymbol{Q}$ is determined by the dynamic form factor $S\left(\boldsymbol{Q},\boldsymbol{Q}'\right)$, thus, the intensity of the diffuse scattering wave is $S(\boldsymbol{Q},\boldsymbol{Q}')\psi(\boldsymbol{Q},\boldsymbol{b},z)\,\psi^*(\boldsymbol{Q}',\boldsymbol{b}',z)$, and the total intensity is an integration over the entire space of momentum transfer.

14.6.4 The Bloch Wave-Multislice Approach

The non-local property of the optical potential is essentially important in recovering the high-order diffuse scattering. The introduction of a non-local potential V' in the equation makes it difficult to be solved using the conventional multislice approach, although it can be done in principle (Kambe and Stampfl 1994). In this case, the Bloch wave theory is likely to be the most suitable choice. Since the average structure is described by a periodic potential V_0, the Bloch wave theory is adequate to calculate Ψ_0, the elastic scattered wave is a linear superposition of Bloch waves $B_i(\boldsymbol{K}_0,\boldsymbol{r})$ (Kambe and Stampfl 1994).

$$\Psi_0(\boldsymbol{r})=\sum_i\alpha_i\left(\boldsymbol{K}_0\right)B_i\left(\boldsymbol{K}_0,\boldsymbol{r}\right), \quad (14.63a)$$

and $B_i(\boldsymbol{K}_0,\boldsymbol{r})$ is the ith eigensolution of (14.35):

$$B_i(\boldsymbol{K},\boldsymbol{r}) = \sum_g C_g^{(i)}(\boldsymbol{K}) \exp\left[2\pi \mathrm{i}(\boldsymbol{K}+\boldsymbol{g})\cdot\boldsymbol{r} + 2\pi \mathrm{i} v_i z\right] \tag{14.63b}$$

where $\alpha_i(\boldsymbol{K}) = C_{I_0}^{(i)} \boldsymbol{K}_0$ are the superposition coefficients, $C_{I0}^{(i)}(\boldsymbol{K}_0)$ are the elements of the first column of the inverse of the matrix whose elements are $C_g^{(i)}$ (row i and column g). The coefficients $C_g^{(i)}(\boldsymbol{K})$ and eigenvalue v_i are determined by an eigenequation in the Bloch wave theory. Substituting (14.63b) into (14.35) and using the inverse Fourier transform, a set of coupled algebraic equations are obtained

$$\left[K^2 - (\boldsymbol{K} + v_i z + \boldsymbol{g})^2\right] C_g^{(i)} + \frac{2m_0 e}{h^2} \sum_h \left[V_{g-h} + V'_{gh}(i)\right] C_h^{(i)} = 0 \tag{14.64}$$

where V_g are the Fourier coefficients of the crystal potential and those of the complex potential V' are

$$V'^{(i)}_{gh} = \frac{e}{V_c} \int \mathrm{d}\boldsymbol{Q} \int \mathrm{d}\boldsymbol{Q}' S\left(\boldsymbol{Q},\boldsymbol{Q}'\right) \widehat{G}\left(\boldsymbol{k}_i + \boldsymbol{g} - \boldsymbol{Q}, \boldsymbol{Q}' - \boldsymbol{k}_i - \boldsymbol{h}\right) \tag{14.65}$$

where V_c is the volume of the crystal and $\widehat{G}(\boldsymbol{U},\boldsymbol{V})$ is the double Fourier transform of $G(\boldsymbol{r},\boldsymbol{r}_1)$, to be calculated below. For a periodically structured crystal and in the Einstein model,

$$\begin{aligned} S\left(\boldsymbol{Q},\boldsymbol{Q}'\right) = \sum_{g''} \delta\left(\boldsymbol{Q}' - \boldsymbol{Q} - \boldsymbol{g}''\right) \Bigg\{ \sum_\beta & \exp\left[2\pi \mathrm{i}\boldsymbol{r}(\beta)\cdot\boldsymbol{g}''\right] f_\beta^e(\boldsymbol{Q}) \left[f_\beta^e\left(\boldsymbol{Q}+\boldsymbol{g}''\right)\right]^* \\ & \times \exp\left[-W_\beta\left(\boldsymbol{g}''\right)\right] - \exp\left[-W_\beta(\boldsymbol{Q}) - W_\beta\left(\boldsymbol{Q}+\boldsymbol{g}''\right)\right] \Bigg\} \end{aligned} \tag{14.66}$$

where the sum of β is over all the atoms in the unit cell, and $\boldsymbol{g}''$ are the reciprocal lattice vectors. Equation (14.65) can be simplified accordingly

$$V'^{(i)}_{gh} = \frac{e}{V_c} \sum_{g''} \int \mathrm{d}\boldsymbol{Q} S\left(\boldsymbol{Q},\boldsymbol{Q}+\boldsymbol{g}''\right) \widehat{G}\left(\boldsymbol{k}_i + \boldsymbol{g} - \boldsymbol{Q}, \boldsymbol{Q} + \boldsymbol{g}'' - \boldsymbol{k}_i - \boldsymbol{h}\right) \tag{14.67}$$

A general method has been proposed based on the Born series for the calculation of $V'^{(i)}_{gh}$ (Wang 1997). The amount for calculating $\widehat{G}(\boldsymbol{U},\boldsymbol{V})$ is usually large, but this calculation is needed only for the reciprocal space in which the mixed dynamic form factor $S(\boldsymbol{Q},\boldsymbol{Q}+\boldsymbol{g}'')$ is appreciable.

Equation (14.64) can be solved numerically using the FORTRAN source code developed by Spence and Zuo (1992). If $\widehat{G}(\boldsymbol{U},\boldsymbol{V})$ is known, the mixed dynamic form factor is also known for phonon excitation, the optical potential can be calculated. We now use Green's function given in (14.58) to calculate $\widehat{G}(\boldsymbol{U},\boldsymbol{V})$. From an identity of

$$\delta\left(\boldsymbol{b} - \boldsymbol{b}_1\right) = \int \mathrm{d}\tau_b \exp\left[2\pi \mathrm{i}\tau_b \cdot \left(\boldsymbol{b} - \boldsymbol{b}_1\right)\right],$$

one has

$$\begin{aligned}\hat{G}(\boldsymbol{U},\boldsymbol{V}) &= -\xi\int \mathrm{d}\boldsymbol{r}\int \mathrm{d}\boldsymbol{r}_1 \exp\left[-2\pi\mathrm{i}\,(\boldsymbol{U}-\boldsymbol{K}_0)\cdot\boldsymbol{r}\right] \\ &\quad\times \exp\left[-2\pi\mathrm{i}\,(\boldsymbol{V}-\boldsymbol{K}_0)\cdot\boldsymbol{r}_1\right] O_p\,(\boldsymbol{b},z-z_1)\left[\delta\,(\boldsymbol{b}-\boldsymbol{b}_1)\right] \\ &= -\xi\int \mathrm{d}\boldsymbol{r}\exp\left[-2\pi\mathrm{i}\,(\boldsymbol{U}-\boldsymbol{K}_0)\cdot\boldsymbol{r}\right] \\ &\quad\times\Big[\int \mathrm{d}z_1 \exp\big[-2\pi\mathrm{i}\,(V_z-K_{0z})\cdot z_1\big]\, O_p\,(\boldsymbol{b},z-z_1) \\ &\quad\times\left\{\exp\left[2\pi\mathrm{i}\,(\boldsymbol{V}_b-\boldsymbol{K}_{0b})\cdot\boldsymbol{b}\right]\right\}\Big]\end{aligned} \tag{14.68}$$

where $O_p(\boldsymbol{b},z-z_1)[\exp(2\pi\mathrm{i}(\boldsymbol{V}_b-\boldsymbol{K}_{0b})\cdot\boldsymbol{b})]$ is the real-space multislice calculation of the incident wave, $\exp\left[2\pi\mathrm{i}\,(\boldsymbol{V}_b-\boldsymbol{K}_{0b})\cdot\boldsymbol{b}\right]$, from thickness $z=z_1$ to $z=z$ (for $z>z_1$ in forward scattering) (Ishizuka 1982; Chen et al. 1997), and the integration of z_1 is to sum the contributions of the waves generated from different crystal thicknesses. The combination of the Bloch wave and multislice theory is likely to solve (14.64). The small-angle forward scattering approximation is the key for computing Green's function using the multislice theory. This section clearly shows that the optical potential has a specific form and it must be computed precisely for both real and imaginary parts, in order to be correctly applied to recover the high-order diffuse scattering.

This section clearly shows that the optical potential has a specific form and it must be computed precisely for both real and imaginary parts, in order to be correctly applied to recover the high-order diffuse scattering.

14.6.5 The Real Component of the Optical Potential

The optical potential is generally a complex function. Its imaginary component accounts for the absorption effect of the elastic wave. Its real component represents the return scattering of the inelastic wave back to the elastic wave, or so-called "virtual" inelastic-scattering process (Wang 1990a, b). To illustrate this point, one uses Yoshioka's approximation by replacing Green's function with its form in free space, which means there is no atom scattering after inelastic excitation. The double Fourier transform of the free space Green's function is

$$\widehat{G}_0(\boldsymbol{U},\boldsymbol{V}) = \frac{m_0}{2\pi^2\hbar^2}\lim_{\varepsilon\to 0}\frac{\delta(\boldsymbol{U}+\boldsymbol{V})}{\left(U^2-K_0^2-\mathrm{i}\varepsilon\right)}, \tag{14.69}$$

thus, the matrix elements of the optical potential are

$$V_{gh}^{\prime(i)} \approx \frac{e\gamma m_0}{2\pi^2\hbar^2 V_c}\left[\int d\tau(\boldsymbol{U})\frac{S(\boldsymbol{k}_i+\boldsymbol{g}-\boldsymbol{U},\boldsymbol{k}_i+\boldsymbol{h}-\boldsymbol{U})}{U^2-K_0^2}\right.$$
$$\left.+i\frac{\pi}{2K_0}\int d\sigma(\boldsymbol{U})S(\boldsymbol{k}_i+\boldsymbol{g}-\boldsymbol{U},\boldsymbol{k}_i+\boldsymbol{h}-\boldsymbol{U})\right] \tag{14.70}$$

where the integral $\tau(\boldsymbol{U})$ is over the entire reciprocal space except a spherical shell defined by $|\boldsymbol{U}| = K_0$; the integral $\sigma(\boldsymbol{U})$ is over the Ewald sphere surface defined by $|\boldsymbol{U}| = K_0$. Equation (14.70) is the most familiar form of the absorption potential usually introduced in dynamical calculations. The imaginary part is the result of an integral over the scattering falling on the surface of the Ewald sphere, the scattering that satisfies the conservation of energy, which is allowed in practice and can occur. On the other hand, the volume integral in the real component covers the entire reciprocal space except the surface of the Ewald sphere, thus, the scattering does not satisfy the conservation of energy and is unlikely to occur. Thus, the virtually scattered wave will return to the elastic wave, but with a phase modulation. Therefore, it can be simply understood that the imaginary component characterizes the lost in intensity, while the real component signifies the gain in elastic wave intensity. There are numerous techniques developed to calculate the imaginary part of (14.70). The real component has, however, always been ignored because it is much smaller than the crystal potential, but it should have the same magnitude as the imaginary component. It is necessary to re-compute the real part and table it as for the imaginary part!

We have proved in Sect. 14.6.2 that a full inclusion of the optical potential in the calculation can recover the multiple diffuse scattering in the calculation made for the first-order scattering. This recovery process is required to meet three conditions: first the real component of the optical potential must be included in the calculation; second Green's function cannot be replaced by its form in free space; and third, the non-local property of the optical potential needs to be retained.

Most of the current calculations reported in the literature were performed considering only the imaginary part of the optical potential given in (14.70). Two effects are missing in these calculations: one, the virtual inelastic scattering is dropped, and the other, the dynamic diffraction after the inelastic scattering is ignored, while the multiple inelastic scattering is still preserved.

14.7 Computation Procedures

Our entire derivations are based on (14.35) and (14.44):

$$\left[\left(-\frac{\hbar^2}{2m_0}\nabla^2 - eV_0 - E\right)\right]\Psi_0(\boldsymbol{r}) = e^2\int d\boldsymbol{r}_1 G(\boldsymbol{r},\boldsymbol{r}_1)\,s(\boldsymbol{r},\boldsymbol{r}_1)\,\Psi_0(\boldsymbol{r}_1), \tag{14.71a}$$

$$\rho\left(\boldsymbol{r},\boldsymbol{r}'\right)=\Psi_0(\boldsymbol{r})\Psi_0^*\left(\boldsymbol{r}'\right)$$
$$+e^2\int \mathrm{d}\boldsymbol{r}_1\int \mathrm{d}\boldsymbol{r}_2\, G\left(\boldsymbol{r},\boldsymbol{r}_1\right)G^*\left(\boldsymbol{r}',\boldsymbol{r}_2\right) \tag{14.71b}$$
$$\times\, \mathrm{s}\left(\boldsymbol{r}_1,\boldsymbol{r}_2\right)\Psi_0\left(\boldsymbol{r}_1\right)\Psi_0^*\left(\boldsymbol{r}_2\right)$$

which were derived under the ground-state approximation. The major task is to calculate Green's function and solving (14.71a). As a summary, the main procedures for calculating the image/diffraction pattern of the multiple elastically and multiple phonon-scattered (or thermal diffusely scattered) electrons, with the consideration of the incoherence among phonon excitations, can be summarized as follows:

1. For a given crystal structure, compute the Fourier transform of Green's function using the multislice theory:

$$\hat{G}(\boldsymbol{U},\boldsymbol{V})=-\xi\int \mathrm{d}\boldsymbol{r}\exp\left[-2\pi\mathrm{i}\left(\boldsymbol{U}-\boldsymbol{K}_0\right)\cdot\boldsymbol{r}\right]$$
$$\times\left[\int \mathrm{d}z_1\exp\left[-2\pi\mathrm{i}\left(V_z-K_{0z}\right)\cdot z_1\right]\right.$$
$$\left.\times\, O_p\left(\boldsymbol{b},z-z_1\right)\left\{\exp\left[2\pi\mathrm{i}\left(\boldsymbol{V}_b-\boldsymbol{K}_{0b}\right)\cdot\boldsymbol{b}\right]\right\}\right] \tag{14.72}$$

   where the small-angle scattering approximation (or high-energy approximation) and the projected potential approximation (for zero-order Laue zone) were made.
2. Calculate the matrix elements of the optical potential:

$$V_{gh}^{\prime(i)}=\frac{e}{V_c}\int \mathrm{d}\boldsymbol{Q}\int \mathrm{d}\boldsymbol{Q}'S\left(\boldsymbol{Q},\boldsymbol{Q}'\right)\widehat{G}\left(\boldsymbol{k}_i+\boldsymbol{g}-\boldsymbol{Q},\boldsymbol{Q}'-\boldsymbol{k}_i-\boldsymbol{h}\right). \tag{14.73}$$

   The Einstein model is optional for this step in order to expedite the calculation.
3. Solve the eigenvalue equation for the Bloch wave coefficients $C_g^{(i)}(\boldsymbol{K})$, and obtain the elastically scattered wave $\Phi_0(\boldsymbol{r})$.
4. Calculate the total amplitude function for an inelastic scattering with momentum transfer $\hbar\boldsymbol{Q}$ generated in the entire thickness of the crystal using the multislice theory:

$$\psi(\boldsymbol{Q},\boldsymbol{b},z)=\int_0^z \mathrm{d}z_1\left\{O_p\left(\boldsymbol{b},z-z_1\right)\right.$$
$$\left.\times\left[\exp\left(2\pi\mathrm{i}\left(\boldsymbol{b}\cdot\boldsymbol{Q}_b+z_1Q_z\right)\right)\Phi_0\left(\boldsymbol{b},z_1\right)\right]\right\}. \tag{14.74}$$

5. Compute the density matrix:

$$\rho\left(\boldsymbol{r},\boldsymbol{r}'\right) = \Phi_0(\boldsymbol{r})\Phi_0^*\left(\boldsymbol{r}'\right) + e^2\xi^2 \int \mathrm{d}\boldsymbol{Q} \int \mathrm{d}\boldsymbol{Q}'\, S\left(\boldsymbol{Q},\boldsymbol{Q}'\right) \times \psi(\boldsymbol{Q},\boldsymbol{b},z)\psi^*\left(\boldsymbol{Q}',\boldsymbol{b}',z\right). \tag{14.75}$$

6. Compute the image/diffraction pattern from $\rho\left(\boldsymbol{r},\boldsymbol{r}'\right)$.

14.8 Quantitative Computation

A many-body quantum-mechanical model for multiple thermal scattering of fast electrons due to phonon excitation has been proposed by Forbes et al. (2010). The distribution of both elastic and inelastically scattered electrons has been received. The model has been extended to include core-shell ionization. Using this model, diffraction patterns were calculated. The diffraction pattern for a 200-Å-thick crystal of $SrTiO_3$ with plane wave illumination is shown in Fig. 14.4a and one can clearly see the Bragg spots and HOLZ ring superimposed on the diffuse background. Electrons in Fig. 14.4b. It contains Bragg peaks and the HOLZ ring but no diffuse background. Subtracting the elastic contribution from the total diffraction pattern removes precisely the Bragg peaks and HOLZ ring, leaving only the diffuse background of inelastically scattered electrons, as shown in Fig. 14.4c.

Using the "frozen lattice" model, quantitative HAADF-STEM images have been quantitatively calculated and compared with experiments (LeBeau et al. 2008), and an excellent agreement has been received. Figure 14.5 (top row) shows experimental HAADF images of a $SrTiO_3$ single crystal recorded along $< 100 >$ for regions with three different thicknesses (25, 55, and 105 nm). Image intensities are represented in terms of fractions of the incident beam signal (see color scale bar on the right). For the largest thickness, 21% of the incident beam intensity is scattered thermally to contribute to the signal when the probe is on the atomic columns compared with 12% when the probe is located between columns. Figure 14.5 also shows image simulations using frozen phonon and Bloch wave calculations (middle and bottom rows). Those shown in the left pane in each case do not take spatial incoherence into account. While a qualitative match exists between simulations and experiments, there are quantitative differences; in particular, the contrast is higher in the simulations. Spatial incoherence (defined here as combined effects from a finite extent of the effective illumination source, instabilities, sample drift, etc.) is known to reduce the contrast and can be modeled by convolving the simulated HAADF images by a Gaussian envelope function. A Gaussian with an FWHM of 0.08 nm (right pane) provided excellent agreement with the experimental data. For thick regions, the discrepancies between the signal strength in Bloch wave simulations and experiments evident in Fig. 14.5 are expected, because this approach does not include further

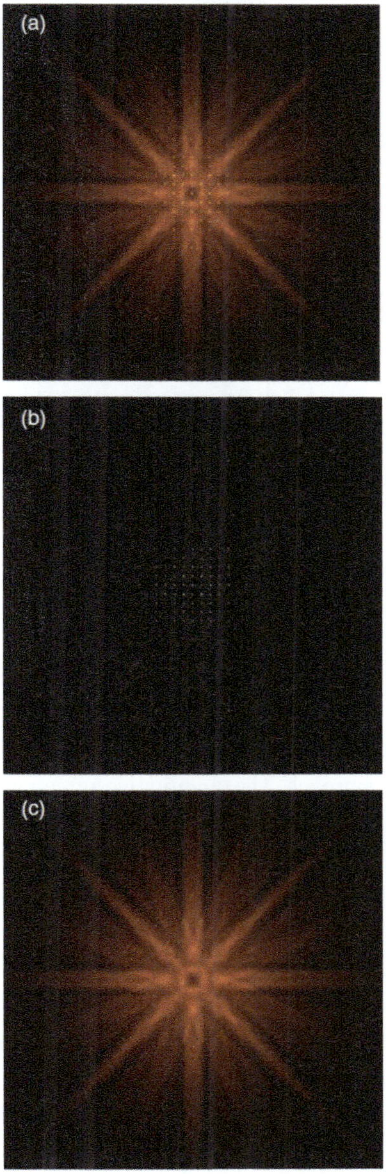

Fig. 14.4 Diffraction pattern components for a 20 nm-thick $SrTiO_3$ specimen formed under plane wave illumination displayed with a log scaling factor: **a** the full intensity, **b** elastically electrons only, and **c** thermally diffusely scattered electrons. Adapted with permission from Forbes et al. (2010)

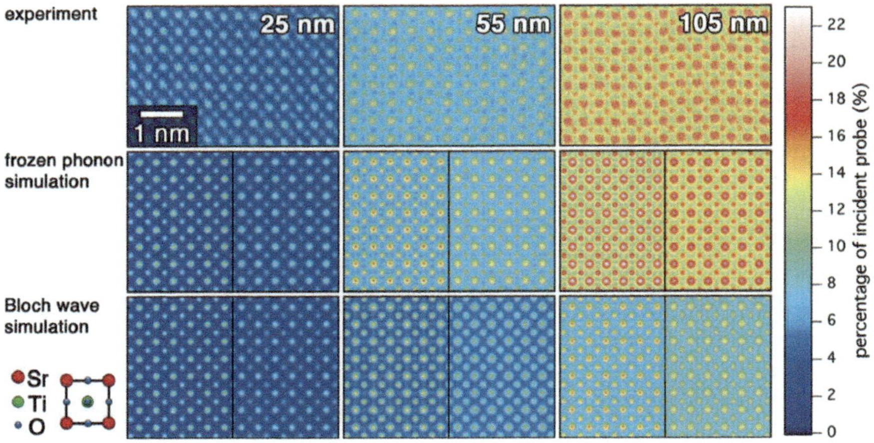

Fig. 14.5 Top row: experimental HAADF images of $SrTiO_3$ along < 100 > with intensity variations normalized to the incident beam intensity (see color scale bar on the right). Regions of three different thicknesses are shown. The Sr columns are the brightest and the Ti–O columns are the second brightest features (see unit cell schematic on the left). The image of the 105 nm region has been drift corrected. Middle row: frozen phonon image simulations. Bottom row: Bloch wave image simulations. In each case, simulations are shown without (left pane) and with convolution with a 0.08 nm FWHM Gaussian (right pane). Adapted with permission from LeBeau et al. (2008)

elastic or inelastic scattering after thermal scattering events. For thicknesses greater than 40 nm these events become important, requiring the use of the frozen phonon approach. The quantitative agreement between the two simulation approaches for thin regions showed that it is tolerable to use a small number of phonon configurations and probe positions in the frozen phonon calculations.

14.9 Summary

1. About the "Frozen Lattice" model

Based on rigorous Green's function and density matrix theories, it has been proved that the calculation based on the "frozen lattice model" for thermal diffuse scattering in electron diffraction gives an identical result as that obtained from the phonon excitation model if the following two conditions are satisfied: (1) the incoherence between different orders of thermal diffuse scattering is considered in the frozen lattice model calculation and (2) the specimen thickness and the mean free path length for phonon excitation both are smaller than the distance traveled by the electron within the lifetime of the phonon. Condition (2) is absolutely satisfied by both low- and high-energy electrons, and satisfying condition (1) is the most critical requirement in the frozen lattice model calculations. More specifically, the diffuse scattering produced by the different orders of diffuse scattering (such as ΔV^2, Δ V^4, and ΔV^6

or multiple diffuse scattering) must be treated incoherent in the frozen lattice model, and the higher order scattering is the result of multiple incoherent first-order scattering events. This incoherence can be precisely accounted for with the introduction of the mixed dynamic form factor $S(\mathbf{Q}, \mathbf{Q}')$, a key quantity for inelastic electron scattering. A new multislice approach has been proposed recently in which the multiple diffuse scattering and the incoherence effects are both accounted for using the $S(\mathbf{Q}, \mathbf{Q}')$ function (Wang 1998c).

The conclusion, reached directly using the integral form of the Schrödinger equation without making the high-energy approximation or ignoring the backscattering, holds for each and all of the orders of diffuse scattering if they are treated incoherently, thus, the quantum mechanical basis of the frozen lattice model is established. The frozen lattice model can also be used to calculate the diffuse diffraction of low-energy electron because their effective penetration depth is relatively small. This conclusion confirms the validity, reliability, and accuracy of using the frozen lattice model in numerous dynamical theories of phonon excitation in electron diffraction and imaging of thin specimens.

2. "Absorption" potential for recovering the high-order diffuse scattering
The "absorption" potential, or the optical potential in general, introduced first by Yoshioka is revisited. The approximation of replacing Green's function by its form in free space is the root of dropping the multiple dynamic thermal diffuse scattering. Using Green's function for a crystal instead of its form in free space, a rigorous theoretical proof is given to show that the inclusion of the optical potential in the dynamical calculation automatically recovers the contributions made by the high-order diffuse scattering although the calculation is done using the equation derived for single diffuse scattering. The conclusion gives the basis for expanding the conventional diffraction theories developed under the first-order diffuse scattering to cases where the specimen thickness is large and/or the degree of disorder is high. This conclusion agrees completely with the results derived previously based on the Frozen lattice model.

The optical potential is no longer a simple potential function, instead it is a non-local function strongly depending on the dynamic diffraction in the crystal because of the involvement of Green's function. It has a much more rich meaning than the conventional interpretation of an absorption effect. These characteristics can be properly taken into account using a Bloch wave-multislice approach, in which the non-local effect is resolved in the Bloch wave matrix diagonalization and the dynamical effect is taken care using Green's function calculated by the multislice theory under the small-angle (or high-energy) approximation. It has also been emphasized that the real component of the potential, which has been ignored in most of calculations, needs to be reconsidered because it has the same magnitude and function as the imaginary component (Wang 1998b).

Under the ground-state approximation, by which we mean that the crystal is in its ground state before each and every inelastic excitation, the introduction of the mixed dynamic form factor and the density matrix automatically produces the incoherence between different orders and different phonon excitation processes, resolving a big problem encountered by many theories, such as the "frozen" lattice model in which the quasi-elastic coherent scattering is a concern.

It is suggested that multiple thermal diffuse scattering (or phonon excitations) is likely to be the dominant source affecting image contrast in quantitative electron microscopy because phonon excitation is a very localized scattering process. To consider this effect properly in HRTEM image calculation, the theory proposed here or the one reported in Wang (1995) is adequate.

In conventional dynamical calculation for elastic wave, the Debye-Waller factor and the optical potential must be included to account for the effects of the phonon excitation on the elastic wave. This type of calculation, however, has missed the contribution from the phonon-scattered electrons. The Debye-Waller factor accounts only for the reduction in Bragg reflected intensity due to atom vibration and it by no way has considered the contribution made by the TDS electrons.

Chapter 15
Electron Diffuse Scattering From Crystals With Correlated Point Defects

Study of point defects is becoming increasingly important in characterization of advanced functional materials. We take $La_{1-x}Ca_xMnO_3$ (LCMO), a material that has been found to exhibit a colossal magnetoresistance (CMR) effect (Jin et al. 1994), as an example. An ordered lattice substitution of La^{3+} by Ca^{2+} creates excess local negative charge, which is balanced partly by the valence conversion of Mn^{3+} into Mn^{4+} and partly by creation of oxygen vacancies (Jonker and van Santen 1953), thus the charge structure of LCMO is

$$La^{3+}_{1-x}Ca^{2+}_{x}Mn^{3+}_{1-x+2y}Mn^{4+}_{x-2y}O^{2-}_{3-y}V^{O}_{y}$$

where V^{O}_{y} stands for the ratio of oxygen vacancies. The conductivity of LCMO is due to the transfer of electrons between Mn^{3+} and Mn^{4+} (Zener 1951), and the transfer probability of the electrons is directly related to the angle Θ_{ij} between the spins of the Mn ions (Anderson and Hasegawa 1955; De Gennes 1960), possibly resulting in the dependence of the electric conductivity of LCMO on the strength of the applied magnetic field (i.e., the magnetoresistance effect).

Determination of short-range order (SRO) structure of oxygen vacancies, however, is a challenge to existing structure analysis techniques. Quantifying SRO structure is rather challenging due to the perturbation produced by SRO on crystal structure can vary from unit cell to unit cell, thus a statistical dynamical diffraction theory is required for considering all possible configurations in dynamic calculations. The statistical dynamical diffraction theory has been proposed (Kato 1980a, b; Kato 1991; Becker and Haddad 1988, 1992) based on the Takagi-Taupin equations (Takagi 1962; Taupin 1964). The theory was developed to deal specifically with the propagation of the average Bragg beam amplitudes at the presence of statistical imperfections in the crystal, but the diffuse scattering produced by the imperfections was not calculated.

Wang (1996a) first presented a new theory that would make the calculation possible. In this chapter, after reviewing the diffuse scattering in electron diffraction, a general theory is proposed to calculate the statistical diffraction intensity of diffusely

Z. L. Wang, *Elastic and Inelastic Scattering in Electron Diffraction and Imaging*, Springer Series in Solid-State Sciences 205, https://doi.org/10.1007/978-3-031-90819-4_15

scattered electrons. This theory is given in the Bloch wave representation best suited for numerical calculations. It is proved that the calculation of diffuse scattering is entirely determined by a dynamic form factor $\mathbf{S}(\mathbf{Q}, \mathbf{Q}')$. The calculations of this factor for two cases involving SRO of point defects are given in Sect. 15.4. Finally, the absorption potential introduced by thermal diffuse scattering (TDS) and SRO is also derived and its role in recovering the high-order diffuse scattering is proved rigorously.

15.1 Diffuse Scattering in Diffraction from Point Defects

Diffuse scattering is produced by structure modulation in a crystalline specimen, and it is usually distributed between Bragg reflected peaks. The Bragg reflections are generated by the periodically structured lattice of the crystal, while the diffuse scattering is produced by the non-periodical components including thermal vibrations of the crystal atoms and SRO of defects. Thermal diffuse scattering (TDS) exists even for a perfect crystal without defects because atomic vibration is a non-periodic perturbation on the crystal potential. Figure 15.1 shows a diffuse scattering pattern recorded from a cubic ZrO_2 stabilized by Y_2O_3 and MgO. The diffuse scattering pattern is produced by the oxygen-vacancy short-range order (Dai et al. 1996a). The distribution of diffuse scattering intensity is rather complex and a three-dimensional geometrical model in reciprocal space is established for describing the equal intensity contours of diffuse scattering observed experimentally. The intensity pattern reflects the correlation and order of oxygen vacancies in the specimen. Electron diffraction is probably the most sensitive and localized technique that can be applied to study SRO of point vacancies, which is an important aspect in studying of oxides functional materials because the variation of cation valences is compensated by creating oxygen vacancies.

In general, the diffuse scattering created by SRO cannot be easily separated experimentally from that produced by TDS. It is thus necessary to include both processes in theoretical calculation. The theory proposed in this chapter is designed for this purpose.

15.2 General Theory

To include in calculation the effects produced by atomic thermal vibrations and structure distortion due to SRO, one uses a general approach in which an *average crystal structure* is introduced (Takagi 1958a, b; Cowley 1995). The crystal potential $V(\mathbf{r}, t)$ is written in the form:

$$V(\mathbf{r}, t) = V_0(\mathbf{r}) + \Delta V(\mathbf{r}, t) \qquad (15.1)$$

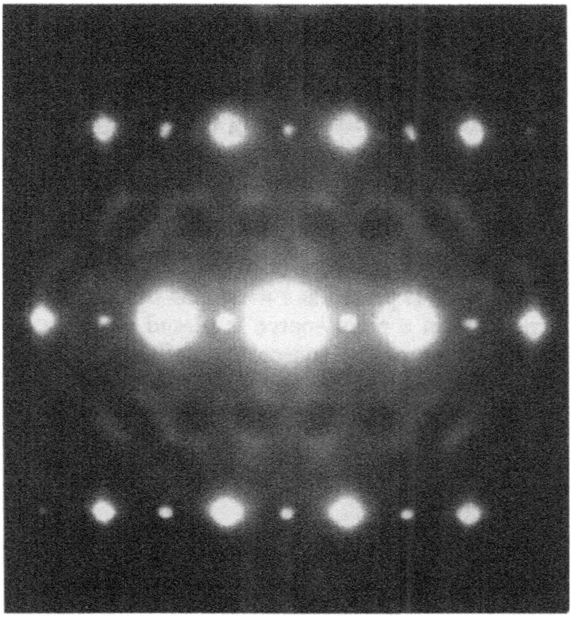

Fig. 15.1 A [013] transmission electron diffraction pattern from cubic ZrO_2 stabilized by Y_2O_3 and MgO showing diffuse scattering created by oxygen vacancies in the crystal. Courtesy of Dr. Z. R. Dai

where $V_0(\mathbf{r}) = \langle V(\mathbf{r}, t)\rangle_{ts}$ is the crystal potential for the average lattice, defined to be time independent and periodic, $\langle\,\rangle_{ts}$ indicates the statistical time and structure average and $\Delta V(\mathbf{r})$ represents the deviation from the average lattice with $\Delta\langle V(\mathbf{r}, t)\rangle_{ts} = 0$, which is non-periodic and time dependent (for TDS). The statistical structure average $\langle\,\rangle_s$ is introduced to take into account the SRO in the crystal structure. The statistical time average is to consider the variation in thermal vibration configurations of crystal lattices. The physical picture contained in (15.1) is schematically shown in Fig. 15.2 using a one-dimensional potential model, where only SRO is considered. For a crystal containing point vacancies, the crystal potential can be separated into a periodic component (V_0) and a deviation term ΔV (Fig. 15.2a). Similar mathematical treatment can be adopted for a binary crystal in which atom substitutions between the A and B types of atom can occur (Fig. 15.2b). The separation of V_0 from ΔV has a powerful application in describing crystals containing point defects using the Patterson function (Cowley 1995).

In high-energy electron scattering, the "frozen" lattice model is assumed in describing TDS. Since the interaction time between an incident electron with a thin specimen is much shorter than the vibration period of an atom, the crystal atoms are seen as stationary by the incident electron. An electron diffraction pattern or image is considered to be a statistical average over the scattering intensities for many instantaneous pictures of the displaced atoms (Hall 1965). The first objective in our theory is to find the scattered electron wave function for a given frozen lattice configuration, then a statistical time average is made on the electron diffraction intensities for a vast number of different thermal vibration configurations. For simplicity, we start from

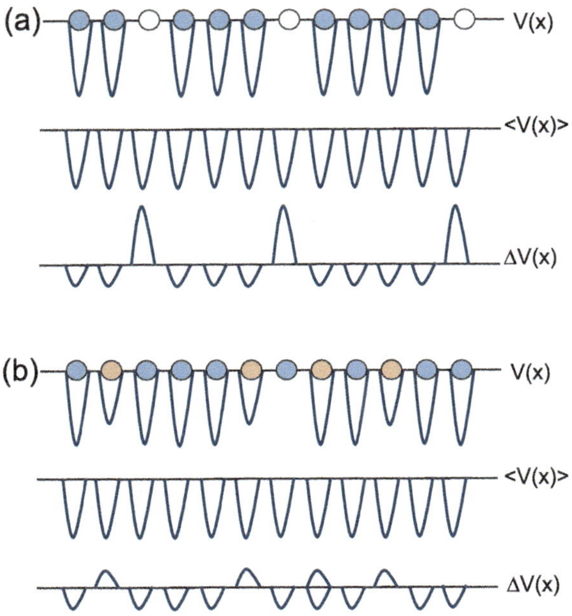

Fig. 15.2 One-dimensional representation of crystal potential V, the structurally averaged potential $V_0 = \langle V \rangle$, and the deviation potential $\Delta V = V - V_0$ for **a** a crystal containing point vacancies and **b** a binary alloy with atom substitutions between the A and B types of atom. V_0 is a periodic function but ΔV is not

the time-independent Schrödinger equation with relativistic correction (Humphreys 1979; Spence 1988a, b):

$$\left[-\left(\hbar^2/2m_0\right)\nabla^2 - e\gamma V_0 - e\gamma\Delta V - E\right]\Psi = 0 \quad (15.2)$$

Shifting the ΔV term to the right-hand side, (15.2) is converted into an integral equation with the use of Green's function $G(\mathbf{r}, \mathbf{r}_1)$:

$$\Psi(\mathbf{r}, t) = \Psi_0(\mathbf{K}_0, \mathbf{r}) + \int \mathrm{d}\mathbf{r}_1 G(\mathbf{r}, \mathbf{r}_1)\left[e\gamma\Delta V(\mathbf{r}_1, t)\Psi(\mathbf{r}_1, t)\right] \quad (15.3)$$

where G is the solution of

$$\left[-\left(\hbar^2/2m_0\right)\nabla^2 - e\gamma V_0 - E\right]G(\mathbf{r}, \mathbf{r}_1) = \delta(\mathbf{r} - \mathbf{r}_1) \quad (15.4)$$

and $\Psi_0(\mathbf{K}_0, \mathbf{r})$ is the elastic wave scattered by the periodic time-independent average potential V_0 due to an incident plane wave with wave vector $\mathbf{K}_0$ and satisfies

$$\left[-\left(\hbar^2/2m_0\right)\nabla^2 - e\gamma V_0 - E\right]\Psi_0 = 0 \quad (15.5)$$

Equation (15.5) can be solved using the Bloch wave or multislice theory. It must be pointed out that the time variable in (15.3) represents the instantaneous lattice configuration of the crystal due to thermal vibration.

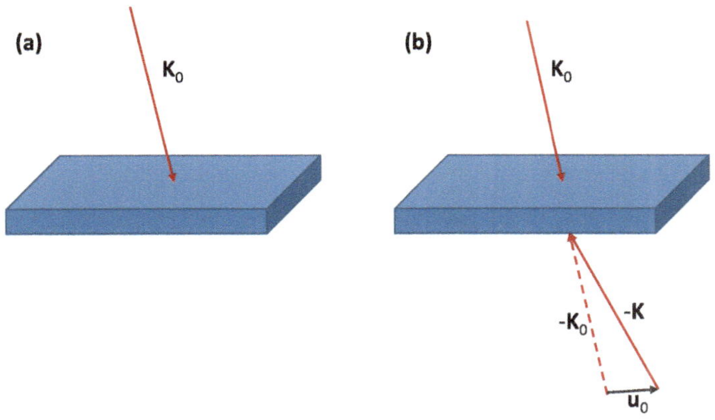

Fig. 15.3 A coordinate system used to describe transmission electron diffraction by a thin slab crystal. The incident beam is nearly parallel to the z-axis, the $\mathbf{b} = xy$ plane is the surface of the crystal slab. $\mathbf{K}_0$ is parallel or nearly parallel to the z- axis. **b** A schematic diagram showing the calculation of $\Psi_0(-\mathbf{K}, \mathbf{r}_1)$ according to the reciprocity theorem

When the observation point is at infinity, the observed diffraction pattern is the squared modulus of the Fourier transform (FT) of $\Psi(\mathbf{r})$. In (15.3), the diffraction amplitude due to the periodic potential V_0 is calculated from the first term:

$$\begin{aligned}\Phi_0(\mathbf{u}_b) &= \int \mathrm{d}x \int \mathrm{d}y \exp\left[-2\pi i\left(u_x x + u_y y\right)\right] \Psi_0(\mathbf{K}_0, x, y, \infty) \\ &= \mathrm{FT}\left[\Psi_0(\mathbf{K}_0, x, y, \infty)\right],\end{aligned} \tag{15.6}$$

where $\mathbf{u}_b = (u_x, u_y)$ is a two-dimensional reciprocal space vector perpendicular or nearly perpendicular to the incident beam direction (see Fig. 15.3a). $\Phi(\mathbf{u}_b)$ is responsible for the intensity of the Bragg reflections. The diffraction due to the non-periodic potential ΔV is represented by

$$\Delta\Phi(\mathbf{u}_b) = \int d\mathbf{r}_1 \hat{G}(\mathbf{u}_b, \infty, \mathbf{r}_1)\left[e\gamma\Delta V(\mathbf{r}_1, t)\Psi(\mathbf{r}_1, t)\right] \tag{15.7}$$

where $\hat{G}(\mathbf{u}_b, \mathbf{r}_1) = \mathrm{FT}\left[G(x, y, z = \infty, \mathbf{r}_1)\right]$. $\Delta\Phi(\mathbf{u}_b)$ is responsible for the diffuse scattering observed in the electron diffraction pattern. The diffraction pattern is calculated under the *first-order diffuse scattering approximation:* $\Psi(\mathbf{r}_1, t)$ is replaced by $\Psi_0(\mathbf{K}_0, \mathbf{r}_1)$ in (15.7) to give

$$\Psi(\mathbf{r}, t) \simeq \Psi_0(\mathbf{K}_0, \mathbf{r}) + \int \mathrm{d}\mathbf{r}_1 G(\mathbf{r}, \mathbf{r}_1) \times \left[e\gamma\Delta V(\mathbf{r}_1, t)\Psi_0(\mathbf{K}_0, \mathbf{r}_1)\right] \tag{15.8a}$$

This approximation holds if the diffuse scattering is much weaker than the Bragg reflections. However, as will be shown in Sect. 15.5, the high-order diffuse scattering terms dropped by this approximation can be recovered by a correction potential to enable (15.8a) to be applied to cases where the disorder is high. This result also applies to the discussions illustrated below. The diffuse scattering intensity is calculated by

$$\begin{aligned} I_D(\mathbf{u}_b) =& \langle |\Delta\Phi(\mathbf{u}_b)|^2\rangle_{ts} \\ =& e^2\gamma^2 \int d\mathbf{r}_1 \int d\mathbf{r}_2 \hat{G}(\mathbf{u}_b, \mathbf{r}_1)\,\hat{G}^*(\mathbf{u}_b, \mathbf{r}_2) \\ & \times \langle \Delta V(\mathbf{r}_1, t)\,\Delta V^*(\mathbf{r}_2, t)\rangle_{ts} \times \Psi_0(\mathbf{K}_0, \mathbf{r}_1)\,\Psi_0^*(\mathbf{K}_0, \mathbf{r}_2) \end{aligned} \tag{15.8b}$$

where the contributions made by different instantaneous crystal lattices due to thermal vibrations are represented by a time average $\langle\,\rangle_t$; the structure average $\langle\,\rangle_s$ is to statistically average on imperfections introduced by SRO. *The most important advantage of this equation is that the time and structure average can be performed analytically before numerical calculations.* We now consider the calculation of $\hat{G}(\mathbf{u}_b, \mathbf{r}_1)$.

From the *reciprocity theorem*, $G(x, y, z = \infty, \mathbf{r}_1) = G(\mathbf{r}_1, x, y, z = \infty)$, provided there is no absorption. This relation means that the wave observed at $z = \infty$ when a point source is placed at $\mathbf{r}_1$ in the specimen is the same as the wave observed at $\mathbf{r}_1$ (in the specimen) when a point source is placed at $z = \infty$ (the image plane). In practice, when a point source is placed at $z = \infty$, the spherical wave of the source that falls on the crystal surface is equivalent to a plane wave, thus, $G(\mathbf{r}_1, x, y, z = \infty)$ is equivalent to the solution of the Schrödinger equation for an incident plane wave. This relation can be proved mathematically (Dudarev et al. 1993a, b) as

$$\hat{G}(\mathbf{u}_b, \mathbf{r}_1) = A_z \Psi_0(-\mathbf{K}_0 - \mathbf{u}_b, \mathbf{r}_1) \tag{15.9}$$

where $\Psi_0(-\mathbf{K}, \mathbf{r}_1)$ is the solution of the Schrödinger equation [Equation (15.5), which will be replaced by (15.48) with the inclusion of the absorption effect] for an incident plane wave vector $(-\mathbf{K})$ $(\mathbf{K} = \mathbf{K}_0 + \mathbf{u}_b)$ and $A_z = -\left[im_0 \exp(2\pi i K_z z)\right]\left(\pi\hbar^2 K_z\right)^{-1}$. The negative sign of the wave vector means that the electron strikes the crystal along the negative z-axis direction, as schematically shown in Fig. 15.3b. Equation (15.9) is a key relation, which makes the calculation of the Green function so convenient. The elastic-scattering wave $\Psi_0(-\mathbf{K},\mathbf{r}_1)$ can be obtained using conventional dynamical approaches, such as the Bethe theory. Thus,

$$\begin{aligned} I_D(\mathbf{u}_b) =& D \int d\mathbf{r}_1 \int d\mathbf{r}_2 \Psi_0(-\mathbf{K}_0 - \mathbf{u}_b, \mathbf{r}_1)\,\Psi_0^*(-\mathbf{K}_0 - \mathbf{u}_b, \mathbf{r}_2) \\ & \times \langle \Delta V(\mathbf{r}_1, t)\,\Delta V^*(\mathbf{r}_2, t)\rangle_{ts} \times \Psi_0(\mathbf{K}_0, \mathbf{r}_1)\,\Psi_0^*(\mathbf{K}_0, \mathbf{r}_2) \end{aligned} \tag{15.10}$$

with $D = e^2\gamma^2 m_0\left[2\pi^2 h^2 E \cos^2\varphi_0\right]^{-1}$, where φ_0 is the angle between $\mathbf{K}$ and the z-axis. In general, $\langle \Delta V(\mathbf{r}_1, t)\,\Delta V^*(\mathbf{r}_2, t)\rangle_{ts}$ is written into a Fourier transform form:

$$\begin{aligned} & \langle \Delta V(\mathbf{r}_1, t)\,\Delta V^*(\mathbf{r}_2, t)\rangle_t \\ & = \int d\mathbf{Q} \int d\mathbf{Q}' \exp\left[2\pi i\left(\mathbf{r}_1 \cdot \mathbf{Q} - \mathbf{r}_2 \cdot \mathbf{Q}'\right)\right] S(\mathbf{Q}, \mathbf{Q}') \end{aligned} \tag{15.11}$$

where $S(\mathbf{Q}, \mathbf{Q}')$ is defined as the diffuse scattering dynamic form factor. Thus, (15.10) is rewritten as

$$\begin{aligned} I_D(\mathbf{u}_b) = & D \int d\mathbf{Q} \int d\mathbf{Q}' S(\mathbf{Q}, \mathbf{Q}') \\ & \times \left\{ \mathrm{FT}\left[\Psi_0^*(-\mathbf{K}_0 - \mathbf{u}_b, \mathbf{r}_1)\,\Psi_0^*(\mathbf{K}_0, \mathbf{r}_1)\right]\Big|_{\mathbf{Q}} \right\}^* \\ & \times \left\{ \mathrm{FT}\left[\Psi_0^*(-\mathbf{K}_0 - \mathbf{u}_b, \mathbf{r}_2)\,\Psi_0^*(\mathbf{K}_0, \mathbf{r}_2)\right]\Big|_{\mathbf{Q}'} \right\} \end{aligned} \quad (15.12)$$

where the subscripts $\mathbf{Q}$ and $\mathbf{Q}'$ mean that the variables of the transformed functions are $\mathbf{Q}$ and $\mathbf{Q}'$, respectively. Since Ψ_0 can be calculated using the existing elastic-scattering theories for the average periodic crystal potential V_0, the key step in the diffuse scattering calculation is to find the $S(\mathbf{Q}, \mathbf{Q}')$ function. The role played by $S(\mathbf{Q}, \mathbf{Q}')$ in diffuse scattering is analogous to that played by the crystal potential in elastic scattering.

Equation (15.12) is a general equation, which has been applied to calculate the diffuse scattering produced by atomic vibrations and point vacancies in a growing surface in reflection high-energy electron diffraction (Wang 1996a). In this paper, this equation is applied to calculate the diffraction of transmitted electrons. We now consider the diffraction of an incident plane wave. An electron wave function in a periodically structured lattice described by the average potential V_0 can be obtained using the Bethe theory. In the Bloch wave representation, the elastic-scattering wave of incident wave vector $\mathbf{K}$ is a linear superposition of Bloch waves $B_i(\mathbf{K}, \mathbf{r})$:

$$\Psi_0(\mathbf{K}, \mathbf{r}) = \sum_i \alpha_i(\mathbf{K}) B_i(\mathbf{K}, \mathbf{r}) \quad (15.13a)$$

where

$$B_i(\mathbf{K}, \mathbf{r}) = \sum_g C_g^{(i)}(\mathbf{K}) \exp\left[2\pi i(\mathbf{K} + \mathbf{g}) \cdot \mathbf{r} + 2\pi i v_i z\right] \quad (15.13b)$$

α_i are the superposition coefficients determined by the boundary conditions, and $C_g^{(i)}$ and ν_i are the Bloch wave coefficients and eigenvalue of the ith Bloch wave $B_i(\mathbf{K}, \mathbf{r})$, respectively. For a periodically structured crystal, the crystal potential can be written in a Fourier series, thus, (15.5) is converted into a set of coupled algebra equations by

$$\left[2KS_g - 2(K_z + g_z)\nu\right] C_g + \left(2\gamma m_0 e/h^2\right) \sum_{\mathbf{h}} V_{\mathrm{g-h}} C_{\mathrm{h}} \simeq 0 \quad (15.14)$$

where V_{g} are the Fourier coefficients of the average crystal potential V_0 and S_g is the excitation error of the g reflection. Equation (15.14) can be solved using the matrix diagonalization technique. For a thin parallel-sided crystal slab, the α_i coefficients are determined by the boundary conditions $\alpha_i = C_0^{(i)*}$ if the absorption potential is zero and $\alpha_i = C_{I0}^{(i)}$ if the absorption effect is included, where $C_{I0}^{(i)}$ are the elements of the first column of the inverse of the matrix whose elements are $C_{\mathbf{g}}^{(i)}$ (row i and column **g**). The Bloch wave theory has shown remarkable success in describing transmission electron diffraction. The theory can be applied to precisely calculate the rocking curves obtained in convergent beam electron diffraction, making it possible to determine the bonding-charge distribution in crystalline specimens (Zuo et al. 1988). The Fortran program developed by Spence (1992) solves (15.14) numerically for any crystallographic system.

For diffuse scattering, substituting (15.13) into (15.12) and performing the integrals gives the TDS intensity at $\mathbf{u}_b$ in reciprocal space:

$$I_D(\mathbf{u}_b) = D\sum_i\sum_j\sum_{i'}\sum_{j'}\sum_{\mathbf{g}}\sum_{\mathbf{h}}\sum_{\mathbf{g}'}\sum_{\mathbf{h}'}\{\text{Bloch waves}\} \times S(\mathbf{Q}, \mathbf{Q}')\,\{\text{thickness}\} \tag{15.15a}$$

where

$$\{\text{Bloch waves}\} = \alpha_i(-\mathbf{K})\alpha_j^*(-\mathbf{K})\alpha_{i'}(\mathbf{K}_0)\,\alpha_{j'}^*(\mathbf{K}_0) \times C_g^{(i)}(-\mathbf{K})C_{\mathbf{h}}^{(j)*}(-\mathbf{K})C_{\mathbf{g}'}^{(i')}(\mathbf{K}_0)\,C_{\mathbf{h}'}^{(j')*}(\mathbf{K}_0) \tag{15.15b}$$

$$\{\text{thickness}\} = \frac{\{\exp[2\pi i(Q_z + g_z + g'_z + v_i + v_i)d] - 1\}}{4\pi^2(Q_z + g_z + g'_z + v_i + v_{i'})} \times \frac{\{\exp[-2\pi i(Q'_z + h_z + h'_z + v_j + v_{j'})d] - 1\}}{(Q'_z + h_z + h'_z + v_j + v_{j'})} \tag{15.15c}$$

$$\mathbf{Q}_b = \mathbf{u}_b - \mathbf{g}_b - \mathbf{g}'_b \tag{15.15d}$$

$$\mathbf{Q}'_b = \mathbf{u}_b - \mathbf{h}_b - \mathbf{h}'_b \tag{15.15e}$$

$\mathbf{K} = \mathbf{K}_0 + \mathbf{u}_b$, **g** and **h** are reciprocal lattice vectors, the subscript b's represent the components parallel to the xy plane and d is the specimen thickness. The sums of i, j, i' and j' are over all the Bloch waves and the sums of **g**, **h**, $\mathbf{g}'$ and $\mathbf{h}'$ are over all the reciprocal lattice vectors. In (15.15a), the {Bloch waves} term characterizes the dynamical diffraction before and after diffuse scattering; S is responsible for the electron angular distribution due to diffuse scattering alone. Equations (15.15d) and (15.15e) are the results of momentum conservation parallel to the surface of the thin slab. The {Bloch waves} term will be modified if the absorption potential is included [see (15.57)]. Based on these discussions, diffuse scattering due to both TDS and SRO of defects can be calculated using the available theoretical approaches, provided the dynamic form factor $S(\mathbf{Q}, \mathbf{Q}')$ is known.

15.3 Calculations of Dynamic form Factors

15.3.1 SRO of Point Vacancies

The distribution of point vacancies varies from cell to cell. Therefore, a statistical structural average on different cell configurations must be made in the calculation. For simplicity, we consider a case in which the crystal structure is dominated by a periodic lattice but with some point vacancies (Fig. 15.4). The positions of the atoms are assumed not to be affected by the point vacancies. Since each atom is also vibrating around its equilibrium position, the instantaneous position of the κth atom site in the crystal is $\mathbf{r}'_\kappa = \mathbf{r}_\kappa + \mathbf{U}_\kappa(t)$, where $\mathbf{U}_\kappa$ is the instantaneous displacement of the atom from its equilibrium position $\mathbf{r}_\kappa$. The calculation can be conveniently performed if the crystal potential is expressed as a Fourier transform of the scattering factors $[f^e_\kappa(\boldsymbol{\tau})]$:

$$
\begin{aligned}
V(\mathbf{r}, t) &= \sum_\kappa \sigma_\kappa V_\kappa \left(\mathbf{r} - \mathbf{r}'_\kappa\right) \\
&= \sum_\kappa \int \mathrm{d}\boldsymbol{\tau} \exp\left[2\pi i \left(\mathbf{r} - \mathbf{r}_\kappa - \mathbf{U}_\kappa\right) \cdot \boldsymbol{\tau}\right] \sigma_\kappa f^e_\kappa(\boldsymbol{\tau})
\end{aligned}
\tag{15.16}
$$

with operator $\sigma_\kappa = 1$ if the site is occupied and $\sigma_\kappa = 0$ if the site is vacant. The structure average is to effectively reduce the scattering power of each atom, thus,

$$
V_0(\mathbf{r}) = \chi_0 \sum_\kappa \int \mathrm{d}\boldsymbol{\tau} \exp\left[2\pi i \left(\mathbf{r} - \mathbf{r}_\kappa\right) \cdot \boldsymbol{\tau}\right] f^e_\kappa(\boldsymbol{\tau}) \exp\left[-W_\kappa(\boldsymbol{\tau})\right] \tag{15.17}
$$

where $\chi_0 = \langle \sigma_\kappa \rangle_s = 1 - \chi_\nu$ and χ_ν is the probability that a site is vacant, i.e., the average density of vacancies in the crystal; χ_0 can be understood as the probability that a site is filled by an atom; a relation of $\langle \exp(X) \rangle_t = \exp\left(\frac{1}{2}\left\langle X^2 \right\rangle_t\right)$ with $\langle X \rangle_t = 0$ was used and $W_\kappa(\boldsymbol{\tau}) = 2\pi^2 \left(\left(\mathbf{U}_\kappa \cdot \boldsymbol{\tau}\right)^2\right)$ is the Debye-Waller factor. Thus, the deviation potential is

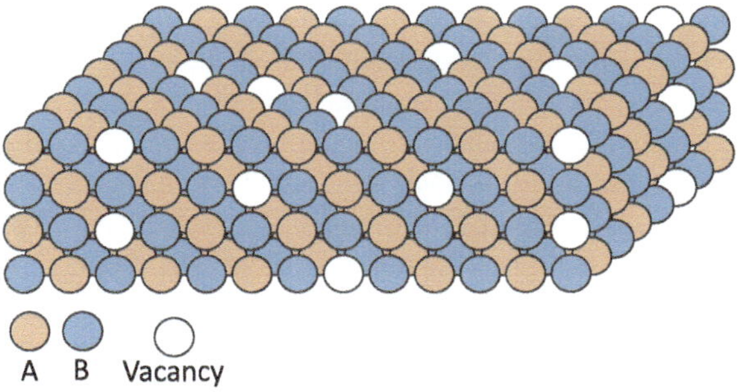

Fig. 15.4 A schematic model showing point vacancies in a crystal. The shadowed spheres are atoms and the open circles represent point vacancies

$$\Delta V(\mathbf{r}_1, t) = \sum_{\kappa} \int d\boldsymbol{\tau} \exp\left[2\pi i (\mathbf{r}_1 - \mathbf{r}_{\kappa}) \cdot \boldsymbol{\tau}\right] f_{\kappa}^{e}(\boldsymbol{\tau}) \times \left\{\sigma_{\kappa} \exp\left(-2\pi i \mathbf{U}_{\kappa} \cdot \boldsymbol{\tau}\right) - \chi_0 \exp\left[-W_{\kappa}(\boldsymbol{\tau})\right]\right\} \tag{15.18}$$

We now calculate the function

$$\begin{aligned} &\langle \Delta V(\mathbf{r}_1, t) \Delta V^*(\mathbf{r}_2, t)\rangle_{ts} \\ &= \sum_{\kappa} \sum_{\kappa'} \int d\boldsymbol{\tau} \int d\mathbf{u} \exp\left[2\pi i (\mathbf{r}_1 - \mathbf{r}_{\kappa}) \cdot \boldsymbol{\tau}\right] \\ &\times \exp\left[-2\pi i (\mathbf{r}_2 - \mathbf{r}_{\kappa'}) \cdot \mathbf{u}\right] f_{\kappa}^{e}(\boldsymbol{\tau}) \left[f_{\kappa'}^{e}(\mathbf{u})\right]^* \\ &\times \left\{\langle \sigma_{\kappa} \sigma_{\kappa'} \rangle_s \langle \exp\left[2\pi i (\mathbf{U}_{\kappa'} \cdot \mathbf{u} - \mathbf{U}_{\kappa} \cdot \boldsymbol{\tau})\right]\rangle_t \right. \\ &\left. - \chi_0^2 \exp\left[-W_{\kappa}(\boldsymbol{\tau}) - W_{\kappa'}(\mathbf{u})\right]\right\} \end{aligned} \tag{15.19}$$

In comparison to (15.11), the dynamic form factor is

$$\begin{aligned} S(\boldsymbol{\tau}, \mathbf{u}) = &\sum_{\kappa} \sum_{\kappa'} \exp\left[2\pi i (\mathbf{r}_{\kappa'} \cdot \mathbf{u} - \mathbf{r}_{\kappa} \cdot \boldsymbol{\tau})\right] f_{\kappa}^{e}(\boldsymbol{\tau}) \left[f_{\kappa'}^{e}(\mathbf{u})\right]^* \\ &\times \exp\left[-W_{\kappa}(\boldsymbol{\tau}) - W_{\kappa'}(\mathbf{u})\right] \\ &\times \left\{\langle \sigma_{\kappa} \sigma_{\kappa'} \rangle_s \exp\left[2F_{\kappa\kappa'}(\boldsymbol{\tau}, \mathbf{u})\right] - \chi_0^2\right\} \end{aligned} \tag{15.20}$$

where

$$F_{\kappa\kappa'}(\boldsymbol{\tau}, \mathbf{u}) = 2\pi^2 \langle (\mathbf{U}_{\kappa} \cdot \boldsymbol{\tau}) (\mathbf{U}_{\kappa'} \cdot \mathbf{u}) \rangle_t \tag{15.21}$$

is defined as a correlation function that is related to the coupling between atom vibrations (Wang 1995). Equation (15.20) includes not only the diffuse scattering produced by point vacancies but also that due to TDS. The structure and time averages can be performed individually.

Based on the harmonic oscillators and adiabatic approximations, in which all the atoms are assumed to interact with harmonic forces and the crystal electrons move as though the ions were fixed in their instantaneous positions, the correlation function $\mathbf{F}_{\kappa\kappa'}(\boldsymbol{\tau}, \mathbf{u})$ has been calculated analytically using the Warren (1990) approximation (Wang 1995). In this approximation, all vibration waves can be considered as either pure longitudinal or pure transverse; the velocities of all longitudinal waves are replaced by an average longitudinal velocity and the velocities of all transverse waves by an average transverse velocity; each average velocity is considered to be a constant independent of the phonon wave vector $\mathbf{q}$, the Brillouin zone is replaced by a sphere of radius q_m, whose volume is equal to that of the Brillouin zone. The result gives a criterion for determining the coherent property of the diffuse scattering waves generated within a local region. In general, the coherent length is approximately 1 nm, about 4–5 interatomic distances. If two atoms are separated by more than 1 nm, the TDS waves generated from the two atom sites are incoherent.

We now consider the statistical structure average of $\langle\sigma_\kappa\sigma_{\kappa'}\rangle_s$. With the Flinn symbols (Flinn 1956),

$$\delta\sigma_\kappa = \sigma_\kappa - \chi_0 = \chi_\nu - \sigma_\kappa^\nu \tag{15.22}$$

where σ_κ^ν is a vacancy operator with $\sigma_\kappa^\nu = 0$ if the site is occupied and $\sigma_\kappa^\nu = 1$ if the site is vacant, then

$$\langle\sigma_\kappa\sigma_{\kappa'}\rangle_s = \langle[\delta\sigma_\kappa + \chi_0]\,[\delta\sigma_{\kappa'} + \chi_0]\rangle_s = \langle\delta\sigma_\kappa\delta\sigma_{\kappa'}\rangle_s + \chi_0^2 \tag{15.23}$$

If the interatomic distance between the κ and κ' atoms is denoted as $\mathbf{r}_{\kappa\kappa'} = \mathbf{r}_\kappa - \mathbf{r}_{\kappa'}$, $\langle\delta\sigma_\kappa\delta\sigma_{\kappa'}\rangle_s$ is directly related to the Cowley short-range order (SRO) parameters (Cowley 1950) defined by

$$\alpha_{\kappa k'} = \langle\delta\sigma_\kappa\delta\sigma_{\kappa'}\rangle_s\,/\chi_0\chi_v \tag{15.24}$$

with $\alpha_{\kappa\kappa} = 1$. In kinematical scattering theory, the SRO parameters are the Fourier coefficients of the diffuse scattering intensity distribution (De Ridder et al. 1976). This is the principle of using the experimentally measured I_D to calculate SRO parameters (Dai et al. 1996a, b). To show the physical meaning of SRO parameters, we now examine the relationship between $\langle\delta\sigma_\kappa\delta\sigma_{\kappa'}\rangle_s$ and the correlation probability. From (22), the probability of finding a vacancy at a given distance $\mathbf{r}_{\kappa\kappa'}$ $(\kappa \neq \kappa')$ from another vacancy at site κ is

$$\begin{aligned} P_{\kappa\kappa'}^{\nu\nu} = \left\langle\sigma_\kappa^\nu\sigma_{\kappa'}^\nu\right\rangle_s &= \langle(\chi_\nu - \delta\sigma_\kappa)\,(\chi_\nu - \delta\sigma_{\kappa'})\rangle_s \\ &= \chi_\nu^2 + \langle\delta\sigma_\kappa\delta\sigma_{\kappa'}\rangle_s \\ &= \chi_\nu^2 + \chi_0\chi_\nu\alpha_{\kappa\kappa'} \end{aligned} \tag{15.25}$$

where the first term χ_ν^2 represents the probability of two vacancies distributed at κ and κ' if there is no correlation (i.e., random distribution). Each vacancy site is a point defect. By the same token, the probability of finding an atom at a given distance $(\mathbf{r}_\kappa - \mathbf{r}_{\kappa'})$ from an atom at site κ' is

$$P_{\kappa\kappa'}^{aa} = \left\langle\sigma_\kappa^a\sigma_{\kappa'}^a\right\rangle = \chi_0^2 + \chi_0\chi_v\alpha_{\kappa\kappa'} \tag{15.26}$$

The probability of finding a vacancy at a given distance $(\mathbf{r}_\kappa - \mathbf{r}_{k'})$ from another atom at site κ' is

$$\begin{aligned} P_{\kappa\kappa'}^{\nu a} = \left\langle\sigma_\kappa^\nu\sigma_{\kappa'}^a\right\rangle &= \langle(\chi_\nu - \delta\sigma_\kappa)\,(\chi_0 + \delta\sigma_{\kappa'})\rangle_s \\ &= \chi_\nu\chi_0 - \langle\delta\sigma_\kappa\delta\sigma_{\kappa'}\rangle_s \\ &= \chi_\nu\chi_0 - \chi_0\chi_\nu\alpha_{\kappa\kappa'} \end{aligned} \tag{15.27}$$

where the first term $\chi_\nu\chi_0$ is the probability of finding a vacancy at site κ and an atom at site κ' if there is no preferred distribution. The whole set of correlation parameters, for all numbers of atoms, can be considered to specify the state of order of the system. The values of $\alpha_{\kappa\kappa'}$ specify the degree to which the neighbors of one sort of atom tend to be preferably of the same sort or of the opposite sort. If $\alpha_{\kappa\kappa'}$ is positive, $P^{\nu\nu}_{\kappa\kappa'} > \chi^2_\nu$, $P^{aa}_{\kappa\kappa'} > \chi^2_0$, the vacancies tend to clump together with vacancies and atoms tend to clump with atoms. If $\alpha_{\kappa\kappa'}$ is negative, $P^{\nu\nu}_{\kappa\kappa'} < \chi^2_\nu$, $P^{a\nu}_{\kappa\kappa'} > \chi_\nu\chi_0$, the vacancies tend to clump together with atoms. Therefore, the measurement of $\alpha_{\kappa\kappa'}$ can reflect the short-range order in the considered system. The decrease of $\alpha_{\kappa\kappa'}$ with the increase of $\mathbf{r}_{\kappa\kappa'}$ gives the range of order.

The dynamic form factor given in (15.20) can be written as

$$\begin{aligned} S(\mathbf{Q},\mathbf{Q}') = &\sum_{\kappa}\sum_{\kappa'} \exp\left[2\pi i\left(\mathbf{r}_{\kappa'}\cdot\mathbf{Q}' - \mathbf{r}_{\kappa}\cdot\mathbf{Q}\right)\right] f^e_{\kappa'}(\mathbf{Q}) \\ &\times \left[f^e_{\kappa'}(\mathbf{Q}')\right]^* \exp\left[-W_\kappa(\mathbf{Q}) - W_{\kappa'}(\mathbf{Q}')\right] \\ &\times \left\{\chi_0\chi_\nu\alpha_{\kappa\kappa'}\exp\left[2F_{\kappa\kappa'}(\mathbf{Q},\mathbf{Q}')\right]\right. \\ &\left.+\chi^2_0\left\{\exp\left[2F_{\kappa\kappa'}(\mathbf{Q},\mathbf{Q}')\right]-1\right\}\right\} \end{aligned} \tag{15.28}$$

This equation covers the diffuse scattering produced by thermal vibration and SRO of point vacancies. The first term in { } is due to point vacancies and the second term is due to TDS. The key in the calculation of (15.28) is the SRO parameters. In the case of SRO in ZrO_2 (Fig. 15.1), the $\alpha_{\kappa\kappa'}$ parameters have been calculated from the experimental diffraction patterns based on kinematical electron diffraction theory (Dai et al. 1996a, b). These parameters can be used in dynamical calculation to quantitatively fit the experimental data.

If there is no order in the distribution of point vacancies, i.e., $\alpha_{\kappa\kappa'} = \delta_{\kappa\kappa'}$, and with the use of the Einstein model, (15.28) is approximated as

$$\begin{aligned} S(\mathbf{Q},\mathbf{Q}') = &\sum_{\kappa} \exp\left[2\pi i\mathbf{r}_\kappa\cdot(\mathbf{Q}'-\mathbf{Q})\right] f^e_\kappa(\mathbf{Q})\left[f^e_\kappa(\mathbf{Q}')\right]^* \\ &\times\left\{\chi_0\exp\left[-W_\kappa(\mathbf{Q}-\mathbf{Q}')\right] - \chi^2_0\exp\left[-W_\kappa(\mathbf{Q}) - W_\kappa(\mathbf{Q}')\right]\right] \end{aligned} \tag{15.29}$$

This is just the result obtained by Dudarev et al. (1993a, b). Equation (15.29) can be further simplified if the position of the κth atom is a sum of the unit-cell position $\mathbf{R}_n$ and the relative position $\mathbf{r}(\beta)$ of the atom with respect to the origin of the unit cell, *i.e.*, $\mathbf{r}_\kappa = \mathbf{R}_n + \mathbf{r}(\beta)$:

$$\begin{aligned} S(\mathbf{Q},\mathbf{Q}') = &\sum_{\mathbf{g}''}\delta(\mathbf{Q}'-\mathbf{Q}-\mathbf{g}'')\left\{\sum_\beta \exp[2\pi i\mathbf{r}(\beta)\cdot\mathbf{g}'']\right. \\ &\times f^e_\beta(\mathbf{Q})[f^e_\beta(\mathbf{Q}+\mathbf{g}'')]'\{\chi_0\exp[-W_\beta(\mathbf{g}'')] \\ &\left.-\chi^2_0\exp[-W_\beta(\mathbf{Q}) - W_\beta(\mathbf{Q}+\mathbf{g}'')]\}\right\}. \end{aligned} \tag{15.30}$$

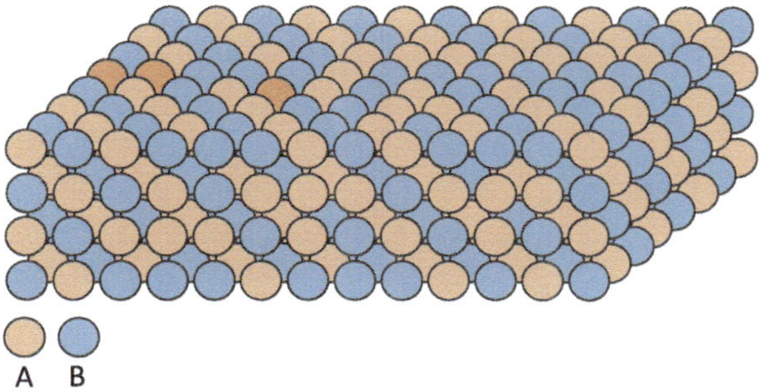

Fig. 15.5 A schematic model showing atom substitution in a binary crystal

where an identity $\sum_n \exp\left[2\pi i \mathbf{R}_n \cdot \mathbf{u}\right] = \sum_{g''} \delta\left(\mathbf{u} - \mathbf{g}''\right)$ for a periodic structure was used and $\mathbf{g}''$ are reciprocal lattice vectors.

15.3.2 Atom Substitution in a Binary Alloy System

A classical example in SRO is a simple binary alloy solution, composed of A and B types of atoms with fractional proportions χ_A and χ_B, such as Cu_3Au and CuAu (Fig. 15.5). These alloys usually have a simple cubic structure. In this section, we calculate the dynamic form factor for this system where the atom substitution can occur. For simplicity, we consider a crystal containing two types of atom. The extension to systems with more than two types of atom follows with elaboration of the algebra but little difference in conception. The ordering in the crystal is defined by order parameters that define correlation between the occupancy of sites, which, for example, is the probability $P^{AB}_{\kappa\kappa'}$ of finding a B atom (at site κ') at a given position $\left(\mathbf{r}_{\kappa'} - \mathbf{r}'_{\kappa}\right)$ from an A atom at site κ. To describe the potential distribution in this system, we first introduce the occupation operators (Flinn 1956; Cowley 1995):

$$\sigma^A_\kappa = \begin{cases} 1 & \text{for an } A \text{ atom at site } \kappa \\ 0 & \text{for a } B \text{ atom at site } \kappa. \end{cases} \tag{15.31a}$$

and

$$\sigma^B_\kappa = \begin{cases} 1 & \text{for a } B \text{ atom at site } \kappa \\ 0 & \text{for an } A \text{ atom at site } \kappa. \end{cases} \tag{15.31b}$$

These parameters are interconnected and we may replace both by a single parameter

$$\delta\sigma_\kappa = \begin{cases} \chi_B & \text{for an } A \text{ atom at site } \kappa \\ -\chi_A & \text{for a } B \text{ atom at site } \kappa. \end{cases} \tag{15.32}$$

Thus,

$$\delta\sigma_\kappa = \sigma^A_x - \chi_A = \chi_B - \sigma^B_x \tag{15.33}$$

with $\langle \delta\sigma_\kappa \rangle_s = 0$ and $\chi_A + \chi_B = 1$. The probability is given by

$$P_{\kappa\kappa'}^{AB} = \langle \sigma_\kappa^A \sigma_{\kappa'}^B \rangle = \chi_A \chi_B - \langle \delta\sigma_\kappa \delta\sigma_{\kappa'} \rangle_s \tag{15.34a}$$

Similarly,

$$P_{\kappa\kappa'}^{AA} = \langle \sigma_\kappa^A \sigma_{\kappa'}^A \rangle = \chi_A^2 + \langle \delta\sigma_\kappa \delta\sigma_{\kappa'} \rangle_s \tag{15.34b}$$

The Cowley SRO parameters are defined by

$$\alpha_{\kappa\kappa'} = \langle \delta\sigma_\kappa \delta\sigma_{\kappa'} \rangle_s / \chi_A \chi_B \tag{15.35}$$

If $\alpha_{\kappa\kappa'}$ is positive, $P_{\kappa\kappa'}^{AA} > \chi_A^2$, $P_{\kappa\kappa'}^{AB} < \chi_A\chi_B$, the same types of atom tend to clump together. If $\alpha_{\kappa\kappa'}$ is negative, $P_{\kappa\kappa'}^{AA} < \chi_A^2$, $P_{\kappa\kappa'}^{AB} > \chi_A\chi_B$, the A atoms tend to clump with B atoms.

After defining these SRO parameters, we now consider the crystal potential distribution with the presence of lattice substitution between the A and B atoms. If the thermal vibrations of crystal atoms are also considered, the instantaneous position of the κth atom in the crystal is $\mathbf{r}'_\kappa = \mathbf{r}_\kappa + \mathbf{U}_\kappa(t)$, where $\mathbf{U}_\kappa$ is the instantaneous displacement of the atom from its equilibrium position $\mathbf{r}_\kappa$. We assume that the lattice position $\mathbf{r}_\kappa$ is not affected by the atom substitution, i.e., lattice relaxation caused by the atom size effect is ignored. The crystal potential may be written as

$$\begin{aligned} V(\mathbf{r}, t) &= \sum_\kappa \left[\sigma_\kappa^A V_A\left(\mathbf{r} - \mathbf{r}_\kappa - \mathbf{U}_\kappa^A\right) + \sigma_\kappa^B V_B\left(\mathbf{r} - \mathbf{r}_\kappa - \mathbf{U}_\kappa^B\right)\right] \\ &= \sum_\kappa \int \mathrm{d}\boldsymbol{\tau} \exp\left[2\pi i\left(\mathbf{r} - \mathbf{r}_\kappa\right)\cdot\boldsymbol{\tau}\right] \\ &\quad \times \left[\sigma_\kappa^A f_A^e(\boldsymbol{\tau}) \exp\left(-2\pi i \mathbf{U}_\kappa^A \cdot \boldsymbol{\tau}\right)\right. \\ &\quad \left. + \sigma_\kappa^B f'_B(\boldsymbol{\tau}) \exp\left(-2\pi i \mathbf{U}_\kappa^B \cdot \boldsymbol{\tau}\right)\right] \end{aligned} \tag{15.36}$$

where $\mathbf{U}_\kappa^A$ (or $\mathbf{U}_\kappa^B$) stands for the vibration displacement of the A (or B atom) located at site κ. The time and structure averages are to effectively reduce the scattering power of each atom, thus,

$$\begin{aligned} V_0(\mathbf{r}) &= \langle V(\mathbf{r}, t)\rangle_{ts} = \sum_\kappa \int \mathrm{d}\boldsymbol{\tau} \exp\left[2\pi i\left(\mathbf{r} - \mathbf{r}_\kappa\right)\cdot\boldsymbol{\tau}\right] \\ &\times \left\{\chi_A f_A^e(\boldsymbol{\tau}) \exp\left[-W_\kappa^A(\boldsymbol{\tau})\right] + \chi_B f_B^e(\boldsymbol{\tau}) \exp\left[-W_\kappa^B(\boldsymbol{\tau})\right]\right\}. \end{aligned} \tag{15.37}$$

where $W_\kappa^A(\boldsymbol{\tau})$ [*or* $W_\kappa^B(\boldsymbol{\tau})$] stands for the Debye-Waller factor of the A (or B) atom located at site κ. The deviation potential is

$$\Delta V(\mathbf{r}_1, t) = \sum_{\kappa} \int \mathrm{d}\boldsymbol{\tau} \exp[2\pi i(\mathbf{r}_1 - \mathbf{r}_{\kappa}) \cdot \boldsymbol{\tau}] \times \Big(f_A^e(\boldsymbol{\tau})\{\sigma_{\kappa}^A \exp(-2\pi i \mathbf{U}_{\kappa}^A \cdot \boldsymbol{\tau}) - \chi_A \exp[-W_{\kappa}^A(\boldsymbol{\tau})]\} + f_B^e(\boldsymbol{\tau})\{\sigma_{\kappa}^B \exp(-2\pi i \mathbf{U}_{\kappa}^B \cdot \boldsymbol{\tau}) - \chi_B \exp[-W_{\kappa}^B(\boldsymbol{\tau})]\}\Big) \tag{15.38}$$

We now calculate the function

$$\begin{aligned}
&\langle \Delta V(\mathbf{r}_1, t)\Delta V^*(\mathbf{r}_2, t)\rangle_{ts} \\
&= \sum_{\kappa}\sum_{\kappa'} \int \mathrm{d}\boldsymbol{\tau} \int \mathbf{d}\mathbf{exp}[2\pi i(\mathbf{r}_1 - \mathbf{r}_{\kappa}) \cdot \boldsymbol{\tau}] \times \exp[-2\pi i(\mathbf{r}_2 - \mathbf{r}_{\kappa'}) \cdot \mathbf{u}] \\
&\times \Big\langle \Big(f_A^e(\boldsymbol{\tau})\{\sigma_{\kappa}^A \exp(-2\pi i \mathbf{U}_{\kappa}^A \cdot \boldsymbol{\tau}) - \chi_A \exp[-W_{\kappa}^A(\boldsymbol{\tau})]\} \\
&+ f_B^e(\boldsymbol{\tau})\{\sigma_{\kappa}^B \exp(-2\pi i \mathbf{U}_{\kappa}^B \cdot \boldsymbol{\tau}) - \chi_B \exp[-W_{\kappa}^B(\boldsymbol{\tau})]\}\Big) \\
&\times \Big([f_A^e(\mathbf{u})]^*\{\sigma_{\kappa'}^A \exp(2\pi i \mathbf{U}_{\kappa'}^A \cdot \mathbf{u}) - \chi_A \exp[-W_{\kappa'}^A(\mathbf{u})]\} \\
&+ [f_B^e(\mathbf{u})]^*\{\sigma_{\kappa'}^B \exp(2\pi i \mathbf{U}_{\kappa'}^B \cdot \mathbf{u}) - \chi_B \exp[-W_{\kappa'}^B(\mathbf{u})]\}\Big)\Big\rangle_{ts}
\end{aligned} \tag{15.39}$$

The statistical structure average can be performed using following relations:

$$\begin{aligned}
\left\langle \sigma_{\kappa}^A \sigma_{\kappa'}^A \right\rangle_s &= \langle (\chi_A + \delta\sigma_{\kappa})\,(\chi_A + \delta\sigma_{\kappa'})\rangle_s \\
&= \chi_A^2 + \langle \delta\sigma_{\kappa}\delta\sigma_{\kappa'}\rangle = \chi_A^2 + \chi_A\chi_B\alpha_{\kappa\kappa'}.
\end{aligned} \tag{15.40a}$$

$$\begin{aligned}
\left\langle \sigma_{\kappa}^B \sigma_{\kappa'}^A \right\rangle_s &= \langle (\chi_B - \delta\sigma_{\kappa})\,(\chi_A + \delta\sigma_{\kappa'})\rangle_s \\
&= \chi_A\chi_B - \langle \delta\sigma_{\kappa'}\delta\sigma_{\kappa'}\rangle = \chi_A\chi_B - \chi_A\chi_B\alpha_{\kappa\kappa'}.
\end{aligned} \tag{15.40b}$$

$$\begin{aligned}
\left\langle \sigma_{\kappa}^A \sigma_{\kappa'}^B \right\rangle_s &= \langle (\chi_A + \delta\sigma_{\kappa})\,(\chi_B - \delta\sigma_{\kappa'})\rangle_s \\
&= \chi_A\chi_B - \langle \delta\sigma_{\kappa}\delta\sigma_{\kappa'}\rangle = \chi_A\chi_B - \chi_A\chi_B\alpha_{\kappa\kappa'}.
\end{aligned} \tag{15.40c}$$

and

$$\begin{aligned}
\left\langle \sigma_{\kappa}^B \sigma_{\kappa'}^B \right\rangle_s &= \langle (\chi_B - \delta\sigma_{\kappa})\,(\chi_B - \delta\sigma_{\kappa'})\rangle_s \\
&= \chi_B^2 + \langle \delta\sigma_{\kappa}\delta\sigma_{\kappa'}\rangle = \chi_B^2 + \chi_A\chi_B\alpha_{\kappa\kappa'}
\end{aligned} \tag{15.40d}$$

In reference to the dynamic form factor defined in (15.11): one has

$$
\begin{aligned}
S(\boldsymbol{\tau},\mathbf{u}) = \sum_{\kappa}\sum_{\kappa'} & \exp[-2\pi i(\mathbf{r}_\kappa\cdot\boldsymbol{\tau}-\mathbf{r}_{\kappa'}\cdot\mathbf{u})] \\
& \times \Big(f_A^e(\boldsymbol{\tau})[f_A^e(\mathbf{u})]^* \exp[-W_\kappa^A(\boldsymbol{\tau}) - W_{\kappa'}^A(\mathbf{u})] \\
& \times \{(\chi_A^2 + \chi_A\chi_B\alpha_{\kappa\kappa'})\exp[2F_{\kappa\kappa'}^{AA}(\boldsymbol{\tau},\mathbf{u})] - \chi_A^2\} \\
& + f_B^e(\boldsymbol{\tau})[f_A^e(\mathbf{u})]^* \exp[-W_\kappa^B(\boldsymbol{\tau}) - W_{\kappa'}^A(\mathbf{u})] \\
& \times \{\chi_A\chi_B(1-\alpha_{\kappa\kappa'})\exp[2F_{\kappa\kappa'}^{BA}(\boldsymbol{\tau},\mathbf{u})] - \chi_A\chi_B\} \\
& + f_A^e(\boldsymbol{\tau})[f_B^e(\mathbf{u})]^* \exp[-W_\kappa^A(\boldsymbol{\tau}) - W_{\kappa'}^B(\mathbf{u})] \\
& \times \{\chi_A\chi_B(1-\alpha_{\kappa\kappa'})\exp[2F_{\kappa k'}^{AB}(\boldsymbol{\tau},\mathbf{u})] - \chi_A\chi_B\} \\
& + f_B^e(\boldsymbol{\tau})[f_B^e(\mathbf{u})]^* \exp[-W_\kappa^B(\boldsymbol{\tau}) - W_{\kappa'}^B(\mathbf{u})] \\
& \times \{(\chi_B^2 + \chi_A\chi_B\alpha_{\kappa\kappa'})\exp[2F_{\kappa\kappa'}^{BB}(\boldsymbol{\tau},\mathbf{u})] - \chi_B^2\} \Big)
\end{aligned}
\tag{15.41}
$$

where $F_{\kappa\kappa'}^{ab}(\boldsymbol{\tau},\mathbf{u}) = 2\pi^2 \left\langle \left[\left(\mathbf{U}_\kappa^a \cdot \boldsymbol{\tau}\right) \left(\mathbf{U}_{\kappa'}^b \cdot \mathbf{u}\right) \right] \right\rangle_t$ is the correlation function between a and b atoms located at sites κ and κ', respectively, and it has been calculated using the Warren approximation (Wang 1995).

If the site substitution is random, so that $\alpha_{\kappa\kappa'} = \delta_{\kappa\kappa'}$, and if the Einstein model is also used for thermal vibration, one has

$$
\begin{aligned}
S(\boldsymbol{\tau},\mathbf{u}) = \sum_{\kappa} & \exp[-2\pi i\mathbf{r}_\kappa\cdot(\boldsymbol{\tau}-\mathbf{u})] \\
& \times \Big(f_A^e(\boldsymbol{\tau})[f_A^e(\mathbf{u})]^* \{\chi_A \exp[-W_\kappa^A(\boldsymbol{\tau}-\mathbf{u})] \\
& - \chi_A^2 \exp[-W_\kappa^A(\boldsymbol{\tau}) - W_\kappa^A(\mathbf{u})]\} \\
& - \chi_A\chi_B\{f_B^e(\boldsymbol{\tau})[f_A^e(\mathbf{u})]^* + f_A^e(\boldsymbol{\tau})[f_B^e(\mathbf{u})]^*\} \\
& \times \exp[-W_\kappa^B(\boldsymbol{\tau}) - W_\kappa^A(\mathbf{u})] \\
& + f_B^e(\tau)[f_B^e(\mathbf{u})]^* \{\chi_B \exp[-W_\kappa^B(\boldsymbol{\tau}-\mathbf{u})] \\
& - \chi_B^2 \exp[-W_\kappa^B(\boldsymbol{\tau}) - W_\kappa^B(\mathbf{u})]\} \Big)
\end{aligned}
\tag{15.42}
$$

Using the same procedure as for deriving (15.30), (15.42) can be rewritten as

$$
\begin{aligned}
S(\boldsymbol{\tau}, \mathbf{u}) = & \sum_{\mathbf{g}''} \delta(\boldsymbol{\tau} - \mathbf{u} - \mathbf{g}'') \Bigg[\sum_{\beta} \exp[2\pi i \mathbf{r}(\beta) \cdot \mathbf{g}''] \\
& \times \Bigg(f_A^e(\boldsymbol{\tau}) [f_A^e(\mathbf{u})]^* \{\chi_A \exp[-W_\beta^A(\mathbf{g}'')] \\
& - \chi_A^2 \exp[-W_\beta^A(\boldsymbol{\tau}) - W_\beta^A(\mathbf{u})]\} \\
& - \chi_A \chi_B \{ f_B^e(\boldsymbol{\tau}) [f_A^e(\mathbf{u})]^* + f_A^e(\boldsymbol{\tau}) [f_B^e(\mathbf{u})]^* \} \\
& \times \exp[-W_\beta^B(\boldsymbol{\tau}) - W_\beta^A(\mathbf{u})] + f_B^e(\boldsymbol{\tau}) [f_B^e(\mathbf{u})]^* \\
& \times \{\chi_B \exp[-W_\beta^B(\mathbf{g}'')] \\
& - \chi_B^2 \exp[-W_\beta^B(\boldsymbol{\tau}) - W_\beta^B(\mathbf{u})]\} \Bigg) \Bigg]
\end{aligned}
\tag{15.43}
$$

To check the consistency of (15.42) with other results, we now consider three extreme cases.
Case 1: If there is no atom substitution, so that $\chi_B = 0$ and $\chi_A = 1$:

$$
\begin{aligned}
S(\boldsymbol{\tau}, \mathbf{u}) = & \sum_{\kappa} \exp[-2\pi i \mathbf{r}_\kappa \cdot (\boldsymbol{\tau} - \mathbf{u})] f_A^e(\boldsymbol{\tau}) [f_A^e(\mathbf{u})]^* \\
& \times \Big(\exp[-W_\kappa^A(\boldsymbol{\tau} - \mathbf{u})] - \exp[-W_\kappa^A(\boldsymbol{\tau}) - W_\kappa^A(\mathbf{u})] \Big)
\end{aligned}
\tag{15.44}
$$

This is just the dynamic form factor for TDS in the Einstein model.
Case 2: If there were no atom vibration at $T = 0$, so that $W_k^A = W_k^B = 0$:

$$
\begin{aligned}
S(\boldsymbol{\tau}, \mathbf{u}) = & \chi_A \chi_B \sum_{\mathbf{g}''} \delta(\boldsymbol{\tau} - \mathbf{u} - \mathbf{g}'') \Bigg\{ \sum_{\beta} \exp[2\pi i r(\beta) \cdot \mathbf{g}''] \\
& \times [f_A^e(\boldsymbol{\tau}) - f_B^e(\boldsymbol{\tau})][f_A^e(\mathbf{u}) - f_B^e(\mathbf{u})]^* \Bigg\}
\end{aligned}
\tag{15.45}
$$

Thus, the diffuse scattering is caused by the difference in scattering factors of the A and B types of atom.
Case 3: If the A and B atoms are the same element, replacing B and A (and $\chi_A = \chi_B = 1/2$) makes (15.42) become

$$
\begin{aligned}
S(\boldsymbol{\tau}, \mathbf{u}) = & \sum_{\kappa} \exp\left[-2\pi i \mathbf{r}_\kappa \cdot (\boldsymbol{\tau} - \mathbf{u})\right] f_A'(\boldsymbol{\tau}) \left[f_A^\ell(\mathbf{u})\right]^* \\
& \times \left\{ \exp\left[-W_\kappa^A(\boldsymbol{\tau} - \mathbf{u})\right] \right. \\
& \left. - \exp\left[-W_\kappa^A(\boldsymbol{\tau}) - W_\kappa^A(\mathbf{u})\right] \right\}
\end{aligned}
\tag{15.46}
$$

This equation is identical to (15.44) for TDS and the SRO effect disappears, just as expected.

It must be pointed out that the calculation of dynamic form factor in this section was performed with the important assumption that the equilibrium position vectors $\mathbf{r}_\kappa$ were the same as for a perfect crystal without defects, i.e., the lattice relaxation caused by point defects was ignored. The relaxation effect can be reasonably taken into account using the results of Hall (1965). If there is no correlation between point defects, the relaxation effect is equivalent to introducing a similar quantity to the Debye-Waller factor that depends on the average displacement of an atom as a result of lattice relaxation. Thus, the Debye-Waller factor $W_\kappa(\mathbf{Q})$ is replaced by $\left[W_\kappa(\mathbf{Q}) + W_\kappa^{(d)}(\mathbf{Q})\right]$, where $W_\kappa^{(d)}(\mathbf{Q}) = 2\pi^2\mathbf{Q}^2\left\langle\nu^2\right\rangle$ and $\left\langle\nu^2\right\rangle$ is the mean square displacement produced by lattice relaxation, the calculation of which can be made based on the random-walk model.

15.4 The "Absorption Potential" and Multiple Diffuse Scattering

Since our calculation is made based on the first-order diffuse scattering approximation, i.e., $\Psi(\mathbf{r}_1, t)$ is replaced by $\Psi_0(\mathbf{K}_0, \mathbf{r}_1)$ in (3), which is

$$\Psi(\mathbf{r}, t) \simeq \Psi_0(\mathbf{K}_0, \mathbf{r}) + \int d\mathbf{r}_1 G(\mathbf{r}, \mathbf{r}_1)\left[e\gamma\Delta V(\mathbf{r}_1, t)\,\Psi_0(\mathbf{K}_0, \mathbf{r}_1)\right] \tag{15.47}$$

where the higher order diffuse scattering terms are dropped. This approximation fails if the specimen is thick and/or the disorder is high. We now modify the solution Ψ_0 of (5) to compensate these high-order diffuse scattering terms so that the theory can be expanded to cases unrestricted by the first-order diffuse scattering approximation. A correction potential $V^{(i)}$ is symbolically introduced:

$$\left[-\left(\hbar^2/2m_0\right)\nabla^2 - e\gamma V_0 - e\gamma V^{(i)} - E\right]\Psi_0 = 0 \tag{15.48}$$

The potential $V^{(i)}$ is chosen so that (47) and (48) exactly satisfy the original Schrödinger equation (2) and the result is

$$\left[V^{(i)}\Psi_0\right] = e\gamma \int d\mathbf{r}_1 \left[G(\mathbf{r}, \mathbf{r}_1)\,\Delta V(\mathbf{r}, t)\Delta V(\mathbf{r}_1, \mathbf{t}) \times \Psi_0(\mathbf{K}_0, \mathbf{r}_1)\right] \tag{15.49}$$

We now prove that the potential $V^{(i)}$ given by (15.49) can be applied to recover the high-order diffuse scattering terms dropped when $\Psi(\mathbf{r}_1, t)$ is replaced by $\Psi_0(\mathbf{K}_0, \mathbf{r}_1)$ in deriving (15.8a) under the first-order diffuse scattering approximation. Starting from the integral form of (15.48) and using Green's function and iterative calculation, one can expand the elastic wave as

$$
\begin{aligned}
\Psi_0(\mathbf{K}_0,\mathbf{r}) =& \Psi_0^{(0)}(\mathbf{K}_0,\mathbf{r}) \\
&+ e\gamma \int d\mathbf{r}_1 G(\mathbf{r},\mathbf{r}_1)\left[V^{(i)}(\mathbf{r}_1)\Psi_0(\mathbf{K}_0,\mathbf{r}_1)\right] \\
=& \Psi_0^{(0)}(\mathbf{K}_0,\mathbf{r}) + (e\gamma)^2 \int d\mathbf{r}_1 G(\mathbf{r},\mathbf{r}_1) \\
&\times \int d\mathbf{r}_2 \left[G(\mathbf{r}_1,\mathbf{r}_2)\Delta V(\mathbf{r}_1,t)\Delta V(\mathbf{r}_2,t)\Psi_0(\mathbf{K}_0,\mathbf{r}_2)\right] \\
=& \Psi_0^{(0)}(\mathbf{K}_0,\mathbf{r}) + (e\gamma)^2 \int d\mathbf{r}_1 \int d\mathbf{r}_2 G(\mathbf{r},\mathbf{r}_1) G(\mathbf{r}_1,\mathbf{r}_2) \\
&\times \Delta V(\mathbf{r}_1,t)\Delta V(\mathbf{r}_2,t)\Psi_0^{(0)}(\mathbf{K}_0,\mathbf{r}_2) + (e\gamma)^4 \\
&\times \int d\mathbf{r}_1 \int d\mathbf{r}_2 \int d\mathbf{r}_3 \int d\mathbf{r}_4 G(\mathbf{r},\mathbf{r}_1) G(\mathbf{r}_1,\mathbf{r}_2) \\
&\times G(\mathbf{r}_2,\mathbf{r}_3) G(\mathbf{r}_3,\mathbf{r}_4)\Delta V(\mathbf{r}_1,t)\Delta V(\mathbf{r}_2,t) \\
&\times \Delta V(\mathbf{r}_3,t)\Delta V(\mathbf{r}_4,t)\Psi_0^{(0)}(\mathbf{K}_0,\mathbf{r}_4) + \ldots
\end{aligned}
\tag{15.50}
$$

where $\Psi_0^{(0)}$ is the Bragg-scattered wave due to the average periodic lattice at the absence of $V^{(i)}$ (e.g., no absorption):

$$
\left[-\left(\hbar^2/2m_0\right)\nabla^2 - e\gamma V_0 - E\right]\Psi_0^{(0)} = 0 \tag{15.51}
$$

This equation can be solved using conventional dynamic electron diffraction theory. Substitution of (15.50) into (15.8a) gives the total scattered wave:

$$
\begin{aligned}
\Psi(\mathbf{r},t) =& \Psi_0^{(0)}(\mathbf{K}_0,\mathbf{r}) \\
&+ (e\gamma) \int d\mathbf{r}_1 G(\mathbf{r},\mathbf{r}_1)\Delta V(\mathbf{r}_1,t)\Psi_0^{(0)}(\mathbf{K}_0,\mathbf{r}_1) \\
&+ (e\gamma)^2 \int d\mathbf{r}_1 \int d\mathbf{r}_2 G(\mathbf{r},\mathbf{r}_1) G(\mathbf{r}_1,\mathbf{r}_2) \\
&\times \Delta V(\mathbf{r}_1,t)\Delta V(\mathbf{r}_2,t)\Psi_0^{(0)}(\mathbf{K}_0,\mathbf{r}_2) \\
&+ (e\gamma)^3 \int d\mathbf{r}_1 \int d\mathbf{r}_2 \int d\mathbf{r}_3 G(\mathbf{r},\mathbf{r}_1) G(\mathbf{r}_1,\mathbf{r}_2) \\
&\times G(\mathbf{r}_2,\mathbf{r}_3)\Delta V(\mathbf{r}_1,t)\Delta V(\mathbf{r}_2,t) \\
&\times \Delta V(\mathbf{r}_3,t)\Psi_0^{(0)}(\mathbf{K}_0,\mathbf{r}_3) \\
&+ (e\gamma)^4 \int d\mathbf{r}_1 \int d\mathbf{r}_2 \int d\mathbf{r}_3 \int d\mathbf{r}_4 G(\mathbf{r},\mathbf{r}_1) G(\mathbf{r}_1,\mathbf{r}_2) \\
&\times G(\mathbf{r}_2,\mathbf{r}_3) G(\mathbf{r}_3,\mathbf{r}_4)\Delta V(\mathbf{r}_1,t)\Delta V(\mathbf{r}_2,t) \\
&\times \Delta V(\mathbf{r}_3,t)\Delta V(\mathbf{r}_4,t)\Psi_0^{(0)}(\mathbf{K}_0,\mathbf{r}_4) + \cdots
\end{aligned}
\tag{15.52}
$$

This Born series is the exact solution of (15.2) without making any approximation. The third term in (15.52) is taken as an example to show its physical meaning, as schematically shown in Fig. 15.1. The Bragg-scattered wave is diffusely scattered at $\mathbf{r}_2$ by $\Delta V(\mathbf{r}_2, t)$. The diffusely scattered wave is elastically scattered by the crystal lattice while propagating from $\mathbf{r}_2$ to $\mathbf{r}_1[G(\mathbf{r}_1, \mathbf{r}_2)]$, then, the second-order diffuse scattering occurs at $\mathbf{r}_1$ $[\Delta V(\mathbf{r}_1, t)]$. Finally, the double diffusely scattered wave exits the crystal at $\mathbf{r}$ after elastic scattering when propagating from $\mathbf{r}_1$ to $\mathbf{r}$ $[G(\mathbf{r}, \mathbf{r}_1)]$. The integrals over $\mathbf{r}_1$ and $\mathbf{r}_2$ sum over the contributions made by all of the possible scattering sources in the crystal.

Therefore, the multiple diffusely scattered waves are comprehensively included in the calculation of (15.8a) if the optical potential $V^{(i)}$ given by (15.49) is introduced in the calculation of Ψ_0 (15.48). This is a key conclusion, which means that, by introducing a proper form of the optical potential, the multiple diffuse scattering terms are automatically included in the calculation using (15.8a), although it was derived for the first-order diffuse scattering. This result has a strong impact on the conventional diffuse scattering theories developed based on the first-order diffuse scattering. Thus, an introduction of a complex potential $V^{(i)}$ in the calculation of the elastic wave makes the existing theories available for calculating the TDS and SRO including all orders of effects. This conclusion is universal for a time-independent system because no assumption and approximation was made in the proof.

We now modify (15.15) to make this Bloch wave theory suitable for calculations beyond the restriction of the first-order diffuse scattering. In practical calculations, a time and structure average is made over (15.49):

$$\left[V^{(i)}\Psi_0\right] = e\gamma \int \mathrm{d}\mathbf{r}_1 \left[G(\mathbf{r}, \mathbf{r}_1) \times \langle \Delta V(\mathbf{r}, t)\Delta V(\mathbf{r}_1, t)\rangle_{ts} \Psi_0(\mathbf{K}_0, \mathbf{r}_1)\right] \tag{15.53}$$

To proceed with the calculation of (15.53), one ignores the diffraction effect of the crystal so that the Green function is replaced by its freespace form (Yoshioka 1957):

$$G_0(\mathbf{r}, \mathbf{r}_1) \simeq \left(2m_0/\hbar^2\right)\left[\exp\left(2\pi i K\,|\mathbf{r} - \mathbf{r}_1|\right)/4\pi\,|\mathbf{r} - \mathbf{r}_1|\right] \tag{15.54}$$

using the dynamic form factor introduced in (15.11), (15.53) is

$$\begin{aligned}\left[V^{(i)}\Psi_0\right] \simeq \left(m_0 e\gamma/2\pi\hbar^2\right) \int \mathrm{d}\mathbf{Q} \int \mathrm{d}\mathbf{Q}' S\left(\mathbf{Q}, \mathbf{Q}'\right) \\ \times \int \mathrm{d}\mathbf{r}_1 \left\{\left[\exp\left(2\pi i K\,|\mathbf{r} - \mathbf{r}_1|\right)/\,|\mathbf{r} - \mathbf{r}_1|\right]\right. \\ \left.\times \exp\left[2\pi i\left(\mathbf{r}\cdot\mathbf{Q} - \mathbf{r}_1\cdot\mathbf{Q}'\right)\right]\Psi_0(\mathbf{K}_0, \mathbf{r}_1)\right\}\end{aligned} \tag{15.55}$$

From (15.55), the correction potential $V^{(i)}$ is a complex function, its imaginary component denotes the "absorption" effect. More importantly, this potential is not a local function because $V^{(i)}$ cannot be separated from wave function Ψ_0. The most accurate representation of this non-local imaginary component is to use its Fourier coefficients given in a matrix form; the $(\mathbf{g}, \mathbf{h})$ matrix elements are given by Yoshioka (1957), Yoshioka and Kainuma (1962), Wang (1995):

$$V_{gh}^{(i)} \simeq -\left(em_0\gamma/2\pi^2\hbar^2 V_c\right)\left\{\int \mathrm{d}\tau(\mathbf{u}) \times \left[S\left(\mathbf{k_g} - \mathbf{u}, \mathbf{k_h} - \mathbf{u}\right) / \left(K_0^2 - u^2\right)\right]\right.$$
$$\left. + i\left(\pi/2K_0\right)\int \mathrm{d}\sigma(\mathbf{u}) S\left(\mathbf{k_g} - \mathbf{u}, \mathbf{k_h} - \mathbf{u}\right)\right\} \tag{15.56}$$

where V_c is the volume of the crystal, the integral $\tau(\mathbf{u})$ is over all reciprocal space $\mathbf{u}$ except a spherical shell defined by $|\mathbf{u}| = K_0$ and the integral $\sigma(\mathbf{u})$ is on the surface of the Ewald sphere defined by $|\mathbf{u}| = K_0$, with $\mathbf{k_g} \simeq \mathbf{K}_0 + \mathbf{g} and \mathbf{k_h} \simeq \mathbf{K}_0 + \mathbf{h}$. The dynamic form factor S is the most important function in this calculation.

If the Green function G in (15.52) is replaced by its form in free space (G_0), this means that the dynamical Bragg diffraction of the electrons is ignored after each order of diffuse scattering. Therefore, Yoshioka's approximation of replacing G by G_0 still recovers the multiple diffuse scattering terms but the dynamic diffraction effects after each event, similar to Høier (1973) multiple inelastic scattering for plasmon excitation. Therefore, the "absorption potential" has a much richer meaning than the conventional interpretation of an absorption effect. In practical calculations, the real component of $V_{\mathrm{gh}}^{(i)}$ may need to be preserved because it may not be small enough for a highly disordered crystal.

Based on the discussion above, we must make the following modifications to the theory presented in Sect. 15.3. Since the reciprocity theorem holds if there is no absorption [i.e., (15.9)], in (15.12), the calculation of $\Psi_0(-K, \mathbf{r})$ should not include absorption, while the calculation of $\Psi_0(\mathbf{K}_0, \mathbf{r})$ must include the absorption potential. Accordingly, the {Bloch waves} term in (15.15a) is modified to

$$\begin{aligned}\{\text{Bloch waves}\} = & C_0^{(i)*}(-\mathbf{K})C_0^{(j)}(-\mathbf{K})C_{l'0}^{(i')}(\mathbf{K}_0) \\ & \times C_{l'0}^{(j')*}(\mathbf{K}_0)\, C_{\mathbf{g}}^{(i)}(-\mathbf{K})C_{\mathbf{h}}^{(j)*}(-\mathbf{K}) \\ & \times C_{\mathbf{g'}}^{(i')}(\mathbf{K}_0)\, C_{\mathbf{h'}}^{(j')*}(\mathbf{K}_0)\end{aligned} \tag{15.57}$$

where the $\left\{C_g\right\}$ coefficients should be calculated with (15.14) without consideration of absorption potential $V_{\mathrm{gh'}}$, while the calculations of $\left\{C_g\right\}$ and $\{C_r\}$ need to include the absorption potential [e.g., V_{g-h} is replaced by $(V_{\mathrm{g-h}} + V_{\mathrm{gh}}{}')$ in (15.14)]. The corrections introduced by the absorption potential on v_i are ignored.

The absorption potentials introduced by TDS have been calculated by Bird and King (1990), Allen and Rossouw (1990) using the Einstein model. The absorption potential introduced by point vacancies has been calculated by Dudarev et al. (1992), who showed the significance of the non-local characteristic of the imaginary potential. Therefore, the conventional assumption that $V^{(i)}$ is a fraction of $V_0[i.e., V^{(i)} = AV_0(\mathbf{r})$, with a proportional constant $A \ll 1]$ is not adequate.

As a summary, the dynamic form factor is directly related to the SRO parameters. These parameters can either be measured experimentally using X-ray or neutron diffraction or be calculated by Monte Carlo simulations. The objective of our dynamical theory is to simulate quantitatively the electron diffraction patterns for deriving the SRO parameters. The calculations using (15.56) need an accurate representation of the electron scattering factor at large scattering angles. The empirical analytical expression proposed by Weickenmeier and Kohl (1991) is expected to give more accurate results than that given by Doyle and Turner (1968) because special consideration was made at large scattering angles.

15.5 Expansion to Include High-Order Thermal Diffuse Scattering

The approach presented in this chapter has been expanded to include the contribution made by multiple TDS (Wang 1996b). By inclusion of a complex potential in dynamical calculation, a rigorous proof is given to show that the high-order diffuse scatterings are fully recovered in the calculations using the equation derived under the distorted wave Born approximation (DWBA), and, more importantly, the statistical time and structure averages over the distorted crystal lattices are evaluated analytically prior numerical calculation. This conclusion establishes the basis for expanding the applications of the existing theories. The exact form of the optical potential is given using a general solution of Green's function, which can be computed numerically using existing dynamical diffraction theories. These conclusions are universal for both low- and high-energy electrons. For transmission electron diffraction, the final result is given in the Bloch wave representation for crystals containing distorted structure. The theory can also be applied to calculate electron images of diffusely scattered electrons in transmission and scanning transmission electron microscopy.

15.6 Expansion to RHEED for Surfaces

In reflection high-energy electron diffraction (RHEED) of growing surfaces in molecular beam epitaxy (MBE), diffuse scattering is generated by both atom vibrations, point vacancies and growth islands (or surface roughness). Most of the existing RHEED theories have been developed under the first-order diffuse scattering approximation, thus, they are restricted for surfaces whose roughness is relatively low. In fact, crystal surfaces grown by MBE are usually rough, the change of surface coverage from 0 to one monolayer counts for the observed RHEED oscillation. Wang (1996c) has developed a formal dynamical theory of RHEED which has been developed to calculate the diffuse scattering produced by both atom vibrations and point vacancies at surfaces. The theory is aimed to recover the *multiple diffuse scatter-*

ings that have been drooped by the distorted wave Born approximation (DWBA). With the inclusion of a complex potential in the dynamical calculation, a rigorous proof is given to show that the high-order diffuse scattering terms are recovered in the calculation using the equation originally derived under the DWBA. This conclusion establishes the basis for expanding the RHEED theories developed under the first-order diffuse scattering to cases where the degree of surface roughness is high, allowing dynamical calculation of RHEED rocking curves for any growing surface. The statistical time and structure averages over the distorted crystal potential are evaluated analytically before numerical calculation. The dynamic form factor is calculated with consideration of anisotropic surface atom vibration and point vacancies at a growing surface.

15.7 Summary

In chapter, a formal dynamical theory is developed to calculate the diffuse scattering produced by atom vibrations and point defects with short-range order (SRO). Diffuse scattering not only produces fine details in diffraction pattern between Bragg beams but also introduces an imaginary potential which reduces the intensities of the Bragg reflected beams. The distribution of diffusely scattered electrons and the absorption potential are directly related to a dynamic form factor $\mathbf{S}(\mathbf{Q}, \mathbf{Q}')$, which have been described for cases where point vacancies and atom substitutions are present, and the final results are directly related to Cowley's SRO parameters. The equation best suited for numerical calculations of diffuse scattering patterns is given in the Bloch wave scheme. The time average on the thermal vibration configurations and statistical structure average over imperfections have been performed analytically before numerical calculation.

A rigorous theoretical proof is given to show that the inclusion of the complex potential in the dynamical calculation automatically recovers the contributions made by the high-order diffuse scattering, although the calculation is done using the equation derived for single diffuse scattering. This conclusion gives the basis for extending the conventional diffraction theories developed under the first-order diffuse scattering to cases where the specimen thickness is large and/or the degree of disorder is high.

Furthermore, the same approach can be utilized to include both TDS and point defects in the diffraction of transmitted and reflected electrons for various scenarios (Wang 1996b, c).

Chapter 16
Multiple Inelastic Electron Scattering

Multiple inelastic electron scattering is significant if the crystal is thick. Contributions from high-order scattering are compatible with that from single scattering if $d/\Lambda > 1$, where d is specimen thickness and Λ is the inelastic electron scattering mean free path length. There are three types of multiple scattering processes. One is plural scattering of the electron by the same inelastic event, such as multiple plasmon excitation. The second type is consecutive multiple electron inelastic scattering by two or more different inelastic events, such as thermal diffuse scattering and valence loss scattering. The third is simultaneous excitation by two or more inelastic events in a single-scattering process, such as double-plasmon excitation (Spence and Spargo 1971; Ashley and Ritchie 1970), in which two plasmons are created during a single inelastic-scattering process. The first two process types are called multiple inelastic scattering, in which inelastic excitations are defined to occur consecutively. The third type is a multi-inelastic scattering process, in which inelastic excitations occur simultaneously during interaction with the same atom (for localized scattering only).

In Chaps. 8 and 10, we described dynamic diffraction theories of double inelastically scattered electrons of Types 2 and 3. However the full dynamic description is very complex if the number of inelastic events increases. It is therefore necessary to seek some methods by which we can qualitatively describe multiple inelastic excitation. In this chapter, the plural scattering process caused by the same event (i.e., Type 1) is discussed in Sects. 16.1 and 16.2; Types 2 and 3 are described in Sects. 16.3 and 16.4. A detailed description of the density matrix theory is given in Sect. 16.3. It is believed that this theory is the best method for describing multiple electron scattering.

Z. L. Wang, *Elastic and Inelastic Scattering in Electron Diffraction and Imaging*, Springer Series in Solid-State Sciences 205, https://doi.org/10.1007/978-3-031-90819-4_16

16.1 Transport Equation Theory

The Bloch wave theory (8.5a) and (8.5b) and the multislice theory (11.21) can, in principle, be applied to treat multiple inelastic scattering in electron diffraction. The scattering of double inelastically scattered electrons by the Bloch wave and Green's function theories was given in Sects. 8.5 and 10.5, respectively. Since the number of calculations becomes huge if we consider high-order scattering, it is desirable to identify simplified theories to take into account multiple scattering effects.

The simplest theory for describing the energy and angular distributions of inelastically scattered electrons is based on an assumption that ignores dynamic diffraction effects of crystals. In this case, scattering of the crystal is treated as if it occurs in an amorphous material, and the angular distribution of the electron is thus considered to be the direct result of inelastic scattering. This is the basic assumption of transport equation theory which is described in the following section.

16.1.1 Energy Distribution of Plural Inelastically Scattered Electrons

We first introduce the transport equation of Landau (1944) based on the detailed balance principle. If $\mathrm{J}(z, E_0 - \Delta E)\, d\Delta E$ is the fraction of the initially monoenergetic beam (of energy E_0) in the energy range $(E_0 - \Delta E) \rightarrow (E_0 - \Delta E - d\Delta E)$ at depth z in the specimen, and $\omega(\varepsilon)d\varepsilon$ [with $\omega(\varepsilon) = \left(d^2P/d\varepsilon \mathrm{d}z\right)$] is the probability that a given electron in the distribution $\mathrm{J}(z, E_0 - \Delta E)$ will lose energy in the range of $\varepsilon \rightarrow \varepsilon + d\varepsilon$ in a path length dz, then J obeys the following equation (Landau 1944):

$$\frac{d\mathrm{J}(z, E_0 - \Delta E)}{dz} = \int_0^\infty d\varepsilon \left[\omega(\varepsilon)\mathrm{J}(z, E_0 - \Delta E + \varepsilon)\right] - \frac{\mathrm{J}(z, E_0 - \Delta E)}{\Lambda} \tag{16.1}$$

where the inelastic mean free path is

$$\frac{1}{\Lambda} = \int_0^\infty d\varepsilon \omega(\varepsilon) \tag{16.2a}$$

and

$$\omega(\varepsilon) = \int d\boldsymbol{\Theta} \frac{d^3P}{d\varepsilon d\Theta \mathrm{d}z} \tag{16.2b}$$

The first term on the right-hand side of (16.1) is the integrated contribution from electrons with energy $(E_0 - \Delta E + \varepsilon)$ after losing additional energy ε; the second term represents the decrease in electron intensity after propagating a distance dz. In general, J is zero for negative energy loss and satisfies the normalization relation.

$$\int_0^\infty d\Delta E \, \mathrm{J}\,(\mathrm{z}, E_0 - \Delta E) = 1 \tag{16.3}$$

It is important to point out that $\mathrm{J}\,(\mathrm{z}, E_0 - \Delta E)$ is the integrated electron intensity distribution of all electrons scattered to different angles but with the same energy loss. Using (16.2a), (16.2b) and (16.1) is rewritten as

$$\frac{d\mathrm{J}\,(\mathrm{z}, E_0 - \Delta E)}{dz} = \int_0^\infty d\varepsilon \omega(\varepsilon) \left[\mathrm{J}\,(\mathrm{z}, E_0 - \Delta E + \varepsilon) - \mathrm{J}\,(\mathrm{z}, E_0 - \Delta E)\right] \tag{16.4}$$

Equation (16.4) can be solved using the Laplace transform **LT**

$$\mathrm{J}(\mathrm{z}, p) = \mathbf{LT}(\mathrm{J}) = \int_0^\infty d\Delta E \exp(-p\Delta E)\mathrm{J}\,(\mathrm{z}, E_0 - \Delta E) \tag{16.5}$$

resulting in

$$\frac{d\mathrm{J}(\mathrm{z}, p)}{dz} = \mathbf{LT}(\omega)\mathrm{J}(\mathrm{z}, p) - \frac{\mathrm{J}(\mathrm{z}, p)}{\Lambda} \tag{16.6}$$

Equation (16.6) can be readily solved, and the inverse Laplace transform ($\mathbf{LT}^{-1}$) of the solution is

$$\begin{aligned}\mathrm{J}\,(\mathrm{z}, E_0 - \varepsilon) &= \exp\left(-\frac{\mathrm{z}}{\Lambda}\right) \mathbf{LT}^{-1}\{\exp[\mathrm{z}\mathbf{LT}(\omega)]\} \\ &= \exp\left(-\frac{\mathrm{z}}{\Lambda}\right) \mathbf{LT}^{-1} \left\{\sum_{n=0} \frac{(z/\Lambda)^n}{n!} [\mathrm{LT}\,(f_E)]^n\right\}\end{aligned} \tag{16.7a}$$

with

$$f_E(\varepsilon) = \frac{\omega(\varepsilon)}{\Lambda} \tag{16.7b}$$

Performing the inverse Laplace transform, (16.7a) becomes (Whelan 1976)

$$\mathrm{J}\,(\mathrm{z}, E_0 - \Delta E) = \exp\left(-\frac{\mathrm{z}}{\Lambda}\right) \sum_{n=0} \frac{(z/\Lambda)^n}{n!} \{f_E\}_n \tag{16.8a}$$

with

$$\{f_E\}_n = \{f_E \otimes \cdots \otimes f_E\}_n = \{f_E\}_{n-1} \otimes f_E \tag{16.8b}$$

The nth power term corresponds to the nth-order plural scattering. Therefore the distribution of electrons as a function of energy loss obeys the Poisson distribution law. This relation is correct if all electrons scattered to different angles are collected. Equation (16.8a) and (16.8b) has been applied to restore the single energy loss function f_E from the experimentally measured electron energy loss distribution function J (Johnson and Spence 1974).

To illustrate the meaning of (16.8a) and (16.8b), we use a delta function approximation to represent the single-plasmon loss function, i.e., $f_E = f\delta\left(\varepsilon - \Delta E_p\right)$, where ΔE_p is the energy of plasmon loss. From (16.8a), the shape of the EELS spectrum is

$$\begin{aligned} \mathrm{J}(\mathrm{d}, E_0 - \varepsilon) &= \exp(-d/\Lambda) \sum_{n=0} \frac{(d/\Lambda)^n}{n!} f^n \delta\left(\varepsilon - n\Delta E_p\right) \\ &= \exp(-d/\Lambda)\delta(\varepsilon) + \exp(-d/\Lambda)(d/\Lambda) f \delta\left(\varepsilon - \Delta E_p\right) \\ &\quad + \exp(-d/\Lambda) \frac{(d/\Lambda)^2}{2!} f^2 \delta\left(\varepsilon - 2\Delta E_p\right) \\ &\quad + \cdots + \exp(-d/\Lambda) \frac{(d/\Lambda)^n}{n!} f^n \delta\left(\varepsilon - n\Delta E_p\right) + \cdots \end{aligned}$$

Therefore the relative intensity of the nth plasmon peak with respect to the zero-loss peak is

$$\frac{(d/\Lambda)^n}{n!} f^n$$

which increases with an increasing (d/Λ). This was observed experimentally, as shown in Fig. 16.1. When $(d/\Lambda) < 1$, the intensity of the zero-loss peak is much stronger than that of the first plasmon peak. With an increase in (d/Λ), the intensity of the zero-loss peak is reduced, and intensities of high-order plasmon peaks increase. For $d/\Lambda = 7.28$ (Fig. 16.1b), almost all electrons experience multiple plasmon losses.

16.1.2 Angular Distribution of Plural Inelastically Scattered Electrons

The angular distribution of plural inelastically scattered electrons can also be derived from the no-diffraction approximation. If $\mathrm{I}(\mathbf{u}, z)$ is the angular distribution function of electrons, which is depth dependent and integrated over energy loss, then the intensity scattered to direction $\mathbf{u}$ by a thin slice of thickness Δz is

$$\Delta \mathrm{I}(\mathbf{u}, z) = \int d\tau\, f_s(\mathbf{u} - \tau)\mathrm{I}(\tau, z)\Delta \mathrm{z} \qquad (16.9)$$

where $f_s(\mathbf{u})$ is the angular distribution function of single loss electrons:

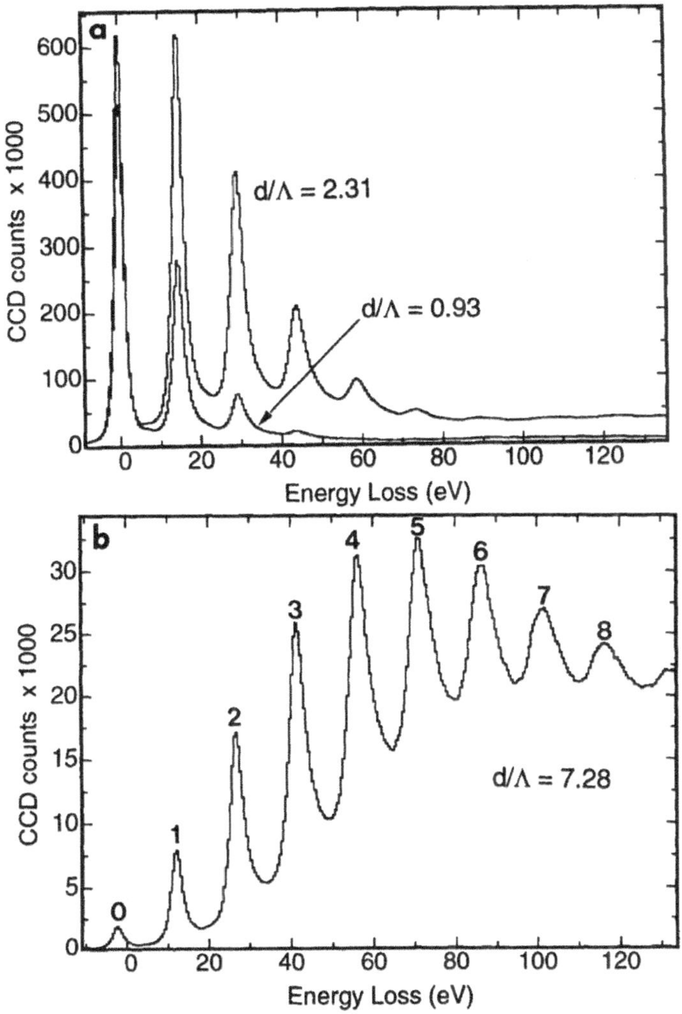

Fig. 16.1 **a** Electron energy loss spectra of an aluminum foil showing intensity distributions of electrons as functions of energy loss and crystal thickness (or d/Λ). **b** Multiple plasmon losses are dominant at large d/Λ

$$f_s(\mathbf{u}) = \int_0^\infty d\varepsilon \frac{d^3\mathrm{P}}{d\varepsilon d\mathbf{u} dz} \tag{16.10}$$

The intensity scattered from **u** is

$$\int d\tau f_s(\tau - \mathbf{u})\mathrm{I}(\mathbf{u}, z)\Delta z = \frac{\mathrm{I}(\mathbf{u}, z)}{\Lambda}\Delta\mathbf{z} \tag{16.11}$$

Combining (16.9) and (16.11), the following transport equation is obtained:

$$\frac{\mathrm{dI}(\mathbf{u},\mathbf{z})}{\mathrm{dz}} = \int \mathrm{d}\tau f_s(\mathbf{u}-\tau)\mathrm{I}(\tau,z) - \frac{\mathrm{I}(\mathbf{u},\mathrm{z})}{\Lambda} \tag{16.12}$$

Taking a two-dimensional Fourier transform of (16.12)

$$\frac{\mathrm{dI}(\mathbf{u},\mathrm{z})}{\mathrm{dz}} = \left[f_s(\mathbf{b}) - \frac{1}{\Lambda}\right]\mathrm{I}(\mathbf{b},\mathrm{z})$$

the solution is

$$\mathrm{I}(\mathbf{b},\mathrm{z}) = \exp\left[\left(f_s(\mathbf{b}) - \frac{1}{\Lambda}\right)\mathrm{z}\right]$$

The inverse Fourier transform of this equation gives the final result (Høier 1973)

$$\mathrm{I}(\mathbf{u},\mathrm{z}) = \exp(-z/\Lambda)\sum_{\mathrm{n}=0}\frac{\mathrm{z}^{\mathrm{n}}}{\mathrm{n}!}\{f_{\mathrm{s}}\}_{\mathrm{n}} \tag{16.13a}$$

with

$$\{f_{\mathrm{s}}\}_{\mathrm{n}} = \{f_{\mathrm{s}}\otimes\cdots f_{\mathrm{s}}\}_{\mathrm{n}} = \{f_{\mathrm{s}}\}_{\mathrm{n}-1}\otimes f_{\mathrm{s}} \tag{16.13b}$$

This equation can conveniently be written as

$$I(\mathbf{u},z) = \sum_{n=0} E_n(z)F_n(\mathbf{u}) \tag{16.14a}$$

with a depth-dependent function

$$\mathrm{E}_{\mathrm{n}}(\mathrm{z}) = \exp(-\mathrm{z}/\Lambda)\frac{(\mathrm{z}/\Lambda)^{\mathrm{n}}}{\mathrm{n}!} \tag{16.14b}$$

and an angular spreading function

$$\mathrm{F}_{\mathrm{n}}(\mathbf{u}) = \Lambda^{\mathrm{n}}\{f_{\mathrm{s}}\}_{\mathrm{n}} \tag{16.14c}$$

or

$$\mathrm{F}_{\mathrm{n}}(\mathbf{u}) = \Lambda\mathrm{F}_{\mathrm{n}-1}(\mathbf{u})\otimes f_{\mathrm{s}} \tag{16.14d}$$

It can be readily proven that

$$\begin{aligned}\int d\mathbf{u}I(\mathbf{u},\mathrm{z}) &= \exp(-\mathrm{z}/\Lambda)\sum_{\mathrm{n}=0}\frac{\mathrm{z}^{\mathrm{n}}}{\mathrm{n}!}\int d\mathbf{u}f_{\mathrm{s}}\otimes\cdots\otimes f_{\mathrm{s}}\\ &= \exp(-\mathrm{z}/\Lambda)\sum_{\mathrm{n}=0}\frac{\mathrm{z}^{\mathrm{n}}}{\mathrm{n}!}\left(\int d\mathbf{u}f_{\mathrm{s}}\right)^{\mathrm{n}} = 1\end{aligned}$$

which means that the number of electrons is conserved at any depth in the crystal.

The electron angular distribution determined from the transport equation completely ignores the diffraction effects of crystals. This is an ideal case. Dynamic diffraction effects become important if the crystal is so thick that Kikuchi pattern effects are significant; thus various approaches have been proposed to incorporate dynamic diffraction effects in the calculation. The dynamic theory of multiple inelastic scattering is so complex that it is very difficult to describe the entire process accurately without making an approximation. Therefore a few simplified theories have been introduced based on appropriate models. The next section illustrates these approaches.

16.2 Improved Theories

Several theories have been introduced to include crystal diffraction effects. Howie (1963) and Goodman et al. (1974) solved (8.5a) under the small-angle inelastic-scattering approximation for plasmon excitations, and some analytical solutions were given. The Poisson distribution law was directly derived (Howie 1963). Yamamoto (1980) also proposed a generalized theory based on the Bloch wave solution given in (8.5a). Rez (1976) attempted to give a more rigorous treatment of the multiple inelastic-scattering theory. The theory introduced by Høier (1973) is based on an approximation in which diffracting effects of the beam before and between inelastic excitation events are ignored and the diffraction effect is considered only after the electron experiences the very last inelastic excitation. This theory is analogous to the transport equation theory except that the final state is a Bloch wave. For simplicity, we illustrate the basic principle of Høier's theory (Høier 1973) here.

To introduce diffraction effects in the multiple inelastic-scattering theory derived in the last section, we first examine the result of the full dynamic single inelastic-scattering theory. The purpose of this step is to find out the proper form of the dynamic scattering operator. The amplitude of electrons to be scattered to $(\mathbf{g} - \mathbf{q})$ at depth z by a single inelastic excitation, based on studies illustrated in Sect. 8.2, can be written as

$$\phi_1(\mathbf{g} - \mathbf{q}) = \sum_{h_1} \sum_{h} [\mathbf{P}_b(d - \mathrm{z})]_{gh_1} \, \mathrm{H}_{h_1 - h}(\mathbf{q}) \, [\mathbf{P}_b(\mathrm{z})]_{h0} \tag{16.15a}$$

where $[\mathbf{P}_\mathrm{b}(\mathrm{z})]_{\mathrm{h}0}$ represents the elastic-scattering matrix from the 0 beam to the **h** beam when the electron travels from $z = 0$ to $z = z$. The inelastic excitation characterized by $\mathrm{H}_{h_1-h}(\mathbf{q})$ occurs at depth z and results in the scattering of the electron for the angle $(\mathbf{h}_1 - \mathbf{h} - \mathbf{q})$. The final elastic electron scattering is represented by $[\mathbf{P}_b(d - z)]_{gh_1}$ which is responsible for scattering the electron from $(\mathbf{h}_1 - \mathbf{q})$ to $(\mathbf{g} - \mathbf{q})$ within a thickness of $d - z$. The $\mathbf{P}_b$ operator, introduced in (4.31), correlates the scattering amplitudes at $z = 0$ with those at $z = z$:

$$[\mathbf{P}_b(z)]_{gh} = \sum_i [\mathbf{P}_{bi}(z)]_{gh} = \sum_i C_g^{(i)} \exp(2\pi i \nu_i z)\, C_h^{(i)} \tag{16.15b}$$

Thus the intensity scattered to $(\mathbf{g} - \mathbf{q})$ by a thin crystal slice is calculated by taking a modulus square of (16.15a), and the result is

$$\begin{aligned} dI_1(\mathbf{g} - \mathbf{q}) =& dz \sum_{h_1} \sum_{h} \sum_{H_1} \sum_{H} [\mathbf{P}_b(d - z)]_{gh_1} [\mathbf{P}_b(d - z)]^*_{gH_1} H_{h_1 - h}(\mathbf{q}) \\ & \times H^*_{H_1 - H}(\mathbf{q}) [\mathbf{P}_b(z)]_{h0} [\mathbf{P}_b(z)]^*{}_{\mathrm{H0}} \end{aligned} \tag{16.16}$$

To generalize (16.16) to the case of multiple inelastic scattering, we define the inelastic-scattering function as

$$T_1^{(I)}(\mathbf{q}, \mathbf{h}, \mathbf{h}_1, \mathbf{H}, \mathbf{H}_1) = H_{h_1 - h}(\mathbf{q}) H^*_{H_1 - H}(\mathbf{q}) \tag{16.17}$$

If we neglect the diffraction effect of the electron before inelastic excitation and assume that the depth dependence of the first-order inelastic scattering is described by a function $D_1(z)$ (to be determined), then the thickness integration in (16.16) is approximated as

$$I_1(\mathbf{g} - \mathbf{q}) = \sum_{h_1} \sum_{h} \sum_{H_1} \sum_{H} [\mathbf{P}_b(d)]_{gh_1} [\mathbf{P}_b(d)]^*_{gH_1} T_1^{(n)}(\mathbf{q}, \mathbf{h}, \mathbf{h}_1, \mathbf{H}, \mathbf{H}_1) \int_0^d dz D_1(z) \tag{16.18}$$

For high-order scattering, $n > 1$, two facts have to be taken into account: one is the incoherence of Bloch waves in different dispersion branches; the other factor is the incoherence of Bragg beams. Therefore the angular distribution of the nth-order plural scattered electrons is

$$I_n(\mathbf{g} - \mathbf{q}) = \sum_h \left| [\mathbf{P}_b(d)]_{gh} \right|^2 T_n^{(I)}(\mathbf{q}, \mathbf{h}) \int_0^d dz D_n(z) \tag{16.19a}$$

where the $T_n^{(I)}$ function is calculated iteratively according to

$$T_n^{(l)}(\mathbf{q}, \mathbf{h}) = T_{n-1}^{(\eta)}(\mathbf{q}, \mathbf{h}) \otimes T_1^{(\eta)}(\mathbf{q}, \mathbf{h}) \tag{16.19b}$$

with the first-order term $T_1^{(l)}(\mathbf{q}, \mathbf{h}) = T_1^{(l)}(\mathbf{q}, \mathbf{h}, \mathbf{h}, \mathbf{h}, \mathbf{h})$ and the convolution calculation is performed on $\mathbf{q}$. The D_n functions in (16.18) and (16.19) can be determined by demanding the sum of (16.18) and (16.19a) to be transformed to the corresponding terms of (16.14) if the diffracting effects diminish, yielding

$$D_{n+1}(z) = \frac{(z/\Lambda)^n}{n!} \frac{\exp(-z/\Lambda)}{\Lambda} \tag{16.20}$$

This is the theory proposed by Høier (1973) for a qualitative description of multiple inelastic scattering. Since the theory considers diffraction effects of the electron only after all inelastic excitation events, that part of the Kikuchi pattern effects has been included.

16.3 Density Matrix Theory of Electron Diffraction

In inelastic electron diffraction, the number of calculations is determined by the number of excited states included. In practice, because many excited states are incoherent with each other, it is desirable to find new approaches that not only take into account dynamic effects but can also incoherently sum over intensities of these states before numerical calculations. One possible approach is to treat electron scattering as a collision process of particles, similar to atomic scattering by molecules in gases. The distribution of atoms is described by a density function, and the evolution of this function represents the interaction process of atoms with gas molecules. An analogous approach has been attempted in describing electron diffraction in crystals. This is the one-particle density matrix theory, introduced as follows.

The density matrix may be conveniently approached from the calculation of electron diffraction patterns. If the electron crystal wave function is $\Phi_s(\mathbf{r}, \mathbf{r}_1, \ldots, \mathbf{r}_M)$, which can be expressed according to (6.2) as

$$\Phi_s(\mathbf{r}, \mathbf{r}_1, \ldots, \mathbf{r}_M) = \sum_n a_n(\mathbf{r}_1, \ldots, \mathbf{r}_M)\, \Psi_n(\mathbf{r}) \tag{16.21}$$

then the diffraction pattern of the outcoming electron is

$$\begin{aligned} I(\mathbf{u}) &= \int d\mathbf{r}_1 \cdots \int d\mathbf{r}_M \left| \sum_n a_n(\mathbf{r}_1, \ldots, \mathbf{r}_M)\, \mathrm{FT}\left[\Psi_n(\mathbf{r})\right] \right|^2 = \sum_n \left|\mathrm{FT}\left[\Psi_n(\mathbf{r})\right]\right|^2 \\ &= \int d\mathbf{b} \int d\mathbf{b}' \exp\left[2\pi i \mathbf{u} \cdot (\mathbf{b}' - \mathbf{b})\right] \sum_n \left[\Psi_n(\mathbf{b}, \infty)\Psi_n^*(\mathbf{b}', \infty)\right] \end{aligned} \tag{16.22}$$

which is a two-dimensional double Fourier transform of $\Sigma_n \left[\Psi_n(\mathbf{r})\Psi_n^*(\mathbf{r}')\right]$. Hence the one-particle density matrix is defined as

$$\rho\left(\mathbf{r}, \mathbf{r}', E\right) = \sum_n \Psi_n(\mathbf{r})\Psi_n^*\left(\mathbf{r}'\right) \delta\left(E - E_0 + \varepsilon_n - \varepsilon_0\right) \tag{16.23}$$

where the δ function stands for the conservation of energy; E is the kinetic energy of the electron; E_0 is the incident electron energy; ε_n is the energy of the nth crystal excited state; and ε_0 is the energy of the crystal ground state. The diagonal elements $\rho(\mathbf{r}, \mathbf{r}, E)$ are proportional to the probability of finding the electron with energy E at point $\mathbf{r}$. Electron scattering is physically described by the evolution of $\rho\left(\mathbf{r}, \mathbf{r}', E\right)$

due to processes of multiple elastic and inelastic scatterings. The fundamental characteristics of the density matrix were described by Blum (1981). The one-particle density matrix was first applied by Dudarev and Ryazanov (1988) to describe multiple inelastic electron scattering. We now outline the basis scheme of this theory.

16.3.1 Kinetic Equation of Multiple Inelastic Electron Scattering

The aim of the one-particle density matrix theory is to derive a kinetic equation to determine the evolution of the density matrix. We start with Yoshioka's coupled equations. For convenience, we define a deviation transition matrix

$$\delta H_{nm}(\mathbf{r}) = H_{nm}(\mathbf{r}) - \delta_{nm} H_{nn}(\mathbf{r}) \tag{16.24}$$

thus $\delta H_{nm}(\mathbf{r}) = H_{nm}(\mathbf{r})$ for $n \neq m$, and $\delta H_{nn}(\mathbf{r}) = 0$. Yoshioka's Equations (6.4a) and (6.4b) are rewritten as

$$\left[\nabla^2 + 4\pi^2\left(U + K_0^2\right)\right]\Psi_0 = \frac{2m_e}{\hbar^2}\sum_m \delta H_{0m}\Psi_m \tag{16.25a}$$

and

$$\left[\nabla^2 + 4\pi^2\left(U + K_n^2\right)\right]\Psi_n = \frac{2m_e}{\hbar^2}\sum_m \delta H_{nm}\Psi_m \tag{16.25b}$$

By using Green's function $G\left(\mathbf{r}, \mathbf{r}', E_n\right)$ satisfying

$$\left[\nabla^2 + 4\pi^2\left(U + K_n^2\right)\right] G\left(\mathbf{r}, \mathbf{r}', E_n\right) = \delta\left(\mathbf{r} - \mathbf{r}'\right) \tag{16.26}$$

Equation (16.25b) is transformed into the integral equation

$$\Psi_n(\mathbf{r}) = \frac{2m_e}{\hbar^2}\sum_m \int d\mathbf{r}_1 G\left(\mathbf{r}, \mathbf{r}_1, E_0 - \varepsilon_n + \varepsilon_0\right) \delta H_{nm}\left(\mathbf{r}_1\right) \Psi_m\left(\mathbf{r}_1\right), \quad n > 0 \tag{16.27}$$

If the density matrix of the elastically scattered electron is denoted by

$$\rho_0\left(\mathbf{r}, \mathbf{r}', E\right) = \Psi_0(\mathbf{r})\Psi_0^*\left(\mathbf{r}'\right)\delta\left(E - E_0\right) \tag{16.28}$$

then substituting (16.27) into (16.23), we have

$$\rho_0\left(\mathbf{r},\mathbf{r}',E\right)=\rho_0\left(\mathbf{r},\mathbf{r}',E\right)+\sum_{n\neq 0}\Psi_n(\mathbf{r})\Psi_n^*\left(\mathbf{r}'\right)\delta\left(E-E_0+\varepsilon_n-\varepsilon_0\right)$$
$$=\rho_0\left(\mathbf{r},\mathbf{r}',E\right)+\frac{4m_e^2}{\hbar^4}\sum_{n\neq 0}\sum_{m}\sum_{m'}\int d\mathbf{r}_1\int d\mathbf{r}_2$$
$$\times G\left(\mathbf{r},\mathbf{r}_1,E_0-\varepsilon_n+\varepsilon_0\right)G^*\left(\mathbf{r}',\mathbf{r}_2,E_0-\varepsilon_n+\varepsilon_0\right)$$
$$\times\,\delta H_{nm}\left(\mathbf{r}_1\right)\delta H_{nm'}^*\left(\mathbf{r}_2\right)\Psi_m\left(\mathbf{r}_1\right)\Psi_{m'}^*\left(\mathbf{r}_2\right)\delta\left(E-E_0+\varepsilon_n-\varepsilon_0\right) \tag{16.29}$$

Since electrons scattered by m and m' states are incoherent, sums over m and m' vanish unless $m=m'$. Using the integral feature of the δ function, (16.29) becomes

$$\rho\left(\mathbf{r},\mathbf{r}',E\right)=\rho_0\left(\mathbf{r},\mathbf{r}',E\right)+\frac{4m_e^2}{\hbar^4}\int d\mathbf{r}_1\int d\mathbf{r}_2G\left(\mathbf{r},\mathbf{r}_1,E\right)G^*\left(\mathbf{r}',\mathbf{r}_2,E\right)$$
$$\sum_{m}\{\Psi_m\left(\mathbf{r}_1\right)\Psi_m^*\left(\mathbf{r}_2\right)\times\sum_{n\neq 0}\left[\delta H_{nm}\left(\mathbf{r}_1\right)\delta H_{nm}^*\left(\mathbf{r}_2\right)\delta\left(E-E_0+\varepsilon_n-\varepsilon_0\right)\right]\} \tag{16.30}$$

The following approximation is made to simplify this equation. In one-particle multiple scattering theory, each time the electron inelastically interacts with crystal atoms, the collision is assumed to take place as though the crystal was at its ground state and the effect of previous collisions was very small. This condition is satisfied if $\tau_0\nu>\Lambda$, where τ_0 is the lifetime of the crystal excitation state. This approximation is mathematically expressed as

$$\sum_{n\neq 0}\delta H_{nm}(\mathbf{r}_1)\delta H_{nm}^*(\mathbf{r}_2)\approx\sum_{n\neq 0}\delta H_{n0}\left(\mathbf{r}_1\right)\delta H_{n0}^*\left(\mathbf{r}_2\right) \tag{16.31}$$

Thus Equation (16.30) is approximated by

$$\rho\left(\mathbf{r},\mathbf{r}',E\right)\approx\rho_0\left(\mathbf{r},\mathbf{r}',E\right)+\frac{4m_e^2}{\hbar^4}\int d\mathbf{r}_1\int d\mathbf{r}_2G\left(\mathbf{r},\mathbf{r}_1,E\right)G^*\left(\mathbf{r}',\mathbf{r}_2,E\right)$$
$$\sum_{m}\{\Psi_m\left(\mathbf{r}_1\right)\Psi_m^*\left(\mathbf{r}_2\right)\times\sum_{n\neq 0}\left[\delta H_{n0}\left(\mathbf{r}_1\right)\delta H_{n0}^*\left(\mathbf{r}_2\right)\delta\left(E-E_0+\varepsilon_n-\varepsilon_0\right)\right]\} \tag{16.32}$$

Using the following identity:

$$\delta\left(E-E_0+\varepsilon_n-\varepsilon_0\right)=\left\{\int d\hbar\omega\delta\left(E+\hbar\omega-E_0-\varepsilon_0+\varepsilon_m\right)\right\}\delta\left(E-E_0+\varepsilon_n-\varepsilon_0\right)$$
$$=\int d\hbar\omega\left[\delta\left(E+\hbar\omega-E_0-\varepsilon_0+\varepsilon_m\right)\delta\left(\hbar\omega+\varepsilon_m-\varepsilon_n\right)\right] \tag{16.33}$$

Equation (16.32) is rewritten as

$$\rho\left(\mathbf{r},\mathbf{r}',E\right)\approx\rho_0\left(\mathbf{r},\mathbf{r}',E\right)+\frac{4m_e^2}{\hbar^4}\int d\mathbf{r}_1\int d\mathbf{r}_2 G\left(\mathbf{r},\mathbf{r}_1,E\right)G^*\left(\mathbf{r}',\mathbf{r}_2,E\right)$$
$$\times\int d\hbar\omega\left\{\sum_m\left[\Psi_m\left(\mathbf{r}_1\right)\Psi_m^*\left(\mathbf{r}_2\right)\delta\left(E+\hbar\omega-E_0-\varepsilon_0+\varepsilon_m\right)\right]\right.$$
$$\left.\times\sum_{n\neq 0}\left[\delta H_{n0}\left(\mathbf{r}_1\right)\delta H_{n0}^*\left(\mathbf{r}_2\right)\delta\left(\hbar\omega+\varepsilon_m-\varepsilon_n\right)\right]\right\} \tag{16.34}$$

Based on the one-particle collision model, the intermediate state m is actually the ground state 0, and ε_m in $\delta\left(\hbar\omega+\varepsilon_m-\varepsilon_n\right)$ is replaced by ε_0. Hence the summation of m and n in (16.34) can be separated. Using the density matrix defined in (16.23) and defining an energy-dependent mixed dynamic form factor

$$\begin{aligned}\bar{\mathbf{s}}\left(\mathbf{r}',\mathbf{r},\hbar\omega\right)&=\frac{4m_e^2}{\hbar^4}\sum_{n\neq 0}\left[\delta H_{n0}(\mathbf{r})\delta H_{n0}^*\left(\mathbf{r}'\right)\delta\left(\hbar\omega+\varepsilon_0-\varepsilon_n\right)\right]\\&=\frac{4m_e^2}{\hbar^4}\sum_{n\neq 0}\left[H_{0n}\left(\mathbf{r}'\right)H_{n0}(\mathbf{r})\delta\left(\hbar\omega+\varepsilon_0-\varepsilon_n\right)\right]\end{aligned} \tag{16.35}$$

a kinetic equation governing the evolution of the density matrix is obtained from (16.34)

$$\begin{aligned}&\rho\left(\mathbf{r},\mathbf{r}',E\right)=\rho_0\left(\mathbf{r},\mathbf{r}',E\right)\\&+\int d\mathbf{r}_1\int d\mathbf{r}_2 G\left(\mathbf{r},\mathbf{r}_1,E\right)G^*\left(\mathbf{r}',\mathbf{r}_2,E\right)\int d\hbar\omega\left[\bar{s}\left(\mathbf{r}_2,\mathbf{r}_1,\hbar\omega\right)\rho\left(\mathbf{r}_1,\mathbf{r}_2,E+\hbar\omega\right)\right]\end{aligned} \tag{16.36}$$

This equation was derived by Dudarev (1991), and it was applied to the study of diffraction effects in Auger electron emission (Gomoyunova et al. 1990), TDS in RHEED (Dudarev et al. 1991 and Dudarev et al. 1993b), multiple plasmon excitation (Dudarev et al. 1992), and diffraction of backscattered electrons (Gorodnichev et al. 1990). A more rigorous derivation of the kinetic equation for a system in thermal equilibrium with the environment was described in detail by Dudarev et al. (1993a).

In general, for consecutive inelastic scattering, the energy-dependent mixed dynamic form factor can be written as a sum of mixed dynamic factors for possible inelastic events

$$\bar{s}=\bar{s}_{\text{valence}}+\bar{s}_{\text{TDS}}+\bar{s}_{\text{single electron}} \tag{16.37}$$

Calculations of mixed dynamic form factors for various inelastic-scattering processes are described in Chap. 6.

16.3.2 Absorption Effect in Calculating Green's Function

We now modify (16.26) to introduce the absorption effect properly

$$\left[\nabla^2 + 4\pi^2 \left(U + K^2\right)\right] G\left(\mathbf{r}, \mathbf{r}', E\right) = \delta\left(\mathbf{r} - \mathbf{r}'\right) + \frac{2m_e}{\hbar^2} \sum_m \delta H_{0m} \Psi_m \quad (16.38)$$

Besides the point source term on the right-hand side of (16.38), terms associated with the effects of inelastic waves were introduced according to (16.25a). We now use the single inelastic-scattering approximation to derive the absorption potential. From (16.27), the first-order solution of the inelastic wave is

$$\Psi_n(\mathbf{r}) \approx \frac{2m_e}{\hbar^2} \int d\mathbf{r}_1 G_0\left(\mathbf{r}, \mathbf{r}_1, E_0 + \varepsilon_n - \varepsilon_0\right) \delta H_{n0}\left(\mathbf{r}_1\right) \Psi_0\left(\mathbf{r}_1\right) \quad (16.39a)$$

where G_0 is determined by

$$\left[\nabla^2 + 4\pi^2 \left(U + K^2\right)\right] G_0\left(\mathbf{r}, \mathbf{r}', E\right) = \delta\left(\mathbf{r} - \mathbf{r}'\right) \quad (16.39b)$$

The introduction of a point source at $\mathbf{r}'$ is equivalent to replacing $\Psi_0\left(\mathbf{r}_1\right)$ by $G\left(\mathbf{r}_1, \mathbf{r}', E\right)$; thus (16.39a) becomes

$$\Psi_n(\mathbf{r}) \approx \frac{2m_e}{\hbar^2} \int d\mathbf{r}_1 G_0\left(\mathbf{r}, \mathbf{r}_1, E_0 + \varepsilon_n - \varepsilon_0\right) \delta H_{n0}\left(\mathbf{r}_1\right) G\left(\mathbf{r}_1, \mathbf{r}', E\right) \quad (16.40)$$

We now consider the following term with the use of (16.40):

$$\begin{aligned}
\frac{2m_e}{\hbar^2} \sum_m \delta H_{0m} \Psi_m \approx & \frac{4m_e^2}{\hbar^4} \sum_m \delta \\
& H_{0m}(\mathbf{r}) \int d\mathbf{r}_1 G_0\left(\mathbf{r}, \mathbf{r}_1, E_0 - \varepsilon_m + \varepsilon_0\right) \delta H_{m0}\left(\mathbf{r}_1\right) G\left(\mathbf{r}_1, \mathbf{r}', E\right) \\
= & \frac{4m_e^2}{\hbar^4} \sum_m \delta H_{0m}(\mathbf{r}) \int d\mathbf{r}_1 \int d\hbar\omega G_0\left(\mathbf{r}, \mathbf{r}_1, E_0 - \hbar\omega\right) \\
& \times \delta\left(\hbar\omega - \varepsilon_m + \varepsilon_0\right) \delta H_{m0}\left(\mathbf{r}_1\right) G\left(\mathbf{r}_1, \mathbf{r}', E\right) \\
= & \int d\mathbf{r}_1 \int d\hbar\omega G_0\left(\mathbf{r}, \mathbf{r}_1, E_0 - \hbar\omega\right) \bar{s}\left(\mathbf{r}, \mathbf{r}_1, \hbar\omega\right) G\left(\mathbf{r}_1, \mathbf{r}', E\right)
\end{aligned} \quad (16.41)$$

where the energy-dependent mixed dynamic form factor defined in (16.35) was used. Substituting (16.41) into (16.38), the equation that determines Green's function is

$$\left[\nabla^2 + 4\pi^2\left(U + K_0^2\right)\right] G\left(\mathbf{r}, \mathbf{r}', E\right)$$
$$- \int d\mathbf{r}_1 \int d\hbar\omega G_0\left(\mathbf{r}, \mathbf{r}_1, E - \hbar\omega\right) \bar{s}\left(\mathbf{r}, \mathbf{r}_1, \hbar\omega\right) G\left(\mathbf{r}_1, \mathbf{r}', E\right) = \delta\left(\mathbf{r} - \mathbf{r}'\right) \tag{16.42}$$

It is obvious that (16.42) is equivalent to (6.19). Again the absorption term is a non-local function.

The kinetic equation (16.36), in principle, includes the contribution from all orders of inelastic scattering. The iterative solution of the kinetic equation is

$$\begin{aligned}
&\rho\left(\mathbf{r}, \mathbf{r}', E\right) = \rho_0\left(\mathbf{r}, \mathbf{r}', E\right) \\
&+ \int d\mathbf{r}_1 \int d\mathbf{r}_2 \int d\hbar\omega \left[G\left(\mathbf{r}, \mathbf{r}_1, E\right) G^*\left(\mathbf{r}', \mathbf{r}_2, E\right) \bar{s}\left(\mathbf{r}_2, \mathbf{r}_1, \hbar\omega\right) \rho_0\left(\mathbf{r}_1, \mathbf{r}_2, E + \hbar\omega\right)\right] \\
&+ \int d\mathbf{r}_1 \int d\mathbf{r}_2 \int d\mathbf{r}_1{}' \int d\mathbf{r}_1' \int d\hbar\omega \int d\hbar\omega' \left[G\left(\mathbf{r}, \mathbf{r}_1, E\right)\right. \\
&\times G^*\left(\mathbf{r}', \mathbf{r}_2, E\right) G\left(\mathbf{r}_1, \mathbf{r}_1', E + \hbar\omega\right) \\
&\times G^*\left(\mathbf{r}_2, \mathbf{r}_2', E + \hbar\omega\right) \bar{s}\left(\mathbf{r}_2, \mathbf{r}_1, \hbar\omega\right) \bar{s}\left(\mathbf{r}_2{}', \mathbf{r}_1', \hbar\omega'\right) \rho_0\left.\left(\mathbf{r}_1{}', \mathbf{r}_2', E + \hbar\omega + \hbar\omega'\right)\right] \\
&+ \ldots
\end{aligned} \tag{16.43}$$

or for the nth-order iterative solution:

$$\begin{aligned}
\rho_n\left(\mathbf{r}, \mathbf{r}', E\right) = \int d\mathbf{r}_1 \int d\mathbf{r}_2 \int d\hbar\omega &\left[G\left(\mathbf{r}, \mathbf{r}_1, E\right) G^*\left(\mathbf{r}', \mathbf{r}_2, E\right)\right. \\
&\times \bar{s}\left(\mathbf{r}_2, \mathbf{r}_1, \hbar\omega\right) \rho_{n-1}\left.\left(\mathbf{r}_1, \mathbf{r}_2, E + \hbar\omega\right)\right]
\end{aligned} \tag{16.44}$$

where the first term is the density matrix of elastically scattered electrons; the second and third terms are density matrices of single and double inelastically scattered electrons, respectively; and the higher order terms correspond to higher order inelastic scattering.

As a summary, (16.36) and (16.42) are two equations of the one-particle density matrix theory. Two approximations were made in deriving these equations. The first approximation is that each time the electron inelastically interacts with the crystal atoms, the collision is assumed to take place as if the crystal was in its ground state and the effect of any previous collision was very small. This assumption is mathematically expressed as $H_{nm} = H_{n0}$ for $m \neq n$. Secondly the single inelastic-scattering approximation is assumed in deriving the absorption potential. The first approximation means that multiple inelastic scattering, each successive inelastic interaction is described by the single inelastic-scattering approximation. Cases in which two inelastic events occur simultaneously during interaction with the same atoms have been ignored. Thus the theory may not be adequate for calculating the scattering cross section of coherent TDS-single-electron, double inelastic excitation, as discussed in Sect. 8.6 (Wang and Fisher 1993). Therefore the kinetic equation is a reasonable approach if all inelastic events occur consecutively, so that no simultaneous excitation of more than one event is allowed. This condition restricts the application of the one-particle density matrix theory for calculating the scattering cross section of double-plasmon excitation (Spence and Spargo 1971; Ashley and Ritchie 1970), in which two plasmons are created during a single inelastic-scattering process. Therefore the kinetic equation cannot be applied to cases of multi-inelastic scattering processes. Finally since the kinetic equation is still an integral equation that is not the final form best suited for numerical calculation, it has to be solved iteratively if we are interested in high-order scattering. The theory is expressed in a tidy form, but its solution is not easy to obtain in comparison to other theories introduced in Chaps. 8 and 9.

There are several advantages associated with the density matrix theory. First dynamic multiple elastic and multiple inelastic scatterings are comprehensively included in the kinetic equation. The incoherence of inelastic waves scattered by different crystal states is evaluated before numerical calculation, and contributions from all excited states are included in a single formula. Second the multiple elastic and multiple inelastic scatterings of an electron are entirely determined by two universal functions, the crystal potential and the mixed dynamic form factor (Dudarev et al. 1993a). Third since we did not assume the shape of the crystal or scattering geometry in deriving the kinetic equation, it can be applied to such cases as RHEED. Finally small-angle scattering was not assumed in the kinetic equation, so this theory can be applied to calculate the diffraction of backscattered electrons. This type of calculation, however, is difficult or impossible using other inelastic-scattering theories introduced in Chaps. 8–12.

16.3.3 Delocalized Multiple Inelastic Scattering

The simplest multiple inelastic scattering is the excitation of plasmons in metals and semiconductors. For valence excitations, the energy-dependent dynamic form factor in real space is written as follows:

$$
\begin{aligned}
\bar{s}\left(\mathbf{r}', \mathbf{r}, \hbar\omega\right) =& \frac{4m_e^2}{\hbar^4} \sum_{n\neq 0} \int d\tau \int d\tau' \exp\left[2\pi i\left(\tau\cdot\mathbf{r} - \tau'\cdot\mathbf{r}'\right)\right] \\
& \times \left[\delta H_{n0}(\tau)\delta H_{n0}^{*}\left(\tau'\right)\right]\delta\left(\hbar\omega + \varepsilon_0 - \varepsilon_n\right) \\
=& \frac{4m_e^2}{\hbar^4} \int d\tau \int d\tau' \exp\left[2\pi i\left(\tau\cdot\mathbf{r} - \tau'\cdot\mathbf{r}'\right)\right] \\
& \times [\sum_{n\neq 0} H_{0n}\left(-\tau'\right) H_{n0}(\tau)]\delta\left(\hbar\omega + \varepsilon_0 - \varepsilon_n\right)
\end{aligned}
\tag{16.45}
$$

Using the relation given in (6.113) for valence excitation, $\bar{s}$ is directly related to the generalized dielectric function by

$$\bar{s}\left(\mathbf{r}', \mathbf{r}, \hbar\omega\right) = \frac{i m_e^2 e^2}{2\pi^3 \hbar^3 \varepsilon_0} \int d\tau \int d\tau' \exp\left[2\pi i \left(\tau \cdot \mathbf{r} - \tau' \cdot \mathbf{r}'\right)\right] \times \left[\frac{1}{\tau^2 \varepsilon_{\tau'\tau}(\omega)} - \frac{1}{\tau'^2 \varepsilon^*_{\tau\tau'}(\omega)}\right] \tag{16.46}$$

For an homogeneous medium with dielectric function $\varepsilon(\omega, \tau)$, (16.46) is simplified as

$$\bar{s}\left(\mathbf{r}', \mathbf{r}, \hbar\omega\right) = \frac{m_e^2 e^2}{\pi^3 \hbar^3 \varepsilon_0} \int d\tau \frac{\exp\left[2\pi i \tau \cdot \left(\mathbf{r} - \mathbf{r}'\right)\right]}{\tau^2} \operatorname{Im}\left[-\frac{1}{\varepsilon(\omega, \tau)}\right] \tag{16.47}$$

Neglecting the dispersion dependence of dielectric function, we have

$$\bar{s}\left(\mathbf{r}', \mathbf{r}, \hbar\omega\right) = \frac{m_e^2 e^2}{\pi^2 \hbar^3 \varepsilon_0} \frac{1}{|\mathbf{r} - \mathbf{r}'|} \operatorname{Im}\left[-\frac{1}{\varepsilon(\omega)}\right] \tag{16.48}$$

This relation and Green's function given in (10.37) for transmission electron diffraction were applied in the numerical calculations of the kinetic equation to assess the contribution of multiple valence excitations in electron diffraction (Dudarev et al. 1993a).

Figure 16.2 compares the calculated and the observed rocking curves of the Si (220) CBED disk. Calculated curves are shown by dotted (11-beam Bloch wave

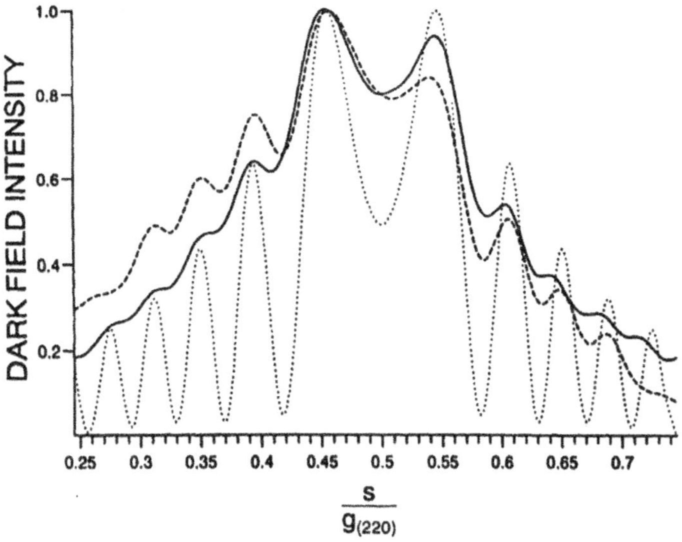

Fig. 16.2 Experimentally observed (*solid line*) and theoretically simulated intensity profiles of the Si(220) convergent beam diffraction disk. Dotted curves were calculated using an 11-beam approximation of dynamic elastic-scattering Bethe theory (see Chap. 3). The dashed curve is the calculated result of the kinetic equation, including multiple plasmon scattering; electron beam energy is 80 keV, crystal thickness is 286 nm. Courtesy of Dr. Dudarev et al. (1993a)

elastic-scattering calculation with absorption) and dashed (numerical solution of the kinetic equation for a one-particle density matrix) lines. The experimental observation is shown by a solid line, which is an intensity line scan without using an energy filter. It is apparent that the calculated result of the kinetic equation gives better fit to the experimental result. In particular, the solution of the kinetic equation reproduces the asymmetry of the rocking curve within the (220) disk, while the Bloch wave calculation results in a high-symmetry profile. This study also indicates that electron multiple scattering leads to the damping of the coherence between states of an electron on different branches of the dispersion surface. The attenuation distance is comparable to the inelastic mean free path (Dudarev et al. 1992).

16.4 Summary

Multiple inelastic electron scattering in TEM becomes significant if the specimen is thick. Theories that describe the angular distribution of the multiple inelastically scattered electron are usually very complex. In this chapter, we introduced a few approaches that can reasonably be applied to the calculation of diffraction patterns of electrons. The energy and angular distributions of inelastic electrons are described by Poisson's distribution law if all diffraction effects are ignored. The one-particle density matrix theory is a tidy formal theory of multiple inelastic scattering. The kinetic equation introduced here covers all scattering characteristics of inelastic electrons, but it cannot be applied to cases involving more than one type of inelastic event, such as coupled-phonon and single-electron excitations. The modified multislice theory introduced in Sect. 14.4 permits a greatly simplified calculation of elastic rescattering of electrons after inelastic excitations.

Chapter 17
Inelastic Excitation of Crystals in Thermal Equilibrium with the Environment

Theories presented in Chaps. 6–16 regarding inelastic electron scattering in crystals were based on the assumption that the incident electron crystal system is isolated from the environment, since the time-independent Schrödinger equation, which is usually employed, is only valid for an isolated system of constant energy and momentum. In practice, the incoming electron crystal system is in contact with the environment, so that the total energy of the system fluctuates due to the exchange of energy with the environment. This fluctuation eventually affects the scattering behavior of the electron. This question was considered by Fanidis et al. (1992, 1993), who proposed a rigorous theory to solve the problem. An alternative approach was proposed by Dudarev et al. (1993a) based on the one-particle density matrix theory. Statistical mechanics calculations predicted that the magnitude of the energy fluctuation due to the thermal equilibrium of a copper crystal of volume 10^6 nm^3 at room temperature is on the order of 400 eV (Mandl 1971).

In this chapter, we outline the first-principles treatment of electron inelastic scattering by a crystal in thermal equilibrium with its environment (Fanidis et al. 1992, 1993).

17.1 Basic Equations

We now consider that the system is in a stationary state with a constant energy E (which may be degenerate). The system is characterized by a wave function that is a superposition of the products of crystal wave functions and incident electron wave functions:

$$\Phi_s(\mathbf{r}, \mathbf{r}_1, \ldots, \mathbf{r}_M, E) = \sum_n a_n(\mathbf{r}_1, \ldots, \mathbf{r}_M, E)\, \Psi_n(\mathbf{r}, E) \qquad (17.1)$$

Z. L. Wang, *Elastic and Inelastic Scattering in Electron Diffraction and Imaging*, Springer Series in Solid-State Sciences 205, https://doi.org/10.1007/978-3-031-90819-4_17

where the system energy is introduced to account for the fluctuation effect. By introducing (17.1) into (6.1), multiplying by a_m, and integrating over $\mathbf{r}_1, \ldots, \mathbf{r}_M$, we obtain a set of equations analogous to Yoshioka's equations:

$$\left[\frac{\hbar^2}{2m_0}\nabla^2 + (E - \varepsilon_n)\right]\Psi_n(\mathbf{r}, E) = \gamma \sum_m H_{nm}(\mathbf{r}, E)\Psi_m(\mathbf{r}, E) \tag{17.2a}$$

where the transition matrix for the system of energy E is

$$H_{nm}(\mathbf{r}, E) = \int a_n^* (\mathbf{r}_1, \ldots, \mathbf{r}_M, E)\, H(\mathbf{r}, \mathbf{r}_1, \ldots, \mathbf{r}_M) \times a_m (\mathbf{r}_1, \ldots, \mathbf{r}_M, E)\, d\mathbf{r}_1, \ldots, d\mathbf{r}_M \tag{17.2b}$$

We now look for the solution to (17.2a) and (17.2b) under the high-energy approximation. By writing

$$\Psi_n(\mathbf{r}, E) = \exp(2\pi i \mathbf{K}_0 \cdot \mathbf{r})\, \Phi_n(\mathbf{r}, E) \tag{17.3}$$

Equation (17.2a) becomes

$$\frac{\partial \Phi_m(\mathbf{r}, E)}{\partial z} \approx \varsigma\left[\Xi + \delta k_n^2\right]\Phi_n(\mathbf{r}, E) + \varsigma \sum_m P_{nm}(\mathbf{r}, E)\Phi_m(\mathbf{r}, E) \tag{17.4}$$

where

$$\delta k_n^2 = 4\pi^2\left(K_n^2 - K_0^2\right) \tag{17.5a}$$

and

$$P_{nm}(\mathbf{r}, E) = -\frac{2m_e}{\hbar^2} H_{nm}(\mathbf{r}, E) \tag{17.5b}$$

The boundary condition is $\Phi_m(\mathbf{b}, 0, E) = \delta_{m0}$.

In statistical mechanics, the probability that a system is of energy E is

$$\rho(E) = \frac{\exp(-E/k_B T)}{\sum_E \exp(-E/k_B T)} \tag{17.6}$$

This is called the weight function of the system. The experimentally measured value of a quantity F is

$$\langle F \rangle_E \equiv \sum_E \rho(E) F(E) \tag{17.7}$$

As stated in (17.4), the electron wave function $\Phi_m(\mathbf{r}, E)$ depends on the energy of the system. For convenience, we split $\Phi_m(\mathbf{r}, E)$ into two components: one part is its averaged value, which is independent of E; the other part is a small deviation that depends on E:

$$\Phi_m(\mathbf{r}, E) = \bar{\Phi}_m(\mathbf{r}) + \delta\Phi_m(\mathbf{r}, E) \tag{17.8}$$

where $\bar{\Phi}_m(\mathbf{r}) = \langle \Phi_m(\mathbf{r}, E)\rangle_E$ and $\langle \delta\Phi_m(\mathbf{r}, E)\rangle_E = 0$. The boundary conditions are $\bar{\Phi}_m(\mathbf{b}, 0) = \delta_{m0}$ and $\delta\Phi_m(\mathbf{b}, 0, E) = 0$. Substituting (17.8) into (17.4), we obtain

$$\frac{\partial}{\partial z}\left(\bar{\Phi}_n + \delta\Phi_n\right) \approx \varsigma\left(\Xi + \delta k_n^2\right)\left[\bar{\Phi}_n(\mathbf{r}) + \delta\Phi_n\right] + \varsigma\sum_m P_{nm}\left(\bar{\Phi}_m + \delta\Phi_m\right) \tag{17.9}$$

Averaging (17.9) over E, using the relations associated with (17.8), and assuming that δK_n^2 is not affected by slight variations in E due to thermal fluctuation, yields

$$\frac{\partial}{\partial z}\bar{\Phi}_n \approx \varsigma\left(\Xi + \delta k_n^2\right)\bar{\Phi}_n + \varsigma\sum_m\left(\bar{P}_{nm}\bar{\Phi}_m + < P_{nm}\delta\Phi_m >_E\right) \tag{17.10}$$

where $\bar{P}_{nm} =< P_{nm}(\mathbf{r}, E) >_E$. Subtracting (17.10) from (17.9) gives

$$\begin{aligned}\frac{\partial}{\partial z}\delta\Phi_n \approx{}& \varsigma\left(\Xi + \delta k_n^2\right)\delta\Phi_n \\ &+ \varsigma\sum_m\left[\left(P_{nm} - \bar{P}_{nm}\right)\bar{\Phi}_m + P_{nm}\delta\Phi_m - < P_{nm}\delta\Phi_m >_E\right]\end{aligned} \tag{17.11}$$

By defining a deviation matrix element

$$\delta P_{nm} = \left(P_{nm} - \bar{P}_{nm}\right) \tag{17.12a}$$

or

$$P_{nm} = \bar{P}_{nm} + \delta P_{nm} \tag{17.12b}$$

which satisfies $< \delta P_{nm} >_E = 0$, (17.10) and (17.11) can be rewritten as

$$\frac{\partial}{\partial z}\bar{\Phi}_n \approx \varsigma\left(\Xi + \delta k_n^2\right)\bar{\Phi}_n + \varsigma\sum_m\left(\bar{P}_{nm}\bar{\Phi}_m + < \delta P_{nm}\delta\Phi_m >_E\right) \tag{17.13}$$

$$\frac{\partial}{\partial z}\delta\Phi_n \approx \varsigma\left(\Xi + \delta k_n^2\right)\delta\Phi_n + \varsigma\sum_m\left(\delta P_{nm}\bar{\Phi}_m + P_{nm}\delta\Phi_m - < P_{nm}\delta\Phi_m >_E\right) \tag{17.14}$$

These are the two basic equations that describe the evolution of different components of the electron wave function. The $\bar{\Phi}_n$ and $\delta\Phi_m$ are coupled by the deviation matrix elements δP_{nm}.

17.2 Electron Images and Diffraction Patterns

The recorded electron image is the result of the incoherent summation of components contributed by electrons scattered by different crystal states

$$
\begin{aligned}
I\mathbf{b}) &= \sum_m \langle |\Phi_m(\mathbf{r}, E) \otimes T_{\text{obj}}(\mathbf{b})|^2 \rangle_E = \sum_m \langle |[\bar{\Phi}_m(\mathbf{r}) + \delta\Phi_m(\mathbf{r}, E)] \otimes T_{\text{obj}}(\mathbf{b}|^2 \rangle_E \\
&= \sum_m \left|\bar{\Phi}_m(\mathbf{r}) \otimes T_{\text{obj}}(\mathbf{b})\right|^2 + \sum_m \langle \left|\delta\Phi_m(\mathbf{r}, E) \otimes T_{\text{obj}}(\mathbf{b})\right|^2 \rangle_E \qquad (17.15)
\end{aligned}
$$

since $< \delta\Phi_m(\text{r}, E) >_E = 0$. For electron diffraction patterns, we similarly have

$$
I(\mathbf{u}) = \sum_m \left|\bar{\Phi}_m(\mathbf{u})\right|^2 + \sum_m < |\delta\Phi_m(\mathbf{u}, E)|^2 >_E \qquad (17.16)
$$

Thus splitting the wave function into an average component and a fluctuating part also separates the image or diffraction pattern intensity into two parts. The sum over m represents the intensity integration over all crystal states. Therefore, strictly speaking, calculations have to be performed for each excited state in inelastic scattering, and there is no easy method to sum over contributions made by all crystal states analytically unless some approximations are made.

17.3 Solution to a Fluctuating Component

We now try to transform (17.14) into a different form to find its first-order solution. Substituting (17.12b) into (17.14) yields

$$
\begin{aligned}
\frac{\partial}{\partial z}\delta\Phi_n &= \varsigma\left(\Xi + \delta k_n^2\right)\delta\Phi_n \\
&\quad + \varsigma \sum_m \left(\delta P_{nm}\bar{\Phi}_m + \bar{P}_{nm}\delta\Phi_m + \delta P_{nm}\delta\Phi_m - < \delta P_{nm}\delta\Phi_m >_E\right) \qquad (17.17)
\end{aligned}
$$

The $\delta P_{nm}\delta\Phi_m$ and $< \delta P_{nm}\delta\Phi_m >_E$ terms represent the higher order effects, and these can be ignored for thin crystals. Also by neglecting the coupling between $\delta\Phi_n$ and $\delta\Phi_m$, (17.17) is approximated as

$$
\frac{\partial}{\partial z}\delta\Phi_n \approx \varsigma\left(\Xi + \bar{P}_{nn} + \delta k_n^2\right)\delta\Phi_n + \varsigma\sum_m \delta P_{nm}\bar{\Phi}_m \qquad (17.18)
$$

If only ZOLZ reflections are considered so that P_{nn} is independent of z, the solution of (17.18) can be directly written as

$$\delta\Phi_n(z) \approx \varsigma \int_0^z dz' \exp\left[\varsigma\left(\Xi + \bar{P}_{nn} + \delta k_n^2\right)(z - z')\right] \sum_m \delta P_{nm}(z')\,\bar{\Phi}_m(z') \tag{17.19}$$

where only variable z is shown for simplicity.

We now consider the following approximations to calculate $\langle \delta P_{n'n}(\mathbf{b}, z)\delta\Phi_n(\mathbf{b}, z)\rangle_E$. For localized scattering, $\delta P_{n'n}$ is expressed as a sum over the scattering function P_{nm}^{κ} centered at each atomic site

$$\begin{aligned}\delta P_{n'n}(\mathbf{b}, z) &= \sum_{\kappa'} \delta P_{n'n}^{\kappa'}\left(\mathbf{b} - \mathbf{b}_{\kappa'}, z - z_{\kappa'}\right) \\ &= \sum_{\kappa'} \left[P_{n'n}^{\kappa'}\left(\mathbf{b} - \mathbf{b}_{\kappa',z} - z_{\kappa'}\right) - \left\langle P_{n'n}^{\kappa'}\left(\mathbf{b} - \mathbf{b}_{\kappa',z} - z_{\kappa'}\right)\right\rangle_E\right]\end{aligned} \tag{17.20}$$

Substituting (17.20) into (17.19), we get

$$\begin{aligned}\delta\Phi_n(z) \approx \varsigma \sum_{\kappa'} \int_0^z dz' \exp\left[\varsigma\left(\Xi + \bar{P}_{nn} + \delta k_n^2\right)(z - z')\right] \\ \times \sum_m \delta P_{n'n}^{\kappa'}\left(\mathbf{b} - \mathbf{b}_{\kappa'}, z - z_{\kappa'}\right)\bar{\Phi}_m(z')\end{aligned} \tag{17.21}$$

For localized scattering, $\delta P_{nm}^{\kappa}(\mathbf{b} - \mathbf{b}_k, z)$ is appreciable only near atom sites. For z' integration within a single atom depth, the only rapid variation function is $\delta P_{nm}^{K}(\mathbf{b} - \mathbf{b}_{\kappa}, z)$; thus we can omit the dynamic scattering operator from the integration (Fanidis et al. 1993). Hence

$$\delta\Phi_n(\mathbf{b}, z) \approx \varsigma \sum_{\kappa} \mathrm{O_P}\left(\mathbf{b}, z - z_{\kappa}\right) \sum_m \delta P_{nm}^{\kappa}\left(\mathbf{b} - \mathbf{b}_{\kappa}\right)\bar{\Phi}_m\left(\mathbf{b}, z_{\kappa}\right) \tag{17.22}$$

where $\delta P_{nm}^{\kappa}(\mathbf{b} - \mathbf{b}_{\kappa}) = \int dz \delta P_{nm}^{\kappa}(\mathbf{b} - \mathbf{b}_{\kappa}, z)$ is the projected scattering function, and

$$\mathrm{O_P}\left(\mathbf{b}, z - z_k\right) = \exp\left[\varsigma\left(\Xi + \bar{P}_{nn} + \delta k_n^2\right)(z - z')\right] \tag{17.23}$$

The numerical calculation of the dynamic scattering operator can follow the conventional multislice theory introduced in Sect. 11.10.3.

We now proceed with the next step for solving (17.13). Multiplying (17.20) by (17.21) and summing over n, we have

$$\begin{aligned}&\sum_n < \delta P_{n'n}(\mathbf{b}, z)\delta\Phi_n(\mathbf{b}, z) >_E \\ &= \varsigma \sum_m \sum_n \sum_{\kappa} \sum_{\kappa'} < \delta P_{n'n}^{\kappa'}(\mathbf{b} - \mathbf{b}_{\kappa'}, z - z_{\kappa'})\mathrm{O_P}\left(\mathbf{b}, z - z_{\kappa}\right) \\ &\times \left[\delta P_{nm}^{\kappa}\left(\mathbf{b} - \mathbf{b}_{\kappa}\right)\bar{\Phi}_m\left(\mathbf{b}, z_{\kappa}\right)\right] >_E\end{aligned} \tag{17.24}$$

For localized scattering, the product of $\delta P_{n'n}^{\kappa'}\delta P_{nm}^{\kappa}$ tends to be very small if the two functions are not centered at the same atom (i.e., $\kappa \neq \kappa'$), and we can discard the terms with $\kappa \neq \kappa'$. Also since the quantity defined in (17.24) is responsible for the absorption effect, the integrated intensity of the absorbed electrons is not affected by the dynamic scattering that causes intensity redistribution. Thus (17.24) is approximated as

$$\sum_n < \delta P_{n'n}(\mathbf{b}, z)\delta\Phi_n(\mathbf{b}, z) >_E$$
$$= \varsigma \sum_m \sum_n \sum_\kappa \langle \delta P_{n'n}^{\kappa} (\mathbf{b} - \mathbf{b}_\kappa, z - z_\kappa)\, \delta P_{nm}^{\kappa} (\mathbf{b} - \mathbf{b}_\kappa)\rangle_E \bar{\Phi}_m (\mathbf{b}, z_k) \tag{17.25}$$

As assumed at the beginning of this section, coupling terms of different states are neglected, so that terms for which $m \neq n'$ are dropped

$$\sum_n \langle \delta P_{n'n}(\mathbf{b}, z)\delta\Phi_n(\mathbf{b}, z)\rangle_E \approx \varsigma \sum_n \sum_\kappa \langle \delta P_{n'n}^{\kappa} (\mathbf{b} - \mathbf{b}_k)|^2\rangle_E \bar{\Phi}_{n'} (\mathbf{b}, z_\kappa)$$
$$= \varsigma \sum_n \sum_\kappa [\langle |P_{n'n}^{\kappa} (\mathbf{b} - \mathbf{b}_\kappa)|^2\rangle_E - |\langle P_{n'n}^{\kappa} (\mathbf{b} - \mathbf{b}_\kappa)\rangle_E|^2]\bar{\Phi}_{n'} (\mathbf{b}, z_\kappa) \tag{17.26}$$

This expression is used to find the solution of $\bar{\Phi}_n$.

17.4 Contributions of Fluctuating Components to Electron Diffraction Pattern and Image

Under the single inelastic-scattering approximation, the contribution of the fluctuating component to the electron diffraction pattern can be calculated by

$$\delta I(\mathbf{u}) = \sum_m \langle |\delta\Phi_m(\mathbf{u}, E)|^2\rangle_E$$
$$\approx |\varsigma|^2 \sum_m \langle | \sum_\kappa \mathrm{FT}[\,\mathrm{O_p} (\mathbf{b}, d - z_\kappa) \left[\delta P_{m0}^{\kappa} (\mathbf{b} - \mathbf{b}_\kappa)\, \bar{\Phi}_0 (\mathbf{b}, z_k)\right]|^2\rangle_E$$
$$= |\varsigma|^2 \sum_m \sum_\kappa \sum_{\kappa'} \int d\mathbf{b} \int d\mathbf{b}' \exp\left[-2\pi i\tau \cdot (\mathbf{b} - \mathbf{b}')\right]$$
$$\times < \left\{\mathrm{O_P} (\mathbf{b}, d - z_\kappa) \left[\delta P_{m0}^{\kappa} (\mathbf{b} - \mathbf{b}_\kappa)\, \bar{\Phi}_0 (\mathbf{b}, z_\kappa)\right] \mathrm{O_P^*} (\mathbf{b}', d - z_{\kappa'})\right.$$
$$\left.\times \left[\left(\delta P_{m0}^{\kappa'} (\mathbf{b}' - \mathbf{b}_{\kappa'})\right)^* \bar{\Phi}_0^* (\mathbf{b}', z_{\kappa'})\right]\right\} >_E \tag{17.27}$$

Since $\mathbf{b}$ and $\mathbf{b}'$ are two distinct variables such that $\mathrm{O_p} (\mathbf{b}, d - z_\kappa)$ and $\mathrm{O_p^*} (\mathbf{b}', d - z_{\kappa'})$ commute, then

$$\delta I(\mathbf{u}) = |\varsigma|^2 \sum_{\kappa} \sum_{\kappa'} \int d\mathbf{b} \int d\mathbf{b}' = \exp[-2\pi i\tau \cdot (\mathbf{b}-\mathbf{b}')] \mathrm{O}_\mathrm{P}(\mathbf{b}, d-z_\kappa) \mathrm{O}_\mathrm{P}^*(\mathbf{b}', d-z_{\kappa'})$$
$$\times \left[\sum_m \langle \delta P_{m0}^{\kappa} (\mathbf{b}-\mathbf{b}_\kappa) \, \delta P_{m0}^{\kappa_0'} (\mathbf{b}'-\mathbf{b}_{\kappa'}) \rangle_E \bar{\Phi}_0 (\mathbf{b}-\mathbf{b}_\kappa) \, \bar{\Phi}_0^* (\mathbf{b}', z_{\kappa'}) \right] \quad (17.28)$$

An analogous expression is derived for calculating the contribution of fluctuating components to the TEM image

$$\delta I(\mathbf{b}) = \sum_m < | \int d\mathbf{b}' \delta\Phi_m (\mathbf{b}', d) \, T_{\mathrm{obj}} (\mathbf{b}'-\mathbf{b}) \, |^2 >_E$$
$$= |\varsigma|^2 \sum_{\kappa} \sum_{\kappa'} \int d\mathbf{b}' \int d\mathbf{b}'' T_{\mathrm{obj}} (\mathbf{b}'-\mathbf{b}) \, T_{\mathrm{obj}}^* (\mathbf{b}''-\mathbf{b}) \, \mathrm{O}_\mathrm{P} (\mathbf{b}', d-z_\kappa) \, \mathrm{O}_\mathrm{P}^* (\mathbf{b}'', d-z_\kappa)$$
$$\times [\sum_m \langle \delta P_{m0}^{\kappa} (\mathbf{b}'-\mathbf{b}_\kappa) \, [\delta P_{m0}^{\kappa'} (\mathbf{b}''-\mathbf{b}_{k'}]^* \rangle_E \bar{\Phi}_0 (\mathbf{b}', z_\kappa) \, \bar{\Phi}_0^* (\mathbf{b}'', z_\kappa)] \quad (17.29)$$

Therefore the essential part of the calculations in (17.27) and (17.29) is to evaluate the quantity

$$PP_{\kappa}^{\kappa'} (\mathbf{b}', \mathbf{b}'') \equiv \sum_m \langle \delta P_{m0}^{\kappa} (\mathbf{b}'-\mathbf{b}_\kappa) \left[\delta P_{m0}^{\kappa'} (\mathbf{b}''-\mathbf{b}_{\kappa'}) \right]^* \rangle_E$$
$$= \sum_m \langle [P_{m0}^{\kappa} (\mathbf{b}'-\mathbf{b}_\kappa) - \bar{P}_{m0}^{\kappa} (\mathbf{b}'-\mathbf{b}_\kappa)]$$
$$\left[P_{m0}^{\kappa'} (\mathbf{b}''-\mathbf{b}_{\kappa'}) - \bar{P}_{m0}^{\kappa'} (\mathbf{b}''-\mathbf{b}_{\kappa'}) \right]^* \rangle_E$$
$$= \sum_m \left\{ \langle P_{m0}^{\kappa} (\mathbf{b}'-\mathbf{b}_\kappa) \left[P_{m0}^{\kappa'} (\mathbf{b}''-\mathbf{b}_{\kappa'}) \right]^* \rangle_E \right.$$
$$\left. - \bar{P}_{m0}^{\kappa} (\mathbf{b}'-\mathbf{b}_\kappa) \left[\bar{P}_{m0}^{\kappa'} (\mathbf{b}''-\mathbf{b}_{\kappa'}) \right]^* \right\} \quad (17.30)$$

This function can be calculated for the specific inelastic-scattering process for which the interaction Hamiltonian is known.

17.5 Nonfluctuating Inelastic Components

We now try to find the solution of $\bar{\Phi}_n$ under the single inelastic-scattering approximation. Equation (17.13) is approximated as

$$\frac{\partial}{\partial z} \bar{\Phi}_n \approx \varsigma \left(\Xi + \bar{P}_{nn} + \delta k_n^2 \right) \bar{\Phi}_n + \varsigma \bar{P}_{n0} \bar{\Phi}_0 + \varsigma \sum_m \langle \delta P_{nm} \delta \Phi_m \rangle_E \quad n \neq 0 \quad (17.31)$$

where the last term is kept to illustrate the effect of energy fluctuation. The last term can be replaced by the expression given in (17.30), so that (17.31) becomes

$$\frac{\partial}{\partial z}\bar{\Phi}_n \approx \varsigma\left[\Xi + \bar{P}_{nn} + i\left(\bar{P}_{nn}\right)_I + \delta k_n^2\right]\bar{\Phi}_n + \varsigma \bar{P}_{n0}\bar{\Phi}_0 \tag{17.32}$$

where the imaginary component of the potential is

$$\left(\bar{P}_{nn}\right)_I = |\varsigma \sum_m \sum_\kappa \langle|\delta P_{nm}^\kappa(\mathbf{b}-\mathbf{b}_\kappa)|^2\rangle_E.$$
$$= |\varsigma| \sum_m \sum_\kappa \left[\langle|P_{nm}^\kappa(\mathbf{b}-\mathbf{b}_\kappa)|^2\rangle_E - \left|\langle P_{mn}^\kappa(\mathbf{b}-\mathbf{b}_k)\rangle_E\right|^2\right] \tag{17.33}$$

The absorption potential $\left(\bar{P}_{nm}\right)_I$ vanishes if the incident electron crystal system is isolated from the environment. It is important to note that the absorption potential $\left(\bar{P}_{nn}\right)_I$ here results from energy fluctuation in the system rather than the transition of electrons to other states, as in conventional electron diffraction introduced in Chap. 6. If HOLZ effects are ignored, the solution of (17.32) is

$$\bar{\Phi}_n \approx \varsigma \int_0^z dz' \exp\left\{\varsigma\left[\Xi + \bar{P}_{nn} + i\left(\bar{P}_{nn}\right)_I + \delta k_n^2\right](z-z')\right\}\left[\bar{P}_{n0}(z')\,\bar{\Phi}_0(z')\right] \tag{17.34}$$

where the numerical calculation of the dynamic scattering operator can follow the conventional multislice theory in Sect. 11.10.3.

The two basic equations governing electron diffraction in a crystal in thermal equilibrium with the environment are (17.19) and (17.34), which have been applied to investigate the effect of phonon scattering in HRTEM (Fanidis et al. 1993).

Analogous to the discussion leading to (17.27) and (17.29), the contribution made by the nonfluctuating inelastic waves to the diffraction pattern is

$$I_i(\mathbf{u}) = \sum_{m=1} \left|\bar{\Phi}_m(\mathbf{u})\right|^2$$
$$\approx |\zeta|^2 \sum_{m=1}\sum_\kappa\sum_{\kappa'} \int d\mathbf{b} \int d\mathbf{b}' \exp[-2\pi i\tau\cdot(\mathbf{b}-\mathbf{b}')]\{\mathrm{O_{Pm}}(\mathbf{b}, d-z_\kappa)\mathrm{O}^*_{\mathrm{Pm}}(\mathbf{b}', d-z_{\kappa'})$$
$$\times \{\bar{P}_{m0}^\kappa(\mathbf{b}-\mathbf{b}_\kappa)\left[\bar{P}_{m0}^{\kappa'}(\mathbf{b}'-\mathbf{b}_{\kappa'})\right]^* \bar{\Phi}_0(\mathbf{b}, z_\kappa)\,\bar{\Phi}_0^*(\mathbf{b}', z_{\kappa'})\}$$
$$\approx |\zeta|^2 \sum_\kappa\sum_{\kappa'} \int d\mathbf{b} \int d\mathbf{b}' \exp[-2\pi i\tau\cdot(\mathbf{b}-\mathbf{b}')]\,\mathrm{O_P}(\mathbf{b}, d-z_\kappa)\,\mathrm{O_P^*}(\mathbf{b}', d-z_{\kappa'})$$
$$\times \{\sum_{m=1} \bar{P}_{m0}^\kappa(\mathbf{b}-\mathbf{b}_\kappa)\left[\bar{P}_{m0}^{\kappa'_0}(\mathbf{b}'-\mathbf{b}_{\kappa'})\right]^* \bar{\Phi}_0(\mathbf{b}, z_k)\,\bar{\Phi}_0^*(\mathbf{b}', z_{\kappa'})\} \tag{17.35}$$

where

$$\mathrm{Op}_m(\mathbf{b}, d-z) = \exp\left\{\varsigma\left[\Xi + \bar{P}_{mm} + i\left(\bar{P}_{mm}\right)_I + \delta k_n^2\right](d-z)\right\} \tag{17.36}$$

An analogous expression is derived for the TEM image

$$\begin{aligned}
I_i(\mathbf{b}) &= \sum_{m=1} \left|\bar{\Phi}_m(\mathbf{r}) \otimes T_{\text{obj}}(\mathbf{b})\right|^2 \\
&\approx |\varsigma|^2 \sum_{m=1}\sum_{\kappa}\sum_{\kappa'} \int d\mathbf{b}' \int d\mathbf{b}'' T_{\text{obj}}(\mathbf{b}'-\mathbf{b})\, T_{\text{obj}}^*(\mathbf{b}''-\mathbf{b})\, \mathrm{Op}_m(\mathbf{b}', d-z_\kappa)\, \mathrm{Op}_m^*(\mathbf{b}'', d-z_\kappa) \\
&\times \left\{\bar{P}_{m0}^{\kappa}(\mathbf{b}'-\mathbf{b}_\kappa)\left[\bar{P}_{m0}^{\kappa'}(\mathbf{b}''-\mathbf{b}_{\kappa'})\right]^* \bar{\Phi}_0(\mathbf{b}', z_\kappa)\, \bar{\Phi}_0^*(\mathbf{b}'', z_\kappa)\right\} \\
&= |\zeta|^2 \sum_{\kappa}\sum_{\kappa'} \int d\mathbf{b}' \int d\mathbf{b}'' T_{\text{obj}}(\mathbf{b}'-\mathbf{b})\, T_{\text{obj}}^*(\mathbf{b}''-\mathbf{b})\, \mathrm{O_P}(\mathbf{b}', d-z_\kappa)\, \mathrm{O_P^*}(\mathbf{b}'', d-z_\kappa) \\
&\times \left\{\sum_{m=1} \bar{P}_{m0}^{\kappa}(\mathbf{b}'-\mathbf{b}_\kappa)\left[\bar{P}_{m0}^{\kappa'}(\mathbf{b}''-\mathbf{b}_{\kappa'})\right]^* \bar{\Phi}_0(\mathbf{b}', z_\kappa)\, \bar{\Phi}_0^*(\mathbf{b}'', z_k)\right\}
\end{aligned} \tag{17.37a}$$

Therefore the essential part of the calculations in (17.35), (17.37a), and (17.37b) is to evaluate

$$\left[\sum_{m=1} \bar{P}_{m0}^{\kappa}(\mathbf{b}'-\mathbf{b}_\kappa)\left[\bar{P}_{m0}^{\kappa_0'}(\mathbf{b}''-\mathbf{b}_{\kappa'})\right]^*\right] \tag{17.37b}$$

17.6 Absorption Effect for Elastic Waves

We now come to examine the elastic wave. From (17.13), the elastic component is

$$\frac{\partial}{\partial z}\bar{\Phi}_0 \approx \varsigma\left(\Xi + \bar{P}_{00}\right)\bar{\Phi}_0 + \varsigma \sum_{m\neq 0}\left(\bar{P}_{0m}\bar{\Phi}_m\right) + \sum_m \left(< \delta P_{0m}\delta\Phi_m >_E\right) \tag{17.38}$$

In calculating the coupling term $\bar{P}_{0m}\bar{\Phi}_m$, we substitute the solution given in (17.34) for $\bar{\Phi}_n$. For localized scattering, delocalized terms are neglected and the exponential operator is dropped, similar to the approximation made in calculating $< \delta P_{0m}\delta\Phi_m >_E$. Thus

$$\varsigma \sum_{m\neq 0} \bar{P}_{0m}\bar{\Phi}_m \approx \varsigma^2 \sum_{m\neq 0}\sum_{\kappa} \bar{P}_{0m}^{\kappa}(\mathbf{b}-\mathbf{b}_\kappa, z-z_\kappa)\, \bar{P}_{m0}^{\kappa}(\mathbf{b}-\mathbf{b}_\kappa)\,\bar{\Phi}_0 \tag{17.39}$$

Substituting (17.30) and (17.39) into (17.38) and neglecting HOLZ reflections, the solution of $\bar{\Phi}_0$ is

$$\bar{\Phi}_0(z) = \exp\left\{\varsigma\left[\Xi + \bar{P}_{00} + i\left(\bar{P}_{00}\right)_I\right]\right\}\bar{\Phi}_0(0) \tag{17.40}$$

where $\left(\bar{P}_{00}\right)_I$ is given by

$$\left(\bar{P}_{00}\right)_I = |\varsigma| \sum_{\kappa} \sum_{m} \langle | P^{\kappa}_{m0} (\mathbf{b} - \mathbf{b}_{\kappa}) |^2 \rangle_E \\ + |\zeta| \sum_{\kappa} \left[\langle | P^{\kappa}_{00} (\mathbf{b} - \mathbf{b}_{\kappa}) \right|^2 \rangle_E - | \bar{P}^{\kappa}_{00} (\mathbf{b} - \mathbf{b}_k) |^2] \tag{17.41}$$

This imaginary potential contains not only the conventional inelastic absorption effect introduced in Chap. 6 [see (6.39)] but also the energy fluctuation of the crystal system. The calculation of (17.41) follows the conventional multislice theory in Sect. 11.10.3.

17.7 Applications in Phonon Scattering

The theory presented in Sections was applied to estimate phonon scattering in HRTEM, since analytic expressions for absorption potentials derived above can be found (Fanidis et al. 1993). For phonon scattering, the interaction Hamiltonian between the incident electron and the crystal atoms can be written as

$$\begin{aligned} H(\mathbf{r}) &= \sum_{n} \sum_{\alpha} H_{\alpha} (\mathbf{r} - \mathbf{R}_n - \mathbf{r}(\alpha)) \\ &= -e \sum_{n} \sum_{\alpha} V_{\alpha} \left(\mathbf{r} - \mathbf{R}_n - \mathbf{r}(\alpha) - \mathbf{u}\binom{n}{\alpha} \right) \\ &= -e \sum_{n} \sum_{\alpha} \Big\{ V_{\alpha} (\mathbf{r} - \mathbf{R}_n - \mathbf{r}(\alpha)) - \mathbf{u}\binom{n}{\alpha} \cdot \nabla V_{\alpha} (\mathbf{r} - \mathbf{R}_n - \mathbf{r}(\alpha)) \\ &+ \frac{1}{2} [\mathbf{u}\binom{n}{\alpha} \cdot \nabla]^n V_{\alpha} (\mathbf{r} - \mathbf{R}_n - \mathbf{r}(\alpha)) + \cdots \Big\} \end{aligned} \tag{17.42}$$

The displacement of the atomic vibration is expressed in the normal coordinates as

$$\begin{aligned} \mathbf{u}\begin{pmatrix} n \\ \alpha \end{pmatrix} &= \frac{\hbar^{1/2}}{(2N_0 M_{\alpha})^{1/2}} \sum_{\mathbf{q}} \sum_{j} \frac{1}{\sqrt{\omega_j(\mathbf{q})}} \mathbf{e} \left(\alpha \middle| {\mathbf{q} \atop j} \right) \\ &\times \exp \left[2\pi i \mathbf{q} \cdot (\mathbf{R}_n + \mathbf{r}(\alpha) \right] \left[a^{+} \binom{\mathbf{q}}{j} + a \binom{\mathbf{q}}{j} \right] \end{aligned} \tag{17.43}$$

For simplicity, our discussion is restricted to the creation of a single phonon, so that the scattering matrix for this process is

$$H_{m0}(\mathbf{r},\mathbf{q},\omega) = e\langle n_s(\mathbf{q},\omega)+1|\sum_n\sum_\alpha\{\mathbf{u}\binom{n}{\alpha}\cdot\nabla V_\alpha(\mathbf{r}-\mathbf{R}_n-\mathbf{r}(\alpha))\}|n_s(\mathbf{q},\omega)\rangle$$
$$= e\sum_n\sum_\alpha A_\alpha\left(\omega_j(\mathbf{q})\right)\mathbf{e}\left(\alpha|{}^{\mathbf{q}}_{j}\right)\cdot\nabla \mathrm{V}_\alpha\left(\mathbf{r}-\mathbf{R}_n-\mathbf{r}(\alpha)\right)$$
$$\times\exp\{2\pi i\mathbf{q}\cdot[\mathbf{R}_n+\mathbf{r}(\alpha)]\} \tag{17.44a}$$

where

$$A_\alpha\left(\omega_j(\mathbf{q})\right) \equiv \frac{\left\{\hbar\left[n_s(\mathbf{q},\omega)+1)\right]\right\}^{1/2}}{\left[2\omega_j(\mathbf{q})M_\alpha N_0\right]^{1/2}} \tag{17.44b}$$

For the κth atom

$$H^{\kappa}_{m0}\left(\mathbf{b}-\mathbf{b}_\kappa, E\right) = eA_\kappa\left(\omega_j(\mathbf{q})\right)\left[\mathbf{e}\left(\kappa|{}^{\mathbf{q}}_{j}\right)\cdot\nabla V_\kappa\left(\mathbf{b}-\mathbf{b}_\kappa\right)\right]\exp\left(2\pi i\mathbf{q}\cdot\mathbf{b}_\kappa\right) \tag{17.45}$$

We now calculate the two quantities defined in (17.27) and (17.37b), which are essential to the numerical calculation

$$PP^{\kappa'}_{\kappa}(\mathbf{b}',\mathbf{b}'') \equiv \sum_m\{\langle P^{\kappa}_{m0}(\mathbf{b}'-\mathbf{b}_\kappa)[P^{\kappa'}_{m0}(\mathbf{b}''-\mathbf{b}_{\kappa'})]^*\rangle_E - \bar{P}^{\kappa}_{mo}(\mathbf{b}'-\mathbf{b}_\kappa)[\bar{P}^{\kappa'}_{m0}(\mathbf{b}''-\mathbf{b}_{\kappa'})]^*\}$$
$$= \frac{4m_e^2}{\hbar^4}\sum_m\{\langle H^{\kappa}_{m0}(\mathbf{b}'-\mathbf{b}_\kappa)[H^{\kappa'}_{m0}(\mathbf{b}''-\mathbf{b}_{\kappa'})]^*\rangle_E$$
$$-\langle H^{\kappa}_{m0}(\mathbf{b}'-\mathbf{b}_\kappa)\rangle_E\cdot\langle[H^{\kappa'_0}_{m0}\left(\mathbf{b}''-\mathbf{b}_{\kappa'}\right)]^*\rangle_E\}$$
$$= \frac{2m_e^2e^2}{\hbar^3N_0\left(M_kM_{\kappa'}\right)^{1/2}}\sum_j\sum_q\frac{1}{\omega_j(\mathbf{q})}\exp\left[2\pi i\mathbf{q}\cdot\left(\mathbf{b}_\kappa-\mathbf{b}_{\kappa'}\right)\right][\mathbf{e}\left(\kappa|{}^{\mathbf{q}}_{j}\right)\cdot\nabla\mathrm{V}_\kappa\left(\mathbf{b}'-\mathbf{b}_\kappa\right)]$$
$$\times[\mathbf{e}^*(\kappa'|{}^{\mathbf{q}}_{j})\cdot\nabla\mathrm{V}_{\kappa'}(\mathbf{b}''-\mathbf{b}_{\kappa'})]\{\langle n_s(\mathbf{q},\omega)\rangle_E+1-\left[\langle[n_s(\mathbf{q},\omega)+1]^{1/2}\rangle_E\right]^2\} \tag{17.46}$$

where

$$< \left[n_s(\mathbf{q},\omega)+1\right]^{1/2} >_E \;=\; \frac{\sum_{n_s=0}(n_s+1)^{1/2}\exp\left[-\left(n_s+1/2\right)\hbar\omega_j/k_BT\right]}{\sum_{n_s=0}\exp\left[-\left(n_s+1/2\right)\hbar\omega_j/k_BT\right]} \tag{17.47}$$

and $\langle n_s\rangle$ is given by (6.69a). To calculate the nonfluctuating component, the quantity appearing in (17.35), (17.37a) and (17.37b) is

$$\sum_{m=1}\bar{P}^{\kappa}_{m0}\left(\mathbf{b}'-\mathbf{b}_k\right)[\bar{P}'_{m0}(\mathbf{b}''-\mathbf{b}_{k'})]^*$$
$$= \frac{2m_e^2e^2}{\hbar^3N_0(M_\kappa M_{\kappa'})^{1/2}}\sum_j\sum_q\frac{1}{\omega_j(\mathbf{q})}\exp[2\pi i\mathbf{q}\cdot(\mathbf{b}_\kappa-\mathbf{b}_{\kappa'})][\mathbf{e}(\kappa|{}^{\mathbf{q}}_{j})\cdot\nabla\mathrm{V}_\kappa(\mathbf{b}'-\mathbf{b}_k)]$$
$$\times[\mathbf{e}^*(\kappa'|{}^{\mathbf{q}}_{j})\cdot\nabla V_{\kappa'}(\mathbf{b}''-\mathbf{b}_k)][\langle[n_s(\mathbf{q},\omega)+1]^{1/2}\rangle_E]^2 \tag{17.48}$$

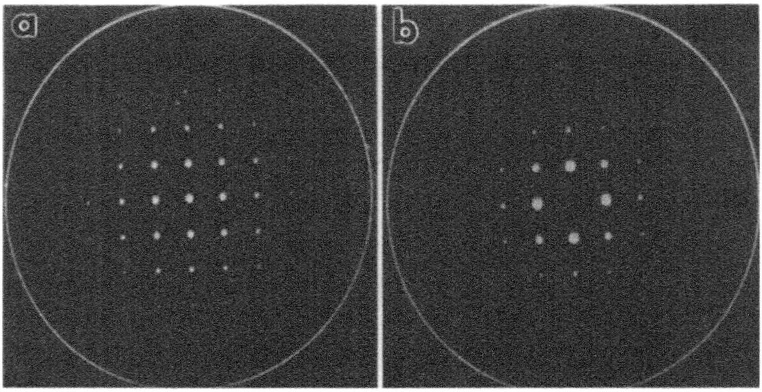

Fig. 17.1 Calculated [001] electron diffraction patterns of Cu at T = 295 K : (a) from the energy fluctuated elastic component $\delta\Phi_0$, intensity maximum 1.898×10^{-3} and (b) from the energy-nonfluctuating elastic component $\bar{\Phi}_0$, intensity maximum 1.365×10^{-1}. Courtesy of Dr. Fanidis et al. (1993)

The absorption potentials $\left(\bar{P}_{nn}\right)_I$ and $\left(\bar{P}_{00}\right)_I$, which have been derived by Fanidis et al. (1993), are rather lengthy and thus are not listed here.

Figure 17.1 compares electron diffraction patterns calculated using the energy-fluctuating elastic component $\delta\Phi_0$ (Fig. 17.1a) and the energy-nonfluctuating elastic component $\bar{\Phi}_0$ (Fig. 17.1b). It is apparent that the Bragg peaks located at high scattering angles are more pronounced in the pattern formed by $\delta\Phi_0$ than that by $\bar{\Phi}_0$.

Further calculations by Fanidis et al. (1993) have shown that even in the case of no energy transfer, thermal equilibrium still influences the image and diffraction pattern intensities. Those authors also illustrated the dependence of image intensity on the polarization of the excited phonon modes and predicted detectable changes in the image and diffraction patterns due to thermal equilibrium with the environment.

Appendix A
Physical Constants, Electron Wavelengths, and Wave Numbers

The fundamental physical constants used in this book are

$$c_0 = 2.99792458 \times 10^8 \text{ m s}^{-1}$$
$$e = 1.6021892 \times 10^{-19}\text{C}$$
$$h = 6.626196 \times 10^{-34} \text{ J s}$$
$$\hbar = h/2\pi = 1.054592 \times 10^{-34} \text{ J s}$$
$$k_B = 1.380622 \times 10^{-23} \text{ J K}^{-1}$$
$$m_0 = 9.109534 \times 10^{-31} \text{ kg}$$
$$\varepsilon_0 = 8.8541878 \times 10^{-12}\text{C V}^{-1} \text{ m}^{-1}$$

The following table gives, as a function of accelerating voltage U_0, the relativistic electron wavelength λ, wave number $K = 1/\lambda$, relativistic factor $\gamma = m_e/m_0$, and velocity $\beta = v/c_0$.

E (kV)	λ (Å)	K (Å^{-1})	m_e/m_0	v/c
0.001	12.26	0.0815	1.000002	0.0020
0.01	3.878	0.2579	1.000019	0.0063
0.1	1.226	0.8154	1.000196	0.0198
0.5	0.5483	1.824	1.000978	0.0806
1	0.3876	2.580	1.00196	0.0625
2	0.2740	3.650	1.00391	0.0882
3	0.2236	4.473	1.00587	0.1079
4	0.1935	5.167	1.00783	0.1244

Z. L. Wang, *Elastic and Inelastic Scattering in Electron Diffraction and Imaging*, Springer Series in Solid-State Sciences 205, https://doi.org/10.1007/978-3-031-90819-4

E (kV)	λ (Å)	K (Å^{-1})	m_e/m_0	v/c
5	0.1730	5.780	1.00978	0.1389
6	0.1579	6.335	1.01174	0.1519
7	0.1461	6.845	1.01370	0.1638
8	0.1366	7.322	1.01566	0.1749
9	0.1287	7.770	1.01761	0.1852
10	0.1220	8.194	1.01957	0.1950
20	0.0859	11.64	1.0391	0.2719
30	0.0698	14.33	1.0587	0.3284
40	0.0602	16.62	1.0783	0.3741
50	0.0536	18.67	1.0978	0.4127
60	0.0487	20.55	1.1174	0.4462
70	0.0448	22.30	1.1370	0.4759
80	0.0418	23.95	1.1566	0.5024
90	0.0392	25.52	1.1761	0.5264
100	0.0370	27.02	1.1957	0.5482
200	0.0251	39.87	1.3914	0.6953
300	0.0197	50.80	1.5871	0.7765
400	0.0164	60.83	1.7828	0.8279
500	0.0142	70.36	1.9785	0.8629
600	0.0126	79.57	2.1742	0.8879
700	0.0113	88.56	2.3698	0.9066
800	0.0103	97.38	2.5655	0.9209
900	0.0094	106.1	2.7912	0.9321
1000	0.0087	114.7	2.9569	0.9411
2000	0.0050	198.3	4.9138	0.9791
4000	0.0028	361.5	8.8277	0.9936

Appendix B
Properties of Fourier Transforms

Function in Real Space	Fourier Transform in Reciprocal Space
$f(x) = \int_{-\infty}^{\infty} du F(u) \exp(2\pi iux)$	$F(u) = \int_{-\infty}^{\infty} dx f(x) \exp(-2\pi iux)$
$c_1 f_1(x) + c_2 f_2(x)$	$c_1 F_1(u) + c_2 F_2(u)$
$f(ax)$	$\frac{F(u/a)}{a}$, with a positive real constant
$f(-x)$	$F(-u)$
$f(x - x_0)$	$F(u) \exp(-2\pi iux_0)$
$f(x) \exp(2\pi iu_0 x)$	$F(u - u_0)$
$f^*(\pm x)$	$F^*(\mp u)$
$f(x) \otimes g(x)$	$F(u)G(u)$
$f(x)g(x)$	$F(u) \otimes G(u)$
$f(x)g^*(x)$	$\int_{-\infty}^{\infty} du' F(u'+u)\, G^*(u') = \int_{-\infty}^{\infty} du' F(u')\, G^*(u'-u)$
$\int dx' f(x' + x)\, g^*(x')$	$F(u)G^*(u)$
df/dx	$2\pi iuF(u)$
$-2\pi ixf(\mathrm{x})$	dF/du

Function in Real Space	Fourier Transform in Reciprocal Space
$\int_{-\infty}^{x} dx' f(x')$	$F(u)/2\pi iu$
$-f(x)/2\pi ix$	$\int_{-\infty}^{u} du' F(u')$
$\exp(2\pi iu_0 x)$	$\delta(u - u_0)$
$\delta(x - x_0)$	$\exp(-2\pi iux_0)$

Z. L. Wang, *Elastic and Inelastic Scattering in Electron Diffraction and Imaging*, Springer Series in Solid-State Sciences 205, https://doi.org/10.1007/978-3-031-90819-4

B.1 Identities

$$\int_{-\infty}^{\infty} dx f(x) g^*(x) = \int_{-\infty}^{\infty} du F(u) G^*(u) \tag{B.1}$$

$$\int_{-\infty}^{\infty} dx f(x) g(x) = \int_{-\infty}^{\infty} du F(u) G(-u) \tag{B.2}$$

$$\int_{-\infty}^{\infty} dx |f(x)|^2 = \int_{-\infty}^{\infty} du |F(u)|^2 \tag{B.3}$$

$$\int_{-\infty}^{\infty} dx |f(x) \otimes g(x)|^2 = \int_{-\infty}^{\infty} du |F(u) G(u)|^2 \tag{B.4}$$

$$\int_{-\infty}^{\infty} dx |f(x) g(x)|^2 = \int_{-\infty}^{\infty} du |F(u) \otimes G(u)|^2 \tag{B.5}$$

Appendix C
Some Properties of Dirac Delta Functions

C.1 Defining Relationships and Normalization Conditions

$$f(\mathbf{r}) = \int d\mathbf{r}' f(\mathbf{r}')\, \delta(\mathbf{r}' - \mathbf{r}) \tag{C.1}$$

$$f(0) = \int d\mathbf{r}' f(\mathbf{r}')\, \delta(\mathbf{r}') \tag{C.2}$$

$$1 = \int d\mathbf{r}' \delta(\mathbf{r}' - \mathbf{r}) \tag{C.3}$$

C.2 Useful Representations of the Delta Function

$$\delta(\mathbf{r} - \mathbf{r}') = \int d\mathbf{u} \exp\left[2\pi i \mathbf{u} \cdot (\mathbf{r} - \mathbf{r}')\right] \tag{C.4}$$

$$\delta(b) = \lim_{\chi \to 0} \left[\frac{1}{\pi\chi} \exp\left(-\frac{b^2}{\chi^2}\right)\right] \tag{C.5}$$

$$\delta(x - x') = \sum_{n=-\infty}^{\infty} \exp\left[2\pi i n (x - x')\right] \tag{C.6}$$

$$\delta(x - x') = \lim_{\varepsilon \to 0} \frac{\varepsilon}{(x - x')^2 + \varepsilon^2} \tag{C.7}$$

Z. L. Wang, *Elastic and Inelastic Scattering in Electron Diffraction and Imaging*, Springer Series in Solid-State Sciences 205, https://doi.org/10.1007/978-3-031-90819-4

$$\lim_{\varepsilon \to 0} \frac{1}{x - x' - i\varepsilon} = \frac{1}{x - x'} + i\delta\left(x - x'\right) \tag{C.8}$$

$$\delta[f(x)] = \sum_{n=1}^{N} \frac{\delta\left(x - x_n\right)}{|df(x)/dx|_{x=x_n}} \tag{C.9}$$

where x_n are the N values of x for which $f(x) = 0$.

$$\delta(x) = \lim_{\Lambda \to \infty} \frac{\sin(x\Lambda)}{\pi x} \tag{C.10}$$

$$\delta\left(x - x'\right) = \frac{1}{\pi} \int\limits_0^\infty dk \cos k\left(x - x'\right) \tag{C.11}$$

$$\pi\delta(x) = \lim_{\Lambda \to \infty} \frac{1 - \cos(x\Lambda)}{\Lambda x^2} \tag{C.12}$$

$$\pi\delta(x) = \lim_{\Lambda \to \infty} \frac{2\sin^2(x\Lambda/2)}{\Lambda x^2} \tag{C.13}$$

$$\delta(ax) = \frac{\delta(x)}{|a|} \tag{C.14}$$

$$\int\limits_0^a dx\delta(x) = 1/2 \text{ if } a > 0$$

$$= -1/2 \text{ if } a < 0 \tag{C.15}$$

$$\int d\mathbf{r}\, \delta(\mathbf{r} - \mathbf{a})\delta(\mathbf{r} - \mathbf{b}) = \delta(\mathbf{b} - \mathbf{a}) \tag{C.16}$$

$$\nabla^2\left[\frac{1}{|\mathbf{r} - \mathbf{r}'|}\right] = 4\pi\delta\left(\mathbf{r} - \mathbf{r}'\right) \tag{C.17}$$

$$\delta(\mathbf{r}) = \frac{\delta(r)}{2\pi r^2} \tag{C.18}$$

$$\delta\left(x - x_0\right) = \frac{d^2}{dx^2}\left(|\, x - x_0 \,|\right) \tag{C.19}$$

$$\delta\left(x - x_0\right) = \frac{d}{dx}\Theta\left(x - x_0\right) \tag{C.20}$$

where $\Theta\left(x - x_0\right)$ is a step function defined by $\Theta\left(x - x_0\right) = 1$ for $x > x_0$, -1 for $x < x_0$.

$$\int\limits_{-\infty}^{\infty} dx f(x) \frac{d}{dx}\delta\left(x - x_0\right) = \left.\frac{df(x)}{dx}\right|_{x=x_0} \tag{C.21}$$

This relation can be proved directly by integration by parts

$$\int_{-\infty}^{\infty} dx f(x)\delta(ax - x_0) = \frac{1}{a}\int_{-\infty}^{\infty} dx f(x/a)\delta(x - x_0) = \frac{f(x_0/a)}{a} \tag{C.22}$$

Appendix D
Integral Form of the Schrödinger Equation

Equation (**??**) can conveniently be written as

$$\left(\nabla^2 + 4\pi^2 K^2\right)\Psi(\mathbf{r}) = -4\pi^2 U(\mathbf{r})\Psi(\mathbf{r}) \tag{D.1}$$

We can transform the term on the right-hand side of this equation into a different form using the Dirac delta function

$$\begin{aligned} U(\mathbf{r})\Psi(\mathbf{r}) &= \int d\mathbf{r}' U\left(\mathbf{r}'\right)\Psi\left(\mathbf{r}'\right)\delta\left(\mathbf{r}-\mathbf{r}'\right) \\ &= \int d\mathbf{r}' U\left(\mathbf{r}'\right)\Psi\left(\mathbf{r}'\right)\int d\mathbf{u}\exp\left[2\pi i\mathbf{u}\cdot\left(\mathbf{r}-\mathbf{r}'\right)\right] \end{aligned} \tag{D.2}$$

where the identity (A.4) was used. We now consider the following relation:

$$\left[\nabla^2 + 4\pi^2 K^2\right]\exp(2\pi i\mathbf{u}\cdot\mathbf{r}) = 4\pi^2\left(K^2 - u^2\right)\exp(2\pi i\mathbf{u}\cdot\mathbf{r})$$

or

$$\frac{\exp(2\pi i\mathbf{u}\cdot\mathbf{r})}{4\pi^2\left(K^2-u^2\right)} = \left(\nabla^2 + 4\pi^2 K^2\right)^{-1}\exp(2\pi i\mathbf{u}\cdot\mathbf{r}) \tag{D.3}$$

We now substitute (D.2) into (D.1), multiply both sides by the inverse of operator $(\nabla^2 + 4\pi^2 K^2)$, and use (D.3)

$$\begin{aligned} \Psi(\mathbf{r}) &= -4\pi^2\left(\nabla^2 + 4\pi^2K^2\right)^{-1} U(\mathbf{r})\Psi(\mathbf{r}) \\ &= -4\pi^2\int d\mathbf{r}' U\left(\mathbf{r}'\right)\Psi\left(\mathbf{r}'\right)\int d\mathbf{u}\exp\left(-2\pi i\mathbf{u}\cdot\mathbf{r}'\right)\left(\nabla^2+4\pi^2K^2\right)^{-1}\exp(2\pi i\mathbf{u}\cdot\mathbf{r}) \\ &= -4\pi^2\int d\mathbf{r}' U\left(\mathbf{r}'\right)\Psi\left(\mathbf{r}'\right)\left\{\int du\frac{\exp\left[2\pi i\mathbf{u}\cdot\left(\mathbf{r}-\mathbf{r}'\right)\right]}{4\pi^2\left(K^2-u^2\right)}\right\} \\ &= -4\pi^2\int d\mathbf{r}' U\left(\mathbf{r}'\right)\Psi\left(\mathbf{r}'\right)G_0\left(\mathbf{r},\mathbf{r}'\right) \end{aligned} \tag{D.4}$$

Z. L. Wang, *Elastic and Inelastic Scattering in Electron Diffraction and Imaging*, Springer Series in Solid-State Sciences 205, https://doi.org/10.1007/978-3-031-90819-4

where Green's function is defined as

$$G_0\left(\mathbf{r},\mathbf{r}'\right)=\int d\mathbf{u}\frac{\exp\left[2\pi i\mathbf{u}\cdot\left(\mathbf{r}-\mathbf{r}'\right)\right]}{4\pi^2\left(K^2-u^2\right)} \tag{D.5}$$

Thus the solution to (**??**) can be written in an integral form as

$$\Psi(\mathbf{r})=\Psi_0(\mathbf{r})-4\pi^2\int d\mathbf{r}'G_0\left(\mathbf{r},\mathbf{r}'\right)U\left(\mathbf{r}'\right)\Psi\left(\mathbf{r}'\right) \tag{D.6}$$

where a term $\Psi_0(\mathbf{r})$ is added, which is a solution of the homogeneous equation $\left(\nabla^2+4\pi^2K^2\right)\Psi(\mathbf{r})=0$. In spherical coordinate system, if $\left(\mathbf{r}-\mathbf{r}'\right)$ is chosen as the polar axis,

$$\begin{aligned}G_0\left(\mathbf{r},\mathbf{r}'\right)&=\int_0^\infty du u^2\int_0^\pi d\theta\sin\theta\int_0^{2\pi}d\varphi\frac{\exp\left(2\pi iu\left|\mathbf{r}-\mathbf{r}'\right|\cos\theta\right)}{4\pi^2\left(K^2-u^2\right)}\\&=\frac{i}{4\pi^2\left|\mathbf{r}-\mathbf{r}'\right|}\int_{-\infty}^{\infty}du\frac{u\exp\left(2\pi iu\left|\mathbf{r}-\mathbf{r}'\right|\right)}{\left(u^2-K^2\right)}\\&=\frac{i}{4\pi^2\left|\mathbf{r}-\mathbf{r}'\right|}\lim_{\varepsilon\to0}\int_{-\infty}^{\infty}du\frac{u\exp\left(2\pi iu\left|\mathbf{r}-\mathbf{r}'\right|\right)}{\left(u^2-K^2-i\varepsilon\right)}\end{aligned} \tag{D.7}$$

where ε was introduced as an aid in evaluating the remaining integral using the standard techniques of complex integration. We transform Eq. (D.7) to the complex plane and integrate around a path, as shown in Fig. D.1 where we integrate from $u=-L$ to $u=L$ along the real axis and then along the path $u=L\exp(i\theta)$ from $\theta=0$ to $\theta=\pi$. The residue theorem in complex variable theory states that the integral around a closed loop of a function containing only regular singularities is simply equal to the sum of residues of all singularities in the loop multiplied by 2π. A residue is what is left of the integrand after the singularity is multiplied and the remaining part evaluated at the singular point. The only singularities in the function we convert are $u_1=K\left(1+i\varepsilon/K^2\right)^{1/2}\approx K+i\varepsilon/2K$ and $u_2=-K\left(1+i\varepsilon/K^2\right)^{1/2}\approx-\mathrm{K}-i\varepsilon/2K$. The path we have chosen involves only the first singularity, whose residue is

$$\frac{1}{2}\exp\left[2\pi i(K+i\varepsilon/2\,K)\left|\mathbf{r}-\mathbf{r}'\right|\right]$$

Thus the residue theorem lead to

$$\begin{aligned}&\int_{-L}^{L}du\frac{u\exp(2\pi iu\left|\mathbf{r}-\mathbf{r}'\right|)}{(u^2-K^2-i\varepsilon)}+iL^2\int_0^\pi d\theta\frac{e^{2i\theta}\exp[(2\pi iL\cos\theta-2\pi L\sin\theta)\left|\mathbf{r}-\mathbf{r}'\right|]}{L^2e^{2i\theta}-K^2-i\varepsilon}\\&\qquad=\pi i\exp\left[2\pi i(K+i\varepsilon/2\,K)\left|\mathbf{r}-\mathbf{r}'\right|\right]\end{aligned} \tag{D.8}$$

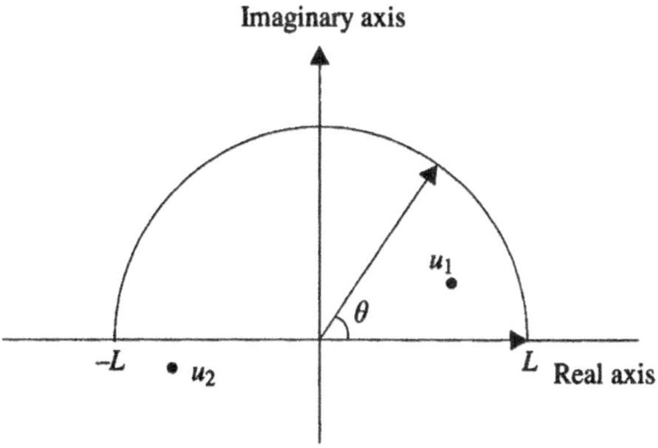

Fig. D.1 The path in the complex plane for evaluating $G(\mathbf{r}, \mathbf{r}')$

For $L \to \infty$, the second integral on the left-hand size of (D.8) vanishes because of the negative real exponential in the integrand and the fact that $\sin\theta$ is positive over $0 < \theta < \pi$. Thus

$$\lim_{\varepsilon \to 0} \int_{-\infty}^{\infty} du \frac{u \exp\left(2\pi i u \left|\mathbf{r} - \mathbf{r}'\right|\right)}{\left(u^2 - K^2 - i\varepsilon\right)} = \pi i \exp\left(2\pi i K \left|\mathbf{r} - \mathbf{r}'\right|\right) \tag{D.9}$$

Green's function is

$$G\left(\mathbf{r}, \mathbf{r}'\right) = -\frac{\exp\left(2\pi i K \left|\mathbf{r} - \mathbf{r}'\right|\right)}{4\pi \left|\mathbf{r} - \mathbf{r}'\right|} \tag{D.10}$$

Therefore the integrate form of (**??**) is

$$\Psi(\mathbf{r}) = \Psi_0(\mathbf{r}) + \pi \int d\mathbf{r}' \frac{\exp\left(2\pi i K \left|\mathbf{r} - \mathbf{r}'\right|\right)}{\left|\mathbf{r} - \mathbf{r}'\right|} U\left(\mathbf{r}'\right) \Psi\left(\mathbf{r}'\right) \tag{D.11}$$

Appendix E
Some Useful Mathematical Relations

$$\int d\mathbf{r} \frac{\exp(-2\pi i \mathbf{u} \cdot \mathbf{r})}{|\mathbf{r} - \mathbf{r}'|} = \frac{\exp\left(-2\pi i \mathbf{u} \cdot \mathbf{r}'\right)}{\pi u^2} \tag{E.1}$$

$$\frac{1}{|\mathbf{r} - \mathbf{r}'|} = \int d\mathbf{u} \frac{\exp\left[2\pi i \mathbf{u} \cdot \left(\mathbf{r} - \mathbf{r}'\right)\right]}{\pi u^2} \tag{E.2}$$

$$\lim_{\varepsilon \to 0} \frac{1}{x - x' \pm i\varepsilon} = P\left[\frac{1}{x - x'}\right] \mp i\pi\delta\left(x - x'\right) \tag{E.3}$$

where P signifies the principal value [i.e., the function $1/\left(x - x'\right)$ is given as $1/\left(x - x'\right)$ for all values of x' except at the point $x = x'$, for which $1/\left(x - x'\right)$ is taken to be identically zero].

$$\lim_{\varepsilon \to 0} \int dx \frac{A(x)}{x - x' \pm i\varepsilon} = P \int dx \frac{A(x)}{x - x'} \mp i\pi A\left(x'\right) \tag{E.4}$$

$$\int_{-\infty}^{\infty} dx \exp\left(-a^2x^2 + bx\right) = \frac{\sqrt{\pi}}{a} \exp\left(\frac{b^2}{4a^2}\right) \tag{E.5}$$

$$\int_{-\infty}^{\infty} dx \int_{-\infty}^{\infty} dy \exp\left[-a^2\left(x^2 + y^2\right) - 2\pi i\left(u_x x + u_y y\right)\right] = \frac{\pi}{a^2} \exp\left(\frac{-\pi^2 u^2}{a^2}\right) \tag{E.6}$$

which means that the Fourier transform of a Gaussian function is a Gaussian function.

$$\int_{0}^{\pi} d\theta \sin\theta \exp(ikr\cos\theta) = \frac{2\sin kr}{kr} = 2j_0(kr) \tag{E.7}$$

Z. L. Wang, *Elastic and Inelastic Scattering in Electron Diffraction and Imaging*, Springer Series in Solid-State Sciences 205, https://doi.org/10.1007/978-3-031-90819-4

References

Afanas'ev, A.M. and Kagan, Yu. (1968) Acta Cryst. A **24**, 163.
Allen, L.J. (1993) Ultramicroscopy **48**, 97.
Allen, L.J. and Rossouw, C.J. (1990) Phys. Rev. B **42**, 11644.
Allen, L.J. and Rossouw, C.J. (1993) Phys. Rev. B **47**, 2446.
Amali, A. and Rez, P. (1992) in Proc. 50th Ann. Meet. Electron Microscopy Soc. America, G.W. Bailey, J. Bentley and J.A. Small, eds., San Francisco Press, 1226.
Amelincks, S. and Van Dyck, D (1993) in Electron Diffraction Techniques, Vol. II, J.M. Cowley ed., International Union of Crystallography, Oxford University Press, 1.
Anstis, G.R. (1989) in Computer Simulation of Electron Microscope Diffraction and Images, W. Krakow and M. O'Keefe eds., The Minerals, Metals & Materials Soc. pp. 229.
Anstis, G.R. (1996) Acta Cryst. A **52**, 450.
Arfken, G. (1970), Mathematical Methods for Physicists, Academic Press (New York), 2nd ed., pp. 485.
Ashcroft, N.W. and Mermin, D.N. (1976) Solid State Physics, Holt Rinehart and Winston (New York).
Ashley, J.C. and Ferrell, T.L. (1976) Phys. Rev. B **14**, 3277.
Ashley, J.C. and Ritchie, R.H. (1970) Phys. Status Solidi **38**, 425.
Ashley, J.C., Ferrell, T.L. and Ritchie, R.H. (1974) Phys. Rev. B **10**, 554.
Bah, M.L., Akjouj, A. and Dobrzynski, L. (1992) Surf. Sci. Rep. **16**, 95.
Bentley, J., Alexander, K.B. and Wang, Z.L. (1990), Proc. of XIIth Intern. Cong. for Electron Microscopy (Seattle), San Francisco Press, Vol. 4, eds. L. D. Peachey and D. B. Williams, 400.
Berry, M.V. (1971) J. Phys. C: Solid State Phys. **4**, 697.
Bethe, H.A. (1928) Ann. Phys. Lpz. **87**, 55.
Bilhorn, D.E., Foldy, L.L., Thaler, R.M. and Tobocman, W. (1964) J. Math. Phys. **5**, 435.
Bird, D.M. (1989) J. Electron Microsc. Technique **13**, 77.
Bird, D.M. and King, Q.A. (1990) Acta Cryst. A **46**, 202.
Bird, D.M. and Wright, A.G. (1989) Acta Cryst. A **45**, 104.
Björkman, G., Lundqvist, B.I. and Sjölander, A. (1967) Phys. Rev. **159**, 551.
Blum, K. (1981) Density Matrix Theory and Applications, Plenum Press, New York.
Bohm, D. (1953) Phys. Rev. **92**, 609.
Borie, B. (1970) Acta Cryst. A **26**, 533.
Borisevich, A.Y., Chang, H.J., Huijben, M., et al., (2010), Phys. Rev. Lett. **105**: 087204.
Borisevich, A., Ovchinnikov, O.S., Chang, H.J., et al. (2010), ACS Nano **4**: 6071–9.

Z. L. Wang, *Elastic and Inelastic Scattering in Electron Diffraction and Imaging*, Springer Series in Solid-State Sciences 205, https://doi.org/10.1007/978-3-031-90819-4

Born, M. (1942a) Rep. Prog. Phys. **9**, 294.
Born, M. (1942b) Proc. R. Soc. Lond. A **180**, 397.
Brüesch, P. (1982) Phonons: Theory and experiments I - Lattice dynamics and models of interatomic forces, Springer-Verlag (Berlin Heidelberg New York).
Buseck, P., Cowley, J.M. and Eyring, L. eds. (1988) High-Resolution Transmission Electron Microscopy, Oxford University Press.
Butler, B.D., Withers, R.L. and Welberry, T.R. (1992) Acta Cryst. A **48**, 737.
Buxton, B.F. (1976) Proc. R. Soc. Series A **350**, 335.
Buxton, B.F., Eades, J.A., Steeds, J.W. and Rackham, G.M. (1976) Phil. Trans. Roy. Soc. Lond. A **281**, 171.
Buxton, B.F., Loveluck, J.E. and Steeds, J.W. (1978) Philos. Mag. A **38**, 259.
Chen, J.H., Van Dyck, D., Op de Beeck, M., Van Landuyt, J. (1997) Ultramicroscopy **69**: 219.
Cherns, D., Howie, A., and Jacobs, M.H. (1973) Z. Naturforsch. **28** a, 565.
Cochran, W. (1963) Rep. Prog. Phys. **26**, 1.
Cochran, W. and Kartha, G. (1956) Acta Cryst. **9**, 941.
Cockayne, D.J.H. (1972) Z. Naturforsch. Teil A **27**, 452.
Cockayne, D.J.H., Ray, I.L.F. and Whelan, M.J. (1969) Philos. Mag. **20**, 1265.
Coene, W. and Van Dyck, D. (1990) Ultramicroscopy **33**, 261.
Coene, W., Van Dyck, D., Van Tendeloo, G. and Van Landuyt, J. (1985) Philos. Mag. A **52**, 127.
Colella, R. (1972) Acta Cryst. A **28**, 11.
Colliex, C., Manoubi, T., Gasgnier, M. and Brown, L.M. (1985). Scanning Electron Microscopy (A.M.F. O'Hare, Chicago), p. 489.
Cowley, J. M. (1988b) in High resolution transmission electron microscopy and associated techniques, Chpt. 1, edited by P. Buseck, J. Cowley and L. Eyring, Oxford University Press (New York).
Cowley, J.M. (1995). Diffraction Physics, 3rd revised ed. (Elsvier Science B.V.). Chpt. 7 and 17.
Cowley, J.M. (1950) Phys. Rev. **77**, 669–675.
Cowley, J.M. (1969) Appl. Phys. Lett. **15**, 58.
Cowley, J.M. (1976) Ultramicroscopy **2**, 3.
Cowley, J.M. (1981) Diffraction Physics, 2nd ed., North-Holland Publishing Co., Amesterdam.
Cowley, J. M. (1988a) Acta Cryst. A **44**, 847.
Cowley, J. M. ed. (1992) Electron Diffraction Techniques, Vol. I, International Union of Crystallography, Oxford University Press.
Cowley, J. M. ed. (1993) Electron Diffraction Techniques, Vol. II, International Union of Crystallography, Oxford University Press.
Cowley, J.M. and Huang, Y. (1992) Ultramicroscopy **40**, 171.
Cowley, J.M. and Moodie, A.F. (1957) Acta Cryst. **10**, 609.
Cowley, J.M. and Pogany, A.P. (1968) Acta Cryst. A **24**, 109.
Craven, A.J., Gibson, J.M., Howie, A. and Spalding, D.R. (1978) Philos. Mag. A **38**, 519.
Crewe, A.V. (1978–79) Chemica Scripta **14**, 17.
Crewe, A.V. and Wall, J. (1970) J. Mol. Biol. **48**, 375.
Dai, Z.R., Wang, Z.L., Chen, Y.R., Wu, H.Z., and Liu, W.X. (1996a) Phil. Mag. A **73**, 415–425.
Dai, Z.R., Wang, Z. L. and Liu, W. X. (1996b) Philos. Mag. A **73**, 1685–1698.
Darwin, C.G. (1914) Philos. Mag. **27**, 315, 675.
De Ridder, R., Van Tendeloo, G. and Amelinckx, S. (1976) Acta Cryst. A **32**, 216–224.
Dederichs, P.H. (1966) Phys. kondens. Mat. **5**, 347.
Disko, M.M., Krivanek, O.L. and Rez, P. (1982) Phys. Rev. B **25**, 4252.
Doyle, P.A. and Turner, P.S. (1968) Acta Cryst. A **24**, 390.
Doyle, P.A. and Berry, M.V. (1973) Z. Naturforsch. **26** a, 571.
Doyle, P.A. (1969) Acta Cryst. A **25**, 569.
Doyle, P.A. (1971) Acta Cryst. A **27**, 109.
Dudarev, S.L. (1991) Sov. Phys. Crystallgr. **35**, 785.
Dudarev, S.L. and Ryazanov, M.I. (1988) Acta Cryst. A **44**, 51.

Dudarev, S.L., Peng, L.M. and Ryazanov, M.I. (1991) Acta Cryst. A **47**, 170.
Dudarev, S.L., Peng, L.M. and Whelan, M.J. (1992) Phys. Lett. A **170**, 111.
Dudarev, S.L., Peng, L.M. and Whelan, M.J. (1993a) Phys. Rev. B **48**, 13408.
Dudarev, S.L., Peng, L.M. and Whelan, M.J. (1993b) Proc. R. Soc. Lond. A **440**, 567.
Eaglesham, D.J. and Berger, S.D. (1994) Ultramicroscopy **53**, 319.
Egerton, R.F. (1986) Electron Energy-Loss Spectroscopy in the Electron Microscope, Plenum Press (New York).
Fanidis, C., Van Dyck, D., Coene, W. and Van Landuyt, J. (1989) in Computer Simulation of Electron Microscope Diffraction and Images, eds. W. Krakow and M. O'Keefe, The Minerals, Metals & Materials Soc. pp. 135.
Fanidis, C., Van Dyck, D. and Van Landuyt, J. (1992) Ultramicroscopy **41**, 55.
Fanidis, C., Van Dyck, D. and Van Landuyt, J. (1993) Ultramicroscopy **48**, 133.
Ferrell, R. (1956) Phys. Rev. **101**, 554.
Ferrell, R. (1957) Phys. Rev. **107**, 450.
Ferrell, T.L. and Echnenique, P.M. (1985) Phys. Rev. Lett. **55**, 1526.
Feynman, R.P. (1948) Rev. Mod. Phys. **20**, 367.
Fisher, P.M.J. (1968) J. Appl. Phys. **7**, 191.
Forbes, B.D., Martin, A.V., Findlay, S.D., D'Alfonso, A.J., Allen, L.J. (2010) Phys. Rev. B **80**: 104103.
Fox, A.G. and Fisher, R.M. (1988) Aust. J. Phys. **41**, 461.
Fujimoto, F. (1959) J. Phys. Soc. Japan **14**, 1558.
Fujimoto, F. and Howie, A. (1966) Phil. Mag. **13**, 1131.
Fujimoto, F. and Kainuma, Y. (1963) J. Phys. Soc. Japan **18**, 1792.
Fujimoto, F. and Komaki, K. (1968) J. Phys. Soc. Japan **25**, 1679.
Fujimoto, F., Takagi, S., Komaki, K., Koike, H. and Uchida, Y. (1972) Radiation Effects **12**, 153.
Fujiwara, K. (1959) J. Phys. Soc. Japan **14**, 1513.
Fujiwara, K. (1961). J. Phys. Soc. Japan **16**, 2226.
Fukuhara, A. (1966) J. Phys. Soc. Japan **21**, 2645.
Gabor, D. (1949) Proc. R. Soc. A, **197**, 454.
Garcia-Molina, R., Gras-Marti, A., Howie, A. and Ritchie, R.H. (1985) J. Phys. C: Solid Stat. Phys. **18**, 5335.
Gevers, R. and David, M. (1982) Phys. Stat. Sol. **113**, 665.
Gjønnes, J. (1962b) J. Phys. Soc. Japan **17**, suppl. B-II, 105.
Gjønnes, J. (1962c) J. Phys. Soc. Japan, **17**, Suppl. B-II, 137.
Gjønnes, J. (1972) Z. Naturforsch **27** a, 434.
Gjønnes, J. (1993) in Electron Diffraction Techniques, Vol. II, J.M. Cowley ed., International Union of Crystallography, Oxford University Press, 223.
Gjønnes, J. (1962a) Acta Cryst. **15**, 703.
Gjønnes, J. (1966) Acta Cryst. **20**, 240.
Gjønnes, J. and Høier, R. (1971) Acta Cryst. A **27**, 166.
Gjønnes, J. and Moodie, A.F. (1965) Acta Cryst. A **19**, 65.
Gjønnes, J. and Taftø, J. (1993) Ultramicroscopy **52**, 445.
Gjønnes, J. and Watanabe, D. (1966) Acta Cryst. **21**, 297.
Gomoyunova, M.V., Dudarev, S.L. and Pronin, I.I. (1990) Surf. Sci. **235**, 156.
Goodman, P., Johnson, A.W.S., Moodie, A.F. and Secomb, T. (1974) Proc. Int. Cryst. Conf. (Melbourne) 349.
Goodman, P. and Moodie, A.F. (1974) Acta Cryst. A **30**, 280.
Gorodnichev, E.E., Dudarev, S.L., Rogozkin, D.B. and Ryazanov, M.I. (1989) Sov. Phys. JETP **69**, 1017.
Gorodnichev, E.E., Dudarev, S.L., Rogozkin, D.B. and Ryazanov, M.I. (1990) Phys. Lett. **144**, 411.
Gunning, J. and Goodman, P. (1992) Acta Cryst. A **48**, 591.
Hall, C.R. (1965) Philos. Mag. **12**, 815.
Hall, C.R. and Hirsch, P.B. (1965) Proc. R. Soc. A **286**, 158.

Hall, C.R., Hirsch, P.B. and Booker, G.R. (1966) Philos. Mag. **14**, 979.
Hashimoto, H., Makita, Y. and Nagoka, N. (1992) Proc. 50th Ann. Meet. of Electron Microsc. Soc. America, G.W. Bailey, J. Bentley and J.A. Small eds., San Francisco Press, pp1194.
Hawkes, P.W. and Spence, J.C.H (Eds.) (2019) Handbook of STEM, Springer.
Heidenreich, R.D. (1962) J. Appl. Phys. **33**, 2321.
Helmholtz, H. von (1886) Crelles J. **100**, 213.
Hillyard, S., Loan, R.F. and Silcox, J. (1993) Ultramicroscopy **49**, 14.
Hirsch, P.B. and Humphreys, C.J. (1970) Proc. 7th Inter. Cong. on Electron Microscopy (IF-SEM, Grenoble) **1**, 229.
Hirsch, P.B., Howie, A., Nicholson, R.B., Pashley, D.W. and Whelan M.J. (1977) Electron Microscopy of Thin Crystals, Roberts E. Krieger Publishing Company (New York).
Hoerni, J.A. (1956) Phys. Rev. **102**, 1534.
Høier, R., (1973) Acta Cryst. A **29**, 663.
Honjo, G., Kodera, S. and Kitamura, K. (1964) J. Phys. Soc. Japn. **19**, 351.
Howie, A. (1962) J. Phys. Soc. Japan, **17** suppl. B-II, 122.
Howie, A. and Stern, R.M. (1972) Z. Naturforsch **27** a, 382.
Howie, A. (1963) Proc. Roy. Soc. (London) **271**, 268.
Howie, A. (1966) Philos. Mag. **14**, 223.
Howie, A. (1979) J. Microsco. **117**, 11.
Howie, A. (1983) Ultramicroscopy **11**, 141.
Howie, A. and Basinski, Z.S. (1968) Philos. Mag. **17**, 1039.
Howie, A and Sworn, C.H. (1970) Philos. Mag. **31**, 861.
Howie, A. and Whelan, M.J. (1961) Proc. R. Soc. A **263**, 217.
Hua, G.L., Welberry, T.R., Withers, R.L. and Thompson, J.G. (1988a) J. Appl. Cryst. **21**, 458.
Hua, G.L., Welberry, T.R. and Withers, R.L. (1988b) J. Phys. C: Solid State Phys. **21**, 3863.
Hua, G.L., Welberry, T.R. and Withers, R.L. (1989) J. Phys. Condens. Matter **1**, 2305.
Huang, K. (1947) Proc. Roy. Soc. London A **190**, 102.
Humphreys, C.J. (1979) Rep. Prog. Phys. **42**, 1825.
Humphreys, C.J. and Hirsch, P.B. (1968) Philos. Mag. **18**, 115.
Humphreys, C. J. and Whelan, M. J. (1969). Philos. Mag. **20**, 165.
Ichimiya, A. (1983) Japan. J. Appl. Phys. **22**, 176.
Ichimiya, A. (1990) Surf. Sci. **235**, 75.
Ichimiya, A. (1993) Ultramicroscopy **48**, 425.
Inokuti, M. (1971) Rev. Mod. Phys. **43**, 297.
Ishida, K. (1970) J. Phys. Soc. Japan **28**, 450.
Ishizuka, K. (1982) Acta Cryst. A **38**, 773.
Ishizuka, K. and Uyeda, N. (1977) Acta Cryst. A **33**, 740.
James, R.W. (1965) The Optical Principles of the Diffraction of X-rays, London: Bell.
Janssens, K.G.F., Vanhellemont, J., De Graef, M. and Van der Biest, O. (1992) Ultramicroscopy **45**, 323.
Jap, B.K. and Glaeser, R.M. (1978) Acta Cryst. A **34**, 94.
Jesson, D.E. and Pennycook, S.J. (1994) Proc. Roy. S. London, in press.
Johnson, D.E. and Spence, J.C.H. (1974) J. Phys. D: Appl. Phys. **7**, 771.
Kainuma, Y. (1955) Acta Cryst. **8**, 247.
Kainuma, Y. and Kogiso, M. (1968) Acta Cryst. A **24**, 81.
Kambe, K. and Moliere, K. (1970) in Advances in structure research by diffraction methods, Vol. 3, R. Brill and R. Mason eds., Pergamon Press (New York) pp. 53.
Kambe, K. (1982) Ultramicroscopy **10**, 223.
Kambe, K. and Stampfl C. (1994) Ultramicroscopy **55**, 221.
Kato, N (1952) J. Phys. Soc. Japan **7**, 897.
Kato, N. (1973) Z. Naturforsch a **28**, 604.
Kato, N. (1976a) Acta Cryst. A **32**, 453.
Kato, N. (1976b) Acta Cryst. A **32**, 458.

Kato, N. (1979) Acta Cryst. A **35**, 9.
Kato, N. (1980a) Acta Cryst. A **36**, 763.
Kato, N. (1980b) Acta Cryst. A **36**, 770.
Kato, N. (1991) Acta Cryst. A **47**, 1.
Kelly, P.M., Jostons, A., Blake, R.G. and Napier, J.G. (1975) Phys. Status Solidi A **31**, 771.
Kilaas, R., O'Keefe, M.A. and Krishnan, K.M. (1987) Ultramicroscopy **21**, 47.
Kim, H.S. and Sheinin, S.S. (1982) Phys. Stat. Sol. (b) **109**, 807.
Kim, H.S. and Sheinin, S.S. (1987) Phys. Stat. Sol. (a) **103**, 349.
Kirkland, E. J. (1998). Advanced Computing in Electron Microscopy., Springer.
Kogiso, M. and Kainuma, Y. (1968) J. Phys. Soc. Japan **25**, 498.
Kohl, H. (1983) Ultramicroscopy **11**, 53.
Kohl, H. and Rose, H. (1985) Advances in Electronics and Electron Physics **65**, 173.
Komatsu, K. and Teramoto, K. (1966) J. Phys. Soc. Japan, **21**, 1152.
Konnert, J. and D'Antonio, P. (1991) Ultramicroscopy **38**, 169.
Konopinski, E.J. (1981) Electromagnetic Fields and Relativistic Particles, McGraw-Hill Book Company, section 10.5.
Korte, U. and Meyer-Ehmsen, G. (1993) Surf. Sci. **298**, 299.
Krakow, W. and O'Keefe, M. eds. (1989) Computer Simulation of Electron Microscope Diffraction and Images, The Minerals, Metals and Materials Society (TMS).
Kreuzer, H.J., Nakamura, K., Wierzbicki, A., Fink, H.W. and Schmid, H. (1992) Ultramicroscopy **45**, 381.
Krivanek, O.L., Ahn, C.C. and Wood, G.J. (1990) Ultramicroscopy, **33**, 177.
Lally, J., Humphreys, C.J., Metherell, A.J. and Fisher, R.M. (1972) Philos. Mag. **25**, 321.
Landau, L.D. and Lifshitz, E.M. (1977) Quantum Mechanics, non-relativistic theory, 3rd ed., Pergamon Press (New York).
Landau, L. (1944) J. Phys. USSR **8**, 201.
Larsen, P.K. and Dobson, P.J. eds. (1988) Reflection High-Energy Electron Diffraction and Reflection Electron Imaging of Surfaces, Plenum Press, NATO ASI Series B: Phys. Vol. 188.
Latyshev, A.V., Krasilnikov, A.B., Sokolov, L.V. and Stenin, S.I. (1991) Surf. Sci. **254**, 90.
Laue, H. von (1935) Ann. Phys. (Leipzig) **23**, 705.
LeBeau, J.M., Findlay, S.D., Allen, L.J., Stemmer, S. (2008) Phys. Rev. Lett. **100**: 206101.
LeBeau, J.M., Findlay, S.D., Wang, X., Jacobson, A.J., Allen, L.J., Stemmer, S. (2009) Phys. Rev. B **79**: 214110.
Lehmann, M., and Lichte, H. (2002) Microsc. Microanal. **8** (Suppl. 2) 34–35.
Lewis, A. and Villagrana, R.E. (1979) Acta Cryst. A **35**, 276.
Lichte, H. (1986) Ultramicroscopy **20**, 293.
Liu, J., Pan, M. and Spinnler, G.E. (1993) Proc. 51st Ann. Meet. of Microsc. Soc. Ame., G.W. Bailey and C.L. Rieder, eds., San Francisco Press, pp.1058.
Liu, J. and Cowley, J.M. (1991) Ultramicroscopy **37**, 50.
Loane, R.F., Xu, P. and Silcox, J. (1991) Acta Cryst. A **47**, 267.
Lu, P. and Smith, D.J. (1991) Surf. Sci. **254**, 119.
Lynch, D.F. and Moodie, A.F. (1972) Surf. Sci. **32**, 422.
Lynch, D.F. and Smith, A.E. (1983) Phys. Stat. Sol. (b) **119**, 355.
Ma, Y. and Marks, L.D. (1989) Acta Cryst. A **45**, 174.
Ma, Y. and Marks, L.D. (1990) Acta Cryst. A **46**, 11.
Ma, Y. and Marks, L.D. (1992) Microsc. Research and Technique **20**, 371.
Ma, Y., Lordi, S., Larsen, P.K. and Eades, J.A. (1993) Surf. Sci. **289**, 47.
Maksym, P.A. and Beeby, J.L. (1981) Surf. Sci. **110**, 423.
Mandl, F. (1971) Statistical Physics, John Wiley & Sons, New York, p. 59.
Maradudin, A.A., Montroll, E.W., Weiss, G.H. and Ipatova, I.P. (1971) Theory of Lattice Dynamics in the Harmonic Approximation, Academic Press, New York.
Marks, L.D. and Ma, Y. (1988) Acta Cryst. A **44**, 392.
Marthinsen, K. and Høier, R. (1986) Acta Cryst. A **42**, 484.

Maslen, V.W. (1987) Philos. Mag. B **55**, 491.
Maslen, V. W. and Rossouw, C. J. (1984) Philos. Mag. **49**, 735.
McCoy, J.M. and Maksym, P.A. (1993) Surf. Sci. **297**, 113.
Melander, A. (1975) Philos. Mag. **31**, 599.
Melander, A. and Sandstrom, R. (1975a) J. Phys. C: Solide State Phys. **8**, 767.
Melander, A. and Sandstrom, R. (1975b) Acta Cryst. A **31**, 116.
Mitra, S.S. and Massa, N.E. (1982) Handbook on Semiconductors, Vol. 1, T.S. Mos and W. Paul eds., North Holland (Amesterdam) pp. 81.
Miyake, S. (1959) J. Phys. Soc. Japan **14**, 1347.
Möllenstedt, G. and Düker, H. (1956) Z. Physik, **145**, 377.
Moon, A.R. (1972) Z. Naturforsch **27** a, 390.
Morse, P.M. (1930) Phys. Rev. **35**, 1310.
Mott, N.F. and Nabarro, F.R.N. (1940) Proc. Phys. Soc. Lond. **52**, 86.
Nagata, F. and Fukuhara, A. (1967) Jap. J. Appl. Phys. **6**, 1233.
Nkoma, J.S. (1987) Surf. Sci. **191**, 595.
Nozieres, P. and Pine, D. (1958) Phys. Rev. **109**, 741.
Ohtsuki, Y.H. (1983) Charge Beam Interactions with Solid, Taylor & Francis Ltd, (New York).
Ohtsuki, Y.H. and Yanagawa, S. (1966) J. Phys. Soc. Japan **21**, 326.
Okamoto, K., Ichinokawa, T. and Ohtsuki, Y. (1971) J. Phys. Soc. Japan **30**, 1690.
Paskin, A. (1958) Acta Cryst. **11**, 165.
Peng, L.M. (1994) Adv. Electronics and Electron Phys., in press.
Peng, L.M. and Cowley, J.M. (1986) Acta Cryst. A **42**, 545.
Peng, L.M. and Cowley, J.M. (1988) Acta Cryst. A **44**, 1.
Peng, L.M. and Whelan, M.J. (1990) Proc. R. Soc. Lond. A **431**, 125.
Peng, L.M., Ren, G., Dudarev S.L. and Whelan, M.J. (1996a) Acta Cryst. A **52** 456.
Peng, L.M., Ren, G., Dudarev S.L. and Whelan, M.J. (1996b) Acta Cryst. A **52**, 257.
Peng, L.M., Dudarev, S.L., Whelan, M.J. (2004) High-energy electron diffraction and microscopy, Oxford University Press.
Pennycook, S.J. and Boatner, L.A. (1988) Nature **336**, 565.
Pennycook, S.J. and Jesson, D.E. (1990) Phys. Rev. Lett. **64**, 938.
Pennycook, S.J., Li, C., Li, M., Tang, C., Okunishi, E., Varela, M., Kim, Y.M. and Jang, J.H. (2018), Journal of Analytical Science and Technology **9**: 11.
Perovic, D.D., Weatherly, G.C., Egerton, R.F., Houghton, D.C. and Jackman, T.E. (1991) Philos. Mag. A **63**, 757.
Perovic, D.D., Howie, A. and Rossouw, C.J. (1993) Philos. Mag. Lett. **67**, 261.
Pine, D. (1953) Phys. Rev. **92**, 626.
Pogany, A.P. and Turner, P.S. (1968) Acta Cryst. A **24**, 103.
Pozzi, G. (1989) Ultramicroscopy **30**, 417.
Qian, W., Spence, J.C.H. and Zuo, J.M. (1993) Acta Cryst. A **49**, 436.
Qin, L.C. and Goodman, P. (1989) Ultramicroscopy **27**, 115.
Qin, L.C. and Urban, K. (1990) Ultramicroscopy **33**, 159.
Radi, G. (1968) Z. Phys. **212**, 146; ibd. **213**, 244.
Radi, G. (1970) Acta Cryst. A **26**, 41.
Raether, H. (1980) Excitation of Plasmons and Interband Transitions by Electrons, Springer Tracts in Mordern Physics, Vol. 88, Springer-Verlag (New York).
Reimer, L. (1984) Transmission Electron Microscopy, Springer-Verlag (New York).
Reimer, L. (1990) Proc. of XIIth Intern. Cong. for Electron Microscopy (Seattle), San Francisco Press, Vol. 2, eds. L. D. Peachey and D. B. Williams, 64.
Reimer, L. and Fromm, I. (1989) Proc. 47th Ann. Meeting Electron Microscopy Society of America, (San Antonio, Texas), ed. G.W. Bailey, San Francisco Press, 382.
Reimer, L., Fromm, I. and Naundorf, I. (1990) Ultramicroscopy **32**, 80.
Rez, P. (1976) The theory of inelastic scattering in the electron microscope of crystals, Ph.D. Thesis, University of Oxford.

Rez, P. (1983) Acta Cryst. A **39**, 697.
Rez, P. (1993) Ultramicroscopy **52**, 260.
Rez, P., Humphreys, C.J. and Whelan, M.J. (1977) Philos. Mag. **35**, 81.
Ritchie, R.H. (1957) Phys. Rev. **106**, 874.
Ritchie, R.H. (1981) Philos. Mag. A **44**, 234.
Ritchie, R.H. and Howie, A. (1988) Philos. Mag. A **58**, 753.
Robertson, I.M. (1991) Philos. Mag. A **64**, 577.
Rose, H. (1976a) Optik **45**, 139.
Rose, H. (1976b) Optik **45**, 187.
Rossouw, C.J. (1981) Ultramicroscopy **7**, 139.
Rossouw, C.J. (1985) Ultramicroscopy **16**, 241.
Rossouw, C.J. and Bursill, L.A. (1985) Acta Cryst. A **41**, 320.
Rossouw, C.J., Miller, P.R., Drennan, J. and Allen, L.J. (1990) Ultramicroscopy **34**, 149.
Rossouw, C.J., Al-Kharaji, M., Cherns, D., Steeds, J.W. and Touaitia, R. (1991) Ultramicroscopy **35**, 229.
Saldin, D.K. and Rez, P. (1987) Philos. Mag. B **55**, 481.
Scherzer, O. (1949) J. Appl. Phys. **20**, 20.
Seale, D.J. and Sheinin, S.S. (1993) Proc. 51st Ann. Meet. of Microsc. Soc. Ame., G.W. Bailey and C.L. Rieder eds., San Francisco Press, pp.1214.
Seale, D.J. and Sheinin, S.S. (1994) to be published.
Sears, V.F. and Shelley, S.A. (1991) Acta Cryst. A **47**, 441.
Self, P., O'Keefe, M.A., Buseck, P.R. and Spargo, A.E.C. (1983) Ultramicroscopy **11**, 35.
Shechtman, D., Hutchison, J.L., Robins, L.H., Farabaugh, E.N. and Feldman, A. (1993) J. Mater. Res. **8**, 473.
Shuman, H., Chang, C.F. and Somlyo, A.P. (1986) Ultramicroscopy **19**, 121.
Sinha, S.K. (1973) Crit. Rev. Solid State Sci. **3**, 273.
Slater, J.C. (1937) Phys. Rev. **51**, 840.
Smith, A.E., Lehmpfuhl, G. and Uchida, Y. (1992) Ultramicroscopy **41**, 367.
Spence, J. C. H. (1988a) Experimental high-resolution electron microscopy (2nd edition), Oxford University Press (New York), p.128.
Spence, J. C. H. (1992) in Electron Diffraction Techniques, Vol. I, J.M. Cowley ed., International Union of Crystallography, Oxford University Press, 465.
Spence, J.C.H. and Zuo, J.M. (1992) Electron Microdiffraction, Plenum Press (New York).
Spence, J.C.H. (1988b) in High-Resolution Transmission Electron Microscopy and Associated Techniques, P. Buseck, J. Cowley and L. Eyring eds., Oxford University Press (New York), Chpt. 5–7.
Spence, J.C.H. and Spargo, A.E. (1971) Phys. Rev. Lett. **26**, 895.
Spence, J.C.H., Zuo, J.M. and Lynch, J. (1989) Ultramicroscopy **31**, 233.
Spencer, J.P., Humphreys, C.J. and Hirsch, P.B. (1972) Philos. Mag. **26**, 193.
Stern, E.A. (1982) Optik, **61**, 45.
Stobbs W.M. and Saxton W.O. (1988) J. Microscopy **151**, 171.
Sturkey, L. (1962) Proc. Phys. Soc. **80**, 321.
Taftø, J. (1983) Phys. Rev. Lett. **51**, 654.
Taftø, J. and Krivanek, O.L. (1982) Nucl. Instrum. Methods **194**, 153.
Takagi, S. (1958a) J. Phys. Soc. Japan **13**, 278.
Takagi, S. (1958b) J. Phys. Soc. Japan **13**, 287.
Takagi, S. (1962) Acta Cryst. **15**, 1311.
Takagi, S. (1969) J. Phys. Soc. Japan **26**, 1239.
Tang D., Brydson R.D., Jefferson D.A. and Thomas J.M. (1989) J. Phys.: Condens. Matter **1**, 1561.
Taupin, D. (1964) Bull. Soc. Franc. Minér. Crist. **87**, 469.
Teo, B.K. and Joy, D.C. (1981) EXAFS Spectroscopy, Plenum Press (New York).
Tonomura, A. (1992) Adv. Phys. **41**, 59.
Tournarie, M. (1962) J. Phys. Soc. Japan **17**, supl. B-II, 98.

Treacy, M.M.J. and Gibson, J.M. (1993) Ultramicroscopy **52**, 31.
Trinkaus, H. (1972) Phys. Stat. Sol. (b) **51**, 307.
Uyeda, R. (1968) Acta Cryst. A **24**, 175.
Van Dyck, D., Danckaert, J., Coene, W., Selderslaghs, E., Broddin, D., Van Landuyt, J. and Amelinckx, S. (1989) in Computer Simulation of Electron Microscope Diffraction and Images, eds. W. Krakow and M. O'Keefe, The Mineral, Metals & Materials Soc., pp.107.
Van Dyck, D. (1975) Phys. Stat. Sol. (b) **72**, 321.
Van Dyck, D. (1976) Phys. Stat. Sol. (b) **77**, 301.
Van Dyck, D. (1980) J. Microsco. **119**, 141.
Van Dyck, D. (1983) J. Microsco. **132**, 31.
Van Dyck, D. (1985) Adv. Electronics and Electron Phys. **65**, 295.
Van Dyck, D. and Coene, W. (1984) Ultramicroscopy **15**, 29.
Van Dyck, D., Lichte, H., Spence, J.C.H. (2000) Ultramicroscopy **81**: 187–194.
Varela, M., Gazquez, J., Pennycook, S.J. (2012) MRS Bull. **37**:29–35.
Viguier, B., Kemker, K.J. and Vabderschaeve, G. (1994) Philos. Mag. A **69**, 19.
Vincent, R., Bird, D.M. and Steeds, J.W. (1984) Philos. Mag. A **50**, 765.
Völkl, E. and Lichte, H. (1990) Ultramicroscopy **32**, 177.
Voss, R., Lempfuhl, G. and Smith, P.J. (1980) Z. Naturforsch. C **35**a, 973.
Walls, M.G. and Howie, A. (1989) Ultramicroscopy **28**, 40.
Wang, Z.L. (1996) Reflection Electron Microscopy and Spectroscopy for Surface Analysis, Cambridge University Press (New York).
Wang, Z. L. and Li, D.C. (1994) Phil. Mag., in press.
Wang, Z.L. and Shapiro, A.J. (1994) Surf. Sci., in press.
Wang, Z.L (1998b) Ultramicroscopy **74**, 7–26.
Wang, Z.L. (1989a) Acta Cryst. A **45**, 636.
Wang, Z.L. (1989b) Acta. Cryst. A **45**, 193.
Wang, Z.L. (1990a) Phys. Rev. B **41**, 12818.
Wang, Z.L. (1990b) Acta Cryst. A **46**, 366.
Wang, Z. L. (1991) Acta Cryst. A **47**, 686.
Wang, Z.L. (1992a) Philos. Mag. B **65**, 559.
Wang, Z.L. (1992b) Acta Cryst. A **48**, 674.
Wang, Z.L. (1993a) Rep. Prog. Phys. **56**, 997.
Wang, Z.L. (1993b) Ultramicroscopy **52**, 504.
Wang, Z.L. (1994) Ultramicroscopy **53**, 73.
Wang, Z.L. (1995) Acta Cryst. A **51**, 569.
Wang, Z.L. (1996a) Acta Cryst. A **52**, 717–729.
Wang, Z.L. (1996b) Surface Sci. **366**, 377–393.
Wang, Z.L. (1996c) Phil. Mag. B **74**, 733–749.
Wang, Z.L. (1998a) Acta Cryst. A **54**, 460–467.
Wang, Z.L. (1998c) Acta Cryst. A **54**, 468–480.
Wang, Z.L. (2003) Micron **34**, 141–155.
Wang, Z.L. and Bentley, J. (1991a) Microsc. Microanal. Microstruct. **2**, 569.
Wang, Z.L. and Bentley, J. (1991b) Ultramicroscopy **38**, 181.
Wang, Z.L. and Bentley, J. (1991c) Inst. Phys. Conf. Ser. **119**, 547.
Wang, Z.L. and Cowley, J.M. (1987a) Ultramicroscopy **21**, 77.
Wang, Z.L. and Cowley, J.M. (1987b) Ultramicroscopy **21**, 335.
Wang, Z.L. and Cowley, J. M. (1989) Ultramicroscopy **31**, 437.
Wang, Z.L. and Cowley, J. M. (1990) Ultramicroscopy **32**, 275.
Wang, Z.L. and Fisher, A. T. (1993) Ultramicroscopy **48**, 183.
Wang, Z.L. and Spence, J.C.H. (1990) Surf. Sci. **234**, 98.
Warren, B.E. (1990) X-Ray Diffraction, Dover Publishing (New York).
Warren, B.E., Averach, B.L. and Roberts, B.W. (1951) J. Appl. Phys. **22**, 1493.
Watenabe, D., Uyeda, R. and Kogiso, M. (1968) Acta Cryst. A **24**, 249.

Weickenmeier, A. and Kohl, H. (1990) Acta Cryst. A **47**.
Weickenmeier, A. and Kohl, H. (1989) Philos. Mag. B **60**, 467.
Weickenmeier, A. and Kohl, H. (1991) Acta Cryst. A **47**, 590.
Whelan, M.J. (1965a) J. Appl. Phys. **36**, 2099.
Whelan, M.J. (1965b) J. Appl. Phys. **36**, 2103.
Whelan, M.J. (1976) J. Phys. C **9**, L195.
Whelan, M.J. (1986) Acta Cryst. A **42**, 493.
Williams, D.B. and Carter, C.B. (2009) Transmission Electron Microscopy - A Textbook for Materials Science, Springer.
Williams, B.G., Sparrow, T.G. and Egerton, R.F. (1984) Proc. R. Soc. Lond. A **393**, 409.
Woll, E.J. and Kohn, W. (1962) Phys. Rev. **126**, 1693.
Wright, A. G. and Bird, D. M. (1989) Acta Cryst. A **45**, 342.
Xu, P., Kirkland, E.J., Silcox, J. and Keyse, R. (1990) Ultramicroscopy **32**, 93.
Yagi, K. (1993) Surf. Sci. Rep. **17**, 305.
Yamamoto, T. (1980) Acta Cryst. A **36**, 126.
Yao, M. and Smith, D.J. (1994) J. Microsc. **175**, 252.
Yoshioka, H. and Kainuma, Y. (1962) J. Phys. Soc. Japan **17**, Suppl. BII, 134.
Yoshioka, H. (1957) J. Phys. Soc. Japan **12**, 618.
Young, A.P. and Rez, P. (1975) J. Phys. C: Solid State Phys. **8**, L1.
Zabala, N. and Echenique, P.M. (1990) Ultramicroscopy **32**, 327.
Zhao, T.C., Poon, H.C. and Tong, S.Y. (1988) Phys. Rev. B **38**, 1172.
Zuo, J.M., Spence, J.C.H. and O'Keefe (1988) Phys. Rev. Lett. **61**, 353.
Zuo, J.M. and Spence, J.C.H. (1991) Ultramicroscopy **35**, 185.
Zuo, J.M., Spence, J.C.H. (2017) Advanced transmission electron microscopy, Springer.

GPSR Compliance

The European Union's (EU) General Product Safety Regulation (GPSR) is a set of rules that requires consumer products to be safe and our obligations to ensure this.

If you have any concerns about our products, you can contact us on ProductSafety@springernature.com

In case Publisher is established outside the EU, the EU authorized representative is:

Springer Nature Customer Service Center GmbH
Europaplatz 3
69115 Heidelberg, Germany

Batch number: 10755461

Printed by Printforce, the Netherlands